T0270789

Textbook of Ion Channels
Volume II

The *Textbook of Ion Channels* is a set of three volumes that provides a wide-ranging reference source on ion channels for students, instructors and researchers. Ion channels are membrane proteins that control the electrical properties of neurons and cardiac cells; mediate the detection and response to sensory stimuli like light, sound, odor, and taste; and regulate the response to physical stimuli like temperature and pressure. In non-excitable tissues, ion channels are instrumental for the regulation of basic salt balance that is critical for homeostasis. Ion channels are located at the surface membrane of cells, giving them the unique ability to communicate with the environment, as well as the membrane of intracellular organelles, allowing them to regulate internal homeostasis. Ion channels are fundamentally important for human health and diseases, and are important targets for pharmaceuticals in mental illness, heart disease, anesthesia, pain and other clinical applications. The modern methods used in their study are powerful and diverse, ranging from single ion-channel measurement techniques to models of ion channel diseases in animals, and human clinical trials for ion channel drugs.

Volume II starts with ion channel taxonomy and features coverage of major ion channel families, and describes the physiological role, structural components, gating mechanisms and biophysics, permeation and selectivity, regulation, pharmacology, and roles in disease mechanisms. Channels in this volume include voltage-activated sodium, calcium and potassium channels, inward-rectifier and two-pore domain potassium channels, calcium-activated potassium channels, cyclic nucleotide-gated channels, pacemaker ion channels, chloride channels, ligand-gated receptors activated by acetylcholine, glutamate, 5-HT$_3$, GABA and glycine, acid-sensing channels, P2X receptors, TRP channels, store-operated channels, pressure-activated piezo channels, ryanodine receptors, and proton channels.

All three volumes give the reader an introduction to fundamental concepts needed to understand the mechanism of ion channels; a guide to the technical aspects of ion channel research; offer a modern guide to the properties of major ion channel families; and include coverage of key examples of regulatory, physiological and disease roles for ion channels.

Textbook of Ion Channels
Volume II

Properties, Function, and Pharmacology of the Superfamilies

Edited by

Jie Zheng
Matthew C. Trudeau

CRC Press
Taylor & Francis Group
Boca Raton London New York

CRC Press is an imprint of the
Taylor & Francis Group, an **informa** business

First edition published 2023
by CRC Press
6000 Broken Sound Parkway NW, Suite 300, Boca Raton, FL 33487-2742

and by CRC Press
4 Park Square, Milton Park, Abingdon, Oxon, OX14 4RN

CRC Press is an imprint of Taylor & Francis Group, LLC

© 2023 selection and editorial matter, Jie Zheng and Matthew C. Trudeau; individual chapters, the contributors

Library of Congress Cataloging-in-Publication Data

Names: Zheng, Jie, 1966- editor. | Trudeau, Matthew C., editor.
Title: Textbook of ion channels / edited by Jie Zheng, Matthew C. Trudeau. Description: First edition. |
Boca Raton: CRC Press, 2023. | Includes bibliographical references and index. |
Contents: volume 1. Basics and methods -- volume 2. Properties, function, and pharmacology of
the superfamilies -- volume 3. Regulation, physiology, and diseases. Identifiers: LCCN 2022043157 (print) |
LCCN 2022043158 (ebook) | ISBN 9780367538156 (v. 1; hardback) | ISBN 9780367560492 (v. 1; paperback) |
ISBN 9780367538163 (v. 2; hardback) | ISBN 9780367560812 (v. 2; paperback) |
ISBN 9780367538194 (v. 3; hardback) |
ISBN 9781032408040 (v. 3; paperback) | ISBN 9781003096214 (v. 1; ebook) | ISBN 9781003096276 (v. 2; ebook) |
ISBN 9781003310310 (v. 3; ebook) Subjects: LCSH: Ion channels--Textbooks.
Classification: LCC QH603.I54 T49 2023 (print) | LCC QH603.I54 (ebook) |
DDC 571.6/4--dc23/eng/20230126 LC record available at https://lccn.loc.gov/2022043157LC ebook
record available at https://lccn.loc.gov/2022043158

ISBN: 978-0-367-53816-3 (hbk)
ISBN: 978-0-367-56081-2 (pbk)
ISBN: 978-1-003-09627-6 (ebk)

DOI: 10.1201/9781003096276

Typeset in Palatino
by Deanta Global Publishing Services, Chennai, India

eResources are available at: HYPERLINK https://www.routledge.com/Textbook-of-Ion-Channels-Volume-II-Properties, Function, and Pharmacology of the Superfamilies/Zheng-Trudeau/p/book/ 9780367538163

Contents

Preface

We view this *Textbook of Ion Channels* as a modern extension of the field's most authoritative textbook, *Ion Channels of Excitable Membranes* by Bertil Hille. It is an update of our previous *Handbook of Ion Channels* (CRC Press 2015). As Hille observed, "The field of ion channels has grown from tentative hypotheses to vigorous mainstream science" (*Progress in Biophysics and Molecular Biology*, 2022, vol. 169–170, p. 18). In order to offer an updated coverage of this vast and rapidly changing field, our approach was to invite leading world experts to write on selected topics that make up 58 different chapters.

The extensive coverage of this textbook makes it necessary to divide the contents into three volumes:

> Volume I, Part 1 deals with the foundational concepts of permeation and gating mechanisms, with a balance of classic theories and latest developments. Volume I, Part 2 focuses on various ion channel techniques. This methods section covers both classic, well-developed techniques and newly developed powerful techniques spanning the basic principle of the method, application to channel research, and practical issues.

> Volume II covers over 25 major ion channel types, with a combination of well-studied ion channels and newly identified channels, and includes in-depth discussion of physiological roles, permeation and ion selectivity, gating mechanisms, biophysics and pharmacology. The organization of this section follows the major superfamilies of ion channels.

> Volume III offers examples of ion channel regulation, the in-depth role of ion channels in key physiological systems and the role of channels in major diseases, discussing genetics to mechanisms and developments in ion channel pharmaceuticals.

As a community-supported project, many colleagues have contributed to the textbook. In addition to the chapter authors, we are deeply indebted to the Editorial Advisory Board for its guidance. The Editorial Advisory Board for the *Textbook of Ion Channels* is Drs. Richard Aldrich, Henry Colecraft, Jianmin Cui, Lucie Delemotte, Teresa Giraldez, Merritt Maduke, Andrea Meredith, Andrew Plested, Gail A. Robertson and William N. Zagotta.

In addition, we thank special topic advisers Drs. Michael Pusch, Laszlo Csanady, Zack Sellers, Zhe Lu and Vivek Garg.

For their professional guidance, we thank our editors at CRC Press: Ms. Carolina Antunes and Ms. Emma Morley. The book project benefited from Ms. Betsy Byers's fantastic and patient assistance, for which we are deeply grateful.

Editors

Jie Zheng, PhD, is a professor at the University of California Davis School of Medicine, where he has served as a faculty member in the Department of Physiology and Membrane Biology since 2004. Zheng earned a bachelor's degree in physiology and biophysics (1988) and a master's degree in biophysics (1991) at Peking University. He earned a PhD in physiology (1998) at Yale University, where he studied with Dr. Fredrick J. Sigworth on patch-clamp recording, single-channel analysis, and voltage-dependent activation mechanisms. He received his postdoctoral training at the Howard Hughes Medical Institute (HHMI) and the University of Washington during 1999 to 2003, working with Dr. William N. Zagotta on the cyclic nucleotide-gated channels activation mechanism and novel fluorescence techniques for ion channel research. Currently, Zheng's research focuses on temperature-sensitive TRP channels.

Matthew C. Trudeau, PhD, is a professor in the Department of Physiology at the University of Maryland School of Medicine in Baltimore, Maryland. He earned a bachelor's degree in biochemistry and molecular biology in 1992 and a PhD in physiology in 1998 while working with Gail Robertson, PhD, at the University of Wisconsin-Madison. His thesis work was on the properties of voltage-gated potassium channels in the human ether-á-go-go related gene (hERG) family and the role of these channels in heart disease. Trudeau was a postdoctoral fellow with William Zagotta, PhD, at the University of Washington and the Howard Hughes Medical Institute (HHMI) in Seattle from 1998 to 2004, where he focused on the molecular physiology of cyclic nucleotide-gated ion channels, the mechanism of their modulation by calcium–calmodulin and their role in an inherited form of vision loss. Currently, Trudeau's work focuses on hERG potassium channels, their biophysical mechanisms, and their role in cardiac physiology and cardiac arrhythmias.

Contributors

Mike Althaus
Department of Natural Sciences/Institute
for Functional Gene Analytics
Bonn-Rhein-Sieg University of Applied
Sciences
Rheinbach, Germany

Diego Alvarez de la Rosa
Biomedical Sciences Department
(Physiology) and Biomedical
Technologies Institute
University of La Laguna
La Laguna, Spain

Jean-Pierre Benitah
Signaling and Cardiovascular
Pathophysiology, UMR-S 1180, Inserm
Université Paris-Saclay
Châtenay-Malabry, France

Cecilia Bouzat
Instituto de Investigaciones Bioquímicas
de Bahía Blanca
Departamento de Biología, Bioquímica y
Farmacia
Universidad Nacional del Sur-Consejo
Nacional de Investigaciones Científicas y
Técnicas
Bahía Blanca, Argentina

Cecilia M. Canessa
School of Medicine
Yale University
New Haven, CT

William A. Catterall
Department of Pharmacology
University of Washington
Seattle, WA

Sara J. Codding
Department of Physiology
University of Maryland School of Medicine
Baltimore, MD

Henry M. Colecraft
Department of Physiology and Cellular
Biophysics
Columbia University Irving Medical Center
New York, NY

Jianmin Cui
Department of Biomedical Engineering
McKelvey School of Engineering
Washington University in St. Louis
St. Louis, MO

Gucan Dai
Doisy Research Center
Department of Biochemistry and
Molecular Biology
Saint Louis University School of Medicine
Saint Louis, MO

Camden Driggers
Department of Chemical Physiology and
Biochemistry
Oregon Health & Science University
Portland, OR

Kate Dunning
University of Strasbourg
Centre National de la Recherche
Scientifique, CAMB UMR 7199
Strasbourg, France

Juan Facundo Chrestia
Instituto de Investigaciones Bioquímicas
de Bahía Blanca
Departamento de Biología, Bioquímica y
Farmacia
Universidad Nacional del Sur-Consejo
Nacional de Investigaciones Científicas
y Técnicas
Bahía Blanca, Argentina

Martin Fronius
Department of Physiology
University of Otago
Dunedin, New Zealand

Marine Gandon-Renard
Signaling and Cardiovascular
    Pathophysiology, UMR-S 1180, Inserm
Université Paris-Saclay
Châtenay-Malabry, France

Steve A.N. Goldstein
Departments of Pediatrics, Physiology &
    Biophysics, and Pharmaceutical
    Sciences
University of California, Irvine
Irvine, CA

A.M. Gomez
Signaling and Cardiovascular
    Pathophysiology, UMR-S 1180, Inserm
Université Paris-Saclay
Châtenay-Malabry, France

Jörg Grandl
Department of Neurobiology
Duke University Medical Center
Durham, NC

Thomas Grutter
University of Strasbourg Institute for
    Advanced Studies, CAMB UMR 7199
Strasbourg, France

H. Criss Hartzell
Department of Cell Biology
Emory University School of Medicine
Atlanta, GA

Josip Ivica
Department of Neuroscience, Physiology
    and Pharmacology
University College London
London, UK

Timothy Jegla
Department of Biology and Huck Institutes
    of the Life Sciences
Penn State University
University Park, PA

Anna K. Koster
Department of Chemistry
Stanford University
Stanford, CA

H. Peter Larsson
Department of Physiology and Biophysics,
    Miller School of Medicine
University of Miami
Miami, FL

Emily R. Liman
Department of Biological Sciences
University of Southern California
Los Angeles, CA

Sarah C.R. Lummis
Department of Biochemistry
University of Cambridge
Cambridge, UK

Merritt Maduke
Department of Molecular and Cellular
    Physiology
Stanford University
Stanford, CA

David D. McKemy
Department of Biological Sciences
University of Southern California
Los Angeles, CA

Jean-Jacques Mercadier
Signaling and Cardiovascular
    Pathophysiology, UMR-S 1180, Inserm
Université Paris-Saclay
Châtenay-Malabry, France

Susanne M. Mesoy
Department of Biochemistry
University of Cambridge
Cambridge, UK

Jacqueline Niu
Department of Physiology and Cellular
    Biophysics
Columbia University Irving Medical
    Center
New York, NY

Laetitia Pereira
Signaling and Cardiovascular
    Pathophysiology, UMR-S 1180, Inserm
Université Paris-Saclay
Châtenay-Malabry, France

Romain Perrier
Signaling and Cardiovascular
  Pathophysiology, UMR-S 1180, Inserm
Université Paris-Saclay
Châtenay-Malabry, France

Colin H. Peters
University of Colorado, Anschutz Medical
  Campus
Aurora, CO

Leigh D. Plant
Department of Pharmaceutical Sciences
Northeastern University
Boston, MA

Andrew Plested
Humboldt University Berlin and Cluster of
  Excellence NeuroCure
Berlin, Germany

Murali Prakriya
Department of Pharmacology
Northwestern University, Feinberg School
  of Medicine
Chicago, IL

Catherine Proenza
University of Colorado, Anschutz Medical
  Campus
Aurora, CO

I. Scott Ramsey
Department of Physiology and Biophysics
Virginia Commonwealth University School
  of Medicine
Richmond, VA

Tamara Rosenbaum
Institute of Cell Physiology
National Autonomous University of
  Mexico
Mexico City, Mexico

Show-Ling Shyng
Department of Chemical Physiology and
  Biochemistry
Oregon Health & Science University
Portland, OR

Benjamin T. Simonson
Department of Biology and Huck Institutes
  of the Life Sciences
Penn State University
University Park, PA

Lucia Sivilotti
Department of Neuroscience, Physiology
  and Pharmacology
University College London
London, UK

Trevor G. Smart
Department of Neuroscience, Physiology
  and Pharmacology
University College London
London, UK

Min-Woo Sung
Department of Chemical Physiology and
  Biochemistry
Oregon Health & Science University
Portland, OR

Jin Bin Tian
Department of Integrative Biology and
  Pharmacology
University of Texas Health Science Center
  at Houston
Houston, TX

Matthew C. Trudeau
Department of Physiology
and
Training Program in Integrative
  Membrane Biology
University of Maryland School of Medicine
Baltimore, MD

Anastasios V. Tzingounis
Department of Physiology and Neurobiology
University of Connecticut
Storrs, CT

Almudena Val-Blasco
Signaling and Cardiovascular
  Pathophysiology, UMR-S 1180, Inserm
Université Paris-Saclay
Châtenay-Malabry, France

Francis I. Valiyaveetil
Department of Chemical Physiology and
    Biochemistry
Oregon Health & Science University
    Portland, OR

Michael D. Varnum
Department of Integrative Physiology and
    Neuroscience
Washington State University College of
    Veterinary Medicine
Pullman, WA

Yangyu Wu
School of Medicine
Tsinghua University
Beijing, China

Heike Wulff
Department of Pharmacology
University of California Davis School of
    Medicine
Davis, CA

Bailong Xiao
State Key Laboratory of Membrane Biology
Tsinghua-Peking Center for Life Sciences
    IDG/McGovern Institute for Brain
    Research at Tsinghua
Tsinghua University
Beijing, China

Liheng Yin
Signaling and Cardiovascular
    Pathophysiology, UMR-S 1180, Inserm
Université Paris-Saclay
Châtenay-Malabry, France

Miao Zhang
Department of Biomedical and
    Pharmaceutical Sciences
Chapman University
Irvine, CA

Michael X. Zhu
Department of Integrative Biology and
    Pharmacology
University of Texas Health Science Center
    at Houston
Houston, TX

1

Taxonomy and Evolution of Ion Channels

Timothy Jegla and Benjamin T. Simonson

CONTENTS

1.1 How Should We Approach Ion Channel Taxonomy?

Making taxonomic sense of the diverse ion channels has been challenging and at times confusing because functional similarities do not always track shared evolutionary inheritance. While channels that are closely related on a molecular level often share many functional features, properties such as the gating mechanism and ion selectivity alone are not sufficient to predict structural similarity over broad evolutionary distance. For example, ionotropic glutamate receptors (Chapter 16) and nicotinic acetylcholine receptors (Chapter 15) are both cation channels activated by neurotransmitters but have no structural similarity and have independent evolutionary origins. Similarly, $GABA_A$ receptors (Chapter 18) are chloride channels but share a common evolutionary origin with the nicotinic acetylcholine receptors independent from voltage-gated and calcium-activated chloride channels (Chapters 13,14). Because an understanding of molecular structure is essential for determining how individual proteins function, ion channel classification schemes based on molecular evolution now provide the most useful lens for taxonomic classification of ion channels. The goal of this chapter is therefore to summarize how we define taxonomic relationships between major classes of animal ion channels in a molecular evolutionary context: What does it mean when we use terms such as ion channel superfamily, family or subfamily? A second goal is to introduce some of the major evolutionary events that have shaped the animal ion channel tool kit, using the large and functionally diverse superfamily of voltage-gated cation channels as an example.

1.2 Ion Channel Superfamilies Have Independent Evolutionary Origins

One of the keys to understanding the functional diversity of our ion channels is that there isn't just one way to make an ion channel protein. It has happened independently multiple times in the evolutionary history of life, and that brings us to the definition of ion channel superfamilies. Here, we will define an ion channel superfamily as the broadest collection of channels that share at least some sequence identity and core structural features indicative of a common evolutionary origin. By this definition, an ion channel superfamily is the broadest collection of channels that can be represented in a single molecular phylogeny. The ion channels explored in this volume, representing the large majority of human ion channel diversity, can be divided into nine structurally unique gene superfamilies based on transmembrane topology of pore-forming subunits and the symmetry/stoichiometric arrangement of subunits required to form the ion pore(s) (Table 1.1). For example, the ligand-gated ion channel (LGIC) superfamily, including the aforementioned nicotinic acetylcholine receptors and $GABA_A$ receptors, encode subunits having four transmembrane domains and form fivefold symmetric channels with five subunits arranged around a single, central ion pore (Table 1.1; Chapters 15, 17–19). In contrast, voltage-gated chloride channels (CLCs) function as dimers with a separate pore in each subunit formed by a complex of 17 transmembrane and membrane-embedded helices (Table 1.1; Chapter 13). Because no structural features or protein sequence is shared across ion channel superfamilies such as these, it is almost certain that each ion channel superfamily has an independent evolutionary origin. Thus, the ion channels described in this volume represent nine de novo instances for evolution of ion channel function in separate structurally distinct lineages of transmembrane proteins.

1.3 Structural Diversification of the Voltage-Gated Cation Channel (VGIC) Superfamily

To examine approaches to classifying channels into families and subfamilies, we will focus on the *v*oltage-*g*ated (cat)*i*on *c*hannel (VGIC) superfamily, which is widespread in both prokaryotes and eukaryotes. It is the largest and most structurally diverse superfamily of ion channels, comprising more than half of all human ion channels (Table 1.1 and Figure 1.1). These channels meet the definition of a channel superfamily because they all share a fourfold symmetric cation pore formed by eight transmembrane helices and four selectivity filter motifs. Four homologous pore domains (PDs) each contribute two of the transmembrane domains and one selectivity filter motif to the pore (Figure 1.1A) (Doyle et al. 1998; Nishida and MacKinnon 2002). This shared pore architecture establishes a common evolutionary origin for all VGICs, and extant animal and prokaryotic Kir inward rectifier K^+ channels (Chapter 9), which are comprised of the core VGIC PD motif alone, are thought to be structural homologs of the original VGIC superfamily channels. Because each Kir subunit contains a single PD, these channels function as tetramers (Figure 1.1A). However, unlike most other channel superfamilies, VGICs vary significantly in overall structure, subunit stoichiometry and function, and the shared pore domain often represents only a small fraction of the channel sequence. The human VGIC superfamily can be divided into nine fundamentally distinct structural classes based on subunit domain

TABLE 1.1

Taxonomy of Human Ion Channel Pore-Forming Subunits

Gene Family	Subunit Topology	Subunit Number	Human Genes	Chapter
Voltage-Gated Cation Channel Superfamily				
Kir	2 TM	4	15	9
K2P	(2 TM) X 2	2	15 or 16	10
Kv	4+2 TM	4	27	4
KCNQ	4+2 TM	4	5	5
BK	4+2 TM	4	4	7
SK	4+2 TM	4	4	8
EAG	4+2 TM	4	8	6
HCN	4+2 TM	4	4	12
CNG	4+2 TM	4	6	11
TRP	4+2 TM	4	32	22–25
RyrR	4+2 TM	4	3	28
IP3R	4+2 TM	4	3	29
Nav	(4+2 TM) X 4	1	11	2
Cav	(4+2 TM) X 4	1	10	3
Hv	4 TM	2	1	30
iGluR	2+1 TM	4	18	16
Ligand-Gated Ion Channel Superfamily				
nACHR	4 TM	5	16	15
5HT-3	4 TM	5	4	17
GABA-A/Gly	4 TM	5	24	18,19
ENaC/P2X Superfamily				
ASIC/Deg	2 TM	3	9	20,21
P2XR	2 TM	3	7	31
VG Chloride Channel Superfamily				
CLC	17 TM*	2	9**	13
Calcium-Activated Chloride Channels (3 Superfamilies)				
CLCA	8 TM	2	4	14
Bestrophin	4 TM	5	4	14
TMEM16	10 TM	2	10***	14
Other Cation Channels Superfamilies				
Orai	4 TM	6	3	26
Piezo	38 TM	3	2	27

* Not all helices span the membrane.
** Only four genes are Cl⁻ channels; the remaining five are Cl^-/H^+ cotransporters.
*** Only two genes are Ca^{2+}-activated Cl⁻ channels.

additions, channel stoichiometry, and the presence or absence of a K⁺-selectivity in at least some members (Figure 1.1). It is relatively common in the literature to refer to these major structural classes of VGIC as superfamilies in their own right, and we won't advocate for or against that usage here. While structural classes are not the largest group of channels with a common evolutionary origin, they are often the largest group of channels that are practical to include in a molecular phylogeny based on length of sequence alignment, and several VGIC structural classes do contain multiple functionally distinct animal channel

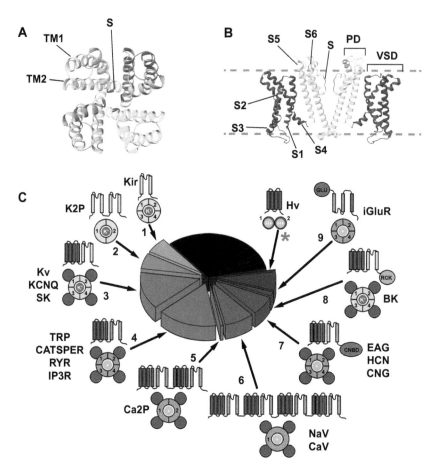

FIGURE 1.1

Structural diversity of the VGIC channel superfamily. (A) VGICs share a common fourfold symmetric cation pore architecture. The ribbon diagram shows an extracellular view of the prokaryotic inward-rectifying potassium channel KcsA (Protein Data Bank: 1BL8) with four subunits forming a single pore. Each subunit contributes two transmembrane domains (TMs) and a selectivity filter sequence (S, red) to the fourfold symmetric pore. Adjacent subunits are differentially shaded. (B) Side view of two diagonally opposed subunits of the tetrameric voltage-gated potassium channel Eag1 (Protein Data Bank: 5K7L) shows the four TM voltage sensor domains (VSD, S1–S4, purple) sitting outside the canonical two TM PDs of the subunits (S5–S6, cyan) with the selectivity filter (S) highlighted in red. Dotted lines show the approximate boundaries of the membrane. (C) The pie chart of the human ion channel set, with slice size indicating gene numbers given in Table 1.1. The VGIC superfamily is highlighted in color and divided by structural class as indicated in the accompanying diagrams labeled 1–9. Hv channels (red asterisk), which consist only of the VSD domain, are shown but not counted as a tenth VGIC structural class because they lack the canonical VGIC PD. The black unlabeled slice collectively represents all other channel superfamilies described in this volume. In the diagrams, VSDs are colored purple, and PDs are colored based on the presence (cyan) or absence (tan) of K+ selectivity. Note however that CNBD-containing VGICs include both K+-selective and non-selective cation channels. Diagrams are provided for the transmembrane topology of channel subunits (extracellular side is up) and stoichiometric arrangement of subunits (numbered) around the central ion channel pore.

families. While it is technically possible and often tempting to construct molecular phylogenies of the entire VGIC superfamily based on the common PD, the sequence is too short to produce statistically robust phylogenies, and such phylogenies should be treated with caution. Therefore, we will limit our discussion here to how we classify VGICs by structure and sequence identity into gene families and subfamilies, and evolutionary relationships between distinct classes of VGICs that are readily inferred from shared structural additions to the VGIC core.

Diversification of the VGIC superfamily into the structural classes depicted in Figure 1.1 has been largely due to two processes that are a major force in the evolution of proteins: intergenic recombination of ancient protein domains into new functional combinations and intragenic duplications of domains into tandem arrays. The most widespread domain addition to the VGIC PD core are the four transmembrane helix voltage-sensor domains (VSDs) from the superfamily that derives its name (Figure 1.1B). VSDs have little contact with each other but can be arranged around the central pore in either a domain-swapped configuration where the VSD of one subunit sits nearest the PD of a neighboring subunit (Long, Campbell, and Mackinnon 2005) (see Kv channels in Chapter 4) or non-swapped configuration where the VSD and PD of the same subunit are in close contact (Whicher and MacKinnon 2016) (see Eag channels in Chapter 5). Other notable ancient domain additions within the VGIC superfamily include the glutamate binding domain of ionotropic glutamate receptors (Panchenko, Glasser, and Mayer 2001; Chen et al. 1999) (Chapter 16), the RCK domains that confer calcium-dependent gating to BK calcium-activated K^+ channels (Yuan et al. 2010) (Chapter 7), and the cyclic nucleotide-binding domain that confers cyclic nucleotide-dependent gating to the CNG (Chapter 11) and HCN (Chapter 12) channels (Zagotta et al. 2003; Li et al. 2017). Homologous cyclic nucleotide-binding domains are found on cyclic nucleotide-sensitive transcription factors (Weber and Steitz 1987) and kinases (Su et al. 1995). Hv voltage-gated proton channels (Chapter 28) present another interesting example for repurposing of a protein domain: they consist only of a proton permeable version of the VSD and function in a dimeric stoichiometry (Lee, Letts, and Mackinnon 2008; Murata et al. 2005; Ramsey et al. 2006). The evolutionary relationship between Hv channels and the VGIC superfamily is uncertain because they lack the defining central pore domain of VGICs (Figure 1.1) and we have therefore not defined them as a tenth structural class of human VGIC here. Hv channels could be extant representatives of an ancestral VSD-like protein that was fused to the VGIC pore via an intergenic recombination or, alternatively, Hvs may have "borrowed" their VSD from a VGIC ancestor via an intergenic recombination event. The more limited phylogenetic spread of Hv channels compared to VSD-containing VGICs favors the latter scenario. Functionally important domain additions via intergenic recombination are common throughout the history of ion channel evolution and will be described in the chapters of this volume where relevant.

Intragenic duplications, in which functional domain(s) of a protein are tandemly duplicated within a single locus to produce a new protein with multiple homologous domains, have also been an important force in VGIC evolution and have produced channels with altered subunit stoichiometry (Figure 1.1). For example, in K2P channels (Chapter 10), an ancestral intragenic duplication linked two Kir-like motifs in tandem (Ketchum et al. 1995). Formation of a canonical fourfold symmetric K^+ channel therefore requires only two K2P subunits and the channels function as dimers instead of tetramers. In voltage-gated Na^+ channels (NaV; Chapter 2) and voltage-gated Ca^{2+} channels (CaV; Chapter 3), two successive intragenic duplications (one VSD/PD motif duplicated to two, and then two VSD/PD motifs duplicated to four) allow these channels to function as fourfold symmetric monomers encoded by a single gene (Mikami et al. 1989; Noda et al. 1984). Duplicated domains within channels can evolve independently leading to functional specialization of each domain. For instance, NaV channels achieve high Na^+ selectivity with an asymmetric selectivity filter (Schlief et al. 1996) derived from the distinct evolutionary history of mutations in each of the four PDs.

1.4 Ion Channel Families and Subfamilies

Taxonomic division of channel superfamilies into gene families and subfamilies has been approached from both evolutionary and functional perspectives, and definitions of family and subfamily have therefore not always been consistent in the literature. Classification at the family and subfamily level is complicated by the fact that some structural classes of channels as described earlier comprise multiple channel types with vastly different functions, while in other cases, functionally similar channels span multiple structural classes or even multiple structurally distinct channel superfamilies (see Ca^{2+}-activated Cl^- channels in Chapter 14). There are nevertheless some helpful taxonomic principles that are widely used in the field, and the gene families described in this volume and listed in Table 1.1 are generally accepted. First, members of a gene family will share a common subunit architecture and most major functional features. Gene family members also share a significantly higher protein sequence identity with each other than with other channels in the larger superfamily or structural class. They will form a unified, separate clade (or branch) in a phylogenetic tree, reflective of a singular evolutionary origin. For example, in animals, the cyclic nucleotide-binding domain (CNBD) structural class (or superfamily) of VGICs has evolved into three distinct gene families (EAG, CNG and HCN) readily identifiable in molecular phylogenies (Figure 1.2). EAG channels are K^+-selective and are gated by depolarization but have lost sensitivity to cyclic nucleotides. HCN channels and CNG channels, on the other hand, retain cyclic nucleotide sensitivity, but have lost K^+ selectivity and are hyperpolarization-gated or voltage-insensitive, respectively.

The EAG family can be further divided into three gene subfamilies based on sequence identity (Eag, Erg and Elk) with more subtle functional differences that underlie distinct physiological roles (Figure 1.2; also see Chapter 5). Gene subfamilies, like gene families, should form one clade within a phylogeny indicative of a single evolutionary origin. This separate, unified branching is key to the definition, not the absolute level of sequence identity. This is because different types of channels can be more or less tolerant of sequence change and thus their sequences can evolve at different rates even when function is tightly conserved. So how do we draw the line between gene families and subfamilies? In truth, much like the division of species into families and genera, the distinctions can be somewhat arbitrary; an evolutionarily distinct group of channels that constitutes a subfamily to one researcher may be called a family by another. However, in both cases, the researcher is referring to a group of channels that have an exclusive evolutionary relationship. In multimeric ion channels, we sometimes have the opportunity to more clearly differentiate what constitutes a gene subfamily versus a family based on functional criteria. Gene subfamilies can be defined functionally as a group of channels that not only branches separately and exclusively in phylogenies but also as a group in which subunits encoded by different genes can co-assemble to form functional heteromeric channels. The potential for heteromeric assembly closely parallels phylogenetic grouping based on sequence, and co-assembly is indeed subfamily-restricted within the EAG family (Zou et al. 2003; Wimmers et al. 2001). For example, subunits encoded by one Erg gene can co-assemble into functional heteromeric channels with subunits encoded by another Erg gene, but they can't co-assemble with Eag or Elk subunits. Channel subfamilies defined in this way can be viewed as functionally independent because genes from different subfamilies will encode separate ion channels with different physiological roles even when expressed in the same cells. Other prominent examples of gene families that can be divided into functionally

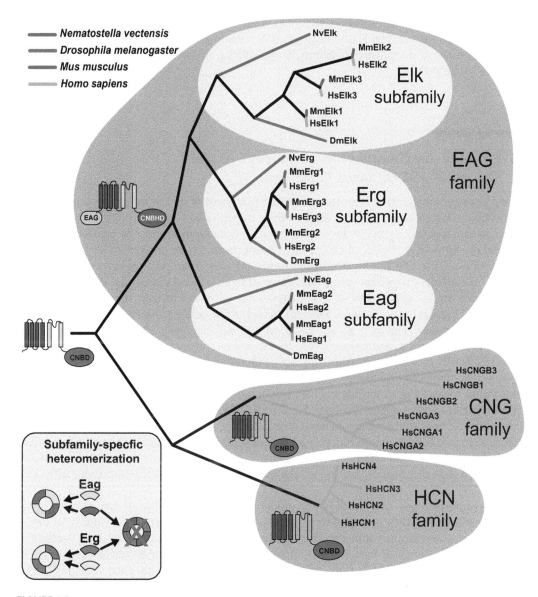

FIGURE 1.2

Bayesian inference phylogeny of animal CNBD-containing VGIC protein sequences illustrates separate clustering of gene families and subfamilies by sequence identity. The EAG, HCN and CNG gene families are shaded in dark gray, and the Eag, Erg and Elk gene subfamilies are shaded in light gray. Subunit structure diagrams are provided to indicate domain structure and are color coded as in Figure 1.1. EAG, HCN and CNG channels are likely to have evolved from a K⁺-selective cyclic nucleotide-modulated ancestor, but the identity of this ancestor and their order of divergence remain undetermined. EAG channels lost cyclic nucleotide modulation but still retain a cyclic nucleotide binding homology domain (CNBHD) (Brelidze, Carlson, and Zagotta 2009). The phylogeny is focused on the EAG family to illustrate the distinct subfamilies and includes EAG channels from human (*Homo sapiens*), mouse (*Mus musculus*), fruit fly (*Drosophila melanogaster*) and sea anemone (*Nematostella vectensis*, Cnidaria). For illustration simplicity, a cnidarian-specific expansion within the Erg subfamily is omitted and only human genes are included for HCN and CNG. Note the clear, separate branching of each gene family and of the three gene subfamilies within EAG. The phylogeny is based on a protein sequence alignment of the structural domains shared across all three gene families: VSD, PD and CNB(H)D. The inset shows and example of rules for subfamily-specific heteromeric assembly for subunits encoded by distinct genes within the EAG family. No subunit mixing occurs between the Erg, Eag and Elk gene subfamilies.

independent subfamilies based on sequence identity and restricted potential for hetero-meric assembly include the Kv voltage-gated K^+ channels (Covarrubias, Wei, and Salkoff 1991) (Chapter 4) and ionotropic glutamate receptors (Ayalon and Stern-Bach 2001; Partin et al. 1993) (Chapter 16).

1.5 Evolutionary History of Ion Channels: Ancient Structures and Dynamic Gene Sets

Prior to the genomic era, it was often assumed that most of our human ion channel types would be restricted to animals because of their unique reliance on rapid electrical signaling in neuromuscular systems. However, genome data from across the tree of life shows that the evolutionary origins of our channel superfamilies lie much farther back in time and often even predate the evolution of eukaryotes. For example, not only did the VGIC superfamily evolve in prokaryotes, but six of nine structural classes of VGICs found in humans (Figure 1.1) can be traced to prokaryotic origins. All nine structural classes were present in ancestral eukaryotes based on broad phylogenetic distribution in the major eukaryotic lineages (Figure 1.3). The reduced channel diversity observed in land plants and fungi (compared to animals) used to be viewed as support for diversification of channels specifically within the animal lineage, but we now know it represents recent loss of ancestral channels selectively in plants and fungi (Figure 1.3). While vascular plants retain only four of the nine major structural classes of VGICs found in animals, all nine are present in various algae lineages. Kir-like, BK-like and CaV-like channels, which are absent in true vascular plants, can even be found in mosses (Edel et al. 2017; Jegla, Busey, and Assmann 2018). Gene loss has played an underappreciated and highly significant role in ion channel evolution and the "channelomes" of living organisms represent a complex interplay of gene duplication and gene loss over vast time frames (Liebeskind, Hillis, and Zakon 2015).

While superfamilies of channels are ancient, many of the animal ion channel gene families and subfamilies described in this volume are indeed uniquely animal, or are at least restricted to animals and choanoflagellates, our closest extant protozoan relatives. Diversification of ancient ion channel superfamilies into the gene families observed in extant organisms has been largely independent in the major eukaryotic lineages (Jegla, Busey, and Assmann 2018). But the expectation that evolutionary innovation of channel types would parallel evolution of nervous system complexity in animals (reaching a pinnacle in the vertebrates) hasn't borne out. Instead, the evolutionary emergence of the major animal ion channel gene families and subfamilies described in this volume was largely complete hundreds of millions of years ago prior to the divergence of cnidarians (including sea anemones, jellyfish and coral) and bilaterians (including vertebrates, insects and mollusks) (Jegla et al. 2009; Li, Liu, et al. 2015; Li, Martinson, et al. 2015). At the time of the cnidarian–bilaterian divergence, animal nervous systems were simple noncentralized nets. Channel evolution has of course continued with independent histories of gene duplications, losses and sequence divergence greatly diversifying the channel sets of each major animal lineage. However, it has occurred within the confines of the channel families and subfamilies we inherited from that simple cnidarian/bilaterian ancestor.

Unraveling that history of gene duplication and loss with genomic and phylogenetic analysis can give us insights into ancestral animal ion channels sets and the nature and

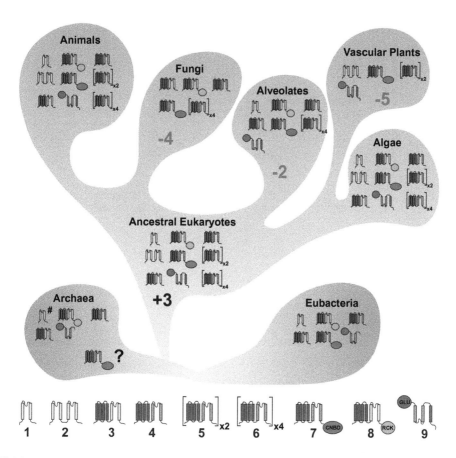

FIGURE 1.3
Phylogenetic distribution of nine human VGIC structural classes in Eubacteria, Archaea and Eukaryota. Icons for structural class are numbered as in Figure 1.1 in the legend at the bottom. Prokaryotic lineages are shaded gray and eukaryotes (shaded tan) are believed to have evolved from the Archaea. Gains and losses of VGIC structural classes within the depicted eukaryotic lineages are noted with a black plus sign (+) or red minus sign (–), respectively. Note all nine VGIC structural classes were present in ancestral eukaryotes and the simplified VGIC sets in Fungi, alveolates (ciliate protozoans, for example) and land plants are the result of gene losses. #: Kir-like channels in the Archaea have RCK domains that are not present in eubacterial and eukaryotic Kir channels. ?: CNBD-containing VGICs are absent in most if not all Archaea genomes, but there are a few putative archaeal sequences that need to be experimentally verified to determine whether this structural class of VGICs is truly absent in the Archaea.

timing of important events in the evolutionary history of our channel families. Evolutionary analysis is especially useful from a biomedical perspective because it can tell us how individual human ion channels are related to the ion channels we genetically manipulate in model organisms. It can help determine which channels are true orthologs (channels that are encoded by the same gene in different species) and which channels are merely paralogs (channels encoded by different genes within the same gene family or subfamily). Compared to orthologs, paralogs have distinct evolutionary origins and histories of selection that could underlie greater differences in physiological roles. This is critical knowledge for interpreting genetic experiments in model species. One interesting reveal of the genomic era is that two whole-genome duplications in ancestral vertebrates played a dominant role in shaping the vertebrate gene set (Ohno 1999; Smith et al. 2013). Figure 1.4 shows

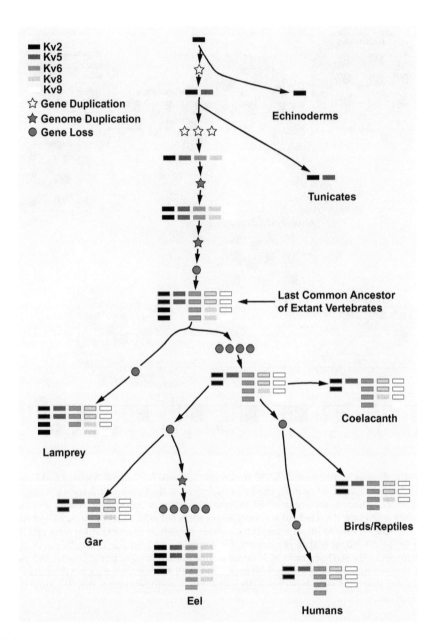

FIGURE 1.4
Evolutionary history of the Kv2 channel subfamily in the deuterostome lineage. Individual genes are depicted as rectangles and shaded according to the legend. Human Kv2 subfamily genes and their 1:1 orthologs (same gene, different species) are outlined in red. The human genes derive from the second of two ancestral vertebrate genome duplications. Teleost fish genes (represented here by the European eel) underwent a third genome duplication, which complicates interpretation of gene orthology relationships with other vertebrates. Note that periods of gene loss follow all three genome duplications. Small-scale duplications of individual genes occurred only prior to the ancestral vertebrate genome duplications in the Kv2 subfamily but have occurred more recently in other channel lineages such a Kv1 subfamily voltage-gated K$^+$ channels and voltage-gated Na$^+$ channels.

an example of how individual duplications of genes combined with these two genome duplications and with gene losses to produce the 12-gene human Kv2 (Shab) subfamily of voltage-gated K⁺ channels (see Chapter 4) from a single Kv2 gene in a deuterostome ancestor. *Note that Kv5,6,8,9 channels depicted in Figure 1.4 belong in the Kv2 subfamily despite their names because they were duplicated from a Kv2 ancestor specifically in the chordate/vertebrate Kv2 lineage, grouped with Kv2 channels in phylogenies and can form functional heteromeric channels when coexpressed with Kv2s* (Li et al. 2015; Bocksteins 2016). The human Kv2 gene subfamily is the product of four single-gene duplications, two genome duplications and seven gene losses (Figure 1.4). Gene loss is frequent following genome duplications presumably due to functional redundancy and has dominated the recent evolution of the vertebrate Kv2 subfamily. Despite gene losses and continued small-scale duplications in some gene families, most vertebrate channels, including those of the human Kv2 subfamily genes shown in Figure 1.4, can trace their origin to the shared ancestral vertebrate genome duplications. Thus, most extant vertebrates share a high degree of 1:1 gene orthology, meaning that most human genes have a clear ortholog in other vertebrate species. However, one should note that teleost fish underwent an additional round of genome duplication (with subsequent gene loss) (Meyer and Van de Peer 2005). The model organism zebrafish thus has a reduced 1:1 orthology with humans (Postlethwait 2006), and is therefore not always the best vertebrate model for study of individual human genes. Humans also share very little 1:1 gene orthology with animals that diverged from us prior to the ancestral vertebrate genome duplications. For example, humans and *Drosophila* share all ion channel families and subfamilies described in this volume, but only a handful of true 1:1 channel orthologs within those families (Jegla et al. 2009). Functional genetic analysis of individual human gene orthologs is therefore best done in vertebrate models, while general shared gene family or subfamily characteristics can be studied in a wide variety of animal models.

Early in the genomic era, it was a common perception that genome complexity increases with increasing physiological and nervous system complexity because the first invertebrate model organism genomes from species such as *Drosophila* and *Caenorhabditis elegans* were genetically simple compared to human and mouse. For example, *Drosophila* has only 5 Kv family voltage-gated K⁺ channels, while humans have 27 (Jegla et al. 2009). While it is true that vertebrates have a comparatively large gene complement due to ancestral genome duplications, animal genome size and ion channel diversity does not closely track the anatomical size and complexity of the nervous system. Now we know that teleost fish are the most gene-rich vertebrates and early diverging metazoans with simple nerve nets (such as cnidarians and ctenophores) have the most Kv family voltage-gated K⁺ channels (40-plus) (Jegla et al. 2012, Li et al. 2015). Similarly, ion channel diversity is much greater in simple algae than in anatomically complex land plants (Figure 1.2). The result of this evolutionary lottery of gene duplication is that the best model species for addressing major questions in ion channel biology will vary, and selection of a model should consider both the physiology and the evolutionary history of the channels that underlie it.

1.6 Much of the Genetic Diversity of Ion Channels Remains Unexplored

One intriguing aspect of the ancient origins and broad phylogenetic distribution of many ion channel superfamilies is that much of the sequence and structural diversity of ion channels remain unexplored on a functional level (Pozdnyakov, Safonov, and Skarlato

2020). Here we focused on looking back in evolutionary time to explore the origins of human ion channel diversity, but there have been many other experiments in channel structure and function throughout the tree of life. For example, multiple additional evolutionarily independent tandem intragenic duplications of VGIC channels exist (like those found in K2P, Ca2P and NaV/CaV channels) in fungal, protozoal and even animal lineages (Pozdnyakov, Safonov, and Skarlato 2020; Lesage et al. 1996). Furthermore, depolarization and hyperpolarization gating phenotypes have evolved independently in both plant and animal CNBD-containing channels (Riedelsberger, Dreyer, and Gonzalez 2015; Jegla, Busey, and Assmann 2018). Most major eukaryotic lineages have unique channel families that collectively represent tens of billions of years of functional evolution and sequence diversification. Yet the vast majority of what we have learned about channel structure and function comes from a few representative animal and plant channels. More broadly, exploring the functional diversity of channels across the phylogenetic tree of life could provide important new insights into the structural and mechanistic basis of channel gating.

Suggested Reading

Ayalon, G., and Y. Stern-Bach. 2001. "Functional assembly of AMPA and kainate receptors is mediated by several discrete protein-protein interactions." *Neuron* 31 (1):103–13. doi:10.1016/s0896-6273(01)00333-6.

Bocksteins, E. 2016. "Kv5, Kv6, Kv8, and Kv9 subunits: no simple silent bystanders." *J Gen Physiol* 147 (2):105–25. doi:10.1085/jgp.201511507.

Brelidze, T. I., A. E. Carlson, and W. N. Zagotta. 2009. "Absence of direct cyclic nucleotide modulation of mEAG1 and hERG1 channels revealed with fluorescence and electrophysiological methods." *J Biol Chem* 284 (41):27989–97. doi:10.1074/jbc.M109.016337.

Chen, G. Q., C. Cui, M. L. Mayer, and E. Gouaux. 1999. "Functional characterization of a potassium-selective prokaryotic glutamate receptor." *Nature* 402 (6763):817–21. doi:10.1038/45568.

Covarrubias, M., A. A. Wei, and L. Salkoff. 1991. "Shaker, Shal, Shab, and Shaw express independent K+ current systems." *Neuron* 7 (5):763–73. doi:10.1016/0896-6273(91)90279-9.

Doyle, D. A., J. Morais Cabral, R. A. Pfuetzner, A. Kuo, J. M. Gulbis, S. L. Cohen, B. T. Chait, and R. MacKinnon. 1998. "The structure of the potassium channel: molecular basis of K+ conduction and selectivity." *Science* 280 (5360):69–77.

Edel, K. H., E. Marchadier, C. Brownlee, J. Kudla, and A. M. Hetherington. 2017. "The evolution of calcium-based signalling in plants." *Curr Biol* 27 (13):R667–R679. doi:10.1016/j.cub.2017.05.020.

Jegla, T., G. Busey, and S. M. Assmann. 2018. "Evolution and structural characteristics of plant voltage-gated K(+) channels." *Plant Cell* 30 (12):2898–909. doi:10.1105/tpc.18.00523.

Jegla, T., H. Q. Marlow, B. Chen, D. K. Simmons, S. M. Jacobo, and M. Q. Martindale. 2012. "Expanded functional diversity of shaker K(+) channels in cnidarians is driven by gene expansion." *PLOS ONE* 7 (12):e51366. doi:10.1371/journal.pone.0051366.

Jegla, T. J., C. M. Zmasek, S. Batalov, and S. K. Nayak. 2009. "Evolution of the human ion channel set." *Comb Chem High Throughput Screen* 12 (1):2–23.

Ketchum, K. A., W. J. Joiner, A. J. Sellers, L. K. Kaczmarek, and S. A. Goldstein. 1995. "A new family of outwardly rectifying potassium channel proteins with two pore domains in tandem." *Nature* 376 (6542):690–5. doi:10.1038/376690a0.

Lee, S. Y., J. A. Letts, and R. Mackinnon. 2008. "Dimeric subunit stoichiometry of the human voltage-dependent proton channel Hv1." *Proc Natl Acad Sci U S A* 105 (22):7692–5. doi:10.1073/pnas.0803277105.

Lesage, F., E. Guillemare, M. Fink, F. Duprat, M. Lazdunski, G. Romey, and J. Barhanin. 1996. "A pH-sensitive yeast outward rectifier K+ channel with two pore domains and novel gating properties." *J Biol Chem* 271 (8):4183–7. doi:10.1074/jbc.271.8.4183.

Li, M., X. Zhou, S. Wang, I. Michailidis, Y. Gong, D. Su, H. Li, X. Li, and J. Yang. 2017. "Structure of a eukaryotic cyclic-nucleotide-gated channel." *Nature* 542 (7639):60–5. doi:10.1038/nature20819.

Li, X., A. S. Martinson, M. J. Layden, F. H. Diatta, A. P. Sberna, D. K. Simmons, M. Q. Martindale, and T. J. Jegla. 2015. "Ether-a-go-go family voltage-gated K+ channels evolved in an ancestral metazoan and functionally diversified in a cnidarian-bilaterian ancestor." *J Exp Biol* 218 (4):526–36. doi:10.1242/jeb.110080.

Li, X., H. Liu, J. Chu Luo, S. A. Rhodes, L. M. Trigg, D. B. van Rossum, A. Anishkin, F. H. Diatta, J. K. Sassic, D. K. Simmons, B. Kamel, M. Medina, M. Q. Martindale, and T. Jegla. 2015. "Major diversification of voltage-gated K+ channels occurred in ancestral parahoxozoans." *Proc Natl Acad Sci U S A* 112 (9):E1010–9. doi:10.1073/pnas.1422941112.

Liebeskind, B. J., D. M. Hillis, and H. H. Zakon. 2015. "Convergence of ion channel genome content in early animal evolution." *Proc Natl Acad Sci U S A* 112 (8):E846–51. doi:10.1073/pnas.1501195112.

Long, S. B., E. B. Campbell, and R. Mackinnon. 2005. "Crystal structure of a mammalian voltage-dependent Shaker family K+ channel." *Science* 309 (5736):897–903. doi:10.1126/science.1116269.

Meyer, A., and Y. Van de Peer. 2005. "From 2R to 3R: evidence for a fish-specific genome duplication (FSGD)." *BioEssays* 27 (9):937–45. doi:10.1002/bies.20293.

Mikami, A., K. Imoto, T. Tanabe, T. Niidome, Y. Mori, H. Takeshima, S. Narumiya, and S. Numa. 1989. "Primary structure and functional expression of the cardiac dihydropyridine-sensitive calcium channel." *Nature* 340 (6230):230–3. doi:10.1038/340230a0.

Murata, Y., H. Iwasaki, M. Sasaki, K. Inaba, and Y. Okamura. 2005. "Phosphoinositide phosphatase activity coupled to an intrinsic voltage sensor." *Nature* 435 (7046):1239–43. doi:10.1038/nature03650.

Nishida, M., and R. MacKinnon. 2002. "Structural basis of inward rectification: cytoplasmic pore of the G protein-gated inward rectifier GIRK1 at 1.8 A resolution." *Cell* 111 (7):957–65. doi:10.1016/s0092-8674(02)01227-8.

Noda, M., S. Shimizu, T. Tanabe, T. Takai, T. Kayano, T. Ikeda, H. Takahashi, H. Nakayama, Y. Kanaoka, and N. Minamino. 1984. "Primary structure of electrophorus electricus sodium channel deduced from cDNA sequence." *Nature* 312 (5990):121–7. doi:10.1038/312121a0.

Ohno, S. 1999. "Gene duplication and the uniqueness of vertebrate genomes circa 1970–1999." *Semin Cell Dev Biol* 10 (5):517–22. doi:10.1006/scdb.1999.0332.

Panchenko, V. A., C. R. Glasser, and M. L. Mayer. 2001. "Structural similarities between glutamate receptor channels and K(+) channels examined by scanning mutagenesis." *J Gen Physiol* 117 (4):345–60. doi:10.1085/jgp.117.4.345.

Partin, K. M., D. K. Patneau, C. A. Winters, M. L. Mayer, and A. Buonanno. 1993. "Selective modulation of desensitization at AMPA versus kainate receptors by cyclothiazide and concanavalin A." *Neuron* 11 (6):1069–82. doi:10.1016/0896-6273(93)90220-l.

Postlethwait, J. H. 2006. "The zebrafish genome: a review and msx gene case study." *Genome Dyn* 2:183–97. doi:10.1159/000095104.

Pozdnyakov, I., P. Safonov, and S. Skarlato. 2020. "Diversity of voltage-gated potassium channels and cyclic nucleotide-binding domain-containing channels in eukaryotes." *Sci Rep* 10 (1):17758. doi:10.1038/s41598-020-74971-4.

Ramsey, I. S., M. M. Moran, J. A. Chong, and D. E. Clapham. 2006. "A voltage-gated proton-selective channel lacking the pore domain." *Nature* 440 (7088):1213–6. doi:10.1038/nature04700.

Riedelsberger, J., I. Dreyer, and W. Gonzalez. 2015. "Outward rectification of voltage-gated K+ channels evolved at least twice in life history." *PLOS ONE* 10 (9):e0137600. doi:10.1371/journal.pone.0137600.

Schlief, T., R. Schönherr, K. Imoto, and S. H. Heinemann. 1996. "Pore properties of rat brain II sodium channels mutated in the selectivity filter domain." *Eur Biophys J* 25 (2):75–91. doi:10.1007/s002490050020.

Smith, J. J., S. Kuraku, C. Holt, T. Sauka-Spengler, N. Jiang, M. S. Campbell, M. D. Yandell, T. Manousaki, A. Meyer, O. E. Bloom, J. R. Morgan, J. D. Buxbaum, R. Sachidanandam, C. Sims, A. S. Garruss, M. Cook, R. Krumlauf, L. M. Wiedemann, S. A. Sower, W. A. Decatur, J. A. Hall, C. T. Amemiya, N. R. Saha, K. M. Buckley, J. P. Rast, S. Das, M. Hirano, N. McCurley, P. Guo, N. Rohner, C. J. Tabin, P. Piccinelli, G. Elgar, M. Ruffier, B. L. Aken, S. M. Searle, M. Muffato, M. Pignatelli, J. Herrero, M. Jones, C. T. Brown, Y. W. Chung-Davidson, K. G. Nanlohy, S. V. Libants, C. Y. Yeh, D. W. McCauley, J. A. Langeland, Z. Pancer, B. Fritzsch, P. J. de Jong, B. Zhu, L. L. Fulton, B. Theising, P. Flicek, M. E. Bronner, W. C. Warren, S. W. Clifton, R. K. Wilson, and W. Li. 2013. "Sequencing of the sea lamprey (*Petromyzon marinus*) genome provides insights into vertebrate evolution." *Nat Genet* 45 (4):415–21, 421.e1–2. doi:10.1038/ng.2568.

Su, Y., W. R. Dostmann, F. W. Herberg, K. Durick, N. H. Xuong, L. Ten Eyck, S. S. Taylor, and K. I. Varughese. 1995. "Regulatory subunit of protein kinase A: structure of deletion mutant with cAMP binding domains." *Science* 269 (5225):807–13. doi:10.1126/science.7638597.

Weber, I. T., and T. A. Steitz. 1987. "Structure of a complex of catabolite gene activator protein and cyclic AMP refined at 2.5 A resolution." *J Mol Biol* 198 (2):311–26. doi:10.1016/0022-2836(87)90315-9.

Whicher, J. R., and R. MacKinnon. 2016. "Structure of the voltage-gated K(+) channel Eag1 reveals an alternative voltage sensing mechanism." *Science* 353 (6300):664–9. doi:10.1126/science.aaf8070.

Wimmers, S., I. Wulfsen, C. K. Bauer, and J. R. Schwarz. 2001. "Erg1, erg2 and erg3 K channel subunits are able to form heteromultimers." *Pflugers Arch Eur J Physiol* 441 (4):450–5.

Yuan, P., M. D. Leonetti, A. R. Pico, Y. Hsiung, and R. MacKinnon. 2010. "Structure of the human BK channel Ca2+-activation apparatus at 3.0 A resolution." *Science* 329 (5988):182–6. doi:10.1126/science.1190414.

Zagotta, W. N., N. B. Olivier, K. D. Black, E. C. Young, R. Olson, and E. Gouaux. 2003. "Structural basis for modulation and agonist specificity of HCN pacemaker channels." *Nature* 425 (6954):200–5. doi:10.1038/nature01922.

Zou, A., Z. Lin, M. Humble, C. D. Creech, P. K. Wagoner, D. Krafte, T. J. Jegla, and A. D. Wickenden. 2003. "Distribution and functional properties of human KCNH8 (Elk1) potassium channels." *Am J Physiol Cell Physiol* 285 (6):C1356–66. doi:10.1152/ajpcell.00179.2003.

2

Voltage-Gated Sodium Channels

William A. Catterall

CONTENTS

2.1 Functional Roles of Voltage-Gated Sodium Channels

2.1.1 Action Potential Generation and Propagation

Electrical signaling in nerve, muscle and endocrine cells depends on initiation of action potentials by voltage-gated sodium channels, as revealed using the voltage-clamp technique (Hodgkin and Huxley 1952). By imposing a rapid depolarization upon the membrane of the giant nerve axon of the squid, it was shown that sodium (Na^+) channels rapidly activate (within 1 ms) and then rapidly inactivate (within 5 ms). The brief pulse of inward Na^+ current produced by activation of channels is responsible for the rapidly rising (depolarizing) phase of the action potential and for its rapid conduction along nerve and muscle fibers. These studies first revealed the transient Na^+ currents as the mechanism of initiation and propagation of action potentials.

DOI: 10.1201/9781003096276-2

2.1.2 Activation, Conductance and Two Phases of Inactivation

The voltage-clamp studies of Hodgkin and Huxley also revealed the three essential functions of Na^+ channels: voltage-dependent activation, fast inactivation and selective Na^+ conductance (Hodgkin and Huxley 1952). They showed that activation of the Na^+ current of squid giant axon is steeply dependent on transmembrane voltage and has sigmoid activation kinetics, consistent with the requirement for movement of three electrically charged "gating particles" across the membrane electric field during activation. In contrast, fast inactivation followed an exponential time course as if a single gating particle controlled this process. Their m^3h formulation of Na^+ channel-gating kinetics has withstood six decades of tests as a quantitative description of the Na^+ current.

The essence of voltage-dependent gating is the movement of gating charges that are part of the Na^+ channel protein across the membrane electric field (Hodgkin and Huxley 1952). This essential gating charge movement was measured directly by the voltage-clamp method (Armstrong and Bezanilla 1973). In the absence of permeant ions, a capacitive current was detected in the squid giant axon that correlated closely with the activation of Na^+ channels. Subsequent studies with higher resolution methods showed that the gating charge movement that drives Na^+ channel activation amounts to transit of 12–16 positive charges across the full membrane electric field (Hirschberg et al. 1995; Kuzmenkin, Bezanilla, and Correa 2004). This voltage-driven movement of gating charge is coupled to a conformational change that opens the pore of the Na^+ channel within 1 ms.

Once open, the Na^+ channel is highly selective for ions. Na^+ is approximately tenfold more permeant than K^+ and fiftyfold more permeant than Ca^{2+} (Hille 1971, 1972). Detailed studies of permeation, saturation, and block by inorganic and organic monovalent cations led Hille to a model of the Na^+ channel selectivity filter as a high-field-strength site with one or more carboxyl groups poised to replace some or all of the waters of hydration of the Na^+ ion in a specific way and coordinate the permeating Na^+ ion as it moves over a series of four potential energy barriers and interacts with a series of three coordination sites on its way through the pore (Hille 1975). This conceptual model of Na^+ permeation and selectivity agrees well with the emerging structural models of the ion selectivity filter of Na^+ channels (see later).

In addition to the fast inactivation process discovered by Hodgkin and Huxley, Na^+ channels have a second slow inactivation process that is engaged during prolonged single depolarizations and trains of repetitive depolarizations on the time scale of hundreds of milliseconds to seconds (Rudy 1978). In a physiological setting, this slow inactivation process is important to limit the frequency of firing and define the length of trains of action potentials in nerve and muscle and to protect cells against excitotoxic injury (Vilin and Ruben 2001). In the central nervous system, the slow inactivation process contributes directly to encoding information in the frequency and duration of action potential trains (Vilin and Ruben 2001).

2.1.3 Persistent and Resurgent Sodium Currents

In addition to the classical transient Na^+ current, Na^+ channels have two additional well-described modes of activity. The transient Na^+ current inactivates incompletely in nerve and muscle cells, leaving a small, persistent Na^+ current in the range of 1% of the amplitude of peak Na^+ current (Crill 1996). This persistent Na^+ current is important in neuronal dendrites and cell bodies where it boosts the size of excitatory postsynaptic potentials and helps to bring the cell to threshold during trains of action potentials. Many classes of neurons also generate resurgent Na^+ currents, which appear as a rebound inward current following a voltage-clamp pulse or an action potential (Raman and Bean 1997). These currents are caused by reopening Na^+ channels from closed or inactivated states, and therefore are distinct from persistent Na+ currents that are caused by prolonged activity of previously open Na+ channels. Resurgent Na+ currents also contribute repetitive firing of action potentials by providing depolarizing drive following inactivation of the transient Na+ current and repolarization of the cell by K+ currents. Persistent and resurgent Na+ currents can work together with slow inactivation to generate complex patterns of firing of action potentials in central neurons (Do and Bean 2003).

2.2 Discovery and Biochemical Properties of the Sodium Channel Protein

2.2.1 Identification, Purification and Reconstitution of Sodium Channels

Because of their essential role in generation of action potentials, Na+ channel-directed neurotoxins have independently evolved in marine dinoflagellates found in plankton, sea anemones and other coelenterates, marine snails, fish, amphibians, spiders, and scorpions. These toxins prevent nerve conduction and induce paralysis through actions at six distinct receptor sites on Na+ channels (Table 2.1). Because of their specificity of interaction with Na+ channels and their ability to bind to Na+ channels with high affinity, these neurotoxins have been exploited as molecular probes of channel structure and function.

Photoreactive derivatives of the polypeptide toxins of scorpion venom were covalently attached to Na+ channels in intact nerve cell membranes from brain, allowing direct identification of the protein components of Na+ channels (Beneski and Catterall 1980). These experiments revealed large α subunits of 260 kDa and smaller β subunits of 30–40 kDa (Hartshorne and Catterall 1984; Figure 2.1 Left). Reversible binding of saxitoxin and tetrodotoxin to their common receptor site was used as a biochemical assay for the channel protein. Solubilization of brain membranes with nonionic detergents released the Na+ channel, and the solubilized channel was purified by chromatographic techniques that separate glycoproteins by size, charge and composition of covalently attached carbohydrate (Catterall 1984; Figure 2.1 Right). Na+ channel α subunits were also purified

TABLE 2.1

Neurotoxin Receptor Sites on Sodium Channels

Receptor Site	Toxin or Drug	Location (Domain/Segment)
Neurotoxin receptor site 1	Tetrodotoxin	I/P, II/P, III/P, IV/P
	Saxitoxin	I/P, II/P, III/P, IV/P
	μ-Conotoxin	
Neurotoxin receptor site 2	Veratridine	I/S6, IIIS6, IV/S6
	Batrachotoxin	I/S6, IIIS6, IV/S6
	Grayanotoxin	
	Aconitine	IS6, IIS6
Neurotoxin receptor site 3	α-Scorpion toxins	I/S5–IS6, IV/S1–S2, IV/S3–S4
	Sea anemone toxins	IV/S3–S4
Neurotoxin receptor site 4	β-Scorpion toxins	II/S1–S2, II/S3–S4
	Tarantula toxins	
Neurotoxin receptor site 5	Brevetoxins	I/S6, IV/S5
	Ciguatoxins	
Neurotoxin receptor site 6	δ-Conotoxins	IV/S3-S4

Source: Adapted from Catterall et al. (2007).

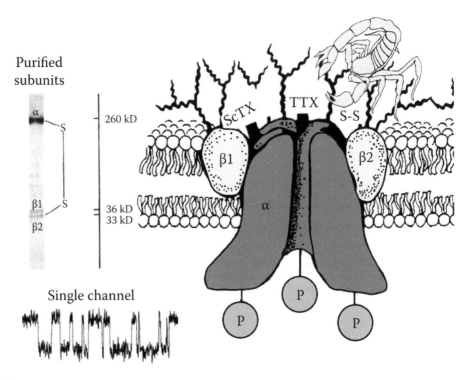

FIGURE 2.1
Subunit structure of voltage-gated sodium channels. Left. SDS polyacrylamide gel electrophoresis patterns illustrating the α and β subunits of the brain Na⁺ channels. (Left) Na⁺ channel purified from rat brain showing the α, β1 and β2 subunits and their molecular weights (Hartshorne and Catterall 1984). As illustrated, the α and β2 subunits are linked by a disulfide bond. Tetrodotoxin and scorpion toxins bind to the α subunits of Na⁺ channels as indicated and were used as molecular tags to identify and purify the Na⁺ channel protein from brain (Beneski and Catterall 1980; Hartshorne and Catterall 1984). (Inset) Single-channel currents conducted by a single purified Na⁺ channel incorporated into a planar bilayer (Hartshorne et al. 1985). Right. Drawing of the subunit structure of the brain Na⁺ channel based on biochemical data (Catterall 1984).

from eel electroplax and a similar complex of α and β subunits was purified from skeletal muscle (Agnew et al. 1980; Barchi 1983). An important step in the study of a purified membrane transport protein is to reconstitute its function in the pure state. This was accomplished by reconstitution of the Na⁺ channel in phospholipid vesicles and in planar phospholipid bilayers (Catterall 1984; Hartshorne et al. 1985). Recordings of Na⁺ currents from single purified brain Na⁺ channels showed that they retain the voltage dependence, ion selectivity and pharmacological properties of native channels, confirming that an active Na⁺ channel protein was purified in functional form (Hartshorne et al. 1985; Figure 2.1, Inset).

2.2.2 Primary Structures of Sodium Channel Subunits

The amino acid sequences of the Na⁺ channel α, β1 and β2 subunits were determined by cloning DNA complementary to their mRNAs, using antibodies and oligonucleotide probes developed from work on purified Na⁺ channels (Noda et al. 1984). The primary

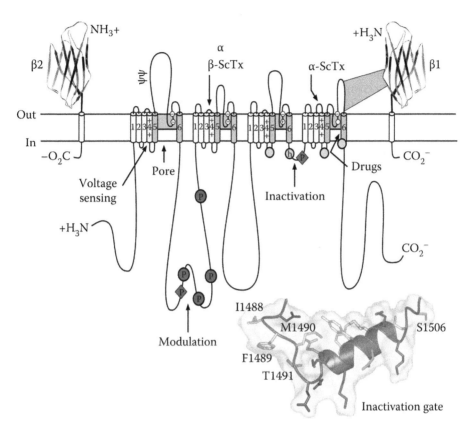

FIGURE 2.2
The primary structures of the subunits of the voltage-gated sodium channels. Cylinders represent probable alpha helical segments. Bold lines represent the polypeptide chains of each subunit with length approximately proportional to the number of amino acid residues in the brain Na^+ channel subtypes. The extracellular domains of the β1 and β2 subunits are shown as immunoglobulin-like folds. Ψ, sites of probable N-linked glycosylation; P in red circles and diamonds, sites of demonstrated protein phosphorylation by PKA (circles) and PKC (diamonds); green, pore-lining segments; white circles, the outer and inner (DEKA) rings of amino residues that form the ion selectivity filter and the tetrodotoxin binding site; yellow, S4 voltage sensors; h in blue circle, inactivation particle in the inactivation gate loop; blue circles, sites implicated in forming the inactivation gate receptor. Sites of binding of α- and β-scorpion toxins and a site of interaction between α and β1 subunits are also shown. Tetrodotoxin is a specific blocker of the pore of Na^+ channels, whereas the α- and β-scorpion toxins block fast inactivation and enhance activation, respectively, and thereby generate persistent Na^+ current that causes depolarization block of nerve conduction. Tetrodotoxin has been used as a tool to probe the pore of the Na^+ channel, whereas the scorpion toxins have been valuable as probes of voltage sensor function. (Inset) Structure of the inactivation gate in solution determined by NMR. (Adapted from Rohl et al. 1999; Catterall 2000.)

structures of these subunits are illustrated as transmembrane folding models in Figure 2.2 (Numa and Noda 1986; Catterall 2000). The large α subunits are composed of 1800–2100 amino acids and contain four repeated domains having greater than 50% internal sequence identity in their transmembrane regions. Each domain contains six segments that form transmembrane α-helices and an additional membrane reentrant loop that forms the outer mouth of the transmembrane pore (see later). In contrast, the smaller β1 and β2 subunits consist of a large extracellular N-terminal segment having a structure similar to antigen-binding regions of immunoglobulin (an Ig-fold), a single transmembrane segment and a short intracellular segment (Isom et al. 1992; Isom et al. 1995; Figure 2.2). Only the principal α subunits of Na$^+$ channels are required for function, but the β subunits increase the level of cell-surface expression and modify the voltage-dependent gating, conferring more physiologically correct functional properties on the expressed α subunits (Isom et al. 1992; Isom et al. 1995). These results indicate that the principal α subunits of the voltage-gated Na$^+$ channels are functionally autonomous, but the auxiliary β subunits improve cell-surface expression, help to determine subcellular localization and modulate physiological properties (Brackenbury and Isom 2011).

2.2.3 Mapping the Molecular Components Required for Sodium Channel Function

Knowledge of the primary structures of the Na$^+$ channel subunits permitted detailed tests of their functional properties. cDNA clones encoding wild-type and mutant channels were expressed in recipient cells and the resulting ion channels were studied by voltage-clamp methods. Much was learned about the molecular basis of Na$^+$ channel function by analyzing the effects of mutations, specific antibodies, drugs and toxins on Na$^+$ channels expressed in recipient cells from their cDNAs. These insights are illustrated in the form of a color-coded molecular map of Na$^+$ channel functions in Figure 2.2 (Catterall 2000). The voltage sensor is composed of the S1–S4 segments in each domain (Figure 2.2, white, yellow; Catterall 2010). The S4 segment contains four to eight repeated motifs having a positively charged amino acid residue (usually arginine) between two hydrophobic residues (Figure 2.2, yellow, +). Mutation of these arginine residues shifts the voltage dependence of gating and reduces the steepness of its voltage dependence (Stühmer et al. 1989; Kontis, Rounaghi, and Goldin 1997), showing that they are the gating charges that sense changes in the electric field across the cell membrane and initiate activation of Na$^+$ channels. Voltage sensors in all four domains contribute to the voltage-sensing process, but the outward movements of the S4 segments are sequential, with domains I, II and III moving rapidly but in sequence, and then domain IV moving more slowly (Cha et al. 1999; Chanda and Bezanilla 2002). The S5 and S6 segments and the P loop between them form the central pore (Figure 2.2, green). The pore motif was first identified by determination of the amino acid residues that are essential for binding the pore-blocking toxin tetrodotoxin (Terlau et al. 1991; Noda et al. 1989; Figure 2.2). Subsequent experiments showed that these amino acid residues also control ion selectivity of Na$^+$ channels (Heinemann, Terlau, and Imoto 1992; Favre, Moczydlowski, and Schild 1996; Sun et al. 1997). The fast inactivation gate, which inactivates Na$^+$ channels within a few milliseconds of opening, is formed by the intracellular loop connecting domains III and IV of the Na$^+$ channel α subunit (Figure 2.2). It was identified by antipeptide antibodies directed against it, which block the fast inactivation process (Vassilev, Scheuer, and Catterall 1988), and further studied by site-directed mutagenesis (Stühmer et al. 1989; West et al. 1992). This loop is thought to fold into the Na$^+$ channel structure and block the pore from the intracellular end during inactivation (Vassilev, Scheuer, and Catterall 1988; West et al. 1992; Figure 2.2). The structure of the fast inactivation gate peptide was determined in solution by nuclear magnetic resonance (NMR) methods (Rohl et al. 1999; Figure 2.2, inset). It consists of a rigid α-helix preceded in the amino acid sequence by two loops that contain the conserved hydrophobic motif iso-leucine–phenylalanine–methionine (IFM) (West et al. 1992). The inactivation gate bends at hinge glycine residues and the crucial IFM motif folds into a receptor site near the intracellular mouth of the pore, binds tightly and serves as a molecular latch to keep the inactivation gate closed (Kellenberger et al. 1997).

2.3 Three-Dimensional Structure of the Sodium Channel

2.3.1 Structure of Bacterial Sodium Channels

Analysis of the human genome revealed that there are 143 ion channel proteins whose pore-forming segments are related to Na^+ channels, and they are associated with at least ten distinct families of auxiliary subunits (Yu and Catterall 2004). The voltage-gated ion channels and their molecular relatives are one of the largest superfamilies of membrane signaling proteins and one of the most prominent targets for drugs used in the therapy of human diseases. The voltage-gated Na^+ channels were the founders of this large superfamily in terms of discovery of their function by Hodgkin and Huxley, and the later discovery of the Na^+ channel protein itself. Surprisingly, the Na^+ channel family is also ancient in evolution (Ren et al. 2001). The bacterial Na^+ channel NaChBac and its prokaryotic relatives are composed of homotetramers of a single subunit whose structure resembles one of the domains of a vertebrate Na^+ channel (Ren et al. 2001). It is likely that these bacterial Na^+ channels are the evolutionary ancestors of the larger, four-domain Na^+ channels in eukaryotes, and a similar bacterial channel may have been the molecular ancestor of Ca^{2+} channels.

Na^+ channel architecture was revealed in three dimensions by determination of the structure of the bacterial channel Na_VAb from *Arcobacter butzleri* at high resolution (2.7 Å) by X-ray crystallography (Payandeh et al. 2011; Figure 2.3). This ancient Na^+ channel has a very simple structure: four identical subunits that each is similar to one homologous domain of a mammalian Na^+ channel without the large intracellular and extracellular loops of the mammalian protein (Payandeh et al. 2011). This structure has revealed a wealth of new information about the structural basis for voltage-dependent gating, ion selectivity and conductance, and the mechanism for blocking of the channel by therapeutically important drugs. As viewed from the top, Na_VAb has a central pore surrounded by four pore-forming modules composed of S5 and S6 segments and the intervening pore loop (Figure 2.3a, blue). Four voltage-sensing modules composed of S1–S4 segments are symmetrically associated with the outer rim of the pore module (Figure 2.3a, green/red). The transmembrane architecture of Na_VAb shows that the adjacent subunits have swapped their functional domains such that each voltage-sensing module is most closely associated with the pore-forming module of its neighbor (Payandeh et al. 2011; Figure 2.3b, c). It is likely that this domain-swapped arrangement enforces concerted gating of the four subunits or domains of Na^+ channels.

2.3.2 The Pore Contains a High-Field-Strength Carboxyl Site and Two Carbonyl Sites

The overall pore architecture includes a large external vestibule, a narrow ion selectivity filter containing the amino acid residues shown to determine ion selectivity in vertebrate Na^+ and Ca^{2+} channels, a large central cavity that is lined by the S6 segments and is water filled, and an intracellular activation gate formed at the crossing of the S6 segments at the intracellular surface of the membrane (Payandeh et al. 2011; Figure 2.3d). The activation gate is tightly closed in the Na_VAb structure (Figure 2.3e), and there is no space for ions or water to move through it. This general architecture resembles voltage-gated K^+ channels (see Chapter 4).

Although the overall pore architecture of Na^+ and K^+ channels is similar, the structures of their ion selectivity filters and their mechanisms of ion selectivity and conductance are

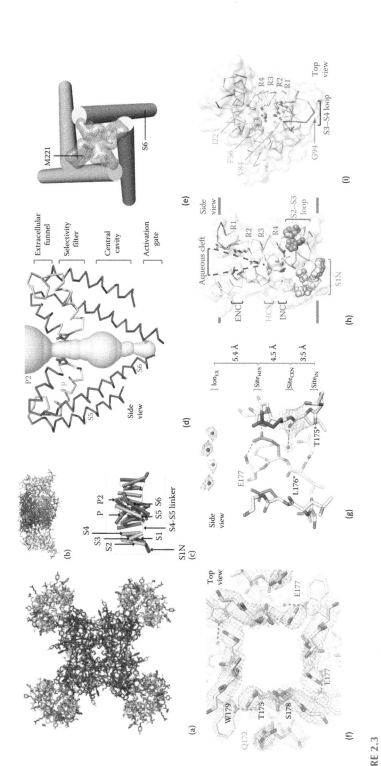

FIGURE 2.3

Structure of the bacterial sodium channel Na$_V$Ab. (a) Top view of Na$_V$Ab channels colored according to crystallographic temperature factors of the main chain (blue <50 Å^2 to red >150 Å^2). The four pore modules in the center are rigid in the crystal structure and therefore are blue. The four voltage-sensing modules surround the pore and are more mobile, as illustrated by warmer colors. (b) Side view of Na$_V$Ab. (c) Structural elements in Na$_V$Ab. The structural components of one subunit are highlighted (1–6, transmembrane segments S1–S6). (Adapted from Payandeh et al. 2011.) (d) Architecture of the Na$_V$Ab pore. Glu177 side chains, purple; pore volume, gray. The P and P2 alpha helices that form the scaffold for the selectivity filter and outer vestibule are shown in green and red, respectively. (e) The closed activation gate at the intracellular end of the pore illustrating the close interaction of side chains the four Met221 residues in closing the pore. (f) Top view of the ion selectivity filter. Symmetry-related molecules are colored white and yellow; P-helix residues are colored green. Hydrogen bonds between Thr175 and Trp179 are indicated by gray dashes. (g) Side view of the ion selectivity filter. Glu177 (E177, purple) interactions with Gln172, Ser178 and the backbone of Ser180 are shown in the far subunit; putative cations or water molecules (red spheres, Ion$_{EX}$). Electron-density around Leu176 (L176, gray) and a bound water molecule are shown in gray. Na$^+$-coordination sites: Site$_{HFS}$, Site$_{CEN}$ and Site$_{IN}$. (Adapted from Payandeh et al. 2011). (h) Side view of the voltage-sensing module of Na$_V$Ab illustrating the conformations of the S1–S4 helices, the extracellular aqueous cleft, the R1–R4 gating charges (blue), extracellular negative cluster (ENC, red), intracellular negative cluster (INC, red) and hydrophobic constriction site (HCS). Residue N49 has been mutated to K in this structure, N49K. (Adapted from Wisedchaisri et al. 2019.)

completely different. K⁺ channels select K⁺ by direct interaction with a series of four ion coordination sites formed by the backbone carbonyls of the amino acid residues that comprise the ion selectivity filter (Chapter 4; see also Volume I, Chapter 1). No charged amino acid residues are involved, and no water molecules intervene between K⁺ and its interacting backbone carbonyls in the ion selectivity filter of K⁺ channels. In contrast, the Na$_V$Ab ion selectivity filter has a high-field-strength site at its extracellular end (Payandeh et al. 2011; Figure 2.3f), which is formed by four glutamate residues in the positions of the key determinants of ion selectivity in vertebrate Na⁺ and Ca²⁺ channels (Figure 2.2). Considering its dimensions of approximately 4.6 Å square, Na⁺ with two planar waters of hydration could fit in this high-field-strength site. This outer site is followed by two ion coordination sites formed by backbone carbonyls (Figure 2.3g). These two carbonyl sites are perfectly designed to bind Na⁺ with four planar waters of hydration but would be much too large to bind Na⁺ directly. In fact, the Na$_V$Ab selectivity filter is large enough to fit the backbone of the K⁺ channel ion selectivity filter inside it. Thus, the chemistry of Na⁺ selectivity and conductance is opposite to that of K⁺: negatively charged residues interact with Na⁺ to remove most (but not all) of its waters of hydration, and Na⁺ is conducted as a hydrated ion interacting with the pore through its inner shell of bound waters. Theoretical considerations of Na⁺ selectivity, saturation and block predicted an outer high-field-strength site that would partially dehydrate the permeating ion and two inner sites that would conduct and rehydrate the permeant Na⁺ ion in the four-barrier, three-site model of selectivity filter function (Hille 1975). This congruence of theory and structure gives clear insight into the chemistry and biophysics of Na⁺ permeation. The ion selectivity filter of mammalian Na⁺ channels has evolved to contain a different high-field-strength site, as described later.

2.3.3 Voltage-Sensor Structure

As pointed out by Hodgkin and Huxley (1952), the steep voltage dependence of activation of Na⁺ channels implies that "electrically charged particles" must move across the membrane in response to changes in membrane potential and thereby provide the driving force for opening of the Na⁺ channel. The predicted transmembrane movement of these gating charges was detected as a small capacitive gating current in high-resolution voltage-clamp studies, first in the squid giant axon and later in other types of cells (Armstrong and Bezanilla 1973). The S4 segments in each of the homologous domains of Na⁺ channels serve as voltage sensors, and the positively charged arginine and lysine residues at intervals of three in the S4 amino acid sequence serve as the gating charges (reviewed in Catterall 2000). The X-ray crystal structure of the Na$_V$Ab channel provides a high-resolution model of an activated voltage sensor (Figure 2.3h; Payandeh et al. 2011). The four transmembrane helices are organized in two helical hairpins composed of the S1–S2 and the S3–S4 transmembrane segments. The four gating-charge-carrying arginine residues in the S4 segment (R1–R4 in Na$_V$Ab) are arrayed in a sequence across the membrane (Figure 2.3h). Just on the intracellular side of the center of the four-helix bundle, a cluster of hydrophobic residues, including a highly conserved phenylalanine residue (Phe56 in Na$_V$Ab), form the hydrophobic constriction site (HCS), which seals the voltage sensor to prevent transmembrane movement of water and ions (Figure 2.3h, green). An analogous Phe residue is crucial for voltage sensor function in K$_V$ channels (Tao et al. 2010). Gating charges R1–R3 are located on the extracellular side of the HCS, and their arginine side chains interact with the negatively charged side chains of the extracellular negative cluster (ENC; Figure 2.3h, red). The array of gating charges facing the aqueous cleft in the voltage sensor on the extracellular side is illustrated in Figure 2.3i. Gating charge R4 is located on the intracellular side of the

HCS and interacts with the intracellular negative cluster (INC; Figure 2.3h, red). Overall, the structure of the voltage sensor seems designed to catalyze movement of the S4 gating charges through the HCS, exchanging ion-pair partners between the INC and ENC. The voltage sensor in the Na$_V$Ab structure has three of its gating charges on the extracellular side of the HCS (Figure 2.3h; Payandeh et al. 2011). This conformation is nearly identical to the conformation of the voltage sensor in the structure of the K$_V$1.2 channel in its open state (Long et al. 2007). Nevertheless, the activation gate of Na$_V$Ab is tightly closed by interaction of the side chains of Ile217 and Met221 (Figure 2.3d, e). Therefore, it is likely that this Na$_V$Ab structure captured the preopen state, which is an expected intermediate in the activation process in which all four voltage sensors have been activated by depolarization and the intracellular activation gate is still closed, but poised to open rapidly in a concerted conformational change of all four subunits.

2.3.4 Voltage-Dependent Activation Involves a Sliding-Helix Mechanism

Much has been learned about the voltage-dependent gating process from structure–function studies and structural modeling. Toxin-binding studies show that the S3–S4 linkers of the voltage sensors in domains II and IV of Na$^+$ channels are available for binding of large, hydrophilic scorpion toxin polypeptides (60–70 residues) in both the resting and activated states (Rogers et al. 1996; Cestèle et al. 1998). These results place the S4 segment in a transmembrane position in both resting and activated states. Outward movement of the S4 segments in Na$^+$ channel voltage sensors was detected in clever experiments that measured the movement of chemically reactive cysteine residues substituted for the native amino acids in S4 by analyzing functional effects of specific chemical reactions at those substituted cysteines (Yang, George, and Horn 1996). These results are consistent with the sliding-helix model of voltage sensing in which gating charges in the S4 segment move outward and exchange ion-pair partners to allow a low energy pathway of gating charge movement (Figure 2.4a, b; Yarov-Yarovoy et al. 2012; Wisedchaisri et al. 2019). This movement of the S4 helix is proposed to initiate a more general conformational change in each domain in the preopen state. After conformational changes have occurred in all four domains, the transmembrane pore can open in a concerted manner and conduct ions (Wisedchaisri et al. 2019). This structural model shows that the S4 segment and its gating charges move through a narrow gating pore that focuses the transmembrane electric field to a distance of approximately 5 Å normal to the membrane (Starace and Bezanilla 2004) and allows the gating charges to move from an intracellular aqueous vestibule to an extracellular aqueous vestibule with a short transit through the HCS (Figure 2.4a, b; Yarov-Yarovoy et al. 2012; Wisedchaisri et al., 2019). This sliding-helix mechanism of outward movement of the gating charges during the activation process has been supported by extensive studies of the ion-pair interactions of gating charges with the ion-pair partners in the INC and ENC using the disulfide locking method (Yarov-Yarovoy et al. 2012). A consensus of laboratories working on this problem in 2012 supported this model for both sodium and potassium channels (Vargas et al. 2012), and recent structural studies have provided further decisive evidence for this mechanism (Wisedchaisri et al. 2019).

A key to understanding the function of the voltage sensor is to determine its structure in the resting state, but this was challenging because the resting state only exists at strongly negative membrane potentials. By inserting mutations that stabilize the resting state at 0 mV and shift the voltage dependence of gating into the positive membrane potential range, it was possible to capture the resting state by disulfide locking and to determine its structure (Wisedchaisri et al. 2019). This surprising structure revealed the molecular basis for the sliding-helix model of voltage sensing in detail (Figure 2.4a, b; Wisedchaisri et al. 2019).

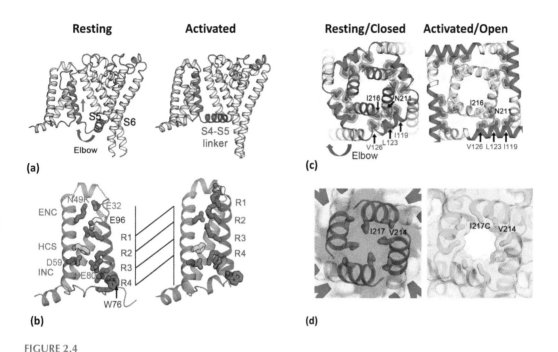

FIGURE 2.4

Sodium channel activation and pore gating models. (a) Structures of the VS are shown as backbone cartoons. S0–S3 are shown in gray, the S3–S4 loop in red and S4 in magenta. A wider and shallower aqueous cleft between the S1–S2 and S3–S4 helix–loop–helix is present in the resting state compared with a deeper cleft in the activated state. (b) Gating charge movement. Four Arg gating charges, R1–R4 (blue); the extracellular negative charge (ENC) cluster of E32 and N49(K) and the intracellular negative charge (INC) cluster of E59 and E80 (red); Phe in the hydrophobic constriction site (HCS, green); and conserved W76 (gray) and E96 (yellow) are shown as sticks. S4 (magenta) moves outward by 11.5 Å , translocating two gating charges through the HCS. Part of S3 is omitted for clarity. (c) Bottom (intracellular) view of the structures in (a), with S0–S3 omitted for clarity. The S4–S5 linker (blue) undergoes a large conformational change that tightens the collar around the S5 and S6 segments of the pore in the resting state (red, closed) and loosens the collar in the activated state (green, open). (d) Spacefilling view of the intracellular mouth of the pore in closed (red) and open (green) states from panel (c). (Adapted from Wisedchaisri et al. 2019.)

In this resting state structure, the S4 segment is pulled nearly 12 Å toward the intracellular side of the membrane, forming a sharply bent elbow at its intracellular end that projects into the cytosol (Figure 2.4a, b, left). This movement of the S4 segment and its attached S4–S5 linker twists the intracellular ends of the S6 segment that form the activation gate and closes the gate tightly (Figure 2.4c, d). In its resting position, the voltage sensor is like a cocked gun with the positive S4 gating charges drawn inward, ready to shoot outward rapidly upon depolarization of the negative internal membrane potential (Figure 2.4a, b, right). The voltage-driven sliding of the S4 helix inward to its resting conformation captures the electrostatic energy of the electric field in the new conformations of the S4, S4–S5 and S6 segments, and the outward movement of the S4 helix upon depolarization couples that energy to concerted opening of the pore.

2.3.5 An Iris-Like Movement Opens the Pore

When the membrane is depolarized, the force holding the S4 segment in its inward position is relieved, allowing the S4 segment to shoot outward, straighten the elbow in the S4–S5 linkers, twist the S4–S5 linkers and S6 segments, and open the pore (Figure 2.4a–d; Movie 1; Lenaeus et al. 2017; Wisedchaisri et al. 2019). The conformational change in the S4–S5 linkers is large, twisting these segments in the plane of the membrane. This large conformational change in the S4–S5 linker causes a more subtle bending and twisting motion of the S6 segments, which opens the pore by moving the hydrophobic side chain of Ile217 (I217) out of the permeation pathway (Figure 2.4c, d; Lenaeus et al. 2017; Wisedchaisri et al. 2019). From these structural models, one can visualize the complete series of conformational changes in the gating of a voltage-gated Na$^+$ channel (Movie 1; Wisedchaisri et al. 2019).

2.4 The Pore Collapses during Slow Inactivation

In addition to the fast inactivation process discovered by Hodgkin and Huxley in their classic work, a separate slow inactivation process operating on the time scale of 100 ms to seconds also terminates the Na$^+$ influx through Na$^+$ channels (Rudy 1978; Vilin and Ruben 2001). This process is engaged during repetitive generation of action potentials in nerve and muscle cells, and limits the length of trains of repetitive action potentials. Bacterial Na$^+$ channels whose structures have been determined have a slow inactivation process (Pavlov et al. 2005), even though their homotetrameric structure means that they do not have a structural component analogous to the intracellular loop connecting domains III and IV of vertebrate channels that mediate fast inactivation. The structure of the slow-inactivated bacterial Na$^+$ channel reveals that the pore has partially collapsed by movement of two opposing S6 segments toward the central axis of the pore and corresponding movement of the two adjacent pairs of S6 segments away from the axis (Figure 2.5a–d; Payandeh et al. 2012). This movement is observed at the selectivity filter at the extracellular end of the pore (Figure 2.5a), in the central cavity (Figure 2.5b) and at the activation gate at the intracellular end of the pore (Figure 2.5c, d). This asymmetric collapse of the pore is accompanied by subtle rotation of the voltage-sensing domain around the cylindrical exterior surface of the pore domain (Payandeh et al. 2012). It is likely that the pore collapse is important for stabilization of the channel in the slow-inactivated state, which requires strong, long-duration hyperpolarization for recovery to the resting state.

2.5 Drugs Block the Pores of Sodium Channels

Many drugs used in therapy act on ion channels. Local anesthetics, which are used to block pain in dental and surgical procedures, bind in the inner pore of Na$^+$ channels in nerves and block them (Hille, 1977; Figure 2.5e–g). In addition, related Na$^+$ channel blocking drugs are used for treatment of epilepsy and cardiac arrhythmia. A combination of site-directed mutagenesis, structural studies and molecular modeling reveals that these drugs bind to a receptor site formed by amino acid residues at specific positions in the S6 segments in domains I, III and IV of the Na$^+$ channel, as illustrated in Figure 2.5e (Ragsdale et al. 1994; Yarov-Yarovoy et al. 2002). The aromatic and hydrophobic side chains of these amino acids contact the aromatic and substituted amino groups of the drug molecules and bind them in their receptor site, where they block ion movement through the pore. The structure of Na$_V$Ab places this drug receptor site in a three-dimensional context (Payandeh

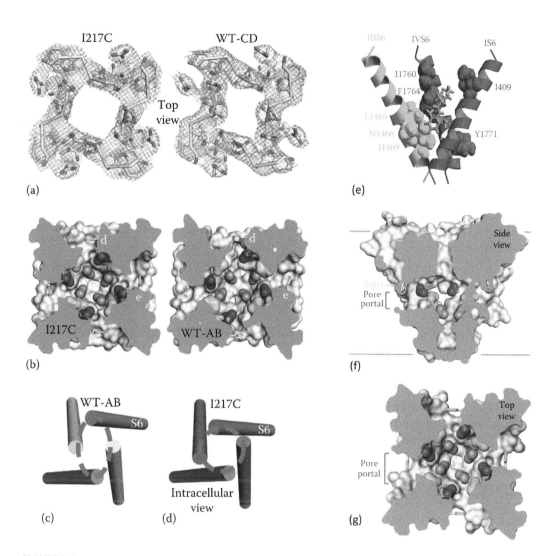

FIGURE 2.5

Slow inactivation and drug receptor site in the pore of Na⁺ channels. (a) Top view of the collapse of the pore during slow inactivation of Na$_V$Ab. Two S6 segments move inward the central axis of the pore and two move outward to produce an asymmetric, partially collapsed conformation. The selectivity filter structure has changed from nearly square in the preopen state of NavAb/I217C to a partially collapsed parallelogram in the inactivated state of Na$_V$Ab/WT-CD. (c) The central cavity is partially collapsed. (c) The activation gate is tightly closed, but collapsed into an asymmetric conformation in WT-AB. (d) Comparison of the distances between equivalent C$_\alpha$ positions for pore-lining residues at the activation gate in the S6 segments in the preopen conformations of Na$_V$Ab/I217C. (Adapted from Payandeh et al. 2012.) (e) A model of the binding site for local anesthetics and related antiepileptic and antiarrhythmic drugs in the IS6, IIIS6 and IVS6 transmembrane segments of a mammalian Na⁺ channel. Amino acid side chains involved in drug binding are shown in space-filling format. Yellow, bound etidocaine, a local anesthetic. (f) Side view through the pore module in the structure of Na$_V$Ab illustrating fenestrations (pore portals) and hydrophobic access to central cavity. Phe203 side chains, yellow sticks. Surface representations of Na$_V$Ab residues aligning with those implicated in drug binding and blocking. Thr206, blue; Met209, green; Val213, orange. Membrane boundaries, gray lines. (g) Top view sectioned below the selectivity filter; colored as in (e). (Adapted from Payandeh et al. 2011.)

et al. 2011; Figure 2.5f, g). The amino acid residues that form the receptor sites for Na⁺ channel blockers line the inner surface of the S6 segments and create a three-dimensional drug receptor site whose occupancy would block the pore (Figure 2.5f, g). Remarkably, fenestrations lead from the lipid phase of the membrane sideways into the drug receptor site, providing a hydrophobic access pathway for drug binding (pore portals; Figure 2.5f, g). This form of drug binding from the membrane phase was predicted in early studies of the mechanism of blocking of Na⁺ channels by local anesthetics of differing size and hydrophobicity (Hille 1977). Access to the drug receptor site in Na$_V$Ab channels from the membrane phospholipid bilayer is limited by the side chain of a single amino acid residue (Phe203 in Na$_V$Ab; Figure 2.5 f, g, yellow). This amino acid residue is located in a position analogous to an amino acid residue in the IVS6 segment of brain Na⁺ channels that controls egress of local anesthetics from their receptor and near amino acid residues that control access of extracellular Na⁺ channel blockers to their receptor site in cardiac and brain Na⁺ channels (Qu et al., 1995).

2.6 Evolutionary Additions to Mammalian Sodium Channels

The structures of mammalian Na⁺ channels from nerve, skeletal muscle and heart determined by cryogenic electron microscopy (Shen et al. 2019; Pan et al. 2018; Jiang et al. 2020) are closely related to bacterial Na⁺ channels in their functional transmembrane regions (root mean square deviation of <3 angstroms, close to the limit of resolution of the structural data). However, the mammalian Na⁺ channels are much larger protein complexes, which have all 24 of their transmembrane segments organized in four homologous domains in a single polypeptide rather than in a homotetramer (Figures 2.2 and 2.6). These mammalian Na⁺ channels also have large unstructured intracellular loops connecting their four domains, and their β subunits are associated in transmembrane positions interacting with both intracellular and extracellular surfaces of the channel protein (Figures 2.2 and 2.6). The structures of the voltage-sensing modules of mammalian Na⁺ channels are very similar to Na$_V$Ab, but there are important evolutionary specializations in the ion selectivity filter and in the fast inactivation gate (Figure 2.6).

The fast inactivation gate is formed by the intracellular linker connecting domains III and IV (III–IV linker). It is locked in its inactivating position in the structures of mammalian Na⁺ channels (Figure 2.6a; Shen et al. 2019; Pan et al. 2018; Jiang et al. 2020). The IFM motif that serves as the latch on the inactivation gate binds in a hydrophobic pocket that serves as its receptor (Figure 2.6a, b, orange). It is released by mutations that convert the IFM motif to QQQ (Jiang et al. 2021). It interacts with all of the amino acid residues previously established as part of the inactivation gate receptor by mutagenesis studies of mammalian Na⁺ channels (see Section 2.2.3).

The high-field-strength site in the ion selectivity filter is formed by four different amino acid residues, Asp–Glu–Lys–Ala, in place of the four Glu residues in Na$_V$Ab (Figure 2.6c, d). Surprisingly, the Lys residue is absolutely required for Na⁺ selectivity, and no other residue can substitute (Favre, Moczydlowski, and Schild 1996; Sun et al. 1997). Structural analysis indicates that this Lys residue forms three hydrogen bonds with strategically placed backbone carbonyls, leaving its lone electron pair pointing its partial negative charge toward the ion permeability pathway in position to interact with Na⁺ (Figure 6e)(Jiang et al. 2020). This unique chemistry may be essential for Na⁺ selectivity.

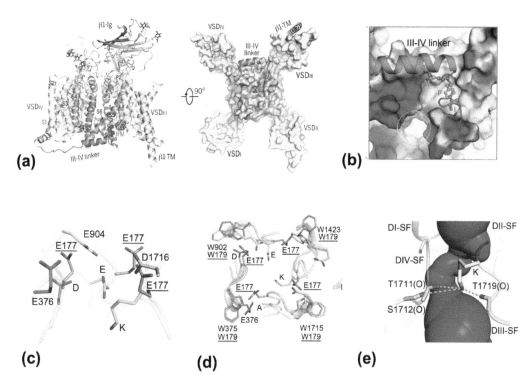

FIGURE 2.6

Structure of mammalian Na$^+$ channels. (a) Structure of Na$_V$1.4 with the β1 subunit. Yellow, voltage sensor; orange, the domain III–IV linker with the fast inactivation gate and IFM motif. (b) High-resolution view of the bound IFM motif. (c, d) High resolution top views of the ion selectivity filter with key amino acid residues at the high field strength that select and conduct sodium labeled and superimposed on Na$_V$Ab. (Adapted from Pan et al. 2018.) (e) Side view of the ion selectivity filter from the cardiac Na$_V$1.5 channel. Note the unique conformation of the ε-amino group of K1421 projecting its lone electron pair into the ion permeation pathway. (Adapted from Jiang et al. 2020.)

2.7 Sodium Channel Diversity

In humans and other mammals, voltage-gated Na$^+$ channel α subunits are encoded by ten genes, which are expressed in different excitable tissues (Table 2.2; Goldin et al. 2000; Yu and Catterall 2004). Na$_V$1.1, Na$_V$1.2, Na$_V$1.3 and Na$_V$1.6 are the primary Na$^+$ channels in the central nervous system. Na$_V$1.7, Na$_V$1.8 and Na$_V$1.9 are the primary Na$^+$ channels in the peripheral nervous system. Na$_V$1.4 is the primary Na$^+$ channel in skeletal muscle, whereas Na$_V$1.5 is primary in the heart. Most of these Na$^+$ channels also have significant levels of expression outside of their primary tissues. The tenth Na$^+$ channel protein is not voltage-gated and is involved in salt-sensing. There is a small family of four Na$_V$β subunits in total. β1 and β3 are associated noncovalently with α subunits and resemble each other most closely in amino acid sequence, whereas β2 and β4 form disulfide bonds with α subunits and also resemble each other closely (Catterall 2000; Brackenbury and Isom 2011). Na$_V$β subunits are promiscuous in their associations with α subunits in transfected cells, but local expression and restricted opportunities for the β2 and β4 to form disulfide bonds

TABLE 2.2

Sodium Channel Protein Family

Channel	Gene	Primary Location	Disease
$Na_V1.1$	SCN1A	CNS	Generalized epilepsy with febrile seizures plus
			Dravet syndrome
			Benign neonatal convulsions
			Familial hemiplegic migraine type III
$Na_V1.2$	SCN2A	CNS	Benign familial neonatal-infantile seizures
			Epilepsy
			Developmental delay and disability
$Na_V1.3$	SCN3A	Embryonic CNS	Developmental delay and disability
$Na_V1.4$	SCN4A	Skeletal muscle	Hypokalemic periodic paralysis type II
			Normokalemic periodic paralysis
			Hyperkalemic periodic paralysis
			Paramyotonia congenita
$Na_V1.5$	SCN5A	Heart	Long QT syndrome type III
			Brugada syndrome
$Na_V1.6$	SCN7A	CNS	Epilepsy
$Na_V1.7$	SCN9A	PNS	Erythromelalgia
			Paroxysmal extreme pain disorder
			Congenital indifference to pain
$Na_V1.8$	SCN10A	DRG	Small fiber neuropathy
$Na_V1.9$	SCN11A	DRG	

Sources: Reviewed in Goldin et al. (2000); Kass (2005); Dib-Hajj and Waxman (2019); Mantegazza et al. (2021).
Abbreviations: CNS, central nervous system; PNS, peripheral nervous system; DRG, sensory neurons in dorsal root ganglia.

with different α subunits may add a level of subtype specificity in native cell environments (Brackenbury and Isom 2011; Shen et al. 2019; Pan et al. 2018; Jiang et al. 2020).

Suggested Reading

This chapter includes additional bibliographical references hosted only online as indicated by citations in blue color font in the text. Please visit https://www.routledge.com/9780367538163 to access the additional references for this chapter, found under "Support Material" at the bottom of the page.

Armstrong, C. M., and F. Bezanilla. 1973. "Currents related to movement of the gating particles of the sodium channels." *Nature* 242 (5398):459–61.

Beneski, D. A., and W. A. Catterall. 1980. "Covalent labeling of protein components of the sodium channel with a photoactivable derivative of scorpion toxin." *Proc Natl Acad Sci U S A* 77 (1):639–43.

Catterall, W. A. 1984. "The molecular basis of neuronal excitability." *Science* 223 (4637):653–61.

Catterall, W. A. 2000. "From ionic currents to molecular mechanisms: the structure and function of voltage-gated sodium channels." *Neuron* 26 (1):13–25.

Crill, W. E. 1996. "Persistent sodium current in mammalian central neurons." *Annu Rev Physiol* 58:349–62.

Goldin, A. L., R. L. Barchi, J. H. Caldwell, F. Hofmann, J. R. Howe, J. C. Hunter, R. G. Kallen, G. Mandel, M. H. Meisler, Y. Berwald Netter, M. Noda, M. M. Tamkun, S. G. Waxman, J. N. Wood, and W. A. Catterall. 2000. "Nomenclature of voltage-gated sodium channels." *Neuron* 28 (2):365–8.

Hartshorne, R. P., and W. A. Catterall. 1984. "The sodium channel from rat brain. Purification and subunit composition." *J Biol Chem* 259 (3):1667–75.

Hartshorne, R. P., B. U. Keller, J. A. Talvenheimo, W. A. Catterall, and M. Montal. 1985. "Functional reconstitution of the purified brain sodium channel in planar lipid bilayers." *Proc Natl Acad Sci U S A* 82 (1):240–4.

Hille, B. 1975. "Ionic selectivity, saturation, and block in sodium channels. A four-barrier model." *J Gen Physiol* 66 (5):535–60.

Hille, B. 1977. "Local anesthetics: hydrophilic and hydrophobic pathways for the drug-receptor reaction." *J Gen Physiol* 69 (4):497–515.

Hodgkin, A. L., and A. F. Huxley. 1952. "A quantitative description of membrane current and its application to conduction and excitation in nerve." *J Physiol* 117 (4):500–44.

Isom, L. L., K. S. De Jongh, D. E. Patton, B. F. X. Reber, J. Offord, H. Charbonneau, K. Walsh, A. L. Goldin, and W. A. Catterall. 1992. "Primary structure and functional expression of the beta-1 subunit of the rat brain sodium channel." *Science* 256 (5058):839–42.

Isom, L. L., D. S. Ragsdale, K. S. De Jongh, R. E. Westenbroek, B. F. Reber, T. Scheuer, and W. A. Catterall. 1995. "Structure and function of the β2 subunit of brain sodium channels, a transmembrane glycoprotein with a CAM motif." *Cell* 83 (3):433–42.

Kontis, K. J., A. Rounaghi, and A. L. Goldin. 1997. "Sodium channel activation gating is affected by substitutions of voltage sensor positive charges in all four domains." *J Gen Physiol* 110 (4):391–401.

Noda, M., S. Shimizu, T. Tanabe, T. Takai, T. Kayano, T. Ikeda, H. Takahashi, H. Nakayama, Y. Kanaoka, N. Minamino, et al. 1984. "Primary structure of electrophorus electricus sodium channel deduced from cDNA sequence." *Nature* 312 (5990):121–7.

Payandeh, J., T. M. Gamal El-Din, T. Scheuer, N. Zheng, and W. A. Catterall. 2012. "Crystal structure of a voltage-gated sodium channel in two potentially inactivated states." *Nature* 486 (7401):135–9.

Payandeh, J., T. Scheuer, N. Zheng, and W. A. Catterall. 2011. "The crystal structure of a voltage-gated sodium channel." *Nature* 475 (7356):353–8.

Ragsdale, D. S., J. C. McPhee, T. Scheuer, and W. A. Catterall. 1994. "Molecular determinants of state-dependent block of sodium channels by local anesthetics." *Science* 265 (5179):1724–8.

Ren, D., B. Navarro, H. Xu, L. Yue, Q. Shi, and D. E. Clapham. 2001. "A prokaryotic voltage-gated sodium channel." *Science* 294 (5550):2372–5.

Rudy, B. 1978. "Slow inactivation of the sodium conductance in squid giant axons. Pronase resistance." *J Physiol (Lond)* 283:1–21.

Starace, D. M., and F. Bezanilla. 2004. "A proton pore in a potassium channel voltage sensor reveals a focused electric field." *Nature* 427 (6974):548–53.

Stuhmer, W., F. Conti, H. Suzuki, X. Wang, M. Noda, N. Yahadi, H. Kubo, and S. Numa. 1989. "Structural parts involved in activation and inactivation of the sodium channel." *Nature* 339 (6226):597–603.

Terlau, H., S. H. Heinemann, W. Stühmer, M. Pusch, F. Conti, K. Imoto, and S. Numa. 1991. "Mapping the site of block by tetrodotoxin and saxitoxin of sodium channel II." *FEBS Lett* 293 (1–2):93–6.

Vargas, E., Yarov-Yarovoy, V., Khalili-Araghi, F., Catterall, W. A., Klein, M. L., Tarek, M., Lindahl, E., Schulten, K., Perozo, E., Bezanilla, F., Roux, B. 2012 "An emerging consensus on voltage-dependent gating from computational modeling and molecular dynamics simulations." *J Gen Physiol* 140(6):587–94. doi: 10.1085/jgp.201210873.

Vassilev, P. M., T. Scheuer, and W. A. Catterall. 1988. "Identification of an intracellular peptide segment involved in sodium channel inactivation." *Science* 241 (4873):1658–61.

Vilin, Y. Y., and P. C. Ruben. 2001. "Slow inactivation in voltage-gated sodium channels: molecular substrates and contributions to channelopathies." *Cell Biochem Biophys* 35 (2):171–90.

West, J. W., D. E. Patton, T. Scheuer, Y. Wang, A. L. Goldin, and W. A. Catterall. 1992. "A cluster of hydrophobic amino acid residues required for fast Na⁺ channel inactivation." *Proc Natl Acad Sci U S A* 89 (22):10910–4.

Yu, F. H., and W. A. Catterall. 2004. "The VGL-chanome: a protein superfamily specialized for electrical signaling and ionic homeostasis." *Sci. STKE* 253 (253):re15.

3

Voltage-Gated Calcium Channels

Jacqueline Niu and Henry M. Colecraft

CONTENTS

3.1 Introduction

The pioneering work of Sydney Ringer in the 1880s probing the importance of ionic composition in the bathing medium for heart contraction in the frog led to the discovery of the necessary role of calcium ions (Ca^{2+}) in a physiological process. It is now known that Ca^{2+} influx into cells controls numerous physiological processes in addition to muscle contraction including neurotransmission, gene transcription, inflammation, cell motility, cell growth and cell death. Voltage-gated Ca^{2+} channels (VGCCs) are a family of ion channel proteins that function primarily as conduits for Ca^{2+} influx principally into excitable cells to regulate their excitability and to convert electrical signals into physiological responses. VGCCs also play important roles in non-excitable cells such as osteocytes and immune cells. To serve their broad physiological roles, VGCCs are molecularly diverse, differentially expressed in distinct tissues and cells, and assembled into discrete macromolecular complexes. Their activity and functional expression are modulated by intracellular signaling molecules to regulate physiology, their dysregulation underlies serious disorders, and they are important pharmacological targets for treating disease. This chapter discusses various aspects of VGCCs including their physiological roles, structure, functional mechanism, regulation, pharmacology, biogenesis and disease.

DOI: 10.1201/9781003096276-3

3.2 VGCC Diversity, Nomenclature and Structural Organization

VGCCs are conventionally classified as being either low voltage-gated (LVGCC) or high voltage-gated (HVGCC) according to the membrane potential threshold at which they are activated (Figure 3.1A). HVGCCs are heteromultimeric protein complexes containing a pore-forming α_1 subunit assembled with auxiliary β, $\alpha_2\delta$ and (in some cases) γ subunits. There are seven distinct HVGCCs α_1 subunits (α_{1A}–α_{1F}, α_{1S}), encoded by distinct genes (*CACNA1A–CACNA1F, CACNA1S*), that are categorized as L-type ($Ca_V1.1$–$Ca_V1.4$) or non-L-type ($Ca_V2.1$–$Ca_V2.3$). Historically, L-type Ca^{2+} channels (LTCCs) are so named because they give rise to large and relatively long-lasting ionic currents. The non-L-type channels are also referred to as P/Q type ($Ca_V2.1$), N-type ($Ca_V2.2$) and R-type ($Ca_V2.3$). By contrast to HVGCCs, LVGCCs ($Ca_V3.1$–$Ca_V3.3$) require only pore-forming α_1 subunits (α_{1G}–α_{1I}) to form functional channels. LVGCCs are historically referred to as T-type because they gave rise to transient currents with tiny single-channel conductances (Catterall et al. 2005; Zamponi et al. 2015; Tsien et al. 1988; Llinas et al. 1992).

VGCC α_1 subunits range from 190 to 250 kDa and have a similar overall topology with four homologous domains (DI–DIV) each possessing six transmembrane segments (S1–S6) (Figure 3.1B). S1–S4 regions from DI–DIV form four distinct voltage-sensor domains (VSDs), while S5–S6 from each domain combine to form the ion-conducting pore and contain the selectivity filter. DI–DIV are joined by three intracellular loops of varying lengths (I–II, II–III and III–IV loops). Cytosolic N- and C-termini lead into DI and follow DIV, respectively. The intracellular loops and termini contain binding sites for various regulatory proteins that control channel trafficking, gating or subcellular localization. Cryogenic electron microscopy (cryo-EM) structures of $Ca_V1.1$ (Figure 3.1C, D) show that DI–DIV adopt a clockwise arrangement, with the four VSDs splayed outside of a central conduction pore (Wu et al. 2016).

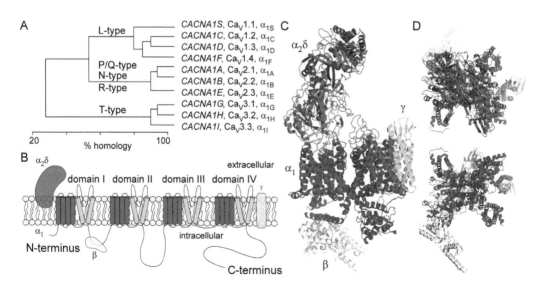

FIGURE 3.1
General structure of voltage-gated Ca^{2+} channels. (A) Classification and nomenclature of voltage-gated Ca^{2+} channels. (B) Cartoon depiction of voltage-gated Ca^{2+} channel subunits. (C and D) Cryo-EM structure of $Ca_V1.1$ (Protein Data Bank: 5GJV). Side view (C), top view (D, top), bottom view (D, bottom).

Auxiliary β subunits of HVGCCs are cytosolic proteins that bind with nanomolar affinity to a conserved region in the I–II loop (the α_1 interacting domain, or AID) (Buraei and Yang 2010). There are four distinct β isoforms, β_1–β_4, each with multiple splice variants. Structurally, $Ca_V\beta$ subunits have two conserved domains – *src* homology 3 (SH3) and guanylate kinase (GK) domains – joined by a HOOK region and with variable N- and C-termini (Van Petegem et al. 2004; Opatowsky et al. 2004; Chen et al. 2004). The tandem SH3-GK module is characteristic of the membrane-associated guanylate kinase (MAGUK) family of scaffold proteins. $Ca_V\beta$s bind the AID using an α_1-binding pocket located in the GK domain. Binding to $Ca_V\beta$ regulates the trafficking and gating trafficking of pore-forming α_1 subunits (Buraei and Yang 2010).

There are four distinct $\alpha_2\delta$ subunit isoforms (α_2-δ–α_2-4), each produced as a precursor protein that is posttranslationally cleaved in the endoplasmic reticulum to generate separate $\alpha_2\delta$ and peptides (Dolphin 2013). The α_2 protein contains a von Willebrand factor-A (VWA) domain as well as four tandem cache domains that are homologous to bacterial chemotaxis factors. Contained within the VWA domain is a metal ion-dependent adhesion site (MIDAS) that coordinates a divalent cation. The α_2 protein is attached by disulfide bonds to the peptide, which is tethered to the membrane on the extracellular side by a glycosyl-phosphatidyl inositol (GPI) anchor. The α_2 subunit interacts with extracellular loops in DI, DII and DIII of α_1 subunits (Wu et al. 2016). Functionally, α_2 subunits increase the trafficking of α_1 subunits in synergy with β subunits, and also modulate channel gating (Dolphin 2013; Zamponi et al. 2015).

A γ subunit was originally identified biochemically as a component of the biochemically purified skeletal muscle $Ca_V1.1$ complex (Takahashi et al. 1987). There are eight distinct γ subunits (γ_1–γ_8) of which γ_1 and γ_6 associate with HVGCCs (Chen et al. 2007). Several of these γ subunits ($\gamma2$, $\gamma3$, $\gamma4$ and $\gamma8$) regulate AMPA receptor trafficking and are collectively known as transmembrane AMPA receptor regulatory proteins (TARPs). Topologically, the γ subunits have four transmembrane spanning regions and cytosolic N- and C-termini. The $Ca_V1.1$ cryo-EM structure shows that $\gamma1$ interacts with the $Ca_V1.1$ DIV VSD (Wu et al. 2016). Functionally, $\gamma1$ stabilizes the inactivated state of $Ca_V1.1$ (Singer et al. 1991).

3.3 Physiological Roles of VGCCs

VGCCs are widely distributed throughout the body, with distinct isoforms being differentially expressed in particular organs, tissues and cell types (Figure 3.2). VGCCs have broad physiological roles that have been identified based on a combination of knockout/knockdown approaches, pharmacological blockers and phenotypes of disease-causing mutations (Zamponi et al. 2015). A unique feature of VGCCs among the family of voltage-gated ion channels is that their activity not only regulates membrane potential, and thus excitability, but also their permeant ion Ca^{2+} is an important second messenger that controls the activity of many Ca^{2+}-dependent proteins in cells. Cells in the body are surrounded by an interstitial fluid with a free Ca^{2+} concentration of 1.1–1.4 mM, while the resting cytosolic Ca^{2+} concentration is tightly controlled in the 100 nM range, resulting in an extracellular to cytoplasmic Ca^{2+} concentration gradient of >10,000. In excitable cells such as neurons, cardiac myocytes, and endocrine cells, action potentials are converted into a downstream physiological response via an increase in intracellular Ca^{2+} concentration mediated through Ca^{2+} influx through VGCCs, release from intracellular stores, or a combination of the two Ca^{2+} sources.

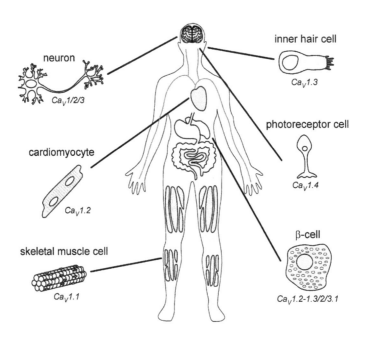

FIGURE 3.2
Differential expression of voltage-gated Ca^{2+} channel α_1-subunit isoforms in distinct organs, tissues and cell types dictates their diverse physiological roles.

In neurons, Ca^{2+} influx through presynaptic HVGCCs (principally $Ca_V2.1$–$Ca_V2.3$) controls neurotransmitter release to enable nerve cells to communicate and to regulate effector organs (Dolphin and Lee 2020; Catterall and Few 2008). Excitation-transcription in neurons is also initiated by Ca^{2+} influx through HVGCCs, with Ca_V1 channels having privileged access to this pathway (Ma et al. 2012). Further, Ca^{2+} influx through HVGCCs can control neuronal excitability by coupling to Ca^{2+}-dependent K^+ channels (Marrion and Tavalin 1998). $Ca_V1.3$ is crucial for synaptic transmission and normal development of inner hairs cells, and is necessary for hearing (Platzer et al. 2000). $Ca_V1.4$ is predominantly expressed in photoreceptor cells where it mediates neurotransmission at ribbon synapses, making it important for normal visual function (Mansergh et al. 2005).

In the heart, $Ca_V1.2$ channels in ventricular cardiomyocytes are localized to dyadic junctions where they lie in proximity to intracellular Ca^{2+} release channels, ryanodine receptors type 2 (RyR2), that gate the release of Ca^{2+} stores from the sarcoplasmic reticulum (SR). During the plateau phase of the ventricular action potential, Ca^{2+} flows in through $Ca_V1.2$ and triggers a larger Ca^{2+} release from the SR via RyR2, producing a Ca^{2+}-induced Ca^{2+} release (CICR) that underlies cardiac excitation–contraction (EC) coupling (Bers 2002). $Ca_V1.3$ is expressed in sinoatrial node cells and contributes to the cardiac pacemaker current to regulate the heart rate (Platzer et al. 2000).

In skeletal muscle, surface $Ca_V1.1$ channels are physically linked to intracellular RyR1 channels. Depolarization of the membrane causes a conformational change in $Ca_V1.1$ that is mechanically transmitted to open RyR1 channels and release Ca^{2+} from intracellular stores to initiate muscle contraction (Bannister and Beam 2013). Ca^{2+} influx through $Ca_V1.1$ is not required for skeletal muscle EC coupling.

Endocrine cells express various VGCCs to control hormone secretion (Zamponi et al. 2015). In pancreatic beta cells, elevations in plasma glucose levels lead to glucose uptake and metabolism to generate ATP. The increased cytosolic ATP binds to and closes ATP-sensitive

K$^+$ channels, leading to membrane depolarization, which opens VGCCs (Ca$_V$1.2 and Ca$_V$2.3) to drive vesicle fusion and release of insulin. Ca^{2+} influx through Ca$_V$1.2 and Ca$_V$1.3 mediates release of catecholamines from chromaffin cells in the adrenal medulla (Marcantoni et al. 2010).

LVGCCs (Ca$_V$3.1–Ca$_V$3.3) are found expressed in many different tissues and cell types in the body including neurons, heart, endocrine, kidney, smooth muscle and sperm (Perez-Reyes 2003). The biophysical properties of LVGCCs result in a window current around the resting potential of distinct cell types. Ca$_V$3.1 channels underlie low threshold Ca^{2+} spikes in neurons, and are necessary for oscillatory burst firing of thalamocortical neurons in deep sleep. In the heart, Ca$_V$3.1 and Ca$_V$3.2 channels contribute to the cardiac pacemaker current.

3.4 Gating of VGCCs

Similar to other voltage-gated ion channels, the opening and closing of VGCCs is regulated by changes in membrane potential. The channels are in a closed state at hyperpolarized potentials and activate to an open state in response to membrane depolarization at low (~–70 mV; Ca$_V$3.1–Ca$_V$3.3), intermediate (~–55 mV; Ca$_V$1.3, Ca$_V$2.3) and high (~–40 mV; Ca$_V$1.1/1.2/1.4; Ca$_V$2.1/2.2) thresholds (Dolphin and Lee 2020). During maintained depolarization, VGCCs enter an inactivated state at different rates depending on the subtype. Ca$_V$3 channels display rapid voltage-dependent inactivation compared to HVGCCs. Some channels (Ca$_V$1.2/1.3; Ca$_V$2.1–2.3) display feedback changes in gating in response to Ca^{2+} influx, evident as either Ca^{2+}-dependent inactivation (CDI) or Ca^{2+}-dependent facilitation (CDF) that is often studied by comparing current waveforms using Ca^{2+} or Ba^{2+} as a charge carrier (Liang et al. 2003; Ben-Johny and Yue 2014). Upon membrane repolarization, VGCCs revert back to the closed state with different deactivation kinetics. LVGCCs deactivate with characteristically slow kinetics compared to HVGCCs.

Voltage sensitivity of VGCCs is conferred by the VSDs. Cryo-EM structures of Ca$_V$1.1 reveal the S4 voltage sensors adopt a 3$_{10}$ helical structure with positively charged amino acids, arginine or lysine, occupying every third or fourth position (Wu et al. 2016). The negative resting membrane potential found in cells holds the S4 voltage sensor in a "down" state where the positively charged residues are below a charge transfer center (CTC) comprised of negative or polar residues, including a conserved aspartate on S3, and a hydrophobic blocking residue on S2. Given the lack of a voltage-gradient under cryo-EM conditions, the Ca$_V$1.1 structures show the S4 voltage sensors in the "up" state, though the levels differ among the distinct domains – the four arginine residues from DI and three from DII are in the "up" state, while those in DIII are unresolved. Cryo-EM structures of Ca$_V$2.2 similarly show S4 voltage sensor from DI, DIII and DIV in an up state, while DII S4 is trapped in a down state by binding to phosphatidylinositol 4,5-bisphosphate (PIP2) (Gao, Yao, and Yan 2021). Voltage-clamp fluorimetry experiments on Ca$_V$1.2 show the four voltage sensors display distinct voltage dependence and kinetics, with an allosteric activation model suggesting DII S4 and DIII S4 most responsible for stabilizing the channel open state, with a small contribution for DI S4 (Pantazis et al. 2014). However, charge neutralization of S4 arginine and lysine residues suggest DI S4 and DIII S4 are rate-limiting for Ca$_V$1.2 activation (Hering et al. 2018). Charge neutralization experiments similarly suggest a dominant role for DI S4 in activation gating of Ca$_V$3.1, with smaller contributions from DII S4 and DIII S4.

The activation gate of VGCCs is present within the channel pore. In the closed state, DI–DIV S6 segments intersect at the intracellular end to form a hydrophobic seal that prevents ionic permeation (Wu et al. 2016). The VSDs are connected to the pore gate via helical S4–S5 linkers that run parallel to the membrane. With depolarization, the upward displacement of the voltage sensors in the electric field is coupled to lateral movement of the S4–S5 linkers that results in disengagement of the S5–S6 helices at the intracellular gate, thereby opening the channel.

In general, determinants of inactivation gating in VGCCs are more widely distributed than in Na_V or K_V channels. Association of intracellular loops with cytosolic proteins (e.g., auxiliary β subunits, syntaxin and CaM) regulates inactivation gating of HVGCCs. Several disease-causing mutations that alter inactivation gating of VGCCs have been mapped to intracellular domains as well as the S6 segments. Single-channel recordings in cardiac $Ca_V1.2$ channels indicates CDI involves a Ca^{2+}-induced switch to a low open probability gating mode (Imredy and Yue 1994). CaM is established as the Ca^{2+} sensor for CDI (Peterson et al. 1999; Zuhlke et al. 1999) and CDF (DeMaria et al. 2001) of HVGCCs. Mutation of a conserved aspartate residue in the selectivity filter of $Ca_V1.2$ selectively eliminated CDI, suggesting the selectivity filter as the endpoint for this inactivation mechanism (Abderemane-Ali et al. 2019). $Ca_V2.2$ channels display preferential closed-state inactivation (CSI). The $Ca_V2.2$ cryo-EM structure revealed an α-helix (termed W-helix) under the intracellular gate formed from residues A764 to S783 in the DII–DIII loop and that contains a residue, W768, which forms hydrophobic interactions with S6 residues to stabilize the intracellular gate in a closed state. Deletion of the W-helix or a W768Q point mutation selectively eliminated $Ca_V2.2$ CSI confirming the necessary role of this unique W-helix in this phenomenon (Dong et al. 2021).

3.5 Ion Permeability of VGCCs

Na^+ and Ca^{2+} ions have a similar ionic radius and extracellular Na^+ outnumbers Ca^{2+} by a hundredfold, yet VGCCs selectively pass Ca^{2+} compared to Na^+ at a ratio of 1000:1. This is accomplished through the selectivity filter of VGCCs, in which each domain contributes one glutamate to form an EEEE locus within the pore that binds Ca^{2+} with micromolar affinity (Yang et al. 1993). In the absence of Ca^{2+} ions, Na^+ and Li^+ ions readily permeate the pore but are blocked by micromolar concentrations of extracellular Ca^{2+} (Hess and Tsien 1984). The concentration dependence of Ca^{2+} block of monovalent cation current through $Ca_V1.2$ channels indicates that Ca^{2+} binds at a high affinity site within the pore with an apparent K_d of ~1 M. Neutralizing E to Q mutations of any of the E residues in the EEEE locus lowered the affinity of Ca^{2+} binding to the $Ca_V1.2$ pore, with double mutations resulting in an up to a thousandfold decreased affinity (Yang et al. 1993).

Assuming that Ca^{2+} binding to the VGCC pore is a diffusion-limited process, if there were only one binding site in the pore, the maximum off rate, k_{off}, of Ca^{2+} would be given by $K_d/k_{on} = 10^{-6}$ M$/10^{-9}$ M $\bullet$ s $= 10^3$ ions per second. However, single-channel recordings reveal picoampere-sized currents indicating VGCCs pass 10^6 ions per second. The thousandfold variance between the expected and observed ionic flux can be bridged by models that assume single-file multi-ion occupancy of the pore and electrostatic repulsions that force ions to exit the pore (Sather and McCleskey 2003). The $Ca_V1.1$ cryo-EM structure identified two densities within the pore modeled as Ca^{2+} ions, including at the EEEE locus. Molecular dynamics modeling indicates the EEEE locus is flexible enough to accommodate multiple Ca^{2+} ions, and predicts three Ca^{2+} ions can exist in the filter at a time where one ion binds to a high affinity site and two ions bind to low affinity sites.

A study combining mutational analyses with modeling in VGCCs identified a ring of nonconserved negatively charged residues above the EEEE locus that forms a divalent cation selectivity site (DCS) responsible for distinctions in divalent cation selectivity among different VGCC types (Cens et al. 2007). In $Ca_V1.1$, an N617D mutation in the DCS generated a channel that does not conduct either monovalent or divalent cations, due to a 4.2-fold increased affinity of the pore for Ca^{2+} (Dayal et al. 2021).

3.6 Pharmacology of VGCCs

VGCCs are targeted by various small molecules and/or toxins that are used either as therapeutics or as research tools (Zamponi et al. 2015; Hockerman et al. 1997) (Table 3.1). L-type Ca^{2+} channels ($Ca_V1.1$–$Ca_V1.4$) are blocked by three distinct classes of drugs: dihydropyridines (e.g., nisoldipine), phenylalkylamine (e.g., verapamil) and benzothiazepines (e.g., diltiazem). Phenylalkylamines and benzothiazepines block the channel pore, whereas dihydropyridines inhibit conduction through an allosteric mechanism. The affinity of block varies for different LTCC types (e.g., IC_{50} for isradipine inhibition of $Ca_V1.2$ is fivefold less than for $Ca_V1.3$) and can also diverge for distinct splice variants of a given type. All three classes display voltage- and state-dependent binding, with binding favored at depolarized potentials and channels in an inactivated state, and also exhibit use-dependent block, where the amount of block increases with the frequency of stimulation, albeit to different degrees (Zamponi et al. 2015). All three classes are used clinically as antihypertensive agents and to treat angina. At therapeutic doses for these clinical indications, block of smooth muscle LTCCs is favored over cardiac LTCCs due to the relatively depolarized resting membrane potential of smooth muscle and expression of distinct $Ca_V1.2$ splice variants with differing sensitivities to block. Verapamil and diltiazem are also utilized to treat supraventricular tachycardias, an indication enabled by their use-dependent block of $Ca_V1.2$ channels in the atrioventricular node of the heart. Photoaffinity labeling and mutagenesis experiments first mapped important determinants of LTCC blockers binding to DIIIS6, DIVS6 and DIIIS5–S6 linker (Hockerman et al. 1997). The binding sites were confirmed and further elaborated by cryo-EM structures of $Ca_V1.1$ in complex with the three distinct classes of LTCC blockers: nifedipine resides in a fenestration between DIII and DIV, while verapamil and diltiazem cross the central cavity in the pore (Zhao et al. 2019). Beyond blockers, there are also DHP agonists (e.g., BAY K 8644) that strongly enhance LTCC current by stabilizing the open state of the channel. The binding site for the DHP agonist, BAYK 8644, overlaps with that of the DHP antagonist nifedipine (Zhao et al. 2019). There are no current clinical indications for DHP agonists. There has been interest in developing blockers that can distinguish among LTCC types, particularly $Ca_V1.2$ and $Ca_V1.3$. This emerged from studies suggesting that $Ca_V1.3$ activity was responsible for the vulnerability of substantial nigra dopaminergic neurons to cell death that causes Parkinson's disease (Chan et al. 2007). However, blockers that can selectively block $Ca_V1.3$ while leaving $Ca_V1.2$ untouched have not yet been successfully developed.

By contrast to LTCCs, small molecule inhibitors for neuronal VGCCs ($Ca_V2.1$–Ca_V3) are relatively lacking. $Ca_V2.2$ channels are important targets for the treatment of pain due in large part to their important role in synaptic transmission from nociceptive neurons in

TABLE 3.1

Structures of Small Molecules and Toxins That Inhibit or Regulate the Various Voltage-Gated Ca²⁺
Channels

VGCC Subunit	Drug/Chemical Compound

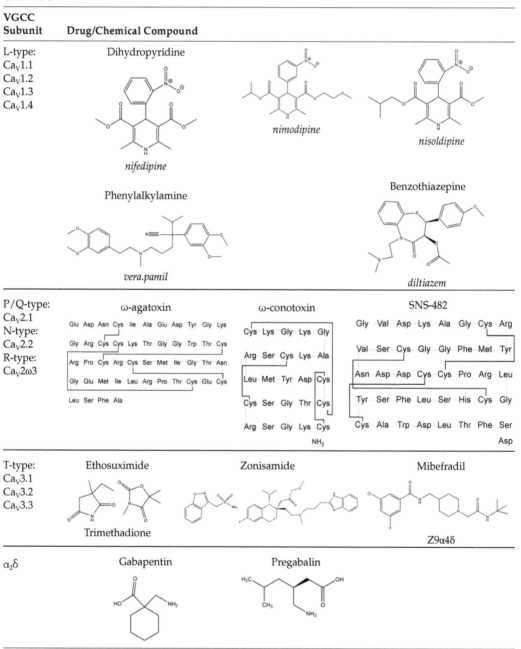

L-type: Ca$_V$1.1 Ca$_V$1.2 Ca$_V$1.3 Ca$_V$1.4	Dihydropyridine ... nifedipine ... nimodipine ... nisoldipine Phenylalkylamine ... vera.pamil ... Benzothiazepine ... diltiazem
P/Q-type: Ca$_V$2.1 N-type: Ca$_V$2.2 R-type: Ca$_V$2ω3	ω-agatoxin ... ω-conotoxin ... SNS-482
T-type: Ca$_V$3.1 Ca$_V$3.2 Ca$_V$3.3	Ethosuximide / Trimethadione ... Zonisamide ... Mibefradil ... Z9α4δ
α$_2$δ	Gabapentin ... Pregabalin

the dorsal horn. Peptide toxins from fish-hunting snails potently block $Ca_V2.2$, including ω-conotoxin GVIA from the cone snail *Conus geographus* and ω-conotoxin MVIIA from *Conus magus* (Olivera et al. 1994). These peptides are pore-blockers that function by physically occluding ionic access to the selectivity and channel pore (Gao, Yao, and Yan 2021). Ziconotide (Prialt) is a synthetic ω-conotoxin MVIIA used clinically as a second-line treatment for severe chronic pain that is refractory to first-line treatments. The drug is delivered by the intrathecal route that limits the therapeutic window due to the potential for central effects (Schmidtko et al. 2010). The surface density of $Ca_V2.2$ in sensory neurons was found to be regulated by association with collapsing response mediated protein-2 (CRMP2) (Chi et al. 2009). A TAT-fused peptide that interferes with the $Ca_V2.2$/CRMP-2 association was found to reduce $Ca_V2.2$ surface density and mediate analgesia in rodent pain models (Brittain et al. 2011). $Ca_V2.1$ channels are strongly inhibited by ω-agatoxin IVA, a peptide isolated from the venom of the funnel web spider *Agenelopsis aperta*, while $Ca_V2.3$ is inhibited by the spider toxin SNX-482. There are no obvious medical indications for blocking wild-type $Ca_V2.1$ channels. By contrast, inhibition of $Ca_V2.3$ may have applications for treatment of seizures and pain (Zamponi et al. 2015).

Beyond the pore-forming α_1 subunits, molecules that target auxiliary proteins of HVGCCs also have clinical utility. Gabapentinoid drugs, gabapentin and pregabalin, which are used clinically for treatment of chronic pain and epilepsy, exert their therapeutic effect through binding α_2-1 and α_2-2 subunits (Gee et al. 1996, Field et al. 2006). While it has been reported that gabapentin binding to α_2-1 reduces the surface trafficking of Ca_V2 channels, other mechanisms not involving HVGCCs have also been proposed for the mechanism underlying relief of pain (Dolphin 2018). Small molecules that inhibit the HVGCC α_1-β subunit interaction have been developed and shown to inhibit $Ca_V2.2$ channel trafficking and display analgesia in rodent models of pain (Chen et al. 2018; Khanna et al. 2019). A genetically encoded inhibitor of HVGCCs was generated by fusing a nanobody that binds auxiliary β subunits to the catalytic HECT domain of the E3 ubiquitin ligase, Nedd4L. This molecule, termed Ca_V-aβlator, potently inhibits HVGCCs by targeting β subunits and promoting ubiquitination of the channel complex, resulting in removal of channels from the cell surface (Morgenstern et al., 2019).

T-type Ca^{2+} channels are blocked by several types of inorganic small molecules, although in most cases they are not selective inhibitors of these channels (Zamponi et al. 2015). The antiepileptic drugs ethosuximide, trimethadione and zonisamide block $Ca_V3.1$–$Ca_V3.3$ channels and also have neuroprotective effects. Several DHP and amphiphilic poly(allylamine) (PAA) LTCC blockers used as antihypertensive agents such as amlodipine and verapamil also block T-type Ca^{2+} channels. Mibefradil is a PAA drug that is a potent T-type channel blocker and used for hypertension, but had to be withdrawn because it inhibited cytochrome P450 3A4 (CYP3A4), causing adverse drug–drug interactions. Z944 is a pan T-type channel blocker that has shown efficacy as an antiseizure agent and also produces analgesia in pain models (Tringham et al. 2012).

3.7 Regulation of VGCCs

The activity of VGCCs is regulated by a host of mechanisms including posttranslational modifications, interactions with G protein subunits released by activation of G protein-coupled receptors (GPCRs), Ca^{2+} sensing proteins and other cell machinery (Figure 3.3).

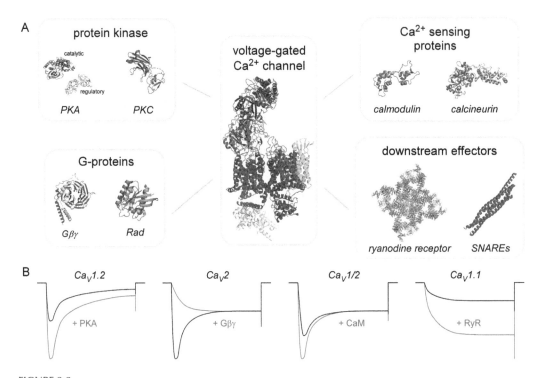

FIGURE 3.3

Regulation of voltage-gated Ca²⁺ channels. (A) Structures of exemplar proteins from various categories that regulate voltage-gated Ca²⁺ channels Ca$_V$1.1 (5GJV), PKA catalytic subunit (4XW5), PKA regulatory subunit (6FLO), protein kinase C (6UWA), Gβ γ (3SN6), Rad (2DPX), calmodulin (2DFS), calcineurin (2P6B), ryanodine receptor (3J8H) and SNAREs (3RK2). (B) Schematic of how various voltage-gated Ca²⁺ channel modulators impact Ca²⁺ current waveforms.

These mechanisms represent powerful means to regulate cellular function and whole organism physiology, and are also targets for therapeutic drugs.

During exercise or in case of danger, norepinephrine acts on β-adrenergic receptors in the heart to initiate a signaling cascade that leads to increased cAMP production and activation of protein kinase A (PKA), resulting in augmentation of heart rate and contractile force. The increased cardiac function is crucial for the physiological fight-or-flight response. Activated PKA strongly increases Ca$_V$1.2 current amplitude in the heart, and this effect is necessary for the positive inotropic effect of β-adrenergic agonists (Bean, Nowycky, and Tsien 1984) (Figure 3.3B). Surprisingly, the major mechanism does not involve direct phosphorylation of Ca$_V$1.2 subunits by PKA. Instead, under basal conditions, a subpopulation of cardiac Ca$_V$1.2 channels are held in a low P_o gating mode by the small monomeric G protein Rad, which binds the auxiliary Ca$_V$β subunit. PKA phosphorylation of Rad decreases its affinity for Ca$_V$β and relieves the inhibition of Ca$_V$1.2 to increase the whole-cell L-type Ca²⁺ current (Liu et al., 2020). Rad belongs to a four-member subfamily of Ras-like monomeric G proteins collectively known as RGK (Rad, Rem, Rem2, Gem/Kir) proteins all of which potently inhibit HVGCCs by binding the β subunit (Beguin et al. 2001; Finlin et al. 2003). The mechanisms of inhibition are diverse and may involve reduced channel surface density in addition to decreased P_o of channels at the cell surface depending on the RGK protein and HVGCC involved (Yang and Colecraft 2013).

Neuronal Ca$_V$2.1–Ca$_V$2.3 channels are regulated by activated GPCRs (Zamponi and Currie 2013). One prominent mechanism involves released Gβ γ subunits binding to pore-forming α_1 subunits and causing a voltage-dependent inhibition of current characterized by slowed activation kinetics (Figure 3.3B) (Ikeda 1996; Herlitze et al. 1996). This inhibition occurs most prominently in $Ca_V2.2$ channels, and involves $G\beta\gamma$ binding to the N-terminus and I–II loop of α_{1B} and shifting channel gating from a willing to reluctant mode characterized by an increased latency to first opening (Agler et al. 2005; Bean 1989; Patil et al. 1996; Colecraft, Brody, and Yue 2001). The inhibition can be relieved by a single strong depolarizing prepulse or a burst of action potential (AP) waveforms, potentially contributing to synaptic plasticity. Opiates acting on -opioid receptors inhibit the presynaptic $Ca_V2.2$ channel in the dorsal horn via this pathway as a contributing mechanism to analgesia (Raingo, Castiglioni, and Lipscombe 2007).

Another important form of regulation for HVGCCs is mediated by Ca^{2+}-binding proteins, most prominently CaM, which is a de facto subunit of these channels. CaM binding provides a mechanism for Ca^{2+} feedback regulation of HVGCCs, but the manner of regulation may be qualitatively different for individual channels. For $Ca_V1.3$, apoCaM binding intrinsically increases the channel P_o and provides the sensor for CDI of the channel (Adams et al. 2014). In the heart, CDI of $Ca_V1.2$ is critical for maintaining normal ventricular action potential duration (APD). Eliminating CDI results in ultralong APDs that would be pathologic (Alseikhan et al. 2002). Diminished CDI contributes to the cardiac pathology of CaM mutations found in human calmodulinopathies (Limpitikul et al. 2014). In tissues where a sustained influx through Ca_V1 channels is required for signaling, such as $Ca_V1.3$ in the inner ear and $Ca_V1.4$ in the retina, CDI is antagonized either by Ca^{2+} binding proteins (CaBP) that compete with CaM ($Ca_V1.3$) or by a module encoded in the distal C-terminus termed of $Ca_V1.4$ (Dolphin and Lee 2020; Singh et al. 2006; Wahl-Schott et al. 2006). In $Ca_V2.1$, CaM mediates both CDI and CDF (DeMaria et al. 2001; Lee, Scheuer, and Catterall 2000). Other notable forms of modulation of HVGCCs include retrograde upregulation of $Ca_V1.1$ by RyR1 in skeletal muscle (Nakai et al. 1996) and syntaxin reducing the availability of $Ca_V2.2$ and $Ca_V2.1$ by stabilizing the inactivated states of these channels (Bezprozvanny, Scheller, and Tsien 1995).

3.8 Biogenesis, Trafficking and Turnover of VGCCs

Like other surface membrane proteins, VGCCs begin their lifecycle with synthesis in the ER membrane, undergo posttranslational modifications in the Golgi, are trafficked to the cell surface where they function, are removed from the cell surface by endocytosis, and are recycled or transported to the lysosome or proteasome for degradation. Though these processes are known or assumed in broad outline. many aspects of the detailed mechanisms and determinants still need to be worked out.

Heterologous expression studies established that forward trafficking of HVGCC α_1 subunits from the ER requires binding to an auxiliary $Ca_V\beta$ subunit (Buraei and Yang 2010). A chimeric channel analysis study in which intracellular loops and termini of $Ca_V1.2$ α_{1C} were systematically swapped into $Ca_V3.1$ (α_{1G}) suggested that the α_{1C} I–II loop contains an ER export signal just downstream of the AID, whereas all other intracellular domains contained ER retention signals. An allosteric model was proposed in which binding of β to the AID causes a conformational change in α_{1C} intracellular domains that shifts the balance between ER retention and export motifs in favor of forward trafficking (Fang and Colecraft 2011). It has also been shown that binding to β protects $Ca_V1.2$ and $Ca_V2.2$ from ubiquitination and proteasomal degradation (Waithe et al. 2011; Altier et al. 2011).

These two mechanisms may act in concert to promote β-dependent forward trafficking of HVGCCs to the cell surface. In adult cardiomyocytes, inducible knockdown of β_2 or expression of a mutant α_{1C}, with a mutated AID that cannot bind β, did not prevent effective $Ca_V1.2$ trafficking to the cell surface, thus breaching the dogma that β binding is always required for HVGCC trafficking to the plasma membrane (Yang et al. 2019; Meissner et al. 2011). Thus, there may be cell-type-specific factors that can substitute for β in promoting HVGCC α_1-subunit forward trafficking, or else provide an environment that makes β binding unnecessary. The surface density of HVGCC α_1 subunits is further markedly augmented by binding to α_2 subunits. The mechanism is proposed to involve increased recycling of endocytosed α_1 subunits as well as stabilization of channels at the cell surface (Zamponi et al. 2015). Beyond the core auxiliary subunits, several other proteins have been shown to regulate surface density of particular HVGCCs, including SH3 and cysteine-rich domain-containing protein 3 (Stac3; $Ca_V1.1$); calmodulin ($Ca_V1.2$); CRMP2 ($Ca_V2.2$); fragile X mental retardation protein (FMRP; $Ca_V2.2$); and RGK GTPases ($Ca_V1.2$).

Unlike HVGCCs, forward trafficking of LVGCCs is not dependent on binding to auxiliary subunits. However, the surface density of LVGCCs is also modulated by protein–protein interactions. The actin-binding protein Kelch-like 1 (KLHL1) augments the surface density of LVGCCs by enhancing recycling of channels to the cell surface (Weiss and Zamponi 2017).

Posttranslational modifications represent an important mechanism for regulating surface density of VGCCs. LVGCCs are modified by asparagine-linked glycosylation (N-glycosylation); elimination of this modification by mutation in $Ca_V3.2$ reduces channel surface density by enhancing their rate of internalization (Weiss and Zamponi 2017). The α_2 subunit is heavily glycosylated, and this modification is important for the ability of this auxiliary subunit to regulate the trafficking of HVGCCs. The surface density of VGCCs is also regulated by ubiquitination, a posttranslational modification that is dependent on a system of ubiquitin writers (E3 ubiquitin ligases) and erasers (deubiquitinases). Functional regulation of the surface density and trafficking of VGCCs by specific E3 ligases have been described, including Nedd 4-1 ($Ca_V1.2$), Parkin ($Ca_V2.2$), RNF138 ($Ca_V2.1$), RFP2 ($Ca_V1.2$) and WWP1 ($Ca_V3.2$). Studies of deubiquitinases that oppose the action of E3 ubiquitin ligases on VGCCs are comparatively limited, with USP5 shown to regulate $Ca_V3.2$ functional expression (Weiss and Zamponi 2017; Zamponi et al. 2015). Overall, many critical questions regarding the molecular players involved and mechanisms by which ubiquitination regulates VGCCs remain outstanding.

3.9 VGCC Channelopathies and Disease

Dysregulation of VGCCs due to either mutations or cellular conditions that alter their functional expression are linked to various diseases (Table 3.2). Distinct mutations in $Ca_V1.1$ are linked to various skeletal muscle diseases (Flucher 2020). Hypokalemic periodic paralysis (HypoPP) is an autosomal dominant genetic disease characterized by flaccid muscle weakness concomitant with low serum potassium levels. Mutations in $Ca_V1.1$ cause HypoPP type 1 (HypoPP1), which represents 60% of HypoPP cases. Most HypoPP1 mutations neutralize S4 gating charges in $Ca_V1.1$ VSDs (e.g., R528G/H and R897S) and disrupt the integrity of the hydrophobic seal in the corresponding VSD. This results in "omega currents" caused by leak of protons or sodium ions through the affected VSD. The presence of omega currents, in conjunction with low serum potassium, promotes a

TABLE 3.2

Distinct Channelopathies, and Alterations in Channel Biophysical Properties, Associated with Mutations in Voltage-Gated Ca^{2+} Channel Subunits

VGCC Subunit	Disease	Biophysical Effects on Ca_V
$Ca_V1.1$	Hypokalemic periodic paralysis type I (HypoPPI)	Loss of function; reduced current density; reduced rate of activation; omega currents through voltage-sensing domain
	Malignant hyperthermia Susceptibility type 5 (MHS5)	Loss of function for channel, but results in increased RyR1 sensitivity for Ca^{2+} release
$Ca_V1.2$ $Ca_V1.3$	Long QT syndrome (LQTS)	Gain of function; increased current density; decreased current inactivation; increased channel expression; usually results in left-shift of activation curve
	Catecholaminergic Polymorphic ventricular tachycardia (CPVT)	Loss of function; decreased current density; decreased channel trafficking; decreased single-channel conductance
	Short QT syndrome (SQTS)	
	Brugada syndrome (BS)	Gain of function; loss of voltage-dependent inactivation
	Timothy syndrome (TS)	Gain of function; increased current density; left-shift in activation curve; increased voltage-dependent inactivation
	Autism spectrum disorders (ASDs)	
$Ca_V1.4$	X-linked congenital Stationary night blindness type 2	Loss of function; left-shift in activation curve; decreased rate of channel inactivation; decreased channel activity (e.g., current density, channel trafficking, channel opening)
$Ca_V2.1$	Familial hemiplegic migraine type I (FHM1)	Gain of function; left-shift in activation curve; increased open probability, increased rate of inactivation
	Episodic ataxia type 2 (EA2)	Loss of function; reduced current density/channel activity; reduced channel trafficking
	Spinocerebellar ataxia type 6 (SCA6)	Gain of function; left-shift in activation curve; increased channel density/perinuclear aggregates
	Epilepsy/absence seizures	Mixed channel effects depending upon channel mutation (e.g., gating shifts, current density)
$Ca_V3.2$	Idiopathic generalized epilepsy (IGE)	Mixed channel effects depending upon channel mutation (e.g., gating shifts, current density)
	Autism spectrum disorders (ASDs)	Loss of function; decreased current density; right-shift in activation curve
β_2 β_4	Short QT syndrome (SQTS)	Loss of function; decreased current density; increased late current
	Brugada syndrome (BS)	
	Epilepsy/ataxia	Gain of function; increased current density; increased rate of inactivation
$\alpha_2\delta_1$	Short QT syndrome (SQTS) Brugada syndrome (BS)	Loss of function; decreased current density

long-lasting depolarization in skeletal muscle that causes $Na_V1.4$ inactivation and attendant muscle weakness. Mutations in $Ca_V1.1$ also cause malignant hypothermia susceptibility type 5 (MHS5), an autosomal dominant disease characterized by excessive Ca^{2+} release in response to volatile anesthetics and depolarizing muscle relaxants that can be fatal if untreated. Most MHS cases are caused by mutations in RyR1. MHS5 mutations alter the coupling between $Ca_V1.1$ and RyR1 in such a manner that there is a hypersensitivity of Ca^{2+} release in response to physiologic and pharmacological activation (Flucher 2020).

Mutations in human α_{1C} subunit are associated with various cardiac arrhythmia syndromes, reflecting the dominant expression of $Ca_V1.2$ in the heart (Zhang et al. 2018). Putative loss-of-function (LoF) mutations in α_{1C} have been linked to short QT (SQTS), Brugada (BrS) and early repolarization syndromes (ERS). Timothy syndrome (TS), a rare autosomal dominant syndrome caused by gain-of-function (GoF) mutations in $Ca_V1.2$, is a multisystemic disease characterized by a prolonged long QT interval (long QT syndrome type 8, LQT8), autism spectrum disorder, syndactyly and facial dysmorphisms (Bauer et al. 2009). The original characterization of the molecular cause of TS identified a G406R mutation in an alternatively spliced exon (exon 8 and 8A) of $Ca_V1.2$ that disrupts the S6 helix in DI of α_{1C} and causes a slowed inactivation of $Ca_V1.2$ (Splawski et al. 2004). The severity and profile of symptoms displayed by TS patients can be quite variable and is dependent on factors such as $Ca_V1.2$ alternative splicing, genetic background and mosaicism. Genome-wide association studies (GWAS) have consistently linked single-nucleotide polymorphisms (SNPs) in *CACNA1C* to various neuropsychiatric disorders including bipolar disorder, schizophrenia, major depressive disorder and autism (Bhat et al. 2012).

Humans with homozygous mutations in CACNA1D that result in a nonconducting $Ca_V1.3$ channel display congenital deafness and bradycardia arising from sinoatrial node dysfunction (SANDD syndrome) (Striessnig, Bolz, and Koschak 2010). By contrast, missense mutations that lead to a GoF in $Ca_V1.3$ cause neurodevelopmental (intellectual impairment, autism spectrum disorder, seizures) and endocrine (hyperaldosteronism and hyperinsulinemic hypoglycemia) disorders. Fitting with the role of $Ca_V1.4$ in mediating sustained neurotransmission at ribbon synapses in photoreceptor cells, mutations in CACNA1F cause X-linked congenital stationary night blindness type 2 (CSNB2) and X-linked cone-rod dystrophy type 3. Electrophysiological studies indicate that mutations that cause either LoF or GoF in $Ca_V1.4$ are linked to disease (Striessnig, Bolz, and Koschak 2010).

Mutations in $Ca_V2.1$ cause a spectrum of neurological diseases including epileptic encephalopathies (EE), familial hemiplegic migraine type 1 (FHM1), episodic ataxia type 2 (EA2), spinocerebellar ataxia type 6 (SCA6) and intellectual disability (ID). FHM1 is associated with GoF missense mutations in $Ca_V2.1$, while EA2 arises from LoF mutations, many of which are nonsense mutations resulting in a premature stop codon. SCA6 results from polyglutamine repeats (>21 CAG) in the C-terminus of $Ca_V2.1$ (Pietrobon 2010).

Beyond the pore-forming subunits, mutations in HVGCC auxiliary subunits have also been linked to disease (Zamponi et al. 2015). Mutations in all four α_2 subunit isoforms have been linked to neurological and/or cardiovascular diseases: α_2-1 – epileptic encephalopathy, Brugada syndrome, short QT syndrome; α_2-2 – epileptic encephalopathy, cerebellar ataxia, schizophrenia; α_2-3 – autism spectrum disorder; α_2-4 – night blindness and cone rod dystrophy. Further, mutations in CACNB2 and CACNB4 have been linked to Brugada syndrome and epilepsy, respectively.

LoF and GoF mutations in $Ca_V3.1$ have been linked to cerebellar ataxias and childhood onset cerebellar atrophy, respectively. Mutations in $Ca_V3.2$ have been associated with idiopathic generalized epilepsy, autism spectrum disorder and hyperaldosteronism. Further, gain of $Ca_V3.2$ functional expression associated with changes in posttranslational modifications including de-ubiquitination and glycosylation have been linked to chronic pain (Weiss and Zamponi 2020).

3.10 Conclusion

There has been tremendous progress over the last four decades in our understanding of the family of VGCCs in terms of their physiological roles, biophysical mechanisms of gating and permeation, pharmacology, structure–function, and involvement in disease. These discoveries have been driven by the application of key enabling techniques and research tools as they have been developed. These include the patch-clamp technique, cloning, heterologous expression, knockout mice and structural biology approaches (X-ray crystallography and cryo-EM). Nevertheless, there are significant gaps in our knowledge of VGCCs that remain. The intracellular loops and termini of VGCCs are critical determinants of channel function and regulation but are mostly unresolved in cryo-EM structures due to their mobility. The mechanisms and determinants underlying biogenesis, trafficking and degradation of VGCCs in their native physiological settings are understudied and not well understood. Mechanistic studies of how mutations in VGCCs cause disease, and bridging the basic insights into developing novel therapies, also remains an important research frontier.

Suggested Reading

This chapter includes additional bibliographical references hosted only online as indicated by citations in blue color font in the text. Please visit https://www.routledge.com/9780367538163 to access the additional references for this chapter, found under "Support Material" at the bottom of the page.

Abderemane-Ali, F., F. Findeisen, N. D. Rossen, and D. L. Minor Jr. 2019. "A selectivity filter gate controls voltage-gated calcium channel calcium-dependent inactivation." *Neuron* 101 (6):1134–49 e3. doi:10.1016/j.neuron.2019.01.011.

Bannister, R. A., and K. G. Beam. 2013. "Ca(V)1.1: the atypical prototypical voltage-gated Ca(2)(+) channel." *Biochim Biophys Acta* 1828 (7):1587–97. doi:10.1016/j.bbamem.2012.09.007.

Bean, B. P., M. C. Nowycky, and R. W. Tsien. 1984. "Beta-adrenergic modulation of calcium channels in frog ventricular heart cells." *Nature* 307 (5949):371–5.

Buraei, Z., and J. Yang. 2010. "The {beta} subunit of voltage-gated Ca2+ channels." *Physiol Rev* 90 (4):1461–506. doi:10.1152/physrev.00057.2009.

Chi, X. X., B. S. Schmutzler, J. M. Brittain, Y. Wang, C. M. Hingtgen, G. D. Nicol, and R. Khanna. 2009. "Regulation of N-type voltage-gated calcium channels (Cav2.2) and transmitter release by collapsin response mediator protein-2 (CRMP-2) in sensory neurons." *J Cell Sci* 122 (23):4351–62. doi:10.1242/jcs.053280.

Colecraft, H. M., D. L. Brody, D. T. Yue. 2001. "G-Protein Inhibition of N- and P/Q-Type Calcium Channels: Distinctive Elementary Mechanisms and Their Functional Impact." *Journal of Neuroscience* 21(4):1137–1147.

Dayal, A., M. L. Fernandez-Quintero, K. R. Liedl, and M. Grabner. 2021. "Pore mutation N617D in the skeletal muscle DHPR blocks Ca(2+) influx due to atypical high-affinity Ca(2+) binding." *eLife* 10. doi:10.7554/eLife.63435.

Dolphin, A. C. 2013. "The alpha2delta subunits of voltage-gated calcium channels." *Biochim Biophys Acta* 1828 (7):1541–9. doi:10.1016/j.bbamem.2012.11.019.

Dolphin, A. C., and A. Lee. 2020. "Presynaptic calcium channels: specialized control of synaptic neurotransmitter release." *Nat Rev Neurosci* 21 (4):213–29. doi:10.1038/s41583-020-0278-2.

Dong, Y., Y. Gao, S. Xu, Y. Wang, Z. Yu, Y. Li, B. Li, T. Yuan, B. Yang, X. C. Zhang, D. Jiang, Z. Huang, and Y. Zhao. 2021. "Closed-state inactivation and pore-blocker modulation mechanisms of human CaV2.2." *Cell Rep* 37 (5):109931. doi:10.1016/j.celrep.2021.109931.

Flucher, B. E. 2020. "Skeletal muscle CaV1.1 channelopathies." *Pflugers Arch* 472 (7):739–54. doi:10.1007/s00424-020-02368-3.

Gao, S., X. Yao, and N. Yan. 2021. "Structure of human Cav2.2 channel blocked by the painkiller ziconotide." *Nature* 596 (7870):143–7. doi:10.1038/s41586-021-03699-6.

Gee, N. S., J. P. Brown, V. U. Dissanayake, J. Offord, R. Thurlow, and G. N. Woodruff. 1996. "The novel anticonvulsant drug, gabapentin (neurontin), binds to the alpha2delta subunit of a calcium channel." *J Biol Chem* 271 (10):5768–76.

Hess, P., R. W. Tsien. 1984. "Mechanism of ion permeation through calcium channels."*Nature* 309(5967):453–6. doi: 10.1038/309453a0.

Hockerman, G. H., B. Z. Peterson, B. D. Johnson, and W. A. Catterall. 1997. "Molecular determinants of drug binding and action on L-type calcium channels." *Annu Rev Pharmacol Toxicol* 37:361–96. doi:10.1146/annurev.pharmtox.37.1.361.

Imredy, J. P., and D. T. Yue. 1994. "Mechanism of Ca(2+)-sensitive inactivation of L-type Ca2+ channels." *Neuron* 12 (6):1301–18.

Liu, G., A. Papa, A. N. Katchman, S. I. Zakharov, D. Roybal, J. A. Hennessey, J. Kushner, L. Yang, B. X. Chen, A. Kushnir, K. Dangas, S. P. Gygi, G. S. Pitt, H. M. Colecraft, M. Ben-Johny, M. Kalocsay, S. O. Marx. 2020. "Mechanism of adrenergic CaV1.2 stimulation revealed by proximity proteomics."*Nature* 577(7792):695–700. doi: 10.1038/s41586-020-1947-z. Epub 2020 Jan 22.

Morgenstern, T. J., J. Park, Q. R. Fan, H. M. Colecraft. 2019. "A potent voltage-gated calcium channel inhibitor engineered from a nanobody targeted to auxiliary CaVβ subunits." *Elife* 8:e49253. doi: 10.7554/eLife.49253.

Pantazis, A., N. Savalli, D. Sigg, A. Neely, and R. Olcese. 2014. "Functional heterogeneity of the four voltage sensors of a human L-type calcium channel." *Proc Natl Acad Sci U S A* 111 (51):18381–6. doi:10.1073/pnas.1411127112.

Pietrobon, D. 2010. "CaV2.1 channelopathies." *Pflugers Arch* 460 (2):375–93. doi:10.1007/s00424-010-0802-8.

Platzer, J., J. Engel, A. Schrott-Fischer, K. Stephan, S. Bova, H. Chen, H. Zheng, and J. Striessnig. 2000. "Congenital deafness and sinoatrial node dysfunction in mice lacking class D L-type Ca2+ channels." *Cell* 102 (1):89–97. doi:10.1016/s0092-8674(00)00013-1.

Sather, W. A., and E. W. McCleskey. 2003. "Permeation and selectivity in calcium channels." *Annu Rev Physiol* 65:133–59. doi:10.1146/annurev.physiol.65.092101.142345.

Schmidtko, A., J. Lotsch, R. Freynhagen, and G. Geisslinger. 2010. "Ziconotide for treatment of severe chronic pain." *Lancet* 375 (9725):1569–77. doi:10.1016/S0140-6736(10)60354-6.

Splawski, I., K. W. Timothy, L. M. Sharpe, N. Decher, P. Kumar, R. Bloise, C. Napolitano, P. J. Schwartz, R. M. Joseph, K. Condouris, H. Tager-Flusberg, S. G. Priori, M. C. Sanguinetti, and M. T. Keating. 2004. "Ca(V)1.2 calcium channel dysfunction causes a multisystem disorder including arrhythmia and autism." *Cell* 119 (1):19–31. doi:10.1016/j.cell.2004.09.011.

Striessnig, J., H. J. Bolz, and A. Koschak. 2010. "Channelopathies in Cav1.1, Cav1.3, and Cav1.4 voltage-gated L-type Ca2+ channels." *Pflugers Arch* 460 (2):361–74. doi:10.1007/s00424-010-0800-x.

Takahashi, M., M. J. Seagar, J. F. Jones, B. F. Reber, and W. A. Catterall. 1987. "Subunit structure of dihydropyridine-sensitive calcium channels from skeletal muscle." *Proc Natl Acad Sci U S A* 84 (15):5478–82. doi:10.1073/pnas.84.15.5478.

Tsien, R. W., D. Lipscombe, D. V. Madison, K. R. Bley, and A. P. Fox. 1988. "Multiple types of neuronal calcium channels and their selective modulation." *Trends Neurosci* 11 (10):431–8. doi:10.1016/0166-2236(88)90194-4.

Weiss, N., and G. W. Zamponi. 2017. "Trafficking of neuronal calcium channels." *Neuronal Signal* 1 (1):NS20160003. doi:10.1042/NS20160003.

Wu, J., Z. Yan, Z. Li, X. Qian, S. Lu, M. Dong, Q. Zhou, and N. Yan. 2016. "Structure of the voltage-gated calcium channel Ca(v)1.1 at 3.6 A resolution." *Nature* 537 (7619):191–6. doi:10.1038/nature19321.

Yang, J., P. T. Ellinor, W. A. Sather, J. F. Zhang, and R. W. Tsien. 1993. "Molecular determinants of Ca2+ selectivity and ion permeation in L-type Ca2+ channels." *Nature* 366 (6451):158–61. doi:10.1038/366158a0.

Yang, L., A. Katchman, J. Kushner, A. Kushnir, S. I. Zakharov, B. X. Chen, Z. Shuja, P. Subramanyam, G. Liu, A. Papa, D. Roybal, G. S. Pitt, H. M. Colecraft, and S. O. Marx. 2019. "Cardiac CaV1.2 channels require beta subunits for beta-adrenergic-mediated modulation but not trafficking." *J Clin Invest* 129 (2):647–58. doi:10.1172/JCI123878.

Yang, T., and H. M. Colecraft. 2013. "Regulation of voltage-dependent calcium channels by RGK proteins." *Biochim Biophys Acta* 1828 (7):1644–54. doi:10.1016/j.bbamem.2012.10.005.

Zamponi, G. W., and K. P. Currie. 2013. "Regulation of Ca(V)2 calcium channels by G protein cou-pled receptors." *Biochim Biophys Acta* 1828 (7):1629–43. doi:10.1016/j.bbamem.2012.10.004.

Zhao, Y., G. Huang, J. Wu, Q. Wu, S. Gao, Z. Yan, J. Lei, and N. Yan. 2019. "Molecular basis for ligand modulation of a mammalian voltage-gated Ca(2+) channel." *Cell* 177 (6):1495–1506 e12. doi:10.1016/j.cell.2019.04.043.

4

Voltage-Gated Potassium Channels

Francis I. Valiyaveetil

CONTENTS

4.1 Introduction

Potassium channels allow the movement of K^+ ions across cellular membranes. In the family of voltage-gated K^+ (K_v) channels, the flux of K^+ through the channel is controlled by the membrane potential (Figure 4.1A). K_v channels constitute one of the largest families of K^+ channels. K_v channels play diverse roles in physiology. In electrically excitable cells such as neurons and muscle cells, K_v channels participate in the repolarization phase of the action potential and therefore regulate the excitability of these cells. In endocrine cells, the action of K_v channels is involved in the controlled release of hormones, while in immune cells, K_v channels play a role in the activation of T cells during the immune response. K_v channels participate in cell volume regulation, and K^+ movement though K_v channels has been linked with cell cycle progression and cell death.

The first K^+ channel gene to be cloned was the *Shaker* channel from Drosophila (Figure 4.1B, C) (Papazian et al. 1987). The cloning of the Shaker K^+ channel heralded the era of

DOI: 10.1201/9781003096276-4

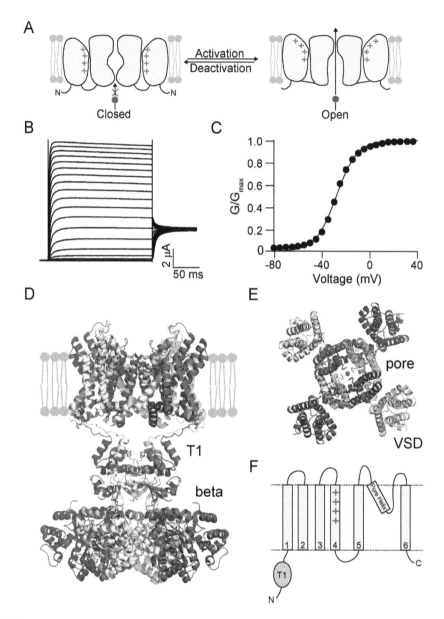

FIGURE 4.1

Structure and function of K_v channels. (A) Illustration of voltage gating in a K_v channel. K^+ ion is depicted as a purple sphere. (B) Current traces for the Shaker K^+ channel elicited with depolarizations from −80 mV to +40 mV in 5 mV steps. (C) Conductance–voltage (G–V) relationship for the Shaker K^+ channel. Solid line represents a Boltzmann fit to the data. (D, E) Side view (D) and top view (E) of the K_v1.2* channel (Protein Data Bank: 2R9R) with each subunit colored differently. (F) Topology of a single subunit of a K_v channel. The transmembrane segments forming the pore domain and the voltage-sensor domain are labeled and are colored differently.

structure function investigations of the K_v channel proteins that has continued with the determination of the crystal structures and recently the cryo-EM structures of K_v channels. There is a vast body of literature on K_v channels and this chapter provides a brief summary of the important aspects of structure, function and the diversity of K_v channels.

4.2 Structure of K_v Channels

Structures have been reported for the archaebacterial K_vAP, the rat $K_v1.2$, and the human $K_v1.3$ and $K_v4.2$ channels (Long et al. 2007; Kise et al. 2021). The highest resolution structure presently available is for a variant of the $K_v1.2$ channel that is commonly referred to as the $K_v1.2$–2.1 chimera or in this chapter as $K_v1.2^*$ channel (Figure 4.1 D, E).

K_v channels are tetramers that are made of identical (homotetramers) or of similar subunits (heterotetramers). Each subunit of the K_v channel consists of three distinct domains (Figure 4.1F). At the cytoplasmic N-terminus is the T1 domain, which is followed by six transmembrane segments (S1–S6) that are arranged in two distinct domains. The S5 and S6 segments from the four subunits form the pore domain, while the S1–S4 segments form the voltage-sensor domain (VSD). The pore domain is present at the center of the K_v channel tetramer, while the VSDs are at the periphery of the pore domain. In the $K_v1.2$ structure, the VSD of each of the subunits interacts with the S5 and S6 segments of the neighboring subunit in a domain-swap arrangement. This domain-swap arrangement is also seen for voltage-gated Na^+ and Ca^{2+} channels, which similar to the K_v channels, belong to the superfamily of voltage-gated ion channels. The domain-swap arrangement is, however, not universal. Within the K_v channel family itself, the domain-swap arrangement is not observed for the Eag ($K_v10.1$), HERG ($K_v11.1$) or the K_vAP channels. The significance of the domain-swap arrangement and the functional consequences of this arrangement are presently not known.

The T1 domain of each subunit comes together in a tetrameric assembly, and tetramerization by the T1 domain is important for directing K_v channel assembly. The T1 domains can form both homo- and heterotetramers. Heterotetramerization can, however, only take place between T1 domains of subunits belonging to the same K_v channel family, K_v1 family subunits for example, and therefore limits assembly of the channel tetramer to subunits of the same family. The T1 domain also forms the binding site for the accessory subunits for the various K_v channels, such as the β subunits and the KChIPs (K^+ channel-interacting proteins).

4.3 The Pore Domain

The K^+ permeation pathway is present along the central fourfold axis of the pore domain (Figure 4.2A). Toward the extracellular side of the ion pathway is the selectivity filter wherein the selection for K^+ takes place (Figure 4.2B). The selectivity filter consists of four ion-binding sites that are constructed using the backbone carbonyl oxygen atoms and the Thr side chain from the sequence TVGYG. This sequence is highly conserved in K^+ channels and the structure of the selectivity filter in K^+ channels is essentially identical.

Toward the cytoplasmic side of the ion permeation pathway is the activation gate. The activation gate is formed by the bundle crossing of the S6 helices. In the closed state, the S6 helices in the bundle cross in close apposition such that the pathway through the bundle crossing is not wide enough for the passage of a hydrated K^+ ion. On voltage activation, the S6 helices undergo a hinged movement that causes a widening of the bundle crossing in an iris-like manner, thereby permitting the flow of K^+ across the membrane. A conserved Gly residue in the S6 helix acts as the hinge point for this movement.

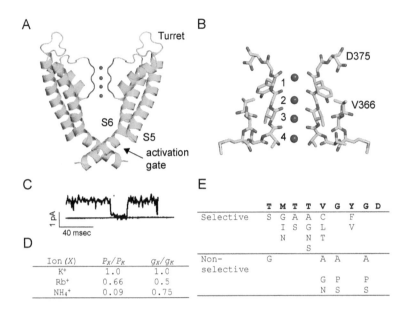

FIGURE 4.2
Ion permeation and the pore domain of K_v channels. (A) Structure of the pore domain of the $K_v1.2^*$ channel (Protein Data Bank: 2R9R). Two opposite subunits of the tetrameric pore are shown with the selectivity filter (residues T370–G375) colored red and the K^+ ions bound shown as purple spheres. (B) Close-up of the selectivity filter. Residues V366-D375 are shown as sticks and the ion-binding sites are labeled. (C) Single-channel recording of the Shaker channel at +80 mV in 100 mM K^+. (D) Ionic permeability and single-channel conductance of the Shaker channel. The ratio of the permeabilities (P_X/P_K) of Rb^+ and $NH4^+$ relative to K^+ calculated from the reversal potential measured under bi-ionic conditions is shown. Single-channel conductances (g) were determined from single-channel recordings. (Data in C and D from Heginbotham and MacKinnon 1993.) (E) Amino acid substitutions in the selectivity filter of the Shaker channel that do not alter ionic selectivity and those that result in a nonselective channel are listed. (Data from Heginbotham et al. 1994.)

4.4 Ionic Selectivity and Permeation

Underlying the physiological roles of K_v channels is the ability to rapidly and selectively conduct K^+ across biological membranes. Single-channel measurements of the Shaker channel indicate a channel conductance of 18 pS (Figure 4.2C) (Heginbotham and MacKinnon 1993). The channel conductance varies depending upon the specific K_v channel (Gutman et al. 2005). K_v channels, similar to other members of the K^+ channel family, show high selectivity for K^+ over Na^+. Measurements of ionic selectivity for the Shaker K_v channel using reversal potential measurements suggest a sequence of $K^+ > Rb^+ > NH_4^+ > Cs^+ >> Na^+$ (Figure 4.2D) (Heginbotham and MacKinnon 1993). The relative conductance of various ions through the Shaker K_v channel, measured under symmetric ionic conditions, follows the sequence $K^+ > NH_4^+ > Rb^+$, while conductance of Li^+, Na^+ and Cs^+ is generally undetectable (Heginbotham and MacKinnon 1993). The divalent ion Ba^{2+} acts as a blocker, as Ba^{2+} can bind to the selectivity filter but cannot permeate through the filter.

Ionic selectivity of the channel is set by the selectivity filter. Extensive mutational studies have been carried out on the selectivity filter residues and these studies show that only a few substitutions are tolerated with a retention of ionic selectivity (Figure 4.2E) (Heginbotham et al. 1994). Nonconservative substitutions result in a nonselective phenotype or result in channels that are either nonconductive or do not form tetrameric assemblies in the membrane. These studies highlight that the role of the precise architecture of the selectivity filter for K⁺ over Na⁺ selectivity.

4.5 Voltage-Sensor Domains and the Mechanism of Voltage Gating

The voltage-dependent opening of the pore, the central property of K_v channels, takes place through the action of VSDs (Yellen 1998, Bezanilla 2005). The VSD is formed from the S1 to S4 segments. The structure of the VSD, visualized in the $K_v1.2$ channel (Long et al. 2007) or in the isolated VSD from the K_vAP channel, shows that the S1–S4 segments are arranged in a loose bundle (Figure 4.3A). The S4 segment shows a characteristic feature of a positively charged residue, mainly Arg, at every third position along the helix (Figure 4.3B). These positively charged residues in the S4 segment sense the changes in the transmembrane potential, which is translated to a movement of the S4 segment that is coupled to the opening or closing of the pore domain.

Measurements in the Shaker K⁺ channel have indicated that during voltage gating 12–14 e_o charges move across the field. Gating charge measurements coupled with charge neutralization studies in the Shaker channel have indicated that the R1–R4 residues in the S4 segment contribute to the gating charges (along with E293 in the S2 segment) (Aggarwal and MacKinnon 1996; Seoh et al. 1996). Changes in the S4 segment during voltage gating have been evaluated using multiple approaches, such as changes in the accessibility of substituted Cys or His residues. Measurements of the S4 movement have also been probed by fluorescence approaches that provide a real-time visualization of S4 movement (Mannuzzu, Moronne, and Isacoff 1996; Cha and Bezanilla 1997). These studies in general support a model in which the four outermost charges transition from an environment wherein they are accessible to the internal solution to an environment wherein they are accessible to the external solution.

The positive charged residues in the S4 segment are stabilized in the membrane by counter charges, Asp and Glu residues, that are present in S1–S3 segments (Long et al. 2007.) These counter charges are arranged in two clusters that are present toward either side of the membrane and referred to as the inner and the outer clusters (Figure 4.3A). These counter charges are conserved in K_v channels. It is anticipated that sequential interactions of the positively charged residues in the S4 segment with the counter charges facilitate the movement of the S4 segment. There is a stretch of hydrophobic residues, referred to as the hydrophobic gasket, that separates the inner and the outer charge clusters. Movement of the S4 segment requires the passage of the positively charged residues through this hydrophobic gasket. Within the hydrophobic gasket is a conserved aromatic residue, Phe290, in the Shaker channel that stabilizes the charges as they transition through the hydrophobic gasket.

There have been different models proposed for the nature of the movement of the S4 segment such as the sliding helix, the helical screw or the paddle movement, and the extent of S4 movement has also been debated (Bezanilla 2005). It has also been suggested that the exact nature of the S4 movement may vary depending on the specific VSD under investigation.

The movement of the S4 segment is coupled to the pore domain (Figure 4.3C). Detailed analysis of voltage gating support the independent outward movement of four voltage

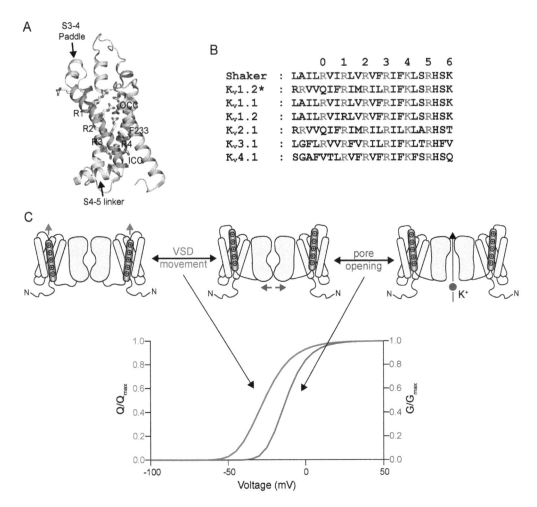

FIGURE 4.3

The voltage-sensor domain and voltage gating in K_v channels. (A) Structure of the voltage-sensor domain in the K_v1.2* channel (Protein Data Bank: 2R9R). The S4 segment is colored salmon. Side chains of the positively charged residues on S4; the negatively charged residues on S1–S3 forming the inner and the outer charge cluster, labeled ICC and OCC; and the conserved Phe residue (F233) are shown in sticks representation. (B) Sequence alignment of the S4 segment of the Shaker, K_v1.2* and the human K_v1.1, K_v1.2, K_v2.1, K_v3.1 and K_v4.1 are shown. Arg residues in the S4 segments are colored red. (C) Voltage gating in K_v channels. Illustration shows the gating mechanism in a K_v channel (top). S4 in the voltage-sensor domain is colored red. Gating charge movement takes place prior to the initiation of the ionic flux through the pore. A schematic representation of the voltage dependence of the normalized gating charge movement (Q/Q_{max}) and conductance (G/G_{max}) for the Shaker channel (bottom). (The schematic is based on data presented in Islas and Sigworth 1999.)

sensors followed by cooperative closed-to-open transitions of the pore domain (Zagotta, Hoshi, and Aldrich 1994). The conventional mechanism proposed for the coupling of the S4 movement to the pore opening is that the outward movement of the S4 segment on depolarization pulls the S4–S5 linker region. The S4–S5 linker region interacts with the C-terminal region of the S6 segment following the activation gate in the pore domain. The movement of the S4–S5 linker leads to a movement of the S6 segment, which results in the opening of the activation gate to initiate the K^+ flux across the membrane (Long et al.

2007). In addition to this conventional model, other coupling mechanisms have also been proposed that involves changes in the interface between the S4 and the segments in the pore domain.

4.6 N-Type and C-Type Inactivation

Activation in a K_v channel is followed by inactivation processes that turn off the flux of K^+ ions through the channel (Figure 4.4A) (Kurata and Fedida 2006). There are two mechanisms of inactivation in K_v channels that are referred to as N-type and C-type inactivation.

In the process of N-type inactivation, the N-terminus of the channel binds to the open pore domain and blocks the flux of K^+ through the channel (Hoshi, Zagotta, and Aldrich 1990). This process of inactivation takes place on the order of milliseconds and is therefore also referred to as fast inactivation (Figure 4.4B). The role of the N-terminus was established by demonstrating that genetic truncation of N-terminus of the channel removes N-type inactivation, while inactivation can be restored by the *trans* addition of a peptide corresponding to the N-terminus. Mutational studies indicated that the binding site of the N-terminus is the cavity region below the selectivity filter in the pore domain. The N-terminus has a number of phosphorylation sites and phosphorylation prevents binding of the N-terminus to the pore indicating that N-type inactivation can be modulated by protein phosphorylation (Kurata and Fedida 2006).

Shaker channels with a truncation at the N-terminus do not undergo N-type inactivation but still display an inactivation process that is referred to a C-type inactivation (Hoshi, Zagotta, and Aldrich 1991). The time course of C-type inactivation is on the order of seconds, much slower than N-type inactivation (Figure 4.4C) (Kurata and Fedida 2006). C-type inactivation is therefore also referred to as slow inactivation. A characteristic feature of the C-type inactivation process is a dependence on the permeant ion (Lopez-Barneo et al. 1993). Functional measurements show that the rate of inactivation increases when the extracellular concentrations of K^+ is lowered and that the rate of inactivation is reduced when the permeant ion is changed from K^+ to Rb^+. Mutations in and around the selectivity filter change the rate of inactivation (Kurata and Fedida 2006). These studies suggested that C-type inactivation involves the selectivity filter. Studies using site-directed labeling, fluorescence measurements and computational studies have been carried out to decipher the structural changes during C-type inactivation (Kurata and Fedida 2006). These studies support a local structural change at the extracellular side of the selectivity filter during C-type inactivation. The studies using site-directed fluorescence labeling have shown that there are also changes in the turret regions of the pore domain on C-type inactivation.

Recent structural studies on the $K_v1.2$ and the Shaker channels have provided the structure of the channel in the C-type inactivated state (Reddi et al. 2022). These structures show a dilation of the outer two ion-binding sites in the selectivity filter, which is caused by conformational changes in the conserved Tyr and Asp side chains (Figure 4.4D). The structure also shows changes in the extracellular mouth of the channel and in the turret region consistent with previous biochemical and spectroscopic studies on C-type inactivation. The structure suggests that the reduction or the loss of ion conduction through the selectivity filter during C-type inactivation is due to the disruption of outer ion-binding sites. Computational studies will be required to decipher why the disruption of the outer ion-binding sites eliminate or reduce ion throughput through the channel.

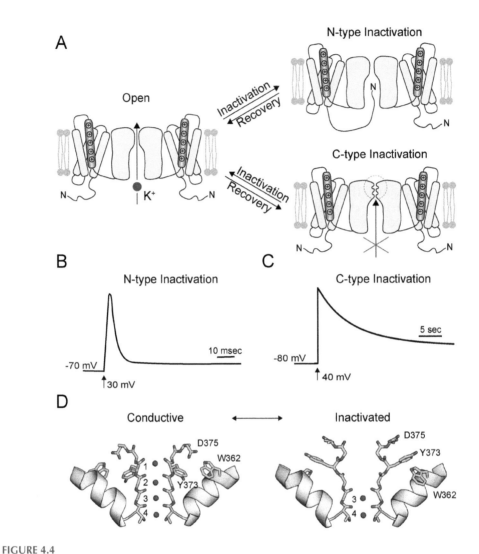

FIGURE 4.4
Inactivation mechanisms in K_v channels. (A) Illustration of the process of N-type (fast) and C-type (slow) inactivation in a K_v channel. (B) N-type inactivation. A schematic representation of macroscopic currents recorded from Shaker channels showing a rapid decay in current following activation due to N-type inactivation. (The schematic is based on current traces presented in Hoshi, Zagotta, and Aldrich 1990.) (C) C-type inactivation. Current traces of the Shaker ($\Delta6$-46, N-type inactivation removed) channels on depolarization show a slow decay in current following activation due to C-type inactivation. (D) Structural changes at the selectivity figure during C-type inactivation. The conductive state of the selectivity filter is based on the $K_v 1.2^*$ channel (Protein Data Bank: 2R9R) and the C-type inactivated state is based on a mutant $K_v 1.2^*$ channel with amino acid substitutions to stabilize the C-type inactivated state (Protein Data Bank: 7SIT). Two opposite subunits are shown with the ion-binding sites labeled and the K^+ ions shown as purple spheres. The side chains for W363 (*W434*), Y373 (*Y445*) and D373 (*D447*) are shown as sticks with the numbers in parenthesis corresponding to the Shaker channel.

4.7 K_v Channel Families

The mammalian genes for the K_v channel α subunits are grouped into twelve distinct families (K_v1–K_v12) (Rudy et al. 2009; Gutman et al. 2005). This chapter describes the K_v1–K_v6, K_v8 and K_v9 families. The K_v7 and K_v10–K_v12 channels are covered in other chapters of this book.

Historically K_v channels have been classified as A-type, when the channel currents are fast activating and inactivating, and as delayed rectifiers, when the channels activate with a delay and are essentially noninactivating. K_v channels that can be classified as A- type or delayed rectifiers are present within the same family.

4.7.1 The K_v1 Family

The K_v1 channels are the mammalian equivalent of the *Drosophila Shaker* channel. K_v1 channels are low-voltage activated channels that open with small depolarizations from the resting potential (Table 4.1) (Dolly and Parcej 1996). They display fast activation kinetics and serve to maintain the membrane potential and modulate electrical excitability. The K_v1 family consists of eight members (K_v1.1 to K_v1.8). All of these K_v1 channels, with the exception of the K_v1.7 subunit, are expressed in the brain at varying levels depending on the specific locus. The K_v1.1 to K_v1.6 subunits form homotetramers on heterologous expression. In native tissue, the formation of heterotetramers of two or more different K_v1 subunits also takes place depending upon the K_v1 subunits that are expressed in the specific cell. The family-specific assembly of the subunits depends upon the T1 domain, which permits the assembly of K_v1 subunits only with other members of the K_v1 family. The functional properties of the assembly depend upon the specific combination of the subunits. For example, the K_v1.1 channel subunit lacks an N-terminal inactivation peptide and therefore N-type inactivation is not observed in homomeric K_v1.1 channels. N-type inactivation in complexes containing the K_v1.1 subunit requires the presence of another K_v1 subunit with an N-terminal inactivation peptide (or assembly with an accessory β subunit that bears an N-terminal inactivation peptide).

Outside the brain, K_v1 subunits are expressed in a variety of excitable tissues such as skeletal muscle, heart, and β cells in the pancreatic islets and also in a wide variety of non-excitable tissues such as immune cells. The K_v1.3 channel is expressed in immune cells such as T lymphocytes, macrophages and microglia (see Volume III, Chapter 7). In T lymphocytes, the outward movement of K^+ through the K_v1.3 channel is required to balance the influx of Ca^{2+}, which is required for T cell activation. Blocking the K_v1.3 channel prevents the calcium influx that blocks T cell activation and proliferation. The K_v1.3 channel has therefore been proposed as a therapeutic target in autoimmune and inflammatory diseases such as multiple sclerosis, psoriasis and type I diabetes (Wulff, Castle, and Pardo 2009). The K_v1.5 channel is expressed in the heart atria and is a component of the ultrarapid outward current I_{Kur} that is important for atrial repolarization during the action potential (see Volume III, Chapter 5). The K_v1.5 channel has been proposed as a target for treating atrial fibrillations, a common cause of cardiac tachycardia (Wulff, Castle, and Pardo 2009), This is because I_{Kur} is not present in ventricles and therefore targeting the K_v1.5 channel provides an atria-specific therapy.

TABLE 4.1

Functional Characteristic of K_v Channels

		Activation	k (Steepness)	Activation Rates	Single-Channel Conductance
		(mV)	(mV)	(ms)	(pS)
Shaker Family					
$K_v1.1$	Fast activation, low threshold, delayed rectifier	−32	8.5	5	10
$K_v1.2$	Fast activation, low threshold, delayed rectifier		13	6	14–18
$K_v1.3$	Fast activation, low threshold, delayed rectifier	−35	6	3	13
$K_v1.4$	Fast activation, low threshold, A-type	−22, −34	5		5
$K_v1.5$	Fast activation, low threshold, delayed rectifier	−14	6, 12		8
$K_v1.6$	Fast activation, low threshold, delayed rectifier	−20	8		9
Shab Family					
$K_v2.1$	Slow activation, high threshold, delayed rectifier	12	3		8
Shaw Family					
$K_v3.1$	Fast activation, high threshold, delayed rectifier	16	10	2	27
$K_v3.2$	Fast activation, high threshold, delayed rectifier	13	7–7.5	4	16–20
$K_v3.4$	Fast activation, high threshold, A-type	3.4, 14	8.4		14
Shal Family					
$K_v4.1$	Fast activation, low threshold, A-type	−47.9	24.2		6

Source: Data from Gutman et al. (2005).

4.7.2 The K_v2 Family

K_v2 channels are the mammalian equivalent of the *Drosophila Shab* channel. The K_v2 family consists of two members, $K_v2.1$ and $K_v2.2$ (Misonou, Mohapatra, and Trimmer 2005). The K_v2 channels have a high voltage threshold for activation. They activate with slow kinetics and provide a sustained K^+ current. They are widely expressed in the brain and in endocrine cells. In neurons, the K_v2 channels underlie the somatic delayed rectifier current, while the K_v2 channels in the pancreatic beta cells play a role in regulating insulin release.

A striking feature of the K_v2 channels is the presence of a large C-terminus. The cytoplasmic C-terminus is long (~440 amino acids in the human $K_v2.1$) and accounts for ~70% of the channel protein. The C-terminus influences the channel activation kinetics. The C-terminal domain shows the presence of multiple phosphorylation sites. The extent and the pattern of phosphorylation at the C-terminus affects the voltage threshold for K_v2 channel activation.

K_v2 channels in neuronal membranes exist in two populations: freely diffusing and in clusters (Sun and De Camilli 2018), The freely diffusing channels act as the classical delayed rectifier channels. The K_v2 channels in clusters are observed at the junction of the endoplasmic reticulum (ER) and the plasma membrane (PM). K_v2 channels in the clusters contribute to the formation of the ER–PM junctions, through the interactions with the VAMP (vesicle associated membrane protein)-associated proteins (VAPs) that are resident in the ER membrane. The physiological significance of these clusters is presently not known. The formation of these clusters by the K_v2 channels is not dependent on ion permeation through the channel illustrating that K_v channels can also play nonelectrical roles in cells.

4.7.3 The K_v3 Family

K_v3 channels are the mammalian equivalent of the *Drosophila Shaw* channel. There are four family members ($K_v3.1$–$K_v3.4$) (Kaczmarek and Zhang 2017). K_v3 channels show a high threshold for activation and activate at voltages more positive than –10 mV. They also display very fast activation and deactivation kinetics. These characteristics permit neurons expressing K_v3 channels to fire action potentials at very fast rates. Neurons show a refractory period following the action potential that is caused by the time for deactivation of K_v channels that open during the action potential. This refractory period is minimal in neurons expressing K_v3 channels due to the rapid deactivation of the K_v3 channels, thereby allowing for high-frequency firing. K_v3 channels are heavily expressed in neurons that show high-frequency activity such as neurons in the auditory nucleus. $K_v3.4$, unlike other family members, is abundantly expressed in nonneuronal tissue, particularly in skeletal muscle and in cells of the pancreas. The $K_v3.1$ and $K_v3.2$ channels display noninactivating currents (over depolarization of 1 s), while the $K_v3.3$ and $K_v3.4$ channels show A-type currents with rapid inactivation. The inactivation is N-type and is modulated by phosphorylation of residues at the N-terminal.

4.7.4 The K_v4 Family

K_v4 channels are the mammalian equivalent of the *Drosophila Shal* gene. There are three family members ($K_v4.1$–4.3) (Birnbaum et al. 2004). On heterologous expression, K_v4 channels show a low threshold for activation and display fast activation kinetics. The channels on heterologous expression undergo N-type inactivation through inactivation peptides at the N-termini of the K_v4 subunits. K_v4 channels are expressed in the nervous system and also in other tissues such as the heart, kidneys and pancreas. K_v4 channels are a major component of the A-type current of neurons (I_{SA}) and the fast component of I_{TO} current in the heart. A unique feature of the K_v4 channels is that they undergo a process referred to as closed-state inactivation (CSI). CSI is distinct from the N- and the C-type inactivation processes in that it does not require pore opening. In CSI, the channel transitions to an inactivated state with the movements of the voltage sensor that precede pore opening. Due to CSI, the K_v4 channels are typically inactivated at resting membrane potentials and require a prior hyperpolarization for the channels to recover from inactivation and to activate.

4.7.5 K_v5, K_v6, K_v8 and K_v9

These subunits do not form functional homotetrameric ion channels and are collectively referred to as electrically silent subunits or K_vS (Bocksteins 2016). These subunits do form heterotetramers with the K_v2 subunits. The K_v2/K_vS heterotetramers show altered biophysical properties compared to the K_v2 homotetramers. The K_v2 channel subunits are ubiquitously expressed, while the K_vS subunits show restricted tissue expression. As the K_v2 channel can co-assemble with K_vS subunits, K_v2 heteromeric channels can show different biophysical properties depending on the nature of the partner K_vS subunits in the heterotetramer complex, thereby allowing for the fine-tuning of electrical excitability.

4.8 Pharmacology of K_v Channels

A classic blocker of K^+ channels is tetraethylammonium (TEA). TEA binds to the K^+ channel pore from either the extracellular or the intracellular side to block ion movement. The K_v1 channels show varying sensitivities to blocking by TEA (Gutman et al. 2005). For external TEA, the $K_v1.1$, $K_v1.3$ and $K_v1.6$ are relatively sensitive to TEA ($IC_{50} = 0.3–10$ mM), while $K_v1.2$, $K_v1.4$, $K_v1.5$ and $K_v1.7$ are relatively insensitive to TEA ($IC_{50} > 100$ mM). K_v3 channels are highly sensitive to blocking by submillimolar concentrations of TEA, while K_v4 channels show low sensitivity to TEA blocking. Another general K^+ channel blocker is 4-aminopyridine (4-AP), which blocks K_v channels from the intracellular side by binding to the cavity in the K_v channel pore. 4-AP blocks most K_v channels with IC_{50} ranging from the micromolar to millimolar range, depending upon the specific K_v channel. There are very few small-molecule blockers that are selective toward specific K_v channels. A challenge in identifying selective small-molecule inhibitors is that the K_v channels, especially members of the same subfamily, show a high degree of sequence identity.

An important class of K_v channel blockers is the peptide toxins that are present in the venoms of scorpions, spiders, snakes, sea anemone and other venomous organisms (Tabakmakher et al. 2019). Toxins active on K_v channels can be broadly classified as "pore-blocking" or "gating-modifier" toxins. Pore-blocking toxins bind to the extracellular mouth of the pore and physically occlude the permeation of K^+ through the channel. Examples of pore-blocking toxins are charybdotoxin and dendrotoxins that block the Shaker family K_v channels. These pore-blocking toxins were extensively used in the prestructural era to deduce structural features of the channel pore. The mechanism of channel block by toxin was structurally elucidated in the crystal structure of a complex of the $K_v1.2$ channel with charybdotoxin. The voltage-sensor toxins bind to the voltage-sensor domain and modify the voltage-gating properties of the channel. An example of a voltage-sensor toxin is hanatoxin, which binds the VSD of the K_v2 channel. An entire catalog of the channel toxins that are active on the various K_v channels are described in the *Kalium* database (Tabakmakher et al. 2019). The pore and the VSD toxins show high specificity for blocking, as they target the turret region and the voltage-sensor paddle region, respectively, and these regions show sequence variability among the various channel subtypes. The peptide toxins also show a larger interaction surface with the channel and therefore have more interactions and thereby greater affinity than observed with small-molecule blockers like TEA or 4-AP.

4.9 K_v Channel Accessory Proteins

The K_v1 channels assemble with accessory proteins that are referred to as the β subunits (Pongs and Schwarz 2010). There are three types of β subunits (β1–β3) that are encoded by three different genes. Further diversity comes from the presence of splice variants (referred to as 1.1, 1.2 and 1.3 for β1) that show an altered sequence at the N-terminus. The β subunit assembles as a tetramer and the crystal structures of the β2 tetramer and in complex with $K_v1.2$ α subunits have been reported (Long et al. 2007). The β subunit shows structural similarity to an aldo-keto reductase enzyme and shows the presence of a bound $NADP^+$ molecule. Biochemical studies have shown weak oxidoreductase activity for the isolated protein. The β1 subunits carry an inactivation peptide at the N-terminus and can induce N-type inactivation when coexpressed with α subunits that do not bear an N-type inactivation peptide. Association with the β subunit can therefore confer A-type behavior to K_v channels that would otherwise behave like delayed rectifier channels. Another established role for the β subunit is that it promotes an increase in the surface expression of the K_v channel, which takes place through a beneficial effect on assembly and/or trafficking of the channel to the cell membrane.

The K_v4 channels associate with two types of accessory subunits: KChIPs and DPPs (dipeptidyl aminopeptidase-like proteins) (Pongs and Schwarz 2010). A structure of the $K_v4.2$ channel in complex with these accessory proteins has been determined (Kise et al. 2021). KChIPs are cytoplasmic calcium sensor proteins that associate with the N-terminal region and the T1 domain of the K_v4 subunits. DPPs are transmembrane proteins with a single transmembrane domain and a large extracellular domain. Interaction of DPPs with the K_v4 channel takes place through the TM domain. Coexpression of KChIPs or DPPs with the K_v4 subunits increases cell-surface expression and changes the gating properties of the channel. The KChIPs and the DPP proteins are present either jointly or separately in the native K_v4 channels. The variant properties of the native K_v4 channels in different cell types are due to the presence of different combinations of these accessory subunits.

4.10 Biogenesis of K_v Channels

During biogenesis of K_v channels, the subunits are cotranslationally integrated into the lipid bilayer of the ER (Deutsch 2002). The targeting to the ER membrane involves a signal sequence in the N-terminal region of the channel polypeptide, and investigations in the $K_v1.3$ channel indicate that the S2 segment acts as the signal sequence. In the ER membrane, the K_v subunits fold, oligomerize to form the tetrameric channel and associate with the accessory subunits that are present in the mature channel complex. The Shaker family K_v subunits carry a glycosylation site in the S1–S2 loop. There are changes in glycosylation detected during the biogenesis. The Shaker polypeptides are detected as an immature, core-glycosylated precursor in the ER. Upon proper folding and assembly, the protein is transferred to the Golgi complex, wherein the oligosaccharide chains are modified. The immature and mature forms of the channel can be differentiated, as they show different electrophoretic mobilities on an SDS–polyacrylamide protein gel. Studies on homotetrameric K_v1 channels have indicated that determinants for the differences in the cell-surface expression of the various K_v1 channel subunits (on heterologous expression) are due to sequence differences in the pore domain. Association with accessory proteins facilitates channel trafficking and cell-surface expression.

4.11 Unresolved Questions and Research Directions

Although there has been much progress in understanding the structure, function and physiological roles of K_v channels, there are multiple questions still unanswered. A major unknown is the structure of a K_v channel in the closed state. This will be a challenging endeavor as K_v channels are generally open at 0 mV, the conditions under which structural studies are carried out. Structure determination of the closed state of a K_v channel will therefore require protein engineering to alter the voltage dependence such that the channel is closed at 0 mV or would require the development of a methodology to carry out structural studies with an applied potential. Comparing the structure of a closed K_v channel to the presently available structures of an open channel will shed light on the action of the VSD, the process of the opening/closing of the pore domain and the mechanism by which the VSD movement is coupled to the pore. Understanding these mechanisms of voltage gating and inactivation will require that the structural studies be complemented by functional measurements and with studies of the protein dynamics that underlie these processes. Structural information is presently only available for the $K_v1.2$, $K_v1.3$ and the $K_v4.2$ channels, and structures of K_v channels belonging to other families is required to highlight the structural basis for the functional differences observed between the various K_v channel families. K_v channels form heteromeric assemblies in the cell membrane. Identification of pharmacology that is specific to the various K_v channel assemblies is required for understanding the physiological roles and for the development of specific therapeutics. There are gaps in our understanding of the mechanism of the assembly of channels in the cell membrane and it is expected that understanding the mechanism of assembly will lead to therapeutic approaches that can correct the faulty assembly of K_v channels, which is a major reason for channel dysfunction in K_v channelopathies. Proteomics approaches have identified a myriad of posttranslation modifications (PTMs) on K_v channels, and future research is required to dissect the physiological significance of these PTMs on K_v channel function.

Acknowledgments

I thank Dr. Kimberly Matulef for providing the electrophysiological data on the Shaker K^+ channel shown in Figures 4.1 and 4.4. Research in my group on K_v channels is supported by National Institutes of Health grant R01GM087546.

Suggested Reading

Aggarwal, S. K., and R. MacKinnon. 1996. "Contribution of the S4 segment to gating charge in the Shaker K^+ channel." *Neuron* 16 (6):1169–77. doi: 10.1016/s0896-6273(00)80143-9.

Bezanilla, F. 2005. "Voltage-gated ion channels." *IEEE Trans Nanobiosci* 4 (1):34–48. doi:10.1109/tnb.2004.842463.

Birnbaum, S. G., A. W. Varga, L. L. Yuan, A. E. Anderson, J. D. Sweatt, and L. A. Schrader. 2004. "Structure and function of Kv4-family transient potassium channels." *Physiol Rev* 84 (3):803–33. doi:10.1152/physrev.00039.2003.

Bocksteins, E. 2016. "Kv5, Kv6, Kv8, and Kv9 subunits: no simple silent bystanders." *J Gen Physiol* 147 (2):105–25. doi:10.1085/jgp.201511507.

Cha, A., and F. Bezanilla. 1997. "Characterizing voltage-dependent conformational changes in the Shaker K^+ channel with fluorescence." *Neuron* 19 (5):1127–40. doi:10.1016/s0896-6273(00)80403-1.

Deutsch, C. 2002. "Potassium channel ontogeny." *Annu Rev Physiol* 64:19–46. doi:10.1146/annurev.physiol.64.081501.155934.

Dolly, J. O., and D. N. Parcej. 1996. "Molecular properties of voltage-gated K⁺ channels." *J Bioenerg Biomembr* 28 (3):231–53. doi:10.1007/BF02110698.

Gutman, G. A., K. G. Chandy, S. Grissmer, M. Lazdunski, D. McKinnon, L. A. Pardo, G. A. Robertson, B. Rudy, M. C. Sanguinetti, W. Stuhmer, and X. Wang. 2005. "International Union of Pharmacology. LIII. Nomenclature and molecular relationships of voltage-gated potassium channels." *Pharmacol Rev* 57 (4):473–508. doi:10.1124/pr.57.4.10.

Heginbotham, L., Z. Lu, T. Abramson, and R. MacKinnon. 1994. "Mutations in the K⁺ channel signature sequence." *Biophys J* 66 (4):1061–7. doi:10.1016/S0006-3495(94)80887-2.

Heginbotham, L., and R. MacKinnon. 1993. "Conduction properties of the cloned Shaker K⁺ channel." *Biophys J* 65 (5):2089–96. doi:10.1016/S0006-3495(93)81244-X.

Hoshi, T., W. N. Zagotta, and R. W. Aldrich. 1990. "Biophysical and molecular mechanisms of Shaker potassium channel inactivation." *Science* 250 (4980):533–8. doi:10.1126/science.2122519.

Hoshi, T., W. N. Zagotta, and R. W. Aldrich. 1991. "Two types of inactivation in Shaker K⁺ channels: effects of alterations in the carboxy-terminal region." *Neuron* 7 (4):547–56. doi:10.1016/0896-6273(91)90367-9.

Islas, L. D., and F. J. Sigworth. 1999. "Voltage sensitivity and gating charge in Shaker and Shab family potassium channels." *J Gen Physiol* 114 (5):723–42. doi:10.1085/jgp.114.5.723.

Kaczmarek, L. K., and Y. Zhang. 2017. "Kv3 channels: enablers of rapid firing, neurotransmitter release, and neuronal endurance." *Physiol Rev* 97 (4):1431–68. doi:10.1152/physrev.00002.2017.

Kise, Y., G. Kasuya, H. H. Okamoto, D. Yamanouchi, K. Kobayashi, T. Kusakizako, T. Nishizawa, K. Nakajo, and O. Nureki. 2021. "Structural basis of gating modulation of Kv4 channel complexes." *Nature* 599 (7883):158–64. doi:10.1038/s41586-021-03935-z.

Kurata, H. T., and D. Fedida. 2006. "A structural interpretation of voltage-gated potassium channel inactivation." *Prog Biophys Mol Biol* 92 (2):185–208. doi:10.1016/j.pbiomolbio.2005.10.001.

Long, S. B., X. Tao, E. B. Campbell, and R. MacKinnon. 2007. "Atomic structure of a voltage-dependent K⁺ channel in a lipid membrane-like environment." *Nature* 450 (7168):376–82. doi:10.1038/nature06265.

Lopez-Barneo, J., T. Hoshi, S. H. Heinemann, and R. W. Aldrich. 1993. "Effects of external cations and mutations in the pore region on C-type inactivation of Shaker potassium channels." *Recept Channels* 1 (1):61–71.

Mannuzzu, L. M., M. M. Moronne, and E. Y. Isacoff. 1996. "Direct physical measure of conformational rearrangement underlying potassium channel gating." *Science* 271 (5246):213–6. doi:10.1126/science.271.5246.213.

Misonou, H., D. P. Mohapatra, and J. S. Trimmer. 2005. "Kv2.1: a voltage-gated K⁺ channel critical to dynamic control of neuronal excitability." *Neurotoxicology* 26 (5):743–52. doi:10.1016/j.neuro.2005.02.003.

Papazian, D. M., T. L. Schwarz, B. L. Tempel, Y. N. Jan, and L. Y. Jan. 1987. "Cloning of genomic and complementary DNA from Shaker, a putative potassium channel gene from Drosophila." *Science* 237 (4816):749–53. doi:10.1126/science.2441470.

Pongs, O., and J. R. Schwarz. 2010. "Ancillary subunits associated with voltage-dependent K⁺ channels." *Physiol Rev* 90 (2):755–96. doi:10.1152/physrev.00020.2009.

Reddi, R., K. Matulef, E. A. Riederer, M. R. Whorton, and F. I. Valiyaveetil. 2022. "Structural basis for C-type inactivation in a Shaker family voltage-gated K⁺ channel." *Sci Adv* 8 (16):eabm8804. doi:10.1126/sciadv.abm8804.

Rudy, B., J. Maffie, Y. Amarillo, B. Clark, E. M. Goldberg, H.-Y.Jeong, I. Kruglikov, E. Kwon, M. Nadal, and E. Zagha. 2009. "Voltage gated potassium channels: structure and function of Kv1 to Kv9 subfamilies." In: *Encyclopedia of Neuroscience*, 397–425. Squire, L. R. (Ed.) Academic Press, Cambridge, MA.

Seoh, S. A., D. Sigg, D. M. Papazian, and F. Bezanilla. 1996. "Voltage-sensing residues in the S2 and S4 segments of the Shaker K⁺ channel." *Neuron* 16 (6):1159–67. doi:10.1016/s0896-6273(00)80142-7.

Sun, E. W., and P. De Camilli. 2018. "Kv2 potassium channels meet VAP." *Proc Natl Acad Sci U S A* 115 (31):7849–51. doi:10.1073/pnas.1810059115.

Tabakmakher, V. M., N. A. Krylov, A. I. Kuzmenkov, R. G. Efremov, and A. A. Vassilevski. 2019. "Kalium 2.0, a comprehensive database of polypeptide ligands of potassium channels." *Sci Data* 6 (1):73. doi:10.1038/s41597-019-0074-x.

Wulff, H., N. A. Castle, and L. A. Pardo. 2009. "Voltage-gated potassium channels as therapeutic targets." *Nat Rev Drug Discov* 8 (12):982–1001. doi:10.1038/nrd2983.

Yellen, G. 1998. "The moving parts of voltage-gated ion channels." *Q Rev Biophys* 31 (3):239–95.

Zagotta, W. N., T. Hoshi, and R. W. Aldrich. 1994. "Shaker potassium channel gating. III: evaluation of kinetic models for activation." *J Gen Physiol* 103 (2):321–62. doi:10.1085/jgp.103.2.321.

5

ERG Family of Potassium Channels

Sara J. Codding and Matthew C. Trudeau

CONTENTS

5.1 Introduction

The ERG (ether á go-go related, Kv11), EAG (ether á go-go, Kv10) and ELK (ether á go-go like, Kv12) voltage-activated potassium channels compose the KCNH family of K channels (Warmke and Ganetzky 1994; Warmke, Drysdale, and Ganetzky 1991). Mammalian ERG channels are composed of three distinct genes: ERG1, ERG2 and ERG3 (KCNH2, KCNH6 and KCNH7; or Kv11.1, Kv11.2 and Kv11.3). EAG is composed of the EAG1 and EAG2 channels (KCNH1 and KCNH5, or Kv10.1 and Kv10.2), and the ELK family contains ELK 1, ELK2 and ELK3 channels (KCNH8, KCNH3 and KCNH4; or Kv12.1, Kv12.2 and Kv12.3) (Figure 5.1A). KCNH channels are distinct from other voltage-activated potassium (Kv) channels because they are more closely related to the pacemaker (HCN) and cyclic nucleotide-gated (CNG) channels than to other Kv channels (Gutman et al. 2003). Two ERG1 isoforms, ERG1a and ERG1b, are the primary alpha subunits that encode the cardiac

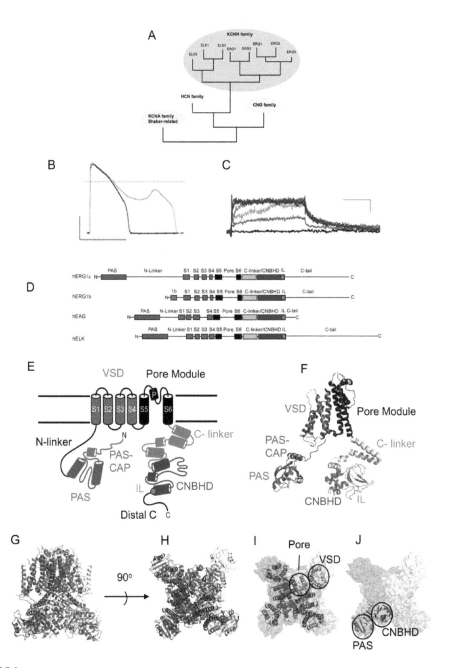

FIGURE 5.1

Organization and introduction to the physiological role of hERG. (A) Cladogram of ERG, EAG and ELK channels within the KCNH family of channels, and relationship to HCN, CNG and other voltage-activated K channels. (B) Cardiac action potential (black trace) from hiPSC-CMs and prolongation of action potential (orange trace) after addition of the hERG and I_{Kr} inhibitory drug E-4031. Scale bar in B is 40 mV and 400 ms. (C) Family of native I_{Kr} currents recorded from hiPSC-CMs. Current in C elicited from –40 to 20 mV in 10 mV steps with repolarization to –40mV. Scale bar in C is 0.5pA/pF and 1 s. (D) Scheme depicting linear organization of hERG1a, hERG1b, hEAG and hELK channels. (E) Schematic of topology of single hERG subunit. (F) Cryo-EM structure of a single hERG subunit. Cryo-EM of hERG tetramer: (G) Side view, (H) top-down view, (I) top-down view with circles to highlight the proximity of VSD and pore domain within the same subunit (non-domain-swapped). (J) Bottom-up view with circles to highlight the intersubunit domain-swapped PAS domain and CNBHD interaction. Each of the four sub-units is a different color. (Panels B and C from Jones et al. 2014. Panels F - J from Protein Data Bank 5VA1.)

I_{Kr} current, which shapes and repolarizes action potentials in mammalian myocardium. Genetic mutations in human ERG1a or ERG1b genes cause the type 2 Long QT Syndrome (LQTS), a predisposition to cardiac arrhythmias and sudden cardiac death. Human ERG (hERG) channels are blocked by a variety of drugs in vitro, and blockade of I_{Kr} in the heart by the off-target effects of pharmaceuticals and other drugs is the major cause of a clinically common acquired form of LQTS . In contrast to ERG, the physiological roles of EAG and ELK are not well established. EAG is a delayed-rectifier potassium channel that is genetically linked to Zimmermann–Laband and Temple–Baraitser multisymptom syndromes and is upregulated in cancer cell lines. ELK channels are also delayed rectifiers, except for ELK2, which has an early inactivating peak, and all have a left-shifted voltage dependence compared to ERG and EAG, which makes them a candidate for encoding a subthreshold potassium current in neurons.

5.2 Physiological Roles

ERG1 channels play a major role in repolarization of the cardiac action potential. This can be demonstrated in experiments where the cardiac action potential (black trace, Figure 5.1B) is prolonged (orange trace, Figure 5.1B) by E-4031, a small molecule that inhibits hERG channels. It is well established that mammalian ERG1 channel subunits form the cardiac I_{Kr} current (Figure 5.1C) in mammalian atrial and ventricular myocytes. A number of key findings have established the relationship between ERG and I_{Kr}. These include the similar kinetics of hERG expressed in heterologous systems and I_{Kr} from guinea pig myocytes, the similar inhibition of hERG channels and I_{Kr} by small molecules (e.g., E-4031), the suppression of I_{Kr} by antisense oligos encoding hERG sequences, animal models in which a transgene encoding a dominant-negative hERG suppresses native rabbit I_{Kr}, and stem cell-derived cardiomyocytes with hERG gene mutations in which human I_{Kr} current is suppressed (Brunner et al. 2008; Itzhaki et al. 2011; Jones et al. 2014; Sanguinetti et al. 1995; Sanguinetti and Jurkiewicz 1990; Trudeau et al. 1995). Two major isoforms of hERG1 co-assemble to form I_{Kr}: ERG1a, the originally identified isoform, and ERG1b, a shorter splice isoform with a divergent N-terminal region (Figure 5.1D) (Jones et al. 2014; Jones et al. 2004; London et al. 1997; McNally, Pendon, and Trudeau 2017; Phartiyal, Jones, and Robertson 2007). In cardiomyocytes, the physiological role of I_{Kr} is to shape and repolarize the late phase (phase 3) of action potentials (Sanguinetti and Jurkiewicz 1990) (Figure 5.1B; see also Volume III, Chapter 5). Unlike ERG1, ERG2 and ERG3 are not detected in the heart.

The physiological role of ERG1 in other systems is not as firmly established as in the heart. Mammalian ERG1a and ERG1b, as well as ERG2 and ERG3, are widely distributed in the central nervous system (CNS) as measured with immunohistochemistry (IHC) and biochemistry (Guasti et al. 2005; Papa et al. 2003; Shi et al. 1997). Neuronal ERG currents were first detected in cerebellar Purkinje neurons where a hERG-like tail current was isolated using a hERG-specific blocker (Sacco et al. 2003). Since then, ERG current has been detected in a number of neuronal cell types, including auditory brainstem and midbrain neurons using pharmacological isolation. Small-molecule drugs, to date, do not distinguish between ERG1, ERG2 and ERG3, so neuronal currents have not been linked to a particular type of ERG subunit. But using toxins that have subunit-specific inhibitory properties, it was determined that ERG1 and ERG3 may form the ERG current in neonatal mouse Purkinje cells (Niculescu et al. 2013). There appear to be three distinct roles for ERG current in neuronal firing, all identified by pharmacological block of the native ERG currents. These are (1) spike frequency adaptation, first described in a neuroblastoma cell line (Chiesa et al. 1997;

Hardman and Forsythe 2009; Pessia et al. 2008); (2) regulation of neuronal firing, where ERG contributes to the negative resting membrane potential (RMP) and blockade elevates the RMP, leading to enhanced firing frequency (Hirdes et al. 2009; Niculescu et al. 2013; Sacco et al. 2003); and (3) a decrease in firing, where ERG blockade depolarizes cells leading to persistent Na channel inactivation and thus a reduction in firing (Hagendorf et al. 2009; Ji et al. 2012). The physiological roles for neuronal ERG channels are discussed in more detail in a review article (Bauer and Schwarz 2018). EAG1 and EAG2 are not detected in the heart, but are instead widely distributed in the CNS as measured with IHC (Ludwig et al. 1994; Saganich et al. 1999). A current from retina, I_{Kx}, exhibits some functional similarities with EAG currents (Frings et al. 1998). But mice that are null for EAG1 do not have major defects in development or behavior, and do not have measurable differences in firing patterns in Purkinje cell neurons (Ufartes et al. 2013). Thus, a physiological role or native cellular correlate for EAG channels has not been established. ELK is detected in the CNS, but not the heart, in IHC and Western blot experiments (Engeland et al. 1998; Trudeau et al. 1999; Zou et al. 2003). Mice genetically null for ELK2 (ELK2$^{-/-}$) exhibit a cellular hyperexcitability phenotype (Bauer and Schwarz 2018; Zhang et al. 2010) and more positive resting potential in CA1 neurons and, in heterologous expression experiments, ELK channels activate at voltages near the reversal potential for K$^+$ ions. Together these results lead to the proposal that ELK2 channels may play a role as a subthreshold K$^+$ current in neurons.

In summary, the physiological role and native correlate is well established for ERG1a and ERG1b channel subunits in the mammalian heart, where they form the native I_{Kr} current. But the role of ERG1-3 in the brain or other tissues is not as well understood. A native correlate for EAG1-2 and ELK1-3 channels is unclear, in part because we lack specific inhibitors or blockers for these channels.

5.3 Subunit Diversity and Basic Structural Organization

All KCNH channels have a similar primary arrangement (Figure 5.1D). Starting at the N-terminus, ERG, EAG and ELK channels have a PAS-CAP domain and a Per-Arnt-Sim (PAS) domain followed by an N-linker region that connects the PAS to the S1 transmembrane domain. Like other voltage-activated channels, KCNHs have six transmembrane domains, S1–S6, in which the S1, S2, S3 and S4 transmembrane domains form the voltage-sensor domain (VSD). The S4 contains positively charged arginine residues like that in other voltage-activated K channels. The S5 and S6 transmembrane domains are connected by a reentrant pore loop domain. An S4–S5 linker connects the VSD to the S5 and is composed of five amino acids, making it much shorter than that in other Kv channels. In the C-terminal region, the S6 is followed by a C-linker domain and cyclic nucleotide binding homology domain (CNBHD). The C-linker and CNBHD are homologous to the corresponding regions of HCN and CNG channels, with some interesting adaptations in how they regulate gating (see Section 5.4.2).

Despite similar architecture, some notable differences among the KCNH channels lead to functional diversity. ERG1a and ERG1b isoforms are encoded by alternate transcripts arising from distinct promoters (London et al. 1997) (Figure 5.1D). As a result, hERG1b has an N-terminal 1b domain and truncated N-linker region in place of the hERG1a PAS domain and long N-linker region. The isoforms are identical starting at amino acid 37 in hERG1b and 377 in hERG1a, and extending to the carboxyl terminus (Figure 5.1D). The N-linker region of hERG is longer than that of other KCNHs and is not well conserved (Figure 5.1D).

Unlike ERG and ELK, EAG channels contain calcium–calmodulin (Ca^{2+}–CaM) binding sites and are inhibited by Ca^{2+}–CaM (Marques-Carvalho et al. 2016; Schonherr, Lober, and Heinemann 2000; see also Volume III, Chapter 2).

A scheme of the topology (Figure 5.1E) and corresponding three-dimensional structure (Figure 5.1F) of a single human ERG subunit indicates the organization of key functional domains (Wang and MacKinnon 2017). The PAS-CAP has two distinct regions: an extended region at the N-terminal end that points upward toward the transmembrane domains and a helix that is adjacent to the PAS domain. The PAS domain structure agrees with that previously determined by X-ray crystallography (Morais Cabral et al. 1998) and is structurally similar to PAS domains found in other proteins, including EAG and ELK (Adaixo et al. 2013). The VSD and Pore domains are adjacent to each other in a single subunit. The S5 helix is positioned near the S1 helix in the same subunit, which is a dissimilar arrangement compared to other Kv channels (see Chapter 4). The S6 helix connects to the C-linker, which is a series of helices that connect to the CNBHD. The CNBHD is structurally similar to the CNBD of the HCN1 channel (Zagotta et al. 2003), but instead of a binding site for a cyclic nucleotide, the sequence contains an "intrinsic ligand." The intrinsic ligand (IL) is formed by a few amino acids and sits inside a binding pocket within the CNBHD, analogous to how a cyclic nucleotide sits in a binding site in the CNBD of HCN and CNG channels (Brelidze et al. 2012; Brelidze et al. 2013). The same basic organization of channel domains reported for hERG is also found in the structure of rat EAG (rEAG) channels (Codding, Johnson, and Trudeau 2020; Whicher and MacKinnon 2016). The N-linker and distal C-terminal region were deleted from the hERG channel used to obtain cryo-EM data and thus structural information for these two regions is not known.

Four individual hERG subunits associate into a fourfold symmetric channel (Figure 5.1G–J) (Wang and MacKinnon 2017). The four S5-P-S6 domains form the central ion-conducting pore and the four VSDs are located on the outer perimeter. In most Kv channels, the VSD is nearby the Pore domain of an adjacent subunit in a domain-swapped configuration (see Chapter 4), but in hERG, the VSD is positioned adjacent to the Pore domain of the same subunit, in a non-domain-swapped configuration (Figure 5.1I). Intracellular N-terminal PAS-CAP and PAS domains, and the C-terminal C-linker and CNBHDs are positioned beneath the transmembrane domains of the channel. The PAS domain interacts with the CNBHD of an adjacent subunit in a domain-swapped configuration (Figure 5.1J).

Like hERG channels, the tetrameric rat EAG channel has a non-domain-swapped VSD and Pore domain orientation, and domain-swapped intersubunit PAS Domain–CNBHD interactions indicating that these domain arrangements are likely common features among KCNH channels (Codding, Johnson, and Trudeau 2020; Whicher and MacKinnon 2016).

5.4 Channel Physiology and Gating Mechanisms

5.4.1 Physiology

5.4.1.1 ERG

ERG1 channels have characteristically unusual currents. In response to a standard family of voltage step commands, hERG channels activate slowly as voltage is stepped in the positive direction and the outward currents increase in size until approximately 0 mV, when, in response to further positive steps, the currents diminish in magnitude due to channel inactivation (Figure 5.2A). Consequently, the current–voltage (I–V) relationship, from currents measured at the end of the voltage command steps, has a characteristic bell-shape with a peak at about 0 mV (Figure 5.2B). Upon repolarization, hERG channels have a characteristic large tail current (Figure 5.2A). The rising phase of the current at the initial repolarization step is due to rapid recovery from inactivation, and the slow decay of the tail current with sustained repolarization is due to slow closing (slow deactivation) of the channels. hERG1a gating can be described by a number (n) of closed (C) states followed by open (O) and inactive (I) states (Scheme 5.1):

$$C_n \ldots C_{n-1} \underset{\beta}{\overset{\alpha}{\rightleftarrows}} O \underset{\delta}{\overset{\gamma}{\rightleftarrows}} I \qquad \text{(Scheme 5.1)}$$

The characteristic hERG currents are consistent with, during depolarization, a relatively slow transition (α) from the C to O state and a tenfold faster transition (γ) from the O to I state. During repolarization, recovery from I to O is tenfold faster (δ) relative to the O to C transition (β). The scheme explains the small outward hERG currents with depolarization (the channels inactivate quickly) and the large tail current with repolarization (channels rapidly reopen and then slowly close).

The steady-state activation curve measured from the tail current has a midpoint of approximately –25 mV (Figure 5.2C). The steady-state inactivation curve has a midpoint of –90 mV, indicating that many hERG channels are inactivated at depolarizing voltages (Smith, Baukrowitz, and Yellen 1996). The calculated product of the steady-state activation and inactivation curves recapitulate the bell-shape of the experimentally measured I–V relationship, indicating that robust inactivation determines the characteristic bell-shaped hERG I–V relationship (Figure 5.2B). The ERG1b isoform has fivefold faster deactivation and slightly slower inactivation than the ERG1a isoform. ERG2 and ERG3 currents are similar in appearance to those of ERG1, with the exception that the voltage-activation curve for ERG3 is left-shifted compared to that of ERG1 or ERG2 (as reviewed in Codding, Johnson, and Trudeau 2020).

5.4.1.2 EAG

EAG1 and EAG2 channels have slowly activating K^+ currents (Figure 5.2D), with almost no measurable inactivation (Robertson, Warmke, and Ganetzky 1996; Saganich et al. 1999) resulting in an outwardly rectifying current–voltage relationship in heterologous expression systems (Figure 5.2E). Thus, EAG channel gating can be explained by a scheme that has a relatively slow C to O transition (α), and has a relatively fast O to C transition (β) and lacks an I state (Scheme 5.2):

$$C_n \ldots C_{n-1} \underset{\beta}{\overset{\alpha}{\rightleftarrows}} O \qquad \text{(Scheme 5.2)}$$

The conductance–voltage relationship for EAG1 has a midpoint at approximately 0 mV (Figure 5.2F), whereas that for EAG2 is left-shifted with a midpoint near –25 mV. A characteristic feature of EAG channels is a slower activation time course with hyperpolarizing prepulse voltage steps similar to the Cole–Moore shift (Robertson, Warmke, and Ganetzky 1996; Cole and Moore 1960). In contrast, ERG and ELK channels do not have a measurable Cole–Moore shift (Trudeau et al. 1999).

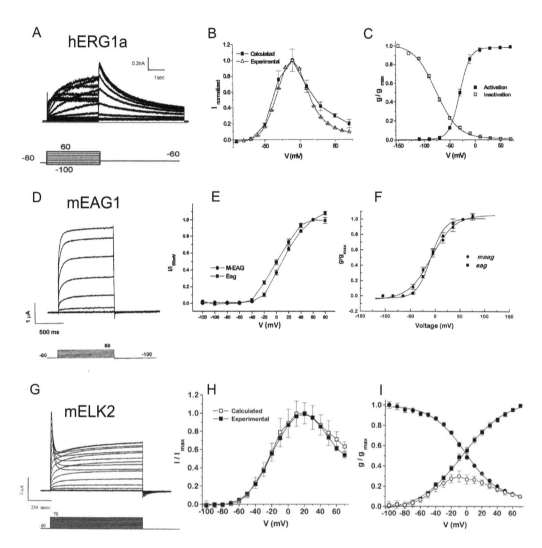

FIGURE 5.2
Ionic currents and steady-state properties of KCNH channels. (A) Current family from hERG1a expressed heterologously in HEK293 cells and recorded with whole-cell patch clamp. (B) Current–voltage plot of hERG (open triangle) and calculated I–V plot (closed square) derived from the product of the (C) steady-state activation curve (closed square) and steady-state inactivation curve (open squares). (D) Mouse EAG1 (mEAG) current recorded from heterologous expression in *Xenopus* oocytes with two-electrode voltage clamp (TEVC). (E) Current–voltage plot and (F) conductance–voltage plot for mEAG (circles) and *Drosophila* EAG (squares). (G) Mouse ELK2 (mELK2) current expressed in *Xenopus* oocytes and recorded with TEVC. (H) Current–voltage plot of mELK2 measured at the end of the 1 s command pulse (closed square) and I–V plot (open square) calculated from the product of activation and inactivation curves in panel I. (I) Voltage-activation curve (closed square) and steady-state inactivation curve (closed circle). The "window conductance" (open circle) is the product of the steady-state activation and inactivation curves. The scale bars are voltage paradigms are indicated in A, D and G. (E and F are from Robertson, Warmke, and Ganetzky 1996. G–I are derived from Trudeau et al. 1999.)

5.4.1.3 ELK

ELK1 and ELK3 both encode an outward, delayed-rectifier type of K current (Engeland et al. 1998; Shi et al. 1998). In contrast, ELK2 currents have an early inactivating peak and a slight reduction in outward current magnitude with sustained depolarization (Figure 5.2G), leading to a partial bell-shaped current–voltage relationship (Figure 5.2H) compared to that of hERG (Trudeau et al. 1999).

ELK2 channels are further characterized by a conductance–voltage relationship with a midpoint near 0 mV and a steady-state inactivation midpoint near 0 mV (Figure 5.2I). The calculated product of the steady-state inactivation and activation curves recapitulate the partial bell-shape of the experimental I–V relationship (Figure 5.2H), indicating that, like for hERG, inactivation determines the distinctive shape of the experimentally measured ELK2 I–V plot.

ELK2 can be described by Scheme 5.3, where the rate of inactivation (γ) is not much faster relative to the rate of activation (α) and the channel recovers from inactivation (δ) and deactivates (β).

$$C \underset{\beta}{\overset{\alpha}{\rightleftharpoons}} \underset{\Delta V}{C} \underset{\delta}{\overset{\gamma}{\rightleftharpoons}} O$$

$$\varepsilon \downarrow VDP$$

$$C^{*} \underset{\beta}{\overset{\alpha}{\rightleftharpoons}} \underset{\Delta V}{C^{*}} \underset{\kappa}{\overset{o}{\rightleftharpoons}} O^{*}$$

(Scheme 5.3)

ELK2 (and zebrafish ELK) channels are further characterized by a distinctive voltage-dependent potentiation (VDP), in which long duration depolarizing pulses markedly shift the midpoint of the conductance–voltage relationship to hyperpolarized potentials by as much as –40 mV. Concomitantly the time course of channel deactivation is slowed, where κ is slower than δ, as indicated by the smaller arrow from O* to C* (Scheme 5.3, red text) (Dai and Zagotta 2017; Li et al. 2015).

5.4.2 Gating Mechanisms

The mechanisms for voltage-dependent gating in KCNH channels have similarities to the mechanisms in other Kv channels, but KCNH channels have fundamental distinguishing differences. From closed confirmations (Figure 5.3A) KCNH channels are activated by a positive change in membrane voltage (Figure 5.3B). Membrane voltage is proposed to cause a rearrangement or movement in the S4 of the VSD (orange arrows, Figure 5.3B). S4 motion is coupled, in a process that is not well understood in KCNHs, to opening of the activation gate (red arrows, Figure 5.3B), which is located in the lower S6 domain. Inactivation in hERG channels and ELK2 channels depends on the identity of pore residues and occurs by a mechanism similar to C-type inactivation (red circle, Figure 5.3B). The intracellular PAS domains and CNBHDs regulate deactivation in hERG, the Cole–Moore shift in EAG and VDP in ELK. Gating in hERG is the most well understood among KCNH channels and here we will mainly focus on hERG mechanisms.

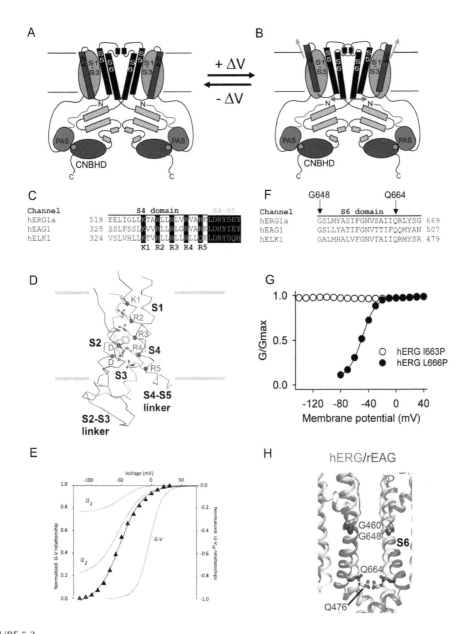

FIGURE 5.3
Mechanisms of activation gating in hERG. Scheme of hERG channel in the (A) closed configuration and (B) open configuration. With depolarizing stimulus, orange arrows depict putative upward S4 motion and red arrows depict lower S6 activation gate motions. Red circle indicates the pore region that contains major determinants of C-type inactivation. (C) Alignment of KCNH channel S4 domains and S4–S5 linker. (D) Cryo-EM structure of the hERG VSD with K1, R2, R3, R4 and R5 in the S4 as indicated and determinants of the charge transfer center aspartic acids (D) and phenylalanine (F463). (Derived from Wang and MacKinnon 2017). (E) Plot of conductance–voltage (G–V) and charge–voltage (Q–V) relationship (triangles). Q1 and Q2 are separate components of charge transfer. (From Wang et al. 2013.) (F) Alignment of S6 domains in KCNHs. (G) Conductance–voltage plot of S6 mutant hERG L666P and "standing open" mutant hERG I663P (Thouta et al. 2014). (H) Overlay of cryo-EM structures of the lower S6 domains of hERG (purple) in the open state and EAG (gray) in the closed state. hERG Q664 and rEAG Q476 residues located at the lowermost point of the S6 gate. hERG G648 and rEAG G460 mark the position of a gating hinge. (Derived from Wang and MacKinnon 2017.)

5.4.2.1 ERG

hERG channel activation is characterized by a relatively slow time course compared to that of other voltage-activated K channels like Shaker. Because hERG inactivation is very rapid and robust an accurate measure of activation from standard hERG currents (Figure 5.2A) is difficult, but activation can be indirectly measured using an "envelope of tails" protocol (Trudeau et al. 1995) and can be measured directly in channels in which inactivation is removed by a point mutation (see Figure 5.4A-D). hERG and other KCNH channels have an S4 domain with a series of five positively charged residues. In the cryo-EM structure of the hERG VSD, the positively charged residues (labeled K1, R2, R3, R4 and R5) are positioned along one face of the S4 helix (Figure 5.3C,D). Opposed to the S4 is a charge transfer center formed in part by two aspartic acid (D) residues and a phenylalanine (F) residue from the S2 and S3 domains (Figure 5.3D). S4 residues K1, R2 and R3 are positioned above F463, and R4 and R5 are positioned below (Figure 5.3D). Thus, S4 is proposed to be in the "up" (i.e., activated or depolarized state) like the S4 state captured in structures of most Kv channels (Volume I, Chapter 2; Chapter 4, this volume) and the downward movement of K1, R2 and R3 relative to F463 is proposed to occur during repolarization and channel deactivation. Measurements of hERG S4 motions with fluorescence or measurements of gating charge movement with gating currents both have now concluded that the S4 movement in hERG is relatively fast compared to the overall time course of activation (Es-Salah-Lamoureux et al. 2010; Goodchild and Fedida 2014). The charge–voltage (Q–V) curve is left-shifted compared to the conductance–voltage curve (Figure 5.3E). These results have led to the proposal for hERG activation that S4 rapidly moves "upward" to the position seen in the cryo-EM structure, and the slow activation time course is likely not due to slow S4 movement but rather to a slow transition in the activation pathway after rapid S4 movement (Islas 2013).

Like other Kv channels, the hERG channel activation gate is located in the lower S6 domain. At S6 residues spanning I655 to Q664 (Figure 5.3F) mutagenesis to cysteine or proline residues disrupted channel activation and channel closing, leading to a standing conductance at negative voltages (e.g., –60 to –100 mV) where wild-type channels were normally closed (Thouta et al. 2014; Wynia-Smith et al. 2008). hERG I663P channels are an example of a standing open channel (Figure 5.3G). A sharp cutoff was reported for the proline mutants where S6 residues R665, L666 and Y667 (Figure 5.3F) had wild-type like activation. hERG L666P was an example of a channel with activation similar to that of wild-type hERG (Figure 5.3G). These data were interpreted to mean that Q664 was located at the lowermost position of the intracellular gate. The role and position of Q664 were directly supported by the cryo-EM structure of hERG. The hERG S6 was solved in the open conformation with a distance of 5 angstroms between Q664 residues in different subunits, which is sufficiently wide to permit the conductance of K^+ ions (Figure 5.3H), but in contrast, the cryo-EM structure of KCNH1 (rat Eag) channels showed that hERG Q664 equivalent residues, rEAG Q476, were less than 0.5 angstroms apart, which would restrict the conductance of K^+ ions (Figure 5.3H) and thus close the S6 gate. The rEAG S6 gate was closed as the structure was solved in complex with Ca^{2+}–CaM, which inhibits rEAG currents (Whicher and MacKinnon 2016; see also Volume III, Chapter 2). Structure–function studies had identified glycine at position G648 (G460 in rEAG) as a hinge region permitting lower S6 movement and this was corroborated by the cryo-EM structure (Figure 5.3H) (Hardman et al. 2007).

In KCNH channels, the coupling between upward S4 motion and lower S6 activation gate opening is not known. In Kv channels, the S4–S5 linker acts as a mechanical lever that directly links VSD motion to the S5 and opening of the activation gate in the S6 (Chapter 4). In KCNH channels, the S4–S5 linker is quite short (Figure 5.3C, D) compared to that of other Kv channels (Chapter 4), and neither hERG nor EAG requires a peptide bond between the S4 and pore to retain voltage-dependent activation (Lorinczi et al. 2015). Thus, VSD to pore coupling in KCNHs may be fundamentally different from that in other Kv channels.

Inactivation is a distinguishing feature of hERG channels. The impact of rapid inactivation and profound steady-state inactivation on macroscopic hERG currents takes the form of an overall suppression of outward currents with an additional pronounced diminution of outward currents at voltages positive to approximately 0 mV (Figure 5.2A), resulting in the characteristic bell-shaped current–voltage relationship (Figure 5.2B). Inactivation in hERG has molecular determinants in and near the pore domain, including point mutations at pore residues S620 and S631 (Figure 5.4A) that abolish or greatly attenuate inactivation (Smith, Baukrowitz, and Yellen 1996; Herzberg, Trudeau, and Robertson 1998). In one example, hERG S631V channels lack the outward current suppression and bell-shaped I–V relationship of wild-type hERG channels (Figure 5.4B, C). hERG inactivation is slowed by elevated external monovalent ions, including K^+ and TEA^+, resulting in paradoxically larger outward hERG K^+ currents. Together, these properties suggest hERG inactivation is similar to the C-type inactivation mechanism originally described in Shaker K channels (Hoshi, Zagotta, and Aldrich 1991). Cryo-EM structures of the inactivation-impaired mutant hERG S631A channel have a similar placement of the central F627 in the GFG of the selectivity sequence compared to the equivalent amino acid in other inactivation-impaired channels, such as rEAG, KcsA or KvChim (Figure 5.4D). In contrast, F627 in control channels with wild-type inactivation is slightly rotated compared to that in hERG S631A and the other channels (Figure 5.4D), which supports the idea that hERG inactivation involves molecular rearrangements in the pore. Point mutations that abolish hERG C-type inactivation also eliminate the rising phase of the hERG tail current, showing that the rising phase is due to a rapid recovery (i.e., the I to O gating transition) from C-type inactivation.

The characteristic slow decay of the hERG tail current (Figure 5.2A) is known as slow deactivation and is due to slow channel closing. In hERG channels, the mechanism deactivation is due to closing of the S6 gate, but deactivation is slowed by a regulatory effect of the intracellular PAS domains and CNBHDs. The PAS domain is a primary determinant of slow deactivation, as hERG channels with an engineered deletion of the PAS domain have five- to tenfold faster deactivation than that of wild-type channels (Figure 5.4E) (Spector, Curran, Zou, et al. 1996; Wang et al. 1998). Likewise, hERG channels with a deletion of the CNBHD also have a five- to tenfold speeding of deactivation, indicating that the CNBHDs are necessary for slow deactivation (Figure 5.4F) (Gianulis, Liu, and Trudeau 2013; Gustina and Trudeau 2011). Mixing PAS-deleted and CNBHD-deleted subunits results in heteromeric channels with a partial recovery of slow deactivation (Figure 5.4G), indicating an intersubunit interaction between PAS domains and CNBHDs that regulate and slow deactivation (Gianulis, Liu, and Trudeau 2013; Gustina and Trudeau 2011). In support of the functional data, a PAS and CNBHD interaction was detected in KCNH1 channels (Figure 5.4H) and an intersubunit PAS and CNBHD interaction was reported in the cryo-EM structure of hERG (Figure 5.1J) and is depicted schematically in Figure 5.3A.

Slow deactivation in hERG is also regulated by the PAS-CAP domain. Deletion of the PAS-CAP or mutations of charged PAS-CAP residues result in faster deactivation (Ng et al. 2011; Wang et al. 1998). The PAS-CAP can be reapplied as a peptide to regulate and slow deactivation in hERG channels with an engineered deletion of the N-terminus (Wang, Myers, and Robertson 2000). The detailed mechanism for PAS-CAP action is not clear,

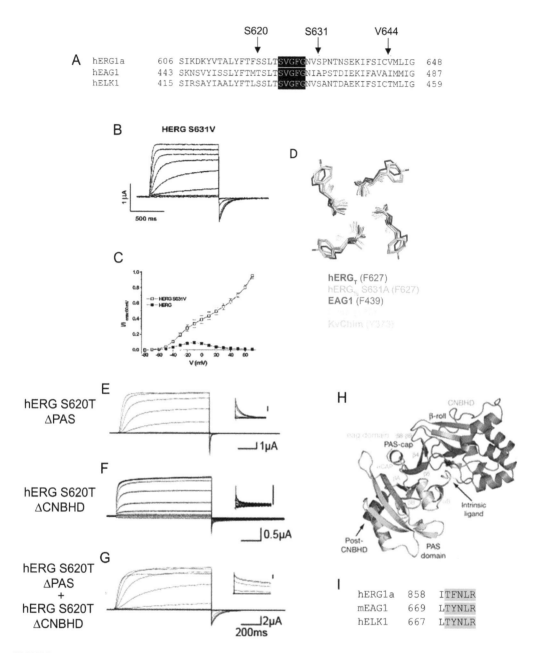

FIGURE 5.4

Inactivation and deactivation mechanisms in hERG channels. (A) Alignment of pore domain of KCNH channels. (B) hERG S631V currents measured with TEVC. (C) Current–voltage plot of hERG S631V (open squares) and wild-type hERG (closed squares). (B and C were derived from Herzberg, Trudeau, and Robertson 1998.) (D) Cryo-EM structure of control hERG F627 residue and equivalent residue as indicated from inactivation deficient channels hERG S631, EAG1, KcsA and KvChim. (From Wang and MacKinnon 2017.) (E) hERG inactivation-removed mutant (hERG S620T) with deletion of the PAS domain. Inset is outward tail current. (F) hERG S620T with a partial deletion of the CNBHD and C-tail (ΔCNBHD). Inset is outward tail current. (G) hERG S620T ΔPAS co-expressed with hERG S620T ΔCNBHD. Deactivation is partially restored, suggesting intersubunit PAS–CNBHD interactions. (Derived from Gustina and Trudeau 2011.) (H) PAS domain and CNBHD interaction in co-crystal structure of KCNH1 (EAG) channels. Intrinsic ligand is indicated by yellow loop. (I) Alignment of intrinsic ligand motifs from KCNH channels.

but in the cryo-EM structure of hERG, the extended PAS-CAP points upward toward the transmembrane domains and the C-linker domain and may regulate gating by interacting with these domains (Figure 5.1E, F).

The intrinsic ligand is another determinant for slow deactivation in hERG. Initially identified in the structures of the CNBHD from zebrafish ELK (Brelidze et al. 2012) and mosquito (ag) ERG (Brelidze et al. 2013), the intrinsic ligand (IL) is formed by a few amino acids at the end of the CNBHD and sit in a binding pocket within the CNBHD analogous to how a cAMP molecule binds to the CNBD in HCN channels (Zagotta et al. 2003). Intrinsic ligand residues in the CNBHD are located nearby the PAS domain in a co-crystal structure (Haitin, Carlson, and Zagotta 2013) (Figure 5.4H). The intrinsic ligand regulates channel deactivation, as mutations in the intrinsic ligand motif (at the F or L of the FNL motif) (Figure 5.4I) disrupts and accelerates hERG deactivation (Brelidze et al. 2013). The FNL motif might regulate deactivation by regulating the CNBHD interaction with the PAS domain, since mutations at the F or L residues perturb a structural interaction with the PAS domain (Codding and Trudeau 2019).

5.4.2.2 EAG

The characteristic Cole–Moore shift in EAG can be explained by several closed states prior to an open state (Scheme 5.2) where hyperpolarization drives channels to a leftward closed state, whereas depolarizing prepulses populate closed states closer to the open state. The Cole–Moore shift in EAG requires the PAS domain and requires amino acids in the intrinsic ligand (Zhao et al. 2017).

5.4.2.3 ELK

The pronounced VDP in ELK channels can be explained by a mechanism whereby long duration depolarizations cause a slow transition to a mode that favors channel opening. In ELK channels, the PAS and CNBHD rearrange relative to each other during VDP gating (Dai and Zagotta 2017). The intrinsic ligand also regulates VDP in ELK channels and undergoes motions during VDP (Dai, James, and Zagotta 2018). In mELK2 channels, partial inactivation occurs by a C-type mechanism similar to that of hERG, since mELK2 inactivation is disrupted by mutations at pore residues S464T and S475A (Figure 5.4A).

The PAS and CNBHDs appear to be specialized to underlie some of the characteristic features of KCNH channel gating, including slow deactivation in hERG, the Cole–Moore shift in EAG and the VDP in ELK.

5.5 Ion Permeability

KCNH channels are all highly selective for K^+ ions over Na^+ ions (Trudeau et al. 1995; Trudeau et al. 1999; Robertson, Warmke, and Ganetzky 1996). There is an interesting difference between the primary sequence of the selectivity filter in the pore of Shaker K channels, which have the characteristic GYG motif that is very highly conserved among Kv channels (Volume I, Chapter 1; Chapter 4, this volume), and KCNH channels, which have a GFG motif lining the ion conduction pathway. The Y versus F does not appear to have a major influence on selectivity.

5.6 Regulation

For hERG, several different types of regulatory effects have been reported. One fundamental type of regulation is co-assembly of two different hERG1 "alpha subunit" isoforms, hERG1a and hERG1b, into heterotetrameric hERG1a/hERG1b channels. Mammalian ERG1b preferentially co-associates with ERG1a subunits in vitro and in heart cells (McNally, Pendon, and Trudeau 2017; Jones et al. 2004; Jones et al. 2014). Heteromeric hERG1a/hERG1b channels have gating characteristics more like those of the cardiac I_{Kr} current than either homomeric hERG1a or homomeric hERG1b channels. Indeed, manipulating native I_{Kr} current in human stem cell-derived cardiomyocytes (hiPSC-CMs) to be more like ERG1a (slower deactivation, more inactivation) leads to AP prolongation and dispersion of repolarization, both of which are arrhythmia precursors, indicating that the precise regulation of hERG1a and hERG1b subunit assembly is critical in the heart (Jones et al. 2014).

External factors including ions and proteins regulate hERG. Elevated concentrations of external monovalent ions, including K^+ and TEA^+, interfere with and reduce C-type inactivation, leading to an increase in the outward hERG current (Wang et al. 1998; Smith, Baukrowitz, and Yellen 1996). In elevated external K^+, the slowed inactivation overcomes the reduction in driving force on K^+ ions and the net effect is an increase outward hERG current. Conversely, a decrease in external K^+ to nominally zero leads to a nonconducting state in hERG and native I_{Kr}, and eventual downregulation of hERG proteins due to ubiquitination by NEDD4-2, and this downregulation effect may underlie low serum K^+-induced acquired cardiac arrhythmias (Guo et al. 2009). hERG is also regulated by elevated external Ca^{2+} ions, which slow hERG activation and speed deactivation rates (Johnson Jr., Mullins, and Bennett 1999), and elevated external protons, which speed the kinetics of deactivation (Shi et al. 2019). The f-actin binding protein TRIOBP-1 co-immunoprecipitates with hERG from mammalian heart and interacts directly with the C-terminal region of hERG1a. TRIOBP-1 downregulates hERG1a currents and I_{Kr} in the heart by about 50% and concomitantly prolongs cardiac cell action potentials or results in a depolarization block to AP firing (Jones et al. 2018).

Ca^{2+}–CaM is a major regulator of EAG channels. Ca^{2+}–CaM associates with and nearly completely inhibits EAG channels within the physiological range of intracellular calcium (Schonherr, Lober, and Heinemann 2000). Biochemical interaction assays using EAG channel intracellular domains and FRET interaction assays using intact channels initially identified Ca^{2+}–CaM binding sites distal to the PAS domain and distal to the CNBHD (Goncalves and Stuhmer 2010; Schonherr, Lober, and Heinemann 2000; Ziechner et al. 2006). The cryo-EM structure of rat EAG in complex with Ca^{2+}–CaM revealed a four EAG subunit:four Ca^{2+}–CaM stoichiometry, where the N-lobe of Ca^{2+}–CaM is associated with the binding site distal to the PAS domain and the C-lobe is associated with the sites distal to the CNBHD (Whicher and MacKinnon 2016; Codding, Johnson, and Trudeau 2020; see also Volume III, Chapter 2). The activation gate of EAG is in a closed confirmation (Figure 5.3H), which is thought to be due to the inhibitory effect of bound Ca^{2+}–CaM (Whicher and MacKinnon 2016).

EAG channel prepulse potentiation and activation time course is regulated by external divalent ions in which elevated concentrations of Ca^{2+} and Mg^{2+} slow the time course of EAG channel activation (Tang, Bezanilla, and Papazian 2000).

In ELK channels, PI(4,5)P2 (PIP2) shifts the conductance–voltage relationship to hyperpolarized potentials and slows channel deactivation. The mechanism is similar to that of the VDP mechanism in ELK, as PIP2 stabilizes an open state of the ELK2 channel (Dai et al. 2019; Li et al. 2015). hERG channels have a form of VDP, but it appears to function on a faster time scale than for ELK channels (Goodchild, Macdonald, and Fedida 2015; Shi, Thouta, and Claydon 2020; Tan et al. 2012).

5.7 Cell Biology (Biogenesis, Trafficking, Turnover)

Some, but not all, of the steps surrounding hERG biogenesis and trafficking to the plasma membrane followed by its eventual retrieval and degradation have been described. Here we will discuss those steps that are the best understood.

Pulse-chase experiments in heterologous systems show that the turnover of hERG1a at the plasma membrane occurs in about 10–12 hours. Glycosylation is one of the most useful measurements to identify hERG subunit biochemical maturation. On a Western blot, hERG subunits are detected as two clearly distinct bands of differing molecular weights: a 150 kD "mature" monomer band that is N-linked glycosylated and a 135 kD "immature" monomer that is O-linked glycosylated (Figure 5.5A, inset) (Zhou et al. 1998). N-glycosylation generally indicates that the protein is destined for the membrane surface, although it is not a measurement of proteins at the plasma membrane per se. The hERG1b isoform is likewise N and O glycosylated resulting in two bands of molecular weights 90 and 75 kD on a Western blot (owing to its smaller size than hERG1a). Following maturation, hERG is delivered to the plasma membrane by COPII machinery (Delisle et al. 2009).

Retention in the endoplasmic reticulum in a variety of ion channels is regulated by a motif in which two arginine residues are separated by any other amino acid (RXR). Owing to an RXR motif in its unique N-terminal region at residues 15-RXR-17, hERG1b traffics inefficiently to the plasma membrane as measured by relatively smaller hERG1b currents, which correlate with a faint mature form of the protein, relative to the immature form, on Western blots. Mutations of the arginine residues in the hERG1b RXR motif to uncharged residues result in larger hERG1b currents, as if hERG1b RXR mutant channels were less retained in the ER. hERG1a enhances the biochemical maturation of hERG1b by a mechanism in which the N-terminal region of hERG1a interacts with and buries the hERG1b RXR motif and the N-linker region of hERG1a makes additional interactions with the hERG1b N-terminal region to upregulate hERG1b (Phartiyal et al. 2008).

hERG1a is degraded by the NEDD4-2/ubiquitin pathway, which includes caveolin and NEDD family interacting proteins. Ubiquitinated hERG proteins are shuttled to the endosome for degradation (Guo et al. 2012).

One of the most remarkable features of hERG biogenesis is its central role in a microtranslatome that shapes repolarization in the heart. hERG mRNA and cardiac Na channel SCN5A mRNA associate cotranslationally on the ribosome in cardiomyocytes, where their expression levels are also coregulated (Eichel et al. 2019). hERG1a and hERG1b mRNAs are also coregulated indicating that the microtranslatome is a complex and critical regulator of cardiac repolarization and likely extends to other ion channels and other systems (Liu et al. 2016).

5.8 Channelopathies and Disease

Genetic mutations in the hERG1 gene cause type 2 long QT syndrome (LQT2), a cardiac arrhythmia syndrome that is a risk factor for sudden cardiac death (Curran et al. 1995).

To date, there are over 450 different mutations reported in the hERG1 (KCNH2) gene and linked to LQT2. Mutations in hERG1 (30%), KCNQ1 (35%) and SCN5A (10%) account for most of the known associations with Long QT Syndrome (LQTS). Mutations in the hERG1 gene result in mutant hERG1a and hERG1b subunits, and studies of hERG1 LQT2

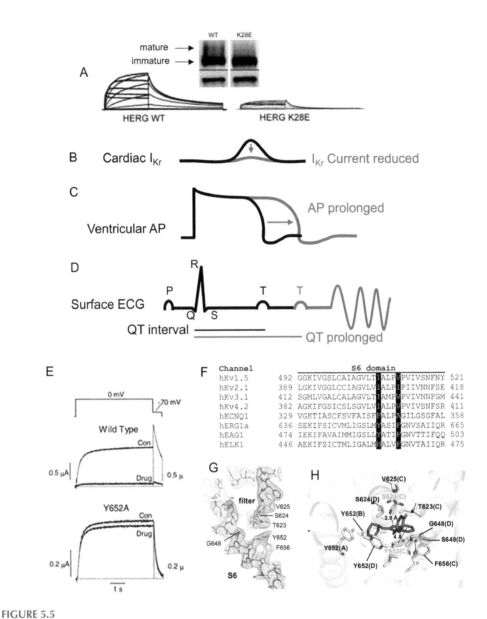

FIGURE 5.5

Physiology, Long QT Syndrome and drug block of hERG. Current family of (A) wild-type hERG and the LQT2-linked mutant hERG K28E. Inset depicts Western blot of cell lysates indicating mature (N-glycosylated) and immature (O-glycosylated) forms of hERG. PDI is the loading control. (Derived from Gianulis and Trudeau 2011.) (B) hERG subunits encode the I_{Kr} current in cardiomyocytes. I_{Kr} produces an outward K^+ current during the late phase of the cardiac action potential that is a major determinant of AP repolarization (black traces). Loss-of-function I_{Kr} current in LQT2 (red trace). (C) Ventricular AP (black trace) and prolonged AP (red trace) in LQT2. (D) Normal surface ECG (black trace) where Q to T interval duration is determined by the length of the ventricular AP and prolonged QT interval (red trace) that degenerates into torsades de pointes arrhythmia. (E) Inhibition of hERG current by MK-499 (top) and loss of inhibition in hERG Y652A mutant channel. (From Mitcheson et al. 2000.) (F) Alignment of KCNH pore and S6 domains and determinants of MK-499 inhibition. (G) Cryo-EM structure of single subunit showing hERG pore labeled with key determinants drug inhibition (Wang and MacKinnon 2017). (H) Cryo-EM structure of hERG pore solved with the inhibitor astemizole (purple). (Derived from Asai et al. 2021.)

mutant channels in heterologous cells reveal loss-of-function phenotypes in which there is a reduced K$^+$ current (Figure 5.5A). In many of these mutants the "mature" N-linked glycosylated band of hERG is reduced or eliminated (Figure 5.5A, inset). Of the LQT2 mutants studied so far, many are trafficking or folding defective, and produce fewer hERG channels at the plasma membrane and concomitantly smaller hERG currents (Anderson et al. 2014). Another class of LQT2 mutants produces gating-defective hERG channels in which the overall outward hERG current is smaller than wild-type currents. Some mutations have dual defects in trafficking and gating, as seen in the hERG K28E LQT2 mutant, which has smaller currents and less maturation, but also faster deactivation gating (Figure 5.5A). The faster deactivation is likely because K28E is located in the PAS domain and interferes with PAS domain regulation of deactivation. Less commonly, LQT2 mutants alter ion permeation or RNA stability.

Most LQT2 mutations have been studied in hERG1a expressed as homomers in heterologous expression systems, which may not accurately reflect disposition of the native mutant channels. A human disease-in-a-dish model of LQT2 provides critical functional connections between a genetic mutation in hERG (A614V) from an LQTS patient with a prolonged QT interval on an ECG and torsades de pointes (an EKG pattern that twists around a point) on an ECG (Itzhaki et al. 2011). Human iPSC-cardiomyocytes from the patient with the hERG A614V mutant had reduced I_{Kr} currents compared to control; a prolonged AP compared to control; and early afterdepolarizations, a cellular precursor to arrhythmias. Thus, there is excellent agreement between the genetic, molecular, cellular, tissue and disease pathophysiology of LQT2. This is further supported by animal models of LQT2 using zebrafish that show prolonged ventricular action potentials and a prolonged QT interval on an ECG with genetic deletion of ERG (Arnaout et al. 2007), and transgenic rabbits with a dominant-negative hERG subunit that has less outward I_{Kr} current, prolonged APs and about 50% rate of sudden death (Brunner et al. 2008).

The loss-of-function phenotype of hERG–LQT2 channels is consistent with LQTS. This can be understood because I_{Kr} is specialized for its physiological role in myocardium by having a small outward current during the AP upstroke and plateau phase, due to rapid inactivation, but I_{Kr} has a large outward current during the AP repolarization phase, due to recovery from inactivation and slow deactivation (Figure 5.5 B, C, black traces) (Zhou et al. 1998; Trudeau et al. 1995; Sanguinetti and Jurkiewicz 1990; Sanguinetti et al. 1995; see also Volume III, Chapter 5). The length of the ventricular AP determines the QT interval on a surface electrocardiogram (Figure 5.5D). A reduction in outward current in an hERG LQT2 mutant, such as hERG K28E (Figure 5.5A), is anticipated to reduce the peak I_{Kr} current in the heart (red trace, Figure 5.5B), which in turn prolongs cardiac action potentials (Figure 5.5C, red trace) and prolongs the QT interval on an electrocardiogram, which can lead to chaotic behavior of the ECG (Figure 5.5D).

Genetic mutations in human EAG1 are associated with Temple–Baraitser syndrome (TBS) and Zimmermann–Laband syndrome (ZLS). TBS and ZLS are characterized by facial and digital dysplasia, intellectual disability and epilepsy (Kortum et al. 2015; Simons et al. 2015). EAG1 channels with TBS or ZLS-associated mutations have gain-of-function phenotypes that subtly alter channel gating, but the mechanisms for how the mutations cause TBS or ZLS are not known (Kortum et al. 2015; Simons et al. 2015). Interestingly, several disease mutations in EAG1, like many for hERG1, occur at the interface of the PAS and CNBHDs. EAG1 and EAG2 channels are upregulated in many different cancer cell lines (Huang et al. 2015; Pardo et al. 1999). Addition of astemizole or shRNAs directed against EAG reduces the proliferation and growth phenotype of these EAG-containing cancer cell lines, suggesting that inhibition of EAG reduces cancer cell growth. Human ELK channels have not been linked to human diseases at this time, but mice null for the ELK2 gene (Elk2$^-$/$^-$) have an increased risk of seizure (Zhang et al. 2010).

5.9 Pharmacology

5.9.1 hERG, I$_{Kr}$ and Acquired LQTS

hERG holds a unique place among ion channels due to its pharmacology. The cardiac I$_{Kr}$ current was initially identified in native guinea pig myocytes because it was inhibited by the drug E-4031 and on this basis I$_{Kr}$ can be separated from the I$_{Ks}$ current in heart cells (Sanguinetti and Jurkiewicz 1990; see Volume III, Chapter 5; Chapter 6, this volume). hERG channels are likewise inhibited by E-4031, establishing the pharmacological link between hERG and I$_{Kr}$ (Trudeau et al. 1995). Figure 5.5E (upper) shows the near-complete inhibition of wild-type hERG by MK-499, another small-molecule inhibitor (Mitcheson et al. 2000).

In addition to E-4031 and MK-499, hERG is inhibited by a wide array of drugs and pharmaceuticals. hERG inhibitors include the Class III antiarrhythmic drugs (E-4031, MK-499, d-sotalol, dofetilide, ibutilide, flecainide, amiodarone), antihistamines (astemizole, terfenadine), antibiotics (erythromycin, ciprofloxacin, grepafloxacin, chloroquine, moxifloxacin, halofantrine), antipsychotics (haloperidol, droperidol, risperidone, fluvoxamine), protease inhibitors (lopinavir, nelfinavir, ritonavir and saquinavir), a gastrointestinal prokinetic drug (cisapride), and illicit drugs including cocaine, as reviewed in detail by Perry et al. (2015). A common, acquired form of LQTS that can result in sudden death arrhythmias is primarily due to the inhibition of hERG1 channels (and native I$_{Kr}$ in the heart) by drugs and pharmaceuticals (Roden et al. 1996; Trudeau et al. 1995; Spector, Curran, Keating, et al. 1996). Notably, the off-target effects of these drugs led the US Food and Drug Administration to issue a Guidance for Industry (1995) that recommended all new research and development drugs undergo a screening process to assess proarrhythmic potential by assessing inhibition of hERG channels.

The mechanism for inhibition of hERG by drugs is blockade of the channel pore by the drug. An alanine scan of the hERG pore showed that hERG Y652A lacked inhibition by MK-499 (Figure 5.5E, lower) as well as channels with alanine mutants G648A and F656A in the S6, and T623A, S624A and V625A in the pore (Mitcheson et al. 2000). The hydrophobic residues F656 and Y652 are nonconserved among most K channels, indicating a possible reason for the specificity of drug molecules for hERG (Figure 5.5F). Putative drug binding sites are nearby one another in three-dimensional space (Figure 5.5G) (Wang and MacKinnon 2017), and the cryo-EM structure of a hERG pore in complex with astemizole shows interactions between residues S624 and F656 and astemizole within the pore of hERG (Figure 5.5H) (Asai et al. 2021). There does not appear to be a strong interaction of astemizole with F656 and thus the role of this site is less clear than before.

Despite the primary and structural homology of hERG and EAG, including the homology of sites Y652 and F656 (Y491 and F495 in hEAG) EAG channels are not inhibited by many drugs. The lack of specific pharmacology has hindered the identification of EAG currents in vivo. Exceptions are astemizole, which inhibits EAG (García-Ferreiro et al. 2004), and chlorpromazine, which is an allosteric inhibitor of EAG channels that binds to the EAG PAS domain (Wang et al. 2020).

ELK channels, like EAG channels, are not blocked by the wide range of drugs that inhibit hERG. ELK channels are inhibited by the small-molecule CX4 in both heterologous expression and in Purkinje neurons (Zhang et al. 2010).

5.10 Conclusion

In summary, it is well established that hERG1a and hERG1b subunits encode the I_{Kr} current in the heart, that mutations in the hERG gene cause LQT2 and drug block of hERG is responsible for an acquired, drug-induced form of LQTS, which is a common clinical problem. Recently EAG channels were identified as being mutated in human diseases, but ELK channels have not yet been linked to a disease. Specific reagents, like those that have been developed and identified for hERG, are needed to help identify the native correlate for EAG and ELK channels in vivo. Molecular specializations that are unique to KCNH channels include the intersubunit domain-swapped PAS–CNBHD interaction, and the non-domain-swapped relationship between the VSD and the pore module that is different from the architecture of most other Kv channels. Future experiments delineating the native current encoding by EAG and ELK channels, and the significance of the non-domain-swapped KCNH pore–VSDs are needed to advance KCNH channel physiology and pathophysiology.

Suggested Reading

This chapter includes additional bibliographical references hosted only online as indicated by citations in blue color font in the text. Please visit https://www.routledge.com/9780367538163 to access the additional references for this chapter, found under "Support Material" at the bottom of the page.

Asai, T., N. Adachi, T. Moriya, H. Oki, T. Maru, M. Kawasaki, K. Suzuki, S. Chen, R. Ishii, K. Yonemori, S. Igaki, S. Yasuda, S. Ogasawara, T. Senda, and T. Murata. 2021. "Cryo-EM structure of K+-bound hERG channel complexed with the blocker astemizole." *Structure* 29 (3):203–12.

Brelidze, T. I., E. C. Gianulis, F. DiMaio, M. C. Trudeau, and W. N. Zagotta. 2013. "Structure of the C-terminal region of an ERG channel and functional implications." *Proc Natl Acad Sci U S A* 110 (28):11648–53. doi:10.1073/pnas.1306887110.

Codding, S. J., Johnson, A. A. and Trudeau, M.C. 2020. "Gating and regulation of KCNH (ERG, EAG and ELK) channels by intracellular domains." *Channels (Austin)* (1):294–309. https://pubmed.ncbi.nlm.nih.gov/32924766/ PMCID: PMC7515569.

Curran, M. E., I. Splawski, K. W. Timothy, G. M. Vincent, E. D. Green, and M. T. Keating. 1995. "A molecular basis for cardiac arrhythmia: HERG mutations cause long QT syndrome." *Cell* 80 (5):795–803. doi:10.1016/0092-8674(95)90358-5.

Dai, G., and W. N. Zagotta. 2017. "Molecular mechanism of voltage-dependent potentiation of KCNH potassium channels." *eLife* 6:e26355. doi:10.7554/eLife.26355.

Gianulis, E. C., Q. Liu, and M. C. Trudeau. 2013. "Direct interaction of eag domains and cyclic nucleotide-binding homology domains regulate deactivation gating in hERG channels." *J Gen Physiol* 142 (4):351–66. doi:10.1085/jgp.201310995.

Gianulis, E. C., and M. C. Trudeau. 2011. "Rescue of aberrant gating by a genetically encoded PAS (Per-Arnt-Sim) domain in several long QT syndrome mutant human ether-a-go-go-related gene potassium channels." *J Biol Chem* 286 (25):22160–9. doi:10.1074/jbc.M110.205948.

Goodchild, S. J., and D. Fedida. 2014. "Gating charge movement precedes ionic current activation in hERG channels." *Channels (Austin)* 8 (1):84–9. doi:10.4161/chan.26775.

Goodchild, S. J., L. C. Macdonald, and D. Fedida. 2015. "Sequence of gating charge movement and pore gating in HERG activation and deactivation pathways." *Biophys J* 108 (6):1435–47. doi:10.1016/j.bpj.2015.02.014.

Gustina, A. S., and M. C. Trudeau. 2011. "hERG potassium channel gating is mediated by N- and C-terminal region interactions." *J Gen Physiol* 137 (3):315–25. doi:10.1085/jgp.201010582.

Haitin, Y., A. E. Carlson, and W. N. Zagotta. 2013. "The structural mechanism of KCNH-channel regulation by the eag domain." *Nature* 501 (7467):444–8. doi:10.1038/nature12487.

Herzberg, I. M., M. C. Trudeau, and G. A. Robertson. 1998. "Transfer of rapid inactivation and sensitivity to the class III antiarrhythmic drug E-4031 from HERG to M-eag channels." *J Physiol* 511(1):3–14. doi:10.1111/j.1469-7793.1998.003bi.x.

Itzhaki, I., L. Maizels, I. Huber, L. Zwi-Dantsis, O. Caspi, A. Winterstern, O. Feldman, A. Gepstein, G. Arbel, H. Hammerman, M. Boulos, and L. Gepstein. 2011. "Modelling the long QT syndrome with induced pluripotent stem cells." *Nature* 471 (7337):225–9. doi:10.1038/nature09747.

Jones, E. M., E. C. Roti Roti, J. Wang, S. A. Delfosse, and G. A. Robertson. 2004. "Cardiac IKr channels minimally comprise hERG 1a and 1b subunits." *J Biol Chem* 279 (43):44690–4. doi:10.1074/jbc. M408344200.

London, B., M. C. Trudeau, K. P. Newton, A. K. Beyer, N. G. Copeland, D. J. Gilbert, N. A. Jenkins, C. A. Satler, and G. A. Robertson. 1997. "Two isoforms of the mouse ether-a-go-go-related gene coassemble to form channels with properties similar to the rapidly activating component of the cardiac delayed rectifier K+ current." *Circ Res* 81 (5):870–8. doi:10.1161/01.res.81.5.870.

Mitcheson, J. S., J. Chen, M. Lin, C. Culberson, and M. C. Sanguinetti. 2000. "A structural basis for drug-induced long QT syndrome." *Proc Natl Acad Sci U S A* 97 (22):12329–33. doi:10.1073/pnas.210244497.

Morais Cabral, J. H., A. Lee, S. L. Cohen, B. T. Chait, M. Li, and R. Mackinnon. 1998. "Crystal structure and functional analysis of the HERG potassium channel N terminus: a eukaryotic PAS domain." *Cell* 95 (5):649–55. doi:10.1016/s0092-8674(00)81635-9.

Phartiyal, P., E. M. Jones, and G. A. Robertson. 2007. "Heteromeric assembly of human ether-a-go-go-related gene (hERG) 1a/1b channels occurs cotranslationally via N-terminal interactions." *J Biol Chem* 282 (13):9874–82. doi:10.1074/jbc.M610875200.

Robertson, G. A., J. M. Warmke, and B. Ganetzky. 1996. "Potassium currents expressed from Drosophila and mouse eag cDNAs in Xenopus oocytes." *Neuropharmacology* 35 (7):841–50. doi:10.1016/0028-3908(96)00113-x.

Sanguinetti, M. C., C. Jiang, M. E. Curran, and M. T. Keating. 1995. "A mechanistic link between an inherited and an acquired cardiac arrhythmia: HERG encodes the IKr potassium channel." *Cell* 81 (2):299–307.

Sanguinetti, M. C., and N. K. Jurkiewicz. 1990. "Two components of cardiac delayed rectifier K+ current. Differential sensitivity to block by class III antiarrhythmic agents." *J Gen Physiol* 96 (1):195–215.

Smith, P. L., T. Baukrowitz, and G. Yellen. 1996. "The inward rectification mechanism of the HERG cardiac potassium channel." *Nature* 379 (6568):833–6.

Thouta, S., S. Sokolov, Y. Abe, S. J. Clark, Y. M. Cheng, and T. W. Claydon. 2014. "Proline scan of the HERG channel S6 helix reveals the location of the intracellular pore gate." *Biophys J* 106 (5):1057–69. doi:10.1016/j.bpj.2014.01.035.

Trudeau, M. C., S. A. Titus, J. L. Branchaw, B. Ganetzky, and G. A. Robertson. 1999. "Functional analysis of a mouse brain Elk-type K+ channel." *J Neurosci* 19 (8):2906–18.

Trudeau, M. C., J. W. Warmke, B. Ganetzky, and G. A. Robertson. 1995. "HERG, a human inward rectifier in the voltage-gated potassium channel family." *Science* 269 (5220):92–5.

Wang, J., M. C. Trudeau, A. M. Zappia, and G. A. Robertson. 1998. "Regulation of deactivation by an amino terminal domain in human ether-a-go-go-related gene potassium channels." *J Gen Physiol* 112 (5):637–47. doi:10.1085/jgp.112.5.637.

Wang, W., and R. MacKinnon. 2017. "Cryo-EM structure of the open human ether-a-go-go-related K(+) channel hERG." *Cell* 169 (3):422–430.e10. doi:10.1016/j.cell.2017.03.048.

Wang, Z., Y. Dou, S. J. Goodchild, Z. Es-Salah-Lamoureux, and D. Fedida. 2013. "Components of gating charge movement and S4 voltage-sensor exposure during activation of hERG channels." *J Gen Physiol* 141 (4):431–43. doi:10.1085/jgp.201210942.

Warmke, J. W., and B. Ganetzky. 1994. "A family of potassium channel genes related to eag in Drosophila and mammals." *Proc Natl Acad Sci U S A* 91 (8):3438–42.

Wynia-Smith, S. L., A. L. Gillian-Daniel, K. A. Satyshur, and G. A. Robertson. 2008. "hERG gating microdomains defined by S6 mutagenesis and molecular modeling." *J Gen Physiol* 132 (5):507–20. doi:10.1085/jgp.200810083.

Zhou, Z., Q. Gong, B. Ye, Z. Fan, J. C. Makielski, G. A. Robertson, and C. T. January. 1998. "Properties of HERG channels stably expressed in HEK 293 cells studied at physiological temperature." *Biophys J* 74 (1):230–41. doi:10.1016/S0006-3495(98)77782-3.

6

KCNQ Channels

Anastasios V. Tzingounis and H. Peter Larsson

CONTENTS

6.1 Introduction

KCNQ channels, also known as Kv7 channels, are voltage-gated K^+ (Kv) channels encoded by the *KCNQ* genes that exhibit unusual properties and serve unique physiological roles. This family includes five members, Kv7.1–7.5, that perform different roles and functions in the body. In addition, some Kv7 channels interact with different KCNE beta subunits, which drastically alters the functional properties of these channels and further increases the diversity of Kv7 channels. This chapter highlights the characteristics and physiological roles of these channels that are distinct from those of other Kv channels. For a general overview of Kv channels, please refer to Chapter 4.

6.2 Physiological Roles

The most well-studied KCNQ channel is the KCNQ1 channel, which, together with the beta subunit KCNE1, forms the I_{Ks} channel in the heart (Sanguinetti et al. 1996; Barhanin et al. 1996) (see also Volume III, Chapter 5). The I_{Ks} channel generates a very slowly activating current, I_{Ks}, which aids in repolarizing the ventricular action potential in the heart (Nerbonne and Kass 2005) (Figure 6.1A, B). The length of the ventricular action potential changes with heart rate, and upregulation of I_{Ks} during sympathetic heart stimulation is one mechanism by which the body shortens the action potential at higher heart rates

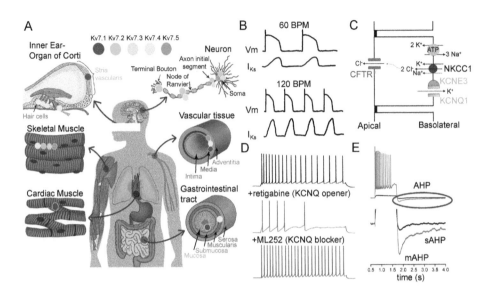

FIGURE 6.1

Physiological role of KCNQ channels. (A) Expression pattern of different KCNQ channels (Adapted by Soldovieri et al., 2011 https://doi.org/10.1152/physiol.00009.2011). (B) Sympathetic upregulation of I_{Ks} contributes to the shorter action potentials at higher heart rates (60 vs. 120 beats per minute). (C) KCNQ1/KCNE3 channels recycle K⁺ at the basolateral membrane to provide driving force for Cl⁻ and water secretion in colon epithelia cells. (Adapted from Preston et al. 2010.) (D) M-currents' activity below threshold puts a break on neuronal excitability, with retigabine, and the channel blocker ML252 decreases and increases the firing rate, respectively. (E) KCNQ2/KCNQ3 channels contribute to the mAHP and sAHP after sustained depolarization in some neurons. Example of voltage responses to two different current injection steps in CA1 pyramidal neurons. Note that as the depolarizing current step gets larger (blue trace), the mAHP and sAHP also become larger.

(Nerbonne and Kass 2005) (Figure 6.1B). A shortened ventricular action potential at higher heart rates is necessary to provide sufficient time for the ventricles to adequately fill with blood between action potentials (Nerbonne and Kass 2005). For more details, please refer to Volume III, Chapter 5. Together with KCNE1, KCNQ1 is also expressed in the inner ear, where KCNQ1/KCNE1 plays an important role in controlling endolymph production in the stria vascularis (Robbins 2001) (Figure 6.1A). KCNQ1 is also expressed with other beta subunits from the KCNE family. KCNQ1 is expressed with KCNE2 and KCNE3, which together form voltage-independent channels in the intestine and kidney (Robbins 2001) (Figure 6.1A). In non-excitable cells, these voltage-independent channels are important for salt and water transport (Preston et al. 2010) (Figure 6.1C).

Together, KCNQ2 and KCNQ3 form the M-channels that generate the M-current in neurons (Wang et al. 1998) (Figure 6.1A). These channels open below the threshold for an action potential and therefore serve as a brake on action potential firing, limiting the firing rate in these neurons (Figure 6.1D). Importantly, KCNQ2 and KCNQ3 channels do not inactivate, allowing them to clamp the membrane potential close to the resting membrane potential. KCNQ2 and KCNQ3 channels are highly enriched at the axon initial segment, the site of action potential induction, and the nodes of Ranvier. In the axon initial segment, and probably to a lesser extent in the nodes of Ranvier, these channels contribute to the resting membrane potential, counteracting the action of the subthreshold persistent sodium current and thereby preventing unwanted excitability (Battefeld et al. 2014).

Besides the M-current, neuronal KCNQ channels also contribute to various afterhyperpolarization (AHP) currents following short bursts or trains of activity (Figure 6.1E). Following high-frequency action potentials or prolonged depolarization, a series of AHP conductances are activated leading to fast, medium and slow AHPs. Ca^{2+}-activated and voltage-activated K^+ (BK) channels, as well as Kv2 channels, typically mediate the fast AHP (<20 ms). Medium AHPs (mAHPs) last 50–200 ms and are typically mediated by KCNQ channels and, depending on the neuron and induction protocol, by SK channels. The main function of the mAHP is to prevent excessive firing by limiting the activity of afterdepolarizations that follow action potentials (Storm 1990). If not limited, afterdepolarizations can lead to a barrage or short burst of action potentials. Indeed, pan-KCNQ channel blockers or deletion of KCNQ2 channels allows CA1 pyramidal neurons to fire in bursts (Soh et al. 2014). In addition to the mAHP, KCNQ channels can also mediate the Ca^{2+}-activated slow AHP (sAHP) in some neurons. The sAHP, which has a duration of 1–20 s, controls the extent of spike frequency adaptation. Sustained neuronal depolarization of excitatory and a subset of inhibitory neurons leads to a reduced neuronal firing rate, i.e., spike frequency adaptation, with both the mAHP and sAHP playing a significant role (Greene and Hoshi 2017). In summary, KCNQ2 and KCNQ3 channels have a dual role in neuronal physiology, acting as a brake to decrease the depolarization rate prior to action potential initiation and limiting runaway activity following neuronal firing. Understanding this dual KCNQ2/3 function is paramount to interpret the impact of pathogenic variants, as KCNQ2 and KCNQ3 variants might not always affect both KCNQ2/3 functions equally.

Research has suggested that KCNQ5 channels (Figure 6.1, orange symbols) may contribute to the M-current, mAHP and sAHP. Most notably, KCNQ5 channels regulate the resting membrane potential of auditory synaptic terminals as well as the excitability of several interneurons. Part of the challenge in elucidating the function of the KCNQ5 channel in the nervous system is the lack of tools available for dissecting its function. Currently, no selective inhibitors for KCNQ5 channels exist to reveal their function in neurons that also express KCNQ2 and KCNQ3 channels (Barrese, Stott, and Greenwood 2018).

Unlike our limited understanding of KCNQ5 channels in neurons, their function has been well described in smooth muscle. Along with KCNQ4 channels, KCNQ5 channels regulate the resting membrane potential in smooth muscle. Several studies have also indicated that KCNQ5 likely forms heteromers with KCNQ4 channels (Barrese, Stott, and Greenwood 2018). Together, they cause hyperpolarization of the membrane potential, preventing activation of voltage-gated Ca^{2+} channels and downstream vasoconstriction. Unlike skeletal muscle-initiated excitation-contraction coupling, smooth muscle contraction requires a Ca^{2+} influx from the extracellular space; thus, small changes in smooth muscle excitability can lead to unintended contractions. Indeed, downstream activation of KCNQ4/5 channels by the cyclic AMP/protein kinase A (PKA) cascade leads to vasodilation, as increasing KCNQ4/5 activity hyperpolarizes the membrane potential, preventing smooth muscle contractions (Barrese, Stott, and Greenwood 2018).

KCNQ4 channels are also expressed in the nervous system, but their expression is limited to a few regions, such as the auditory system and the raphe nuclei. Most notably, KCNQ4 channels are expressed in the inner ear and cochlear hair cells. Work from multiple labs using *Kcnq4* genetic manipulations has shown that KCNQ4 contributes to the hair cell resting membrane potential and the low-voltage-activated K^+ conductance in outer hair cells (Barrese, Stott, and Greenwood 2018; Jentsch 2000).

6.3 Channel Diversity/Structural Organization

Researchers first identified KCNQ1 channels in 1996 through positional cloning (Sanguinetti et al. 1996; Barhanin et al. 1996). In a span of two years (1998–2000), multiple groups identified and described the biophysical and pharmacological properties of the other four KCNQ channels (KCNQ2–KCNQ5) and their involvement in human disease (Jentsch 2000). Beyond mammals, KCNQ channels exist in multiple organisms. It has been suggested that the diversity of KCNQ channel members arose from duplication early in vertebrate evolution, with some vertebrate KCNQ channels undergoing an additional duplication in fish to form orthologues. Another independent cycle of duplication in worms gave rise to KCNQ channel genes, known as KQT-1 and KQT-3 in nematodes. Unlike other animals, insects have only one KCNQ channel gene. See Chapter 1 for details. Despite their diversity, all KCNQ channels share a similar basic structure.

KCNQ channels consist of six transmembrane (S1–S6) segments, a short N-terminus and a long intracellular C-terminus that accounts for half of the protein (Sun and MacKinnon 2020) (Figure 6.2A). As with most Kv channels, KCNQ channels have a swapped-domain configuration in which the voltage-sensing domain (S1–S4) of one subunit interacts with the pore domain (S5–S6) of an adjacent subunit (Sun and MacKinnon 2020). However, KCNQ channels also display several distinguishing properties. For instance, KCNQ channels have a relatively small single-channel conductance (femtosiemens to low picosiemens) and contain fewer positively charged amino acids in the voltage sensor S4 than other Kv channels (Figure 6.2B). In the pore region, instead of a methionine, a lysine follows the Kv signature sequence GYGD (Figure 6.2C), which contributes to the unique ion conductance sequence of these channels. The S6 gate has a PAG sequence in its distal end, rather than the conserved Kv PVP sequence (Figure 6.2D). The distal part of S6 bends at the PVP/PAG site, allowing for dilation of the cytoplasmic pore entrance (Seebohm et al. 2006). The replacement of PVP with PAG suggests that KCNQ S6 may not dilate to the same extent as other Kv channels. Indeed, recent KCNQ1 structures suggest that the cytosolic gate expands less than other Kv channels (3.5 Å during gate opening compared with >10 Å). KCNQ channels are also unique in that they do not contain an N-terminus tetramerization domain; rather, tetramerization is mediated through the C-terminus. The C-terminus is very large and

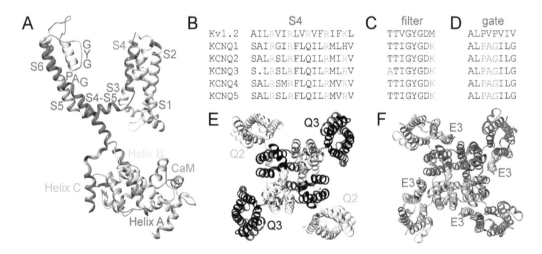

FIGURE 6.2
KCNQ structural organization. (A) Side view of one KCNQ1 subunit (Protein Data Bank: 6uum) (Sun and MacKinnon 2020). (B–D) Sequence alignment of Kv1.2 with KCNQ1–5 for (B) S4, (C) selectivity filter, (D) lower S6. (E and F) Top view of (E) a model of a KCNQ2 (Q2: gray)/KCNQ3 (Q3: black) channel and (F) a KCNQ1/KCNE3 (E3: gray) channel (Protein Data Bank: 6uv1) (Sun and MacKinnon 2020). The four KCNQ1 subunits shown in four different colors.

is further subdivided into four helices, A–D. Helices A–B contain calmodulin, A-kinase anchoring protein (AKAP), and phosphatidylinositol 4,5-bisphosphate (PIP_2) binding sites, whereas helices C–D form two coiled-coil domains (Sun and MacKinnon 2020; Sachyani et al. 2014; Greene and Hoshi 2017) (Figure 6.2A). The two coiled-coil domains are important for the tetramerization of KCNQ channels, with the helix-C coiled-coil domain dictating oligomerization.

Research has clearly demonstrated that KCNQ2–5 channels can form either as homomers or heteromers. In particular, KCNQ3 subunits can form homotetramers or heterotetramers with KCNQ2, KCNQ4 or KCNQ5 subunits. Similarly, KCNQ4 can form either homotetramers or heterotetramers with KCNQ3 or KCNQ5 subunits (Barrese, Stott, and Greenwood 2018). KCNQ2 subunits primarily form homotetramers or heterotetramers with KCNQ3 subunits (Figure 6.2E) and possibly triheterotetramers with KCNQ3 and KCNQ5. Because of the heterogeneity and numerous possibilities of KCNQ channel multimers, the native composition of KCNQ channels is not fully known. However, pharmacological and genetic evidence indicates that the majority of neuronal KCNQ channels consist of KCNQ2/3 heterotetramers and possibly KCNQ3/5 heterotetramers (Jentsch 2000). Ablation of KCNQ2 subunits eliminates the M-current in sympathetic neurons and reduces it by more than 80% in hippocampal neurons. Similarly, ablation of KCNQ3 subunits substantially reduces the M-current (50%) in cortical and hypothalamic neurons. Additionally, native M-currents are typically sensitive to low millimolar concentrations of tetraethylammonium (TEA), as are KCNQ2/3 heterotetramers (Wang et al. 1998). In contrast to neuronal tissue, smooth muscle cells express a combination of KCNQ4, KCNQ5 and KCNQ4/5 channels (Barrese, Stott, and Greenwood 2018). In the heart and in many epithelial cells, KCNQ1 channels are the major KCNQ channel. KCNQ1 does not heteromerize with any other members of the KCNQ family; rather, it forms homotetramers that complex with members of the KCNE family, KCNE1–5 (see next section on gating) (Figure 6.2F). All five KCNEs are single-transmembrane-segment proteins. Results from heterologous expression systems suggest that the ratio of KCNQ1 to KCNE subunits ranges from 4:1 to 4:4. The KCNQ1/KCNE ratio in vivo is not known, and it is unclear whether a single KCNQ1 channel assembles with several different KCNE subunits simultaneously.

6.4 Gating

KCNQ channels exhibit some properties that differ from those of most Kv channels. It is assumed that most Kv channels do not open until all four S4 voltage sensors in the tetrameric Kv channel have activated (see Volume I, Chapter 2). In contrast, in KCNQ1 channels expressed alone (without KCNE subunits), the open probability increases as each voltage sensor activates (Osteen et al. 2010). Even in the absence of any activated voltage sensors, the KCNQ1 channel open probability is a few percent. In contrast, for Shaker K channels, the open probability in the absence of any activated voltage sensors is $<10^{-7}$. Therefore, it has been proposed that KCNQ1 may have a weak coupling between the voltage sensor and the gate, which may explain why the gating of KCNQ1 channels is so easily affected by different KCNE beta subunits.

Coexpression of KCNQ1 with KCNE1 shifts the voltage dependence of opening by >50 mV, dramatically slows the activation kinetics, and eliminates any open probability in the

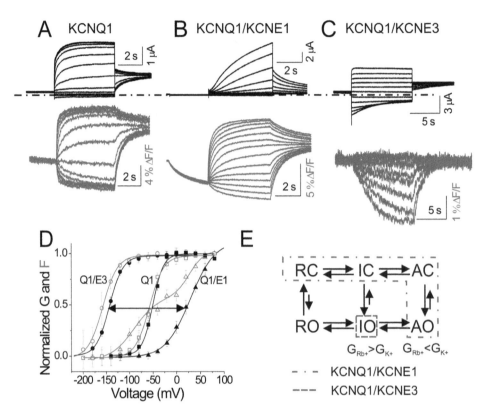

FIGURE 6.3

KCNQ1 channel gating. (A–C) Currents (top) and S4 movement measured by fluorescence (bottom) from (A) KCNQ1, (B) KCNQ1/KCNE1 and (C) KCNQ1/KCNE3 channels in response to voltage steps. (D) Conductance–voltage and fluorescence–voltage curves for KCNQ1, KCNQ1/KCNE1 and KCNQ1/KCNE3 channels. (E) Simplified six-state model of KCNQ1, KCNQ1/KCNE1 (only states in red box) and KCNQ1/KCNE3 channels (only state in blue box at physiological voltages). (Adapted from Barro-Soria et al. 2017.)

absence of activated voltage sensors (Figure 6.3A–D). The rightward shift in voltage dependence and the slow kinetics of KCNQ1/KCNE1 channels are key for generating the slowly activating I_{Ks} in the heart (Nerbonne and Kass 2005) (Figure 6.1B). The molecular mechanism by which KCNE1 causes these dramatic changes in function is not completely clear. However, a triad of residues in the middle of the transmembrane segment of KCNE1 is necessary for most of these effects (Melman, Krumerman, and McDonald 2002).

Coexpression of KCNQ1 with KCNE3 eliminates voltage-dependent closing in the physiological voltage range, basically rendering KCNQ1/KCNE3 channels as voltage-independent channels in physiological settings (Figure 6.3A–D). It has been proposed that KCNE3 causes these functional changes by electrostatic interactions between two negatively charged residues at the top of the transmembrane segment of KCNE3 and the positive charges in the voltage sensor in KCNQ1, trapping the voltage sensors of KCNQ1 in their activated state (Barro-Soria et al. 2017).

Coexpression of KCNQ1 with KCNE2 also results in voltage-independent KCNQ1/KCNE2 channels, whereas coexpression with KCNE4 and KCNE5 positively shifts the voltage dependence out of the physiological range. The physiological role of KCNE4 and KCNE5 is not clear, but it is possible that these beta subunits assemble with KCNQ1 channels with one to three KCNE1–3 subunits to further increase the diversity of KCNQ1 channels.

In most Kv channels, the S4 voltage sensor is assumed to have two stable states: an inward state at negative voltages and an outward state at positive voltages. However, studies have shown that S4 sensors in KCNQ1 channels exhibit three stable states: an inward state at very negative voltages, an intermediate state at less negative voltages and an outward state at positive voltages (Zaydman et al. 2014) (Figure 6.3E). KCNQ1 channels are primarily closed when the S4 sensors are in the inward state, but the open probability increases significantly when the S4 sensors are in the intermediate state. Coexpression with KCNE1 eliminates this open probability when the S4 sensors are in the intermediate state, with KCNQ1/KCNE1 channels opening only when the S4 sensors are in the outward state (Zaydman et al. 2014) (Figure 6.3E). In contrast, KCNQ1/KCNE3 channels open when the S4 sensors are in the intermediate state. The change in S4-to-gate coupling caused by KCNE1 is thought to underlie many of the functional effects on KCNQ1 caused by the coexpression with KCNE1, such as the >50 mV shift in opening and the slowing of activation kinetics (Zaydman et al. 2014). However, the molecular mechanism by which KCNE1 causes these effects is not completely understood.

KCNQ2–5 channels open at voltages below threshold for action potential firing and are noninactivating, both properties that are important for their function to regulate action potential firing and generate AHPs (Greene and Hoshi 2017). Compared to KCNQ1, fewer biophysical/functional studies have been done on KCNQ2-5 channel gating. These studies suggest that KCNQ2-5 channels gate similar to KCNQ1 channels without KCNE subunits.

6.5 Ion Permeability

Like the native M-current, neuronal KCNQ channels possess permeability properties that distinguish them from other members of the Kv channel superfamily. Like most K^+ channels, KCNQ2/3 channels display a type-IV Eisenman permeability sequence ($Tl^+ > K^+ > Rb^+ > NH_4^+ > Cs^+ > Na^+$) (Prole and Marrion 2004). However, the conductance sequence does not follow the same pattern; rather, the conductance has a sequence of $K^+ > Tl^+ > NH_4^+/Rb^+ > Cs^+$. The relatively low Rb^+ conductance of KCNQ2/3 channels over K^+ (0.24), which is even lower for the native M-current (0.1), is more akin to previous measurements for inwardly rectifier Kir2.1 channels (0.1) rather than previous findings for Kv channels (i.e., Kv1, 0.5) (Prole and Marrion 2004). The unique conductance sequence of KCNQ2 channels is partly due to the presence of a lysine following the Kv signature sequence GYGD in the selectivity filter (Figure 6.2C). Members of the Kv family typically have a methionine following the GYGD sequence. Indeed, mutations in which the lysine is replaced by a methionine lead to Kv-like Rb^+ conductance in KCNQ2 channels (Prole and Marrion 2004). This finding indicates that the outer-pore lysine is partly responsible for the low Rb^+ conductance; however, other mechanisms likely play a role because all members of the KCNQ family share the sequence GYGDK (Figure 6.2C), whereas not all members have low Rb^+ conductance. KCNQ3 channels are also unique in that their selectivity filter is unstable. This instability arises because KCNQ3 channels have an ATX sequence preceding the GYGDK sequence, rather than the usually conserved TTX sequence (Figure 6.2C). This alanine (A315) has also been suggested to decrease trafficking of KCNQ3 to the plasma membrane. KCNQ1 channels display different K^+/Rb^+ conductance ratios depending on the presence or absence of the KCNE1 subunit. This difference has been attributed to the presence of different opening states in KCNQ1 versus KCNQ1/KCNE1 channels with the

S4 voltage sensor in different conformations (Zaydman et al. 2014) (Figure 6.3E). Thus, there is substantial heterogeneity among KCNQ members in regard to their conductance sequences, preventing generalizations.

6.6 Pharmacology

One of the great challenges in identifying the cellular functions of KCNQ channels is their insensitivity to canonical K^+ channel blockers and toxins. For instance, with the exception of KCNQ2 channels, KCNQ channels are insensitive to extracellular TEA (Wang et al. 1998). Block by extracellular TEA typically requires the presence of a tyrosine at the outer-pore vestibule, which is only present in KCNQ2 channels. Besides TEA, KCNQ channels are also insensitive to extracellular 4-AP, Cs^+ and divalent ions such as Mg^{2+} that are known to inhibit some K^+ channels (however, KCNQ channels are sensitive to extracellular Ba^{2+}) (Robbins 2001). Thus, for many years, isolating KCNQ-mediated currents in cells required voltage-clamped protocols that capitalized on the noninactivating and slow deactivation properties of the M-current and KCNQ channels. Although these protocols enabled isolation of the M-current from the many K^+ currents found in different cell types, they prevented a determination of the full biophysical gating properties and physiological impacts across different tissues and cell types. The identification of linopirdine and its more potent analog XE991 alleviated these drawbacks. These two compounds both block the majority of KCNQ channels with high affinity, thus allowing detailed investigations of KCNQ channel properties in different cell types (Miceli et al. 2018).

Although XE991 and linopirdine are powerful tools for probing KCNQ channel physiology, they do have some limitations. These compounds are state-dependent blockers, i.e., they block open KCNQ channels (Greene and Hoshi 2017). Consequently, for cells or subcompartments with fairly hyperpolarized membrane potentials (i.e., axons), XE991 must be applied for up to 30 minutes in order to fully block KCNQ channels. Additionally, XE991 cannot distinguish between different KCNQ channel subtypes, and its affinity for blocking depends on the presence or absence of auxiliary KCNE subunits for KCNQ1 channels (Zaydman et al. 2014). The development of ML-252, another pan-KCNQ blocker, may mitigate some of the aforementioned issues (Miceli et al. 2018; Barrese, Stott, and Greenwood 2018). Currently, there exists a limited number of subtype-specific blockers, with the exception of compounds that distinguish between KCNQ1 channels and the other family members, most notably chromanol 293B, HMR 1556 and JNJ-303 (Naffaa and Al-Ewaidat 2021). Most of these compounds are pore blockers, which access the pore from the cytosolic end of the pore. However, it has been suggested that JNJ-303 may access the channel pore through a fenestration in the membrane-spanning portion of the pore (Barrese, Stott, and Greenwood 2018).

In parallel to the identification of XE991 as a KCNQ channel blocker, researchers identified retigabine as a KCNQ channel activator (Barrese, Stott, and Greenwood 2018; Miceli et al. 2018). Retigabine was first categorized as a potent anticonvulsant and later as a neuronal KCNQ channel activator. Application of retigabine increases the maximum KCNQ channel open probability and shifts the voltage dependence of activation toward hyperpolarized membrane potentials. Retigabine binds to the pore domain wedged between the S5 and S6 transmembrane domains (Figure 6.4A). KCNQ2–KCNQ5 channels, but not the retigabine-insensitive KCNQ1, have a tryptophan in S5, allowing for the formation

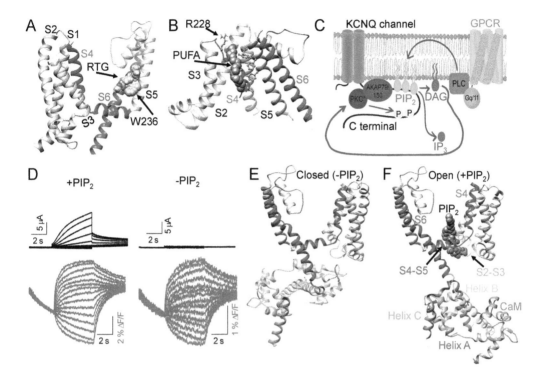

FIGURE 6.4

Pharmacology and regulation of KCNQ channels. (A) Side view of retigabine binding to KCNQ2 channels pore domain (Protein Data Bank: 7cr2) (Li et al. 2021). (B) Side view of PUFA binding to hKCNQ1 voltage sensor S4 of one subunit and S5 from the neighboring subunit (Yazdi et al. 2021). (Model based on Protein Data Bank: 5vms.) (C) Inhibition of KCNQ2 channels is due to downstream activation of PLC that (i) hydrolyzes PIP_2 to DAG and IP_3, and (ii) subsequent activation of PKC by DAG which phosphorylates the KCNQ2 C terminus leading to lower affinity of KCNQ2 channels for the remaining plasma membrane PIP_2. (D) Currents (top) and S4 movement measured by fluorescence (bottom) from KCNQ1/KCNE1 channels in response to voltage steps in the presence (left) and absence (right) of PIP_2. (Adapted from Barro-Soria et al. 2017.) (E and F) Side view of KCNQ1/KCNE3 channels in (E) the absence and (F) the presence of PIP_2 (Protein Data Bank: 6uzz and 6v01) (Sun and MacKinnon 2020).

of a hydrogen bond between the indole nitrogen atom and retigabine. Although retigabine potentiates KCNQ2–KCNQ5 channels, KCNQ3 channels have the highest affinity for retigabine. Retigabine was a first-in-class Kv channel opener approved by the US Food and Drug Administration. However, it has been discontinued due to a series of adverse effects, including, but not limited to, urinary retention and blue–gray mucocutaneous discoloration. In an effort to minimize side effects, researchers have recently developed several compounds that exhibit higher KCNQ channel specificity than retigabine, such as SF0034, RL-81 and KB-3061. However, none of these compounds have been approved for clinical use (Miceli et al. 2018). One of the most unexpected recent developments in KCNQ channel pharmacology is the recognition that naturally occurring compounds also target KCNQ channels. For instance, the neurotransmitter GABA and various ketones can bind to KCNQ3 and KCNQ5 channels, increasing their activity. Thus, naturally derived anticonvulsants previously identified might exert their effects by activating KCNQ channels (Abbott 2020).

In addition to pore KCNQ activators such as retigabine, several voltage-sensor activators have also been developed. Compounds such as ICA-27243, ICA-069673 and Ztz240 bind to the activated conformation of the voltage-sensing domain (VSD) in KCNQ2 channels, driving the voltage activation to highly hyperpolarized membrane potentials (Wang et al. 2018; Li et al. 2021). Researchers have also identified agonists of KCNQ1 channels, such as R-L3, ML277 and CP1 (mimicking PIP_2), and polyunsaturated fatty acids (Naffaa and Al-Ewaidat 2021; Yazdi et al. 2021), some which also target the VSD (Figure 6.4B). The efficacy of many KCNQ1 agonists and antagonists varies with the expression of KCNE1 subunits; in some cases, the compounds are ineffective against KCNQ1/KCNE1 channels with four KCNE1 subunits. It has been suggested that some of these differences in efficacy result from the different open states in KCNQ1 and KCNQ1/KCNE1 channels (Zaydman et al. 2014) (Figure 6.3E).

6.7 Regulation

Neuronal KCNQ channels mediate the M-current. The M-current was the first Kv-related activity found to be inhibited by neurotransmitters. Specifically, Brown and Adams (1980) demonstrated that M-current inhibition through cholinergic receptor activation increases sympathetic neuron excitability and partially controls the pronounced spike frequency adaptation of bullfrog sympathetic neurons. Consequently, acetylcholine through the atropine-sensitive muscarinic receptors blocks the M-current and causes sympathetic neurons to fire repetitively.

Soon after the initial discovery of the M-current, several groups detected the M-current and its modulation by acetylcholine in multiple regions of the nervous system. In the ensuing years, researchers showed that M-channels are the downstream effectors of a plethora of neuromodulators and neurotransmitters, including ATP, bradykinin, glutamate, serotonin and substance P (Delmas and Brown 2005). Together, these findings suggest that M-current modulation is a widespread phenomenon in the nervous system. However, for many years it was unknown how muscarinic receptors, and Gq/11 protein-coupled receptors more generally, inhibit the M-current. The molecular identification of KCNQ2/3 channels as the channels underlying the canonical M-current provided clues toward answering this question. In subsequent studies, the groups of B. Hille and D. Logothetis showed that PIP_2 depletion is responsible for the neuromodulation of the M-current by Gq/11 protein-coupled receptors (Delmas and Brown 2005) (Figure 6.4C). More importantly, studies demonstrated that translocatable enzymes and *Ciona intestinalis* voltage-sensor-containing phosphatase could deplete PIP_2 from the plasma membrane without activating PLC directly, demonstrating that PIP_2 is required for the proper function of KCNQ channels. Thus, PIP_2 depletion helps explain how a myriad of neurotransmitters and neuromodulators converge and inhibit the M-current across different cell types.

All KCNQ channels require PIP_2 to function (Loussouarn et al. 2003; Gamper and Shapiro 2007) (see also Volume III, Chapter 4). However, the affinity for PIP_2 among KCNQ2–5 channels differs substantially. KCNQ3 channels have the highest PIP_2 affinity and KCNQ5 have the lowest. The almost tenfold differences in PIP_2 affinity among KCNQ2–5 channels, along with the finding that the M-channel subunit composition is not always fixed, suggests that muscarinic modulation of KCNQ channels can vary across development and cell types (Gamper and Shapiro 2007).

In KCNQ1 channels, in the absence of PIP_2 the voltage sensor S4 still responds to voltage changes, but the channel gate does not open, showing that PIP_2 is necessary for the S4-to-gate coupling (Zaydman et al. 2013) (Figure 6.4D). Mutagenesis studies suggested that PIP_2

is sandwiched between the S4–S5 linker and the lower end of S6, as if PIP_2 couples the S4 to the gate by bridging the S4–S5 linker with S6 (Zaydman et al. 2013). However, in the KCNQ1/KCNE3 cryogenic electron microscopy structure, one PIP_2 is located between the S2–S3 and S4–S5 linkers, and not between the S4–S5 linker and S6 (Sun and MacKinnon 2020) (Figure 6.4F). Comparing the KCNQ1/KCNE3 structures with and without PIP_2, there is a very large conformational change of the C-terminal cytosolic region, which rotates calmodulin (CaM) almost 180 degrees and, in the presence of PIP_2, makes a continuous helix all the way from S6 far into the cytosolic domain (Sun and MacKinnon 2020) (Figure 6.4E, F). In addition, S6 slightly bends at the PAG motif, widening the constricted part of the pore from 1 Å to 3 Å. However, further widening of the pore might be necessary for K⁺ to permeate (Sun and MacKinnon 2020). How CaM and PIP_2 modulate the voltage-dependent opening of KCNQ1 is not completely clear, but a recent study suggested that CaM and PIP_2 compete for binding to the S2–S3 linker in the voltage-sensing domain of KCNQ1 with PIP_2 stabilizing the open state with S4 in the fully-activated state.

ATP is also necessary for KCNQ1 channels to open. In the absence of PIP_2, the voltage sensor S4 still responds to voltage changes, but in the absence of ATP the pore does not open. An ATP-binding site has been proposed in the cytosolic domain at the extension of S6, perhaps holding the gate closed while still allowing S4 to move. KCNQ1 channels have a weak coupling between the voltage sensor and the gate, thereby allowing regulating molecules, such as KCNE subunits, PIP_2 and ATP, to easily alter the voltage sensor-to-gate coupling drastically. The regulation of KCNQ1 channels by ATP and PIP_2 might be important during ischemia to preserve the function of cells until the ATP and PIP_2 levels are restored following adequate blood flow to the cells.

The identification of PIP_2 as the critical nexus for KCNQ channel neuromodulation does not preclude additional signaling entities from controlling the activity of KCNQ channels. Indeed, phorbol esters, which activate protein kinase C (PKC), can inhibit the M-current and KCNQ channels. However, the possibility of a role for PKC in M-current modulation was controversial for many years due to conflicting reports. This was finally resolved following the recognition that KCNQ2 channels associate with an AKAP and more specifically AKAP5 (AKAP79/AKAP150) (Greene and Hoshi 2017). AKAP5 tethers PKC to KCNQ2 channels, allowing PKC to phosphorylate KCNQ2 channels. PKC binding to AKAP5 prevents access of PKC inhibitors that target the PKC catalytic domain, explaining earlier conflicting results. PKC can phosphorylate a couple of serines of the rat KCNQ2 C-terminus (S534, S541) (Figure 6.4C). S541 is critical, as its mutation to alanine prevents modulation of KCNQ2 channels by muscarinic receptor agonists. Consistent with the in vitro findings, *Kcnq2* knock-in mice with S559 (mouse homolog; 558 human homolog) replaced with alanine are protected from seizures induced by muscarinic partial agonists (Greene, Kosenko, and Hoshi 2018). This result led to the question: If PIP_2 depletion underlies neuromodulation, how does the absence of PKC sites in KCNQ2 prevent KCNQ2 neuromodulation? It turns out that the PKC site controls the sensitivity of KCNQ2 channels to PIP_2. Thus, PKC phosphorylation weakens the interaction of the KCNQ2 C-terminus with CaM, which in turn lowers the affinity of KCNQ2 channels for PIP_2 and makes them more amenable to neuromodulation by muscarinic receptor activation. Thus, neuromodulation of KCNQ2 channels by Gq/11 protein-coupled receptors is due to the synergy of three players: PIP_2, CaM and PKC (Greene and Hoshi 2017).

In addition to PKC, the KCNQ2 C-terminus is the site of phosphorylation for several kinases, such as CamKII, Cdk2 and p38 MAPK. As with PKC, phosphorylation of the KCNQ2 C-terminus by the various kinases decreases the affinity of KCNQ2 channels for PIP_2, leading to decreased probability of opening, inhibition of current and greater susceptibility to neuromodulation. Unlike the aforementioned kinases, phosphorylation of

KCNQ2 channels by CK2 increases the interaction of KCNQ2 with ankyrin-G, a protein that tethers KCNQ2 channels to the axon. Additionally, CK2 controls the interaction of CaM with KCNQ2 channels, which also regulates the affinity of KCNQ2 channels to membrane-bound PIP_2 (Salzer et al. 2017).

PKA triggers upregulation of KCNQ1 channels following sympathetic stimulation, which is an important response in the heart when there is a need for increased blood flow to the body. PKA shifts the voltage dependence of opening, speeds opening and slows closing of KCNQ1/KCNE1 channels by phosphorylating S27 in the N-terminus of KCNQ1. The presence of the AKAP protein Yotiao, which binds to the C-terminus of KCNQ1, is necessary for the effect of PKA on KCNQ1. The molecular mechanism of the upregulation of KCNQ1/KCNE1 activity by S27 phosphorylation is not known (Barrese, Stott, and Greenwood 2018). The N-terminus of KCNQ2 channels can also be phosphorylated by PKA, although the impact on neurons is unclear. PKA phosphorylation of KCNQ5 channels in vascular smooth muscle leads to vasodilation. A series of PKA sites both at the N-terminus and C-terminus of KCNQ5 are critical for the modulation of KCNQ5 channels by PKA. The N-terminus PKA site is analogous to the PKA site in KCNQ1 channels. Earlier work has also shown that PKA phosphorylation of KCNQ1 channels regulates their sensitivity to PIP_2; thus, PKA, like other kinases, might control the activity of KCNQ channels by regulating their PIP_2 sensitivity. In summary, KCNQ channel activity is under the control of multiple kinases and signaling molecules that primarily control the sensitivity of KCNQ channels to PIP_2 (Barrese, Stott, and Greenwood 2018).

6.8 Cell Biology (Biogenesis and Trafficking)

There have been apparent conflicting conclusions from studies on the biogenesis and assembly of KCNQ1/KCNE1 channels, with some studies suggesting that KCNQ1 and KCNE1 assemble in the plasma membrane as transient complexes and others suggesting an assembly early in the secretory pathway necessary for trafficking to the plasma membrane. These findings could be reconciled by recent data supporting that KCNQ1/KCNE1 assembly occurs in endoplasmic reticulum–plasma membrane (ER–PM) junctions, such as at t-tubules, thereby bypassing the Golgi to reach the plasma membrane via an unconventional pathway (Oliveras et al. 2020). In this study, both KCNQ1 and KCNE1 expressed on their own predominately reside in the classic secretory pathway, but some KCNE1 reach the surface via the Golgi. When KCNQ1 and KCNE1 are coexpressed, both KCNQ1 and KCNE1 surface expression increases, which may be through an unconventional pathway at ER–PM junctions. Our knowledge about KCNQ2 and KCNQ3 is limited, but multiple studies have shown that KCNQ2 and KCNQ3 trafficking requires the presence of calmodulin (Greene and Hoshi 2017).

6.9 Channelopathies

Several hundred mutations have been identified in KCNQ1 that are associated with long QT syndrome (Nerbonne and Kass 2005). This disease is manifested by a prolonged QT interval seen on an electrocardiogram evaluating the heart (Figure 6.5A). Patients with long QT syndrome are at increased risk of developing fatal cardiac arrhythmias, such as torsade

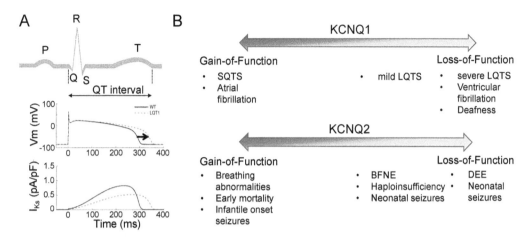

FIGURE 6.5

KCNQ channelopathies. (A) A decreased I_{Ks} (bottom) leads to a prolonged ventricular action potential (middle) and QT interval (top). (B) Spectra of loss-of-function and gain-of-function mutations in KCNQ1 and KCNQ2.

de pointes and sudden cardiac death (Nerbonne and Kass 2005). These loss-of-function (or decrease-of-function in many cases) mutations in KCNQ1 and KCNE1 cause long QT syndrome by prolonging the ventricular action potential due to a reduced outward I_{Ks} at the end of the action potential (Figure 6.5A). Homozygous or compound heterozygous loss-of-function mutations in KCNQ1 cause Jervell and Lange-Nielsen syndrome, which results in hearing loss or deafness, most likely by affecting the endolymph production in stria vascularis in the cochlea, in addition to causing a prolonged QT interval and a high risk of sudden cardiac death. Loss-of-function mutations in KCNQ1 have also been associated with epilepsy and diabetes. A few gain-of-function mutations have also been found in KCNQ1 (Figure 6.5B). These gain-of-function mutations cause short QT syndrome, which can lead to atrial fibrillation by shortening the ventricular and atrial action potentials due to an increased outward I_{Ks} at the end of the action potential. A shorter action potential in the atria increases the risk of generating reentrant excitation pathways in the atria that are substrates for atrial fibrillation. See Volume III, Chapter 5 for more details.

Hundreds of KCNQ2 and, to a lesser extent, KCNQ3 channel variants have been associated with epilepsy and neurodevelopmental disorders (see Volume III, Chapter 8) (Figure 6.5B). KCNQ2 and KCNQ3 channels were first identified in patients with self-limiting benign familial neonatal epilepsy (BFNE) (Nappi et al. 2020). In these patients, seizures occur within a few days post-birth, but resolve fairly rapidly. Most patients show no lingering long-term symptoms. However, over the last ten years, it has become clear that KCNQ2 dysfunction can also lead to very severe forms of epilepsy. Researchers now recognize that KCNQ2 channels lead to not only self-limiting BFNE but also developmental and epileptic encephalopathy (DEE) (Figure 6.5B). As with BFNE, DEE patients experience seizures soon after birth, which might resolve as the children get older. However, developmental cognitive delays persist despite the absence of seizures. A main distinction between KCNQ2 BFNE and DEE is the nature of the KCNQ2 mutations. KCNQ2 DEE variants are typically dominant-negative loss-of-function variants, that is, one dysfunctional subunit can lead to a decrease in KCNQ2/3 activity. The pore module, voltage sensor, and intracellular segments that bind calmodulin are hotspots for DEE variants. The majority of the variants are loss of function; however, a subset of variants are gain-of-function mutations. Loss of function and gain of function lead to very different clinical presentations. Unlike loss-of-function variants, patients with gain-of-function variants do not exhibit seizures when they are born but develop seizures later in life. Most gain-of-function KCNQ2 variants result in respiratory problems and myoclonus soon after birth. KCNQ2 gain of function is much more severe than KCNQ2 loss of function in patients, leading to very poor prognosis and death. In contrast to KCNQ2 DEE, self-limiting BFNE is due to haploinsufficiency, which might explain its resolution later in life, as KCNQ2 and KCNQ3 wild-type levels might increase, overcoming the early KCNQ dysfunction (Nappi et al. 2020).

Variants of KCNQ4 and KCNQ5 also lead to neurological and neuropsychiatric disorders. KCNQ4 was first identified in patients with autosomal dominant-negative deafness (DFNA2A) that results in high-frequency sensorineural hearing loss. KCNQ4 variants are loss of function, thus decreasing the low-voltage-activated K^+ current found in hair cells in a dominant-negative manner (Jentsch 2000). KCNQ5 is the only KCNQ family member that was not associated with a disorder when it was first identified. However, recent work has implicated KCNQ5 channels in epileptic encephalopathy. In particular, several patients have been identified that have either loss- or gain-of-function variants in KCNQ5 channels (Nappi et al. 2020).

6.10 Conclusions

KCNQ channels are a diverse group of K^+ channels with a wide variety of properties and physiological functions. Moreover, some members are further regulated by KCNE beta subunits that drastically alter the functional properties of the channels. Mutations in KCNQ channels are associated with a range of diseases, such as cardiac arrhythmia, deafness and epilepsy, further emphasizing the important roles this family of K channels plays in the body.

Suggested Reading

This chapter includes additional bibliographical references hosted only online. Please visit https://www.routledge.com/9780367538163 to access the additional references for this chapter, found under "Support Material" at the bottom of the page.

Abbott, G. W. 2020. "KCNQs: ligand- and voltage-gated potassium channels." *Front Physiol* 11:583. doi:10.3389/fphys.2020.00583.

Barrese, V., J. B. Stott, and I. A. Greenwood. 2018. "KCNQ-encoded potassium channels as therapeutic targets." *Annu Rev Pharmacol Toxicol* 58:625–48. doi:10.1146/annurev-pharmtox-010617-052912.

Barro-Soria, R., R. Ramentol, S. I. Liin, M. E. Perez, R. S. Kass, and H. P. Larsson. 2017. "KCNE1 and KCNE3 modulate KCNQ1 channels by affecting different gating transitions." *Proc Natl Acad Sci U S A* 114 (35):E7367–E76. doi:10.1073/pnas.1710335114.

Battefeld, A., B. T. Tran, J. Gavrilis, E. C. Cooper, and M. H. Kole. 2014. "Heteromeric Kv7.2/7.3 channels differentially regulate action potential initiation and conduction in neocortical myelinated axons." *J Neurosci* 34 (10):3719–32. doi:10.1523/JNEUROSCI.4206-13.2014.

Brown, D. A., and P. R. Adams. 1980. "Muscarinic suppression of a novel voltage-sensitive K+ current in a vertebrate neurone." *Nature* 283 (5748):673–6. doi:10.1038/283673a0.

Delmas, P., and D. A. Brown. 2005. "Pathways modulating neural KCNQ/M (Kv7) potassium channels." *Nat Rev Neurosci* 6 (11):850–62. doi:10.1038/nrn1785.

Greene, D. L., and N. Hoshi. 2017. "Modulation of Kv7 channels and excitability in the brain." *Cell Mol Life Sci* 74 (3):495–508. doi:10.1007/s00018-016-2359-y.

Greene, D. L., A. Kosenko, and N. Hoshi. 2018. "Attenuating M-current suppression in vivo by a mutant Kcnq2 gene knock-in reduces seizure burden and prevents status epilepticus-induced neuronal death and epileptogenesis." *Epilepsia* 59 (10):1908–18. doi:10.1111/epi.14541.

Jentsch, T. J. 2000. "Neuronal KCNQ potassium channels: physiology and role in disease." *Nat Rev Neurosci* 1 (1):21–30.

Li, X., Q. Zhang, P. Guo, J. Fu, L. Mei, D. Lv, J. Wang, D. Lai, S. Ye, H. Yang, and J. Guo. 2021. "Molecular basis for ligand activation of the human KCNQ2 channel." *Cell Res* 31 (1):52–61. doi:10.1038/s41422-020-00410-8.

Melman, Y. F., A. Krumerman, and T. V. McDonald. 2002. "A single transmembrane site in the KCNE-encoded proteins controls the specificity of KvLQT1 channel gating." *J Biol Chem* 277 (28):25187–94.

Miceli, F., M. V. Soldovieri, P. Ambrosino, L. Manocchio, I. Mosca, and M. Taglialatela. 2018. "Pharmacological targeting of neuronal Kv7.2/3 channels: a focus on chemotypes and receptor sites." *Curr Med Chem* 25 (23):2637–60. doi:10.2174/0929867324666171012122852.

Naffaa, M. M., and O. A. Al-Ewaidat. 2021. "Ligand modulation of KCNQ-encoded (KV7) potassium channels in the heart and nervous system." *Eur J Pharmacol* 906: 174278. doi:10.1016/j.ejphar.2021.174278.

Nappi, P., F. Miceli, M. V. Soldovieri, P. Ambrosino, V. Barrese, and M. Taglialatela. 2020. "Epileptic channelopathies caused by neuronal Kv7 (KCNQ) channel dysfunction." *Pflugers Arch* 472 (7):881–98. doi:10.1007/s00424-020-02404-2.

Nerbonne, J. M., and R. S. Kass. 2005. "Molecular physiology of cardiac repolarization." *Physiol Rev* 85 (4):1205–53.

Oliveras, A., C. Serrano-Novillo, C. Moreno, A. de la Cruz, C. Valenzuela, C. Soeller, N. Comes, and A. Felipe. 2020. "The unconventional biogenesis of Kv7.1-KCNE1 complexes." *Sci Adv* 6 (14):eaay4472. doi:10.1126/sciadv.aay4472.

Osteen, J. D., C. Gonzalez, K. J. Sampson, V. Iyer, S. Rebolledo, H. P. Larsson, and R. S. Kass. 2010. "KCNE1 alters the voltage sensor movements necessary to open the KCNQ1 channel gate." *Proc Natl Acad Sci U S A* 107 (52):22710–5. doi:10.1073/pnas.1016300108.

Preston, P., L. Wartosch, D. Gunzel, M. Fromm, P. Kongsuphol, J. Ousingsawat, K. Kunzelmann, J. Barhanin, R. Warth, and T. J. Jentsch. 2010. "Disruption of the K+ channel beta-subunit KCNE3 reveals an important role in intestinal and tracheal Cl– transport." *J Biol Chem* 285 (10):7165–75. doi:10.1074/jbc.M109.047829.

Prole, D. L., and N. V. Marrion. 2004. "Ionic permeation and conduction properties of neuronal KCNQ2/KCNQ3 potassium channels." *Biophys J* 86 (3):1454–69. doi:10.1016/S0006-3495(04)74214-9.

Sachyani, D., M. Dvir, R. Strulovich, G. Tria, W. Tobelaim, A. Peretz, O. Pongs, D. Svergun, B. Attali, and J. A. Hirsch. 2014. "Structural basis of a Kv7.1 potassium channel gating module: studies of the intracellular c-terminal domain in complex with calmodulin." *Structure* 22 (11):1582–94. doi:10.1016/j.str.2014.07.016.

Salzer, I., F. A. Erdem, W. Q. Chen, S. Heo, X. Koenig, K. W. Schicker, H. Kubista, G. Lubec, S. Boehm, and J. W. Yang. 2017. "Phosphorylation regulates the sensitivity of voltage-gated Kv7.2 channels towards phosphatidylinositol-4,5-bisphosphate." *J Physiol* 595 (3):759–76. doi:10.1113/JP273274.

Seebohm, G., N. Strutz-Seebohm, O. N. Ureche, R. Baltaev, A. Lampert, G. Kornichuk, K. Kamiya, T. V. Wuttke, H. Lerche, M. C. Sanguinetti, and F. Lang. 2006. "Differential roles of S6 domain hinges in the gating of KCNQ potassium channels." *Biophys J* 90 (6):2235–44. doi:10.1529/biophysj.105.067165.

Soh, H., R. Pant, J. J. LoTurco, and A. V. Tzingounis. 2014. "Conditional deletions of epilepsy-associated KCNQ2 and KCNQ3 channels from cerebral cortex cause differential effects on neuronal excitability." *J Neurosci* 34 (15):5311–21. doi:10.1523/JNEUROSCI.3919-13.2014.

Sun, J., and R. MacKinnon. 2020. "Structural basis of human KCNQ1 modulation and gating." *Cell* 180 (2):340–347.e9. doi:10.1016/j.cell.2019.12.003.

Wang, C. K., S. M. Lamothe, A. W. Wang, R. Y. Yang, and H. T. Kurata. 2018. "Pore- and voltage sensor-targeted KCNQ openers have distinct state-dependent actions." *J Gen Physiol* 150 (12):1722–34. doi:10.1085/jgp.201812070.

Wang, H. S., Z. Pan, W. Shi, B. S. Brown, R. S. Wymore, I. S. Cohen, J. E. Dixon, and D. McKinnon. 1998. "KCNQ2 and KCNQ3 potassium channel subunits: molecular correlates of the M-channel." 282 (5395):1890–3.

Yazdi, S., J. Nikesjo, W. Miranda, V. Corradi, D. P. Tieleman, S. Y. Noskov, H. P. Larsson, and S. I. Liin. 2021. "Identification of PUFA interaction sites on the cardiac potassium channel KCNQ1." *J Gen Physiol* 153 (6):e202012850. doi:10.1085/jgp.202012850.

Zaydman, M. A., M. A. Kasimova, K. McFarland, Z. Beller, P. Hou, H. E. Kinser, H. Liang, G. Zhang, J. Shi, M. Tarek, and J. Cui. 2014. "Domain-domain interactions determine the gating, permeation, pharmacology, and subunit modulation of the IKs ion channel." *eLife* 3:e03606. doi:10.7554/eLife.03606.

Zaydman, M. A., J. R. Silva, K. Delaloye, Y. Li, H. Liang, H. P. Larsson, J. Shi, and J. Cui. 2013. "Kv7.1 ion channels require a lipid to couple voltage sensing to pore opening." *Proc Natl Acad Sci U S A* 110 (32):13180–5. doi:10.1073/pnas.1305167110.

7

BK channels

Jianmin Cui

CONTENTS

7.1 Functional Properties of BK Channels

Large conductance, voltage, and Ca^{2+}-activated K$^+$ channels are known as BK or MaxiK channels to denote the large single-channel conductance of 100–300 pS (Figure 7.1A). BK channel currents were first discovered in the early 1980s when the newly invented patch clamp techniques were used to record various cell types, and their large single channel currents stood out from patch clamp recordings with intracellular solutions containing Ca^{2+} (Pallotta et al., 1981). A decade later, the gene that encodes the pore-forming α subunit of BK channels, *slo1* or *KCNMA1*, was first identified in *Drosophila*, and then in mouse and human (Butler et al., 1993). BK channels activate in response to membrane depolarization and elevation of intracellular Ca^{2+} concentrations: the open probability of single BK channels increases with voltage and intracellular Ca^{2+} concentrations, while the macroscopic conductance (G) of many BK channels increases with voltage and the conductance–voltage (G–V) relation shifts toward more negative voltages with increasing intracellular Ca^{2+} concentrations (Figure 7.1). The dependence of BK channel activation on both voltage and Ca^{2+} makes the channel function versatile in two unique aspects. First, BK channel function is not defined by any single voltage or intracellular Ca^{2+} concentrations; the G–V relation of BK channels varies depending on the intracellular Ca^{2+} concentrations at which it is measured, while the dose–response curve of channel activation on intracellular Ca^{2+} concentrations also changes with voltage (Cui et al., 1997). Second, the channel function varies

DOI: 10.1201/9781003096276-7

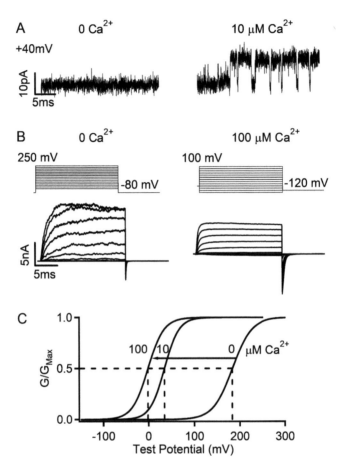

FIGURE 7.1

Voltage and Ca²⁺ dependent activation of BK channels. A) Single-channel currents of a BK channel with Slo1 and β4 subunits showing activation by intracellular Ca²⁺. Lower current: 0. B) Macroscopic currents of Slo1 channels in response to the voltage pulses and intracellular Ca²⁺ concentrations. C) Increasing intracellular Ca²⁺ concentrations shifts the conductance–voltage (*G–V*) relation to more negative voltages.

with voltages changing from –200 to 300 mV and intracellular Ca²⁺ concentration variations from 0 to 10 mM that cover all possible physiological and pathological conditions.

Besides Ca²⁺, various physiologically important intracellular ions and molecules also alter BK channel activation. Intracellular Mg^{2+} of millimolar physiological concentrations activates BK channels (Shi and Cui, 2001). The probability of single-channel openings increases, and the *G–V* relation of macroscopic currents shifts to more negative voltage ranges with increasing $[Mg^{2+}]_i$. Although the effects of Mg^{2+} on BK channel activation are similar to that of Ca²⁺, the underlying mechanisms of Ca²⁺- and Mg^{2+}-dependent activation are distinct. Other divalent cations also activate BK channels with the effectiveness Cd^{2+} > Sr^{2+} > Mn^{2+} > Fe^{2+} > Co^{2+} (Oberhauser et al., 1988). Ba^{2+}, an ion that effectively blocks BK channels, was found to also activate BK channels. Divalent ions, proton H^+, and intracellular molecules other than ions such as alcohol and PIP_2 also enhance BK channel activation.

In addition to activating BK channels, intracellular divalent cations also block BK channels. Single channel current–voltage relation of BK channels deviates from a straight line and the current becomes smaller at positive voltages (inwardly rectifying), which is more prominent in intracellular Ca²⁺ concentrations ≥ 100 μM (Cox et al., 1997b). Likewise,

Mg^{2+} also blocks the channel. These results suggest that Ca^{2+} and Mg^{2+} may reach a site in the channel pore or intracellular vestibule from the intracellular side; the blocking and unblocking events happen fast and cannot be resolved by the recording system. On the other hand, Ba^{2+} blocks BK channels by occupying sites for conducting ions (Neyton and Miller, 1988) to cause blocking events longer than the close time of normal gating. Since the blocking of BK channels by Ca^{2+} is prominent only at intracellular Ca^{2+} concentrations higher than 100 µM and above +50 mV, at physiological conditions, Ca^{2+} increases BK channel current by activation. On the other hand, Mg^{2+} activates as well as blocks BK channels in physiological conditions so that the influence of Mg^{2+} on BK channel currents is more complex.

7.2 Physiological Roles of BK Channels

BK channels are expressed throughout major tissues of the body and play important roles in controlling many physiological processes such as neuronal excitability, neurotransmitter release, muscle contraction, secretion, and endothelial functions. BK channels have also been found to regulate membrane potential and Ca^{2+} stores of intracellular organelles, including mitochondria, nucleus, and lysosome. These diverse roles originated from the unique large conductance and dual sensitivity to membrane potentials and intracellular Ca^{2+} concentration of BK channels. BK channels in the plasma membrane are located spatially proximal to various cell surface or intracellular Ca^{2+} channels including voltage-gated Ca^{2+} (Ca_V) channels, N-methyl-d-aspartate (NMDA) receptors, transient receptor potential (TRP) channels, inositol trisphosphate receptors (IP3R), and ryanodine (RyR) receptors (Figure 7.2). Thus, BK channels can quickly sense the elevation of intracellular Ca^{2+} level and membrane depolarization, and the large single-channel conductance of BK channels

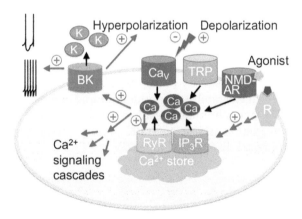

FIGURE 7.2
Physiological roles of BK channels. In cells, BK channels are activated by membrane depoarlzation and intracellular Ca^{2+} that enters cytosol via various sources. The activation of BK channels regulates neuronal excitability and action potential firing frequency, intracellular Ca^{2+} by deactivating the Ca_V channels with a negative-feedback mechanism, and transporting K^+. BK: BK channels; Ca_V: voltage-gated Ca^{2+} channels; TRP: transient recepror potential channels; NMDAR: N-methyl-D-aspartate receptor; RyR: ryanodine receptor; IP$_3$R: inositol 1,4,5-triphosphate receptor; R: membrane receptors. The red circled "+" and blue circled "–" indicate increasing and decreasing, respectively.

enables rapid K^+ efflux, thereby effectively hyperpolarizing the membrane. Because of these properties BK channels contribute to physiological processes with three major mechanisms (Figure 7.2). First, BK channels regulate excitability and firing behaviors in neurons. BK channels rapidly activate during the upstroke of an action potential when the membrane potential is depolarized and intracellular Ca^{2+} concentration is raised due to the opening of Ca_V channels, and quickly deactivate when the membrane potential drops to a more negative range. In this way, BK channels help repolarize the action potential and contribute to the generation of fast after-hyperpolarization (fAHP) that usually lasts 1–10 ms. Since AHP is a major determinant of the refractory period and therefore the firing frequency of a neuron, BK channels are important in controlling the shape and frequency of action potentials in various neurons (Hu et al., 2001). With this mechanism BK channels are involved in the control of movements (Dong et al., 2022) and circadian rhythms (Harvey et al., 2020). Second, BK channels reduce intracellular Ca^{2+} concentration via a negative feedback mechanism to regulate Ca^{2+} dependent physiological processes. In neurons BK channels activate by Ca^{2+} entry from Ca_V channels, which hyperpolarizes the membrane. Membrane hyperpolarization in turn deactivates Ca_V channels to stop further Ca^{2+} entry and thereby reduce neurotransmitter release. In smooth muscle cells Ca^{2+} sparks released from intracellular Ca^{2+} store via ryanodine receptors (RyRs) activate BK channels to elicit spontaneous transient outward currents (STOCs), which opposes pressure-induced depolarization and contraction in arterial, urinary, and tracheal smooth muscle (Nelson et al., 1995). Third, BK channels serve as K^+ transporters in various types of cells including epithelial, endothelial, and glial cells. In airway epithelium BK channels play a significant role in mucociliary clearance by regulating membrane potential and Cl^- secretion (Kis et al., 2016). In the kidney BK channels mediate flow-induced K^+ secretion (FIKS) in the distal nephron in response to increased fluid movements (Carrisoza-Gaytan et al., 2016).

Consistent with their important roles in many physiological processes, BK channels are involved in various pathological conditions. Based on recent studies of human patients and rodent models, BK channels are implicated in the pathogenesis of several diseases including epilepsy, cerebellar ataxia, autism and mental retardation, stroke, hypertension, asthma, diabetes, obesity, and tumor progression.

The first human BK channelopathy due to a missense mutation of Asp-to-Gly (D434G) in the cytosolic RCK1 domain (Figure 7.3) of Slo1 channel was identified from patients who have a syndrome of coexistent generalized epilepsy and paroxysmal dyskinesia (GEPD) (Du et al., 2005). By reducing the flexibility of a key region in the RCK1 domain to enhance its apparent Ca^{2+} sensitivity (Yang et al., 2010), this gain-of-function mutation leads to increased neuronal excitability by inducing rapid repolarization of action potential and fast firing frequency, ultimately resulting in the GEPD syndrome (Dong et al., 2022). Recently, more congenital and de novo BK channel mutations have been identified to associate with neurological disorders (Cui, 2021).

7.3 BK Channel Structure

Four Slo1 subunits form a functional BK channel. Auxiliary β and γ subunits associate with the Slo1 channel to modify gating mechanisms (Figure 7.3). A Slo1 subunit is composed of an N-terminal membrane-spanning domain (MSD) and a large cytosolic domain (CTD) at the C-terminus that consist of almost two-thirds of the entire sequence (Figure 7.3A). The MSD includes the transmembrane (TM) segments S0–S6, with S1–S4 being conserved as

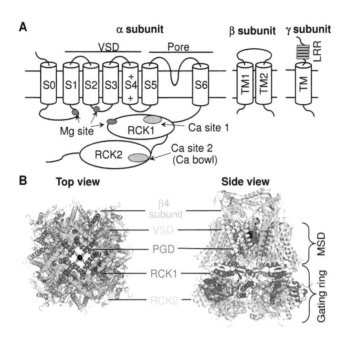

FIGURE 7.3

BK channel structure. A) Diagram of the Slo1, β, and γ subunits of BK channels. B) Structure of the BK channel with human Slo1 and β4 subunits (PDB ID: 6V22). Top view as seen from the extracellular side. Orange and majenta dots are Ca^{2+} and Mg^{2+} ions, respectively. TM, transmembrane segment; RCK, regulators of K^+ conductance; LRR, leucine-rich-repeat; VSD, voltage sensor domain; PGD, pore-gate domain; MSD, membrane-spanning domain.

the voltage sensor domain (VSD) and S5–S6 as the pore-gate domain (PGD). Unlike most 6-TM voltage-gated K^+ (Kv) channels, the MSD of BK channels has an additional S0 segment that is linked to the VSD through a long intracellular S0–S1 linker, positioning the N-terminus of Slo1 to the extracellular side. S5–S6 from all four subunits forms a central pore, which is surrounded by four S0-and-VSDs at the periphery (Tao and MacKinnon, 2019) (Figure 7.3B). In most Kv channels the VSD and the PGD are domain swapped, i.e., the VSD of one subunit is in contact with the PGD of a neighboring subunit, but in BK and a number of other Kv channels the VSD makes contact with the PGD of the same subunit without domain swapping. Four β subunits are found in each BK channel in association with the VSD of Slo1 (Figure 7.3B).

X-ray crystallography and cryo-EM structures (Tao and MacKinnon, 2019) reveal that the CTD of each Slo1 subunit contains two regulators of K^+ conductance (RCK1 and RCK2) domains, and the eight RCK domains from four subunits form a ring-like structure known as the gating ring (Figure 7.3B). In the gating ring, each RCK1 domain makes contact with two RCK2 domains. The RCK1 of each subunit also interacts non-covalently with the bottom of the VSD of a neighboring subunit with a domain swap (Tao and MacKinnon, 2019; Yang et al., 2008). In addition, a peptide linker (C linker) of 16 amino acids covalently connects the end of S6 in the MSD to the beginning of RCK1 in each Slo1 subunit. The RCK1 and RCK2 domains in each Slo1 subunit are connected by a peptide linker that is not conserved among BK channel orthologs. A cut could be made in the linker between RCK1 and RCK2, and the resulting two pieces of the Slo1 subunit coexpressed to form functional channels, indicating that the free RCK2 can associate with RCK1 to endow correct structure and function.

The structures either with Ca^{2+} and Mg^{2+} bound or metal free were solved using cryo-EM (Tao and MacKinnon, 2019). A Ca^{2+}-binding site, known as the Ca^{2+} bowl, is identified to reside in the RCK2 domain. The bound Ca^{2+} is coordinated by side-chain carboxylates from residues D898 and D900, as well as main chain carbonyl groups from residue Q892 and D895 (residue numbers in mSlo1, GenBank accession number 347143). Another Ca^{2+}-binding site is formed by residues D367 and E535 and the main chain carbonyl group from R514 in the RCK1 domain. The Mg^{2+}-binding site is formed by residues in the MSD of one subunit, N172, and the RCK1 of a neighboring subunits, E374 and E399. The metal-binding sites in the structure are consistent with that identified by mutagenesis studies in which neutralization of acidic residues reduced Ca^{2+} or Mg^{2+} sensitivity (Yang et al., 2008; Schreiber and Salkoff, 1997; Zhang et al., 2010). A large conformational rearrangement is observed at the interface between the CTD and the VSD comparing the Ca^{2+}/Mg^{2+} bound and the metal-free structures.

7.4 Allosteric Mechanisms of BK Channel Activation

Although voltage-dependent activation of BK channels is affected by Ca^{2+} and vice versa, the mechanisms of voltage- and Ca^{2+}-dependent activation are distinct (Cui et al., 1997). Voltage can activate BK channels in the absence of Ca^{2+} binding, and likewise, Ca^{2+} can also activate BK channels when the voltage sensors are at the resting state. In the absence of Ca^{2+} binding, BK channels are activated only when voltage is extremely positive. For instance, the macroscopic currents of BK channels formed by homogeneous mSlo1 are measurable at about +110 mV, and the $G-V$ relation reaches half-maximum at the voltage ($V_{1/2}$) of ~180 mV (Figure 7.1C). On the other hand, intracellular Ca^{2+} concentrations increase from 0 to ~100 µM that is saturating for the two Ca^{2+}-binding sites in the gating ring can enhance BK channel open probability by four orders of magnitude from ~10^{-7} to 10^{-3} when the voltage sensors are at the resting state.

For *Shaker* type Kv channels, activation is strictly dependent on voltage; the logarithm of the probability of channel opening, $logP_o$, exhibits a linear dependence on voltage at very negative voltages where P_o is ~10^{-7}. Such a relationship suggests that the coupling between the voltage sensor and the pore is obligatory, that is, the channel opens when the voltage sensor is activated, and the channel closes when the voltage sensor stays at resting state. Contrary to *Shaker*, in BK channels, the coupling of voltage sensor or Ca^{2+} binding with channel opening is not obligatory. The BK channel has an intrinsic open probability of ~10^{-7} even in the absence of voltage sensor activation and Ca^{2+} binding. Because of this intrinsic open probability, at negative potentials, $logP_o$ of BK channels deviates from a linear relation with voltage, becoming flat as P_o reaches 10^{-7} while voltage further hyperpolarizes (Horrigan et al., 1999). Therefore, instead of being obligatory, the coupling of voltage or Ca^{2+} with the activation gate in BK channels is allosteric. Voltage sensor activation and Ca^{2+} binding promote opening of the channel, but the probability of channel opening is not bound to the state of the voltage sensor or the occupancy of Ca^{2+} binding. It is because of such an allosteric mechanism that either voltage or Ca^{2+} can activate BK channels independently. In a channel with obligatory coupling between the VSD and the activation gate, Ca^{2+} may not alter open probability at any given voltage unless it alters voltage-dependent activation of the VSD.

The allosteric mechanism of voltage- and Ca^{2+}-dependent activation of BK channels can be quantitatively described by Schemes I and II, respectively (Horrigan et al., 1999; Cox et

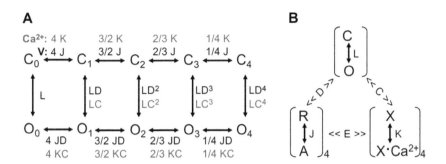

A

B

FIGURE 7.4

Allosteric gating mechanisms of BK channels. A) An allosteric model of voltage (Scheme I, parameters in black)- or Ca^{2+}(Scheme II, parameters in blue)-dependent gating. B) A general allosteric gating mechanism (Scheme III) including allosteric interactions between Ca^{2+} binding, voltage sensor activation, and channel opening. C and O: closed and open conformations of the channel, respectively; L: the equilibrium constant for the C–O transition; J: the equilibrium constant for voltage sensor activation; D: allosteric factor describing interaction between channel opening and voltage sensor activation; K_C: equilibrium constant for Ca^{2+} binding to closed channels; C: allosteric factor describing interaction between channel opening and Ca^{2+} binding; E: allosteric factor describing interaction between voltage sensor activation and Ca^{2+} binding.

al., 1997a) (Figure 7.4A). In Scheme I, voltage sensor activation occurs in both the closed (C) and open (O) states, and the channel can open when 0, 1, 2, 3, or all 4 voltage sensors are activated. The voltage sensor activation, however, is more favored in the open state, with an equilibrium constant for activation larger than that in the closed state, differing with an allosteric factor D. Similarly, channel opening is favored by voltage sensor activation; with each additional voltage sensor activated, the equilibrium constant for channel opening is increased by a D factor. Such an allosteric mechanism can well fit the gating current data that measure voltage sensor activation as well as the ionic current data that measure channel opening in the voltage range of –450 to 300 mV. Likewise, in Scheme II, Ca^{2+} can bind to both open and closed states with different affinities, and the ratio of these is called the allosteric factor C. Ca^{2+} binding favors channel opening by the factor C. This scheme could fit the data of BK channel currents with various intracellular Ca^{2+} concentrations. More careful studies reveal that voltage sensor activation is influenced by Ca^{2+} binding, and vice versa, also with an allosteric mechanism. Thus, BK channel activation by voltage and Ca^{2+} is governed by all three sets of allosteric mechanisms integrated in Scheme III (Figure 7.4B) (Horrigan and Aldrich, 2002). Scheme III can fit the results of gating current and ionic current measurements at all intracellular Ca^{2+} concentrations and voltages within the experimental limitations. It is noteworthy that the allosteric model of Ca^{2+}-dependent activation assumes only four identical Ca^{2+}-binding sites. The model would be more complex if two different Ca^{2+}-binding sites in each Slo1 subunit are considered. Nevertheless, the principle of allosteric mechanism should not change with increased complexities.

7.5 Structural Basis of BK Channel Function

7.5.1 Large Single-Channel Conductance

BK channels have the largest single-channel conductance of all K^+ channels although the sequence of the selectivity filter is highly conserved within the K^+ channel superfamily.

Part of such a large conductance is derived from two clusters of acidic residues that are located at the intracellular and extracellular vestibules within the K^+ permeation pathway. These clusters of negative charges thus serve as electrostatic traps to attract and concentrate local K^+, thereby increasing single-channel conductance (Brelidze et al., 2003). At the intracellular vestibule, E321 and E324, which are conserved and specific to Slo1 located in the cytosolic end of the S6 segment, form a ring of eight negative charges. Reducing the negative charges and increasing positive charges by mutations gradually reduce single channel conductance by up to 50%. The effects of these charge changes on single-channel conductance are diminished in the presence of high $[K^+]_i$ (~1 Molar), and these mutations only reduce the conductance of the outward current but not the inward current, supporting the mechanism that the negative charges attract K^+ to enhance the probability of K^+ entering the channel. Likewise, conserved D261/E264 in the extracellular vestibule of Slo1 also contributes to ~18% of BK channel single channel conductance, and the contribution of these negative charged residues to BK channel conductance in both intracellular and extracellular vestibules can be explained by their contributions to the electrostatic potentials at the intracellular and extracellular entrance of the pore, respectively. Nevertheless, these charges in the vestibules are not the sole determinant of the large single-channel conductance as the single-channel conductance of the mutant BK channel in the absence of these negative charges is still about six times larger than that of Shaker K^+ channels. This suggests that other structural aspects specific to the BK channel pore may also be important for ion conduction.

7.5.2 Voltage- and Ca²⁺-Dependent Activation

Ion channels sense physiological stimuli to open and close, thereby regulating physiological processes. Channel activation generally involves three principal steps: sensor of stimuli changes conformation, the conformational change is propagated to the activation gate, and then the gate opens. Similar to other Kv channels, the voltage sensor of BK channels contains S1–S4 TM segments. Basic residues R167 in S2 and R213 in S4 and acidic residues D153 in S2, D186 in S3, and E219 in S4 are gating charges (Ma et al., 2006). This differs from other Kv channels where the gating charges are primarily basic residues found in S4. BK channels also contain an additional S0 that associates and interacts with the VSD. These features implicate unique movements of BK channel voltage sensors during activation (Ma et al., 2006; Pantazis et al., 2010). Ca^{2+} binding to either the RCK1 or RCK2 site can activate the channel; when mutations destroy one of the Ca^{2+}-binding sites the remaining site still activates the channel with a Ca^{2+} sensitivity of 40%–60% of the wildtype channel (Yang et al., 2010), measured by the voltage of *G–V* shifting in response to saturating intracellular Ca^{2+} concentrations (≥100 μM). However, the two sites may not be completely independent in activating the channel: the Ca^{2+} sensitivities contributed by each individual site do not add up to that of the wild-type channel, indicating that the two sites are cooperative in activating the channel.

   Indentical to other Kv channels, the pore of BK channels is formed by S5–S6 and contains a highly conserved K^+ selectivity filter. For most K^+ channels, the intracellular bundle crossing formed by the cytosolic side of four inner helices acts as the activation gate that restricts K^+ ion flux when the channels are closed. However, when the BK channel is closed, the entrance of the K^+ ion to the pore cavity from inside of the cell may not be restricted at the cytosolic side of S6, because even blockers and methanethiosulfonate (MTS) reagents with larger sizes can enter the cavity (Li and Aldrich, 2004). Cysteine scanning and modification results suggest that the S6 α-helix in BK channels may be disrupted at the unique diglycine motif G310/G311, resulting in a wider opening of the inner pore compared to other Kv channels that usually have a single glycine at the *glycine hinge* position and narrower opening at the bundle crossing. Therefore, with the disrupted S6, the permeation gate for K[+] in BK channels was suggested to locate at the selectivity filter. More recently, a molecular dynamic simulation of the BK channel structures in the Ca^{2+} and Mg^{2+} bound and free states suggested that the closed state of BK channels lacks a physical blockage, but seems to stop ionic flow with a vapor barrier generated by hydrophobic dewetting of the pore. Movement of S6 helices upon removal of Ca^{2+} and Mg^{2+} binding results in conformational changes that mainly involve an amphipathic segment of S6, $V_{319}PEIIE_{324}$, to expose hydrophobic residues V319 and I323 to the pore inner surface. The pore thus becomes not only more hydrophobic, but also narrower and more elongated, promoting dewet transitions that completely deplete liquid water. The dry pore stops ionic flow through the channel as a hydrophobic gate although it remains physically open with an average diameter of ~6 Å. This explains the experimental observation that moderately sized quaternary ammonium compounds can access the deep-pore region to block the channel even in the closed state during membrane repolarization (Jia et al., 2018). Both voltage sensor movement and Ca^{2+} binding regulate the opening and closing of this gate.

In Kv channels with the VSD and the PGD domain swapped the S4–S5 linker interacts with the cytosolic part of S6 that propagates the S4 movement to open the activation gate. BK channels do not show a VSD-PGD domain swap, while the S4-S5 linker in BK channel structures is not physically close to S6. Instead, the S4 and S5 helices in the same subunit show an extensive contact in BK channel structures. These interactions may be important in the coupling between the VSD and the PGD in the voltage dependent opening of the pore, and vice versa, in restricting the VSD movements in BK channels.

The cytosolic gating ring is important for coupling the activation gate with both voltage sensor movements and Ca^{2+} binding. Conformational changes of the gating ring upon Ca^{2+} binding are proposed to pull the gate open (Niu et al., 2004) via the C-linker or to alter the interactions between the C-linker with the CTD or the membrane. In addition, noncovalent interactions between the MSD and the cytosolic gating ring may be also involved in the coupling of the activation gate with voltage sensor movements and Ca^{2+} binding (Tao and MacKinnon, 2019; Geng et al., 2020; Zhang et al., 2022). The mechanism of coupling remains an active area of research.

7.5.3 Activation by Other Intracellular Ions

Both Mg^{2+} and Ca^{2+} at millimolar concentrations can bind to the Mg^{2+} site and activate BK channels, though at physiological conditions, this site is more likely occupied by Mg^{2+}. The bound Mg^{2+} is located in close proximity to the gating charge R213 at the cytoplasmic side of S4 in the VSD (Yang et al., 2008). The electrostatic interaction between Mg^{2+} and R213 stabilizes the voltage sensor at the activated state. Thus, Mg^{2+} binding activates the channel primarily by promoting voltage sensor activation. Mg^{2+} also binds to the high-affinity Ca^{2+}-binding sites to compete with Ca^{2+}, thereby reducing the Ca^{2+}-dependent activation of the channel (Shi and Cui, 2001). However, the binding of Mg^{2+} to the Ca^{2+} sites per se does not seem to activate the channel.

Other divalent cations including Sr^{2+}, Cd^{2+}, Mn^{2+}, Co^{2+}, and Ni^{2+} can all bind to the Mg^{2+}-binding site to activate the channel. Sr^{2+} also binds to the two Ca^{2+}-binding sites to activate BK channels. On the other hand, Ba^{2+} activates BK channels by specifically binding to the Ca^{2+} bowl but not the RCK1 site or the Mg^{2+} site. Cd^{2+} is different from other divalent cations, which seem to activate the channel by binding to a unique site that differs from any of the Ca^{2+}- and Mg^{2+}-binding sites, although the site for Cd^{2+} may involve D367 in the RCK1 Ca^{2+} site. Proton activates the channel by binding H365 and H394 in the RCK1 domain and electrostatically interacting with D367 in the RCK1 high-affinity Ca^{2+}-binding site. Thus, both Cd^{2+} and H^+ may open the activation gate via the same mechanism as Ca^{2+} via the RCK1 Ca^{2+} site.

7.6 β and γ Subunits Modulate BK Channel Function

Four β (β1–β4) subunits (Orio et al., 2002) and five γ (γ1–γ5) subunits (Dudem et al., 2020; Yan and Aldrich, 2012) of BK channels have been identified. These auxiliary subunits show distinct tissue distribution and modulate BK channel function differently, thus providing a major mechanism for the diverse BK channel phenotypes observed in various tissues. All β (KCNMB) subunits share 20%–53% of sequence identity and a similar membrane topology with two TM segments (TM1 and TM2), a large extracellular loop (116–128 amino acid residues), and two cytosolic termini. All γ subunits belong to a large leucine-rich-repeat-containing (LRRC) protein family that contains a single TM segment with a large extracellular leucine-rich-repeat (LRR) motif, and a short cytosolic COOH tail (Figure 7.3A). It is not known how γ subunits associate with Slo1.

All β subunits except for β3 alter voltage sensor movements by stabilizing the activated state; the β1 and β2 subunits shift the voltage dependence of gating charge movement, the $Q–V$ relation, toward less positive voltages, while β4 reduces the slope of $Q–V$ relation. The β1, β2, and β4 subunits change Ca^{2+} sensitivity of BK channel activation, such that the $G–V$ relation shifts more toward negative voltages in response to Ca^{2+} concentration increase. In the absence of voltage sensor activation and Ca^{2+} binding, the β1, β2, and β4 subunits also reduce the intrinsic open probability of the channel by as much as an order of magnitude.

Apparently, the β1, β2, and β4 subunits all alter BK channel gating with similar mechanisms. However, the effects of these β subunits on BK current differ significantly. For instance, neither the β1 nor β2 subunit shifts the $G–V$ relation by a large voltage at zero intracellular Ca^{2+} concentrations. Due to the increase of Ca^{2+} sensitivity, both the β1 and β2 subunits increase BK activation at physiological voltages and intracellular Ca^{2+} concentrations. On the other hand, since the β4 subunit shifts the $G–V$ relation to more positive voltages at 0 intracellular Ca^{2+} concentrations it inhibits BK currents at intracellular Ca^{2+} concentrations ≤ 10 μM, while enhancing BK currents at intracellular Ca^{2+} concentrations ≥ 10 μM due to a higher Ca^{2+} sensitivity. In addition, β4 subunit slows down channel activation. Therefore, at physiological voltages and intracellular Ca^{2+} concentrations, the β4 subunit reduces the function of BK channels. Besides enhancing Ca^{2+} sensitivity, the β2 subunit also inactivates BK channels with a *ball and chain* mechanism such that the N-terminus of β2 blocks the pore during channel opening to reduce BK currents.

All γ subunits alter voltage dependence by shifting $G–V$ relation to more negative voltages. The amount of the shift caused by various γ subunits is γ1(LRRC26) > γ2 (LRRC52) > γ3 (LRRC55) ~ γ5 (LINGO-1) > γ4 (LRRC38) (Dudem et al., 2020; Yan and Aldrich, 2012). The shift of the $G–V$ relation caused by the association of the γ1 subunit is so large (approximately –140 mV) that the BK channel opens at the physiological voltages even when the intracellular Ca^{2+} concentrations are at the resting level. The γ subunits do not alter Ca^{2+} sensitivity of channel activation, and the overall effect of these auxiliary subunits is to enhance BK channel currents except for γ5 (LINGO-1), which reduces BK channel currents by causing a rapid inactivation.

7.7 Pharmacology of BK Channels

BK channels are modulated by various toxins, peptides, and small molecules. BK channels are inhibited by the K^+ channel blocker Tetraethylammonium (TEA) and charybdotoxin (ChTX), which is a peptide toxin in scorpion venom. While ChTX is a potent BK channel inhibitor, it also inhibits other K^+ channels. More specific inhibitors of BK channels have been identified and served as powerful tools in studying tissue distribution and physiological roles of BK channels as well as the structure and molecular mechanisms of BK channel function. For example, iberiotoxin (IbTX) (Galvez et al., 1990), another polypeptide isolated from the venom of the scorpion, is a commonly used potent and specific blocker of BK channels. Paxilline (PAX), an alkaloid from the tremorgenic fungi, is also a commonly used specific BK channel inhibitor, with a half inhibition concentration (IC_{50}) of the Slo1 channel around 10 nM. PAX inhibits other ion channels or pumps at higher concentrations. A single PAX molecule has been shown in the inner pore cavity and binding at the crevice between the pore helix and S6 of one of the four subunits just beneath the selectivity filter. PAX binding is dependent on the conformation around G311 at the closed state. It no longer binds to the site when the channel is open, consistent with the observation that PAX is a closed-channel blocker (Zhou et al., 2020). The mechanism of PAX inhibition corroborates with the structure of the closed BK channel, in which the inner pore is physically open. Some inhibitors are selective to BK channels associated with specific auxiliary subunits. ChTX and IbTX association to BK channels formed with the Slo1 and the β4 subunits is slowed by ~1,000 fold as compared to the Slo1 channels (Meera et al., 2000). On the other hand, another scorpion venom toxin, martentoxin, inhibits BK channels containing the β4 subunits with high potency but the BK channels formed by Slo1 subunits alone are not sensitive. Such inhibitors also include the conopeptide, Vt3.1, that preferentially inhibits BK channels associated with the β4 subunit. Vt3.1 interacts with residues both from the Slo1 and the β4 subunits, and the identification of these interactions predicted the structural feature of the β4 subunit with the extracellular loop extending from the periphery of the channel to central pore (Li et al., 2014) (Figure 7.3B).

BK channel openers (or activators), such as the benzimidazalone derivative NS1619 and the biarylthiourea derivative NS11021, are also powerful tools in studying physiological roles and molecular mechanisms of function (Cui, 2020). BK channels are activated by voltage and intracellular Ca^{2+} by allosteric mechanisms (Figure 7.4). BK channel openers may interact with the channel protein and alter the voltage or Ca^{2+} sensors, the allosteric coupling between sensors and the pore/gate, or the pore/gate to enhance channel activation. NS11021 may enter the BK channel pore to enhance the intrinsic opening of BK channels (Rockman et al., 2020). NS1619 may act like another BK channel opener, the dehydroabietic acid derivative Cym04 that alters the interaction between the pore and the RCK1 domain via the C-linker (Figure 7.3) to stabilize voltage sensor activation as well as intrinsic pore opening (Gessner et al., 2012). Similarly, GoSlo-SR-5-6, a member of the anthraquinone analogs known as GoSlo BK channel openers, may interact with the cytosolic end of S6 to stabilize voltage sensor activation and pore opening (Webb et al., 2015). The arginine derivative BC5 is an allosteric agonist of BK channels recently identified using in silico docking and experimental screening. BC5 binds to the interface between the VSD and RCK1 domain (Figure 7.3) to enhance the coupling between Ca^{2+} binding and pore opening (Zhang et al., 2022). Some BK channel openers also exhibit subunit specificity. The long-chain polyunsaturated omega-3 fatty acid docosahexaenoic acid (DHA) activates BK channels composed of Slo1 with β1 or β4 subunits more than the channels composed of

Slo1 with β2, γ1 or Slo1 only (Hoshi et al., 2013). Mallotoxin (also known as rottlerin), isolated from the powder on the fruit of the kamala tree, is a potent BK opener, but its effect is attenuated by the association of γ1-3 subunits (Guan et al., 2017).

7.8 Concluding Remarks

Great progress in the understanding of BK channels, from molecular structure, working mechanism, cellular, and *in vivo* functions, to human diseases, has been achieved since the initial discovery of BK currents four decades ago. We now know the structure of the Slo1 subunit of BK channels with three structural domains, the PGD, the voltage sensor domain (VSD), and the large cytosolic Ca^{2+}-binding domain (CTD). The structure–function relations of each of these domains are quite well understood, and the biophysical principles of allosteric coupling among these domains have been shown. BK channels have become one of the most studied ion channels owing to their large single-channel conductance, dual activation by membrane depolarization and intracellular Ca^{2+}, and diverse physiological functions. Emerging from these structure–function studies, mechanisms of BK channel gating including allosteric activation, non-domain swap between the VSD and PGD, and the hydrophobic gate, which differ from canonical Kv channel activation mechanisms, make BK channels a unique model for understanding the principles of ion channel activation.

BK channels play various roles in physiology mainly owing to their sensitivity to voltage and intracellular Ca^{2+}. In addition, BK channel function is also modulated by various physiological stimuli, posttranslational modifications (phosphorylation, methylation, heme), different splicing isoforms, and association of auxiliary β and γ subunits. All these factors fine-tune the properties of BK channels, thereby rendering BK channels tissue-specific functions. Mouse models of BK channel mutations and diseases associated with aberrant BK channel function advanced our understanding of the physiological and pathophysiological roles of BK channels. Therefore, further understanding of the molecular gating mechanism and biology of BK channels will greatly facilitate rational design of BK channel modulators to treat BK channel–related diseases, such as epilepsy, hypertension, asthma, stroke, and urinary incontinence.

Acknowledgments

The author would like to thank Dr. Huanghe Yang for contributing to an earlier edition of this chapter. Dr. Guohui Zhang helped make the figures and references.

Suggested Reading

Brelidze, T. I., X. Niu, and K. L. Magleby. 2003. "A ring of eight conserved negatively charged amino acids doubles the conductance of BK channels and prevents inward rectification." *Proc Natl Acad Sci U S A* 100 (15):9017–22.

Butler, A., S. Tsunoda, D. P. McCobb, A. Wei, and L. Salkoff. 1993. "mSlo, a complex mouse gene encoding "maxi" calcium-activated potassium channels." *Science* 261 (5118):221–4.

Carrisoza-Gaytan, R., M. D. Carattino, T. R. Kleyman, and L. M. Satlin. 2016. "An unexpected journey: Conceptual evolution of mechanoregulated potassium transport in the distal nephron." *Am J Physiol Cell Physiol* 310 (4):C243–59. doi:10.1152/ajpcell.00328.2015.

Cox, D. H., J. Cui, and R. W. Aldrich. 1997a. "Allosteric gating of a large conductance Ca-activated K$^+$ channel." *J Gen Physiol* 110 (3):257–81.

Cox, D. H., J. Cui, and R. W. Aldrich. 1997b. "Separation of gating properties from permeation and block in mslo large conductance Ca-activated K$^+$ channels." *J Gen Physiol* 109 (5):633–46.

Cui, J., D. H. Cox, and R. W. Aldrich. 1997. "Intrinsic voltage dependence and Ca^{2+} regulation of mslo large conductance Ca-activated K$^+$ channels." *J Gen Physiol* 109 (5):647–73.

Cui, J. 2020. "The action of a BK channel opener." *J Gen Physiol* 152 (6):e202012571. doi:10.1085/jgp.202012571.

Cui, J. 2021. "BK channel gating mechanisms: Progresses toward a better understanding of variants linked neurological diseases." *Front Physiol* 12. doi:10.3389/fphys.2021.762175.

Dong, P., Y. Zhang, A. S. Hunanyan, M. A. Mikati, J. Cui, and H. Yang. 2022. "Neuronal mechanism of a BK channelopathy in absence epilepsy and dyskinesia." *Proc Natl Acad Sci U S A* 119 (12):e2200140119. doi:10.1073/pnas.2200140119.

Du, W., J. F. Bautista, H. Yang, A. Diez-Sampedro, S.-A. You, L. Wang, P. Kotagal, H. O. Luders, J. Shi, J. Cui, G. B. Richerson, and Q. K. Wang. 2005. "Calcium-sensitive potassium channelopathy in human epilepsy and paroxysmal movement disorder." *Nat. Genet.* 37 (7):733–8.

Dudem, S., R. J. Large, S. Kulkarni, H. McClafferty, I. G. Tikhonova, G. P. Sergeant, K. D. Thornbury, M. J. Shipston, B. A. Perrino, and M. A. Hollywood. 2020. "LINGO1 is a regulatory subunit of large conductance, Ca(2+)-activated potassium channels." *Proc Natl Acad Sci U S A* 117 (4):2194–200. doi:10.1073/pnas.1916715117.

Galvez, A., G. Gimenez-Gallego, J. P. Reuben, L. Roy-Contancin, P. Feigenbaum, G. J. Kaczorowski, and M. L. Garcia. 1990. "Purification and characterization of a unique, potent, peptidyl probe for the high conductance calcium-activated potassium channel from venom of the scorpion Buthus tamulus." *J Biol Chem* 265 (19):11083–90.

Geng, Y., Z. Deng, G. Zhang, G. Budelli, A. Butler, P. Yuan, J. Cui, L. Salkoff, and K. L. Magleby. 2020. "Coupling of Ca(2+) and voltage activation in BK channels through the alphaB helix/voltage sensor interface." *Proc Natl Acad Sci U S A* 117 (25):14512–21. doi:10.1073/pnas.1908183117.

Gessner, G., Y.-M. Cui, Y. Otani, T. Ohwada, M. Soom, T. Hoshi, and H. Heinemann Stefan. 2012. "Molecular mechanism of pharmacological activation of BK channels." *Proc Natl Acad Sci U S A* 109 (9):3552–7. doi:10.1073/pnas.1114321109.

Guan, X., Q. Li, and J. Yan. 2017. "Relationship between auxiliary gamma subunits and mallotoxin on BK channel modulation." *Sci Rep* 7 (1):42240. doi:10.1038/srep42240.

Harvey, J. R. M., A. E. Plante, and A. L. Meredith. 2020. "Ion channels controlling circadian rhythms in suprachiasmatic nucleus excitability." *Physiol Rev* 100 (4):1415–54. doi:10.1152/physrev.00027.2019.

Horrigan, F. T., and R. W. Aldrich. 2002. "Coupling between voltage sensor activation, Ca^{2+} binding and channel opening in large conductance (BK) potassium channels." *J Gen Physiol* 120 (3):267–305.

Horrigan, F. T., J. Cui, and R. W. Aldrich. 1999. "Allosteric voltage gating of potassium channels I. Mslo ionic currents in the absence of Ca^{2+}." *J Gen Physiol* 114 (2):277–304.

Hoshi, T., Y. Tian, R. Xu, S. H. Heinemann, and S. Hou. 2013. "Mechanism of the modulation of BK potassium channel complexes with different auxiliary subunit compositions by the omega-3 fatty acid DHA." *Proc Natl Acad Sci U S A* 110 (12):4822–7. doi:10.1073/pnas.1222003110.

Hu, H., L. R. Shao, S. Chavoshy, N. Gu, M. Trieb, R. Behrens, P. Laake, O. Pongs, H. G. Knaus, O. P. Ottersen, and J. F. Storm. 2001. "Presynaptic Ca2+-activated K+ channels in glutamatergic hippocampal terminals and their role in spike repolarization and regulation of transmitter release." *J Neurosci* 21 (24):9585–97.

Jia, Z., M. Yazdani, G. Zhang, J. Cui, and J. Chen. 2018. "Hydrophobic gating in BK channels." *Nat Commun* 9 (1):3408. doi:10.1038/s41467-018-05970-3.

Kis, A., S. Krick, N. Baumlin, and M. Salathe. 2016. "Airway hydration, apical K(+) secretion, and the large-conductance, Ca(2+)-activated and voltage-dependent potassium (BK) channel." *Ann Am Thorac Soc* 13 (Suppl 2):S163–8. doi:10.1513/AnnalsATS.201507-405KV.

Li, M., S. Chang, L. Yang, J. Shi, K. McFarland, X. Yang, A. Moller, C. Wang, X. Zou, C. Chi, and J. Cui. 2014. "Conopeptide Vt3.1 preferentially inhibits BK potassium channels containing beta4 subunits via electrostatic interactions." *J Biol Chem* 289 (8):4735–42. doi:10.1074/jbc.M113.535898.

Li, W., and R. W. Aldrich. 2004. "Unique inner pore properties of BK channels revealed by quaternary ammonium block." *J Gen Physiol* 124 (1):43–57.

Ma, Z., X. J. Lou, and F. T. Horrigan. 2006. "Role of charged residues in the S1-S4 voltage sensor of BK channels." *J Gen Physiol* 127 (3):309–28. doi:10.1085/jgp.200509421.

Meera, P., M. Wallner, and L. Toro. 2000. "A neuronal beta subunit (KCNMB4) makes the large conductance, voltage- and Ca2+-activated K+ channel resistant to charybdotoxin and iberiotoxin." *Proc Natl Acad Sci U S A* 97 (10):5562–7. doi:10.1073/pnas.100118597.

Nelson, M. T., H. Cheng, M. Rubart, L. F. Santana, A. D. Bonev, H. J. Knot, and W. J. Lederer. 1995. "Relaxation of arterial smooth muscle by calcium sparks." *Science* 270 (5236):633–7. doi:10.1126/science.270.5236.633.

Neyton, J., and C. Miller. 1988. "Potassium blocks barium permeation through a calcium-activated potassium channel." *J Gen Physiol* 92 (5):549–67. doi:10.1085/jgp.92.5.549.

Niu, X., X. Qian, and K. L. Magleby. 2004. "Linker-gating ring complex as passive spring and Ca^{2+}-dependent machine for a voltage- and Ca^{2+}-activated potassium channel." *Neuron* 42 (5):745–56.

Oberhauser, A., O. Alvarez, and R. Latorre. 1988. "Activation by divalent cations of a Ca^{2+}-activated K^+ channel from skeletal muscle membrane." *J Gen Physiol* 92 (1):67–86.

Orio, P., P. Rojas, G. Ferreira, and R. Latorre. 2002. "New disguises for an old channel: MaxiK channel beta-subunits." *News Physiol Sci* 17:156–61.

Pallotta, B. S., K. L. Magleby, and J. N. Barrett. 1981. "Single channel recordings of Ca^{2+}-activated K^+ currents in rat muscle cell culture." *Nature* 293 (5832):471–4.

Pantazis, A., V. Gudzenko, N. Savalli, D. Sigg, and R. Olcese. 2010. "Operation of the voltage sensor of a human voltage- and Ca2+-activated K+ channel." *Proc Natl Acad Sci U S A* 107 (9):4459–64. doi:10.1073/pnas.0911959107.

Rockman, M. E., A. G. Vouga, and B. S. Rothberg. 2020. "Molecular mechanism of BK channel activation by the smooth muscle relaxant NS11021." *J Gen Physiol* 152 (6). doi:10.1085/jgp.201912506.

Schreiber, M., and L. Salkoff. 1997. "A novel calcium-sensing domain in the BK channel." *Biophys J* 73 (3):1355–63.

Shi, J., and J. Cui. 2001. "Intracellular Mg^{2+} enhances the function of BK-type Ca^{2+}-activated K^+ channels." *J Gen Physiol* 118 (5):589–606.

Tao, X., and R. MacKinnon. 2019. "Molecular structures of the human Slo1 K(+) channel in complex with beta4." *Elife* 8. doi:10.7554/eLife.51409.

Webb, T. I., A. S. Kshatri, R. J. Large, A. M. Akande, S. Roy, G. P. Sergeant, N. G. McHale, K. D. Thornbury, and M. A. Hollywood. 2015. "Molecular mechanisms underlying the effect of the novel BK channel opener GoSlo: Involvement of the S4/S5 linker and the S6 segment." *Proc Natl Acad Sci U S A* 112 (7):2064–9. doi:10.1073/pnas.1400555112.

Yan, J., and R. W. Aldrich. 2012. "BK potassium channel modulation by leucine-rich repeat-containing proteins." *Proc Natl Acad Sci U S A* 109 (20):7917–22. doi:10.1073/pnas.1205435109.

Yang, H., J. Shi, G. Zhang, J. Yang, K. Delaloye, and J. Cui. 2008. "Activation of Slo1 BK channels by Mg2+ coordinated between the voltage sensor and RCK1 domains." *Nat Struct Mol Biol* 15 (11):1152–9. doi:10.1038/nsmb.1507.

Yang, J., G. Krishnamoorthy, A. Saxena, G. Zhang, J. Shi, H. Yang, K. Delaloye, D. Sept, and J. Cui. 2010. "An epilepsy/dyskinesia-associated mutation enhances BK channel activation by potentiating Ca2+ sensing." *Neuron* 66 (6):871–83. doi:10.1016/j.neuron.2010.05.009.

Zhang, G., S.-Y. Huang, J. Yang, J. Shi, X. Yang, A. Moller, X. Zou, and J. Cui. 2010. "Ion sensing in the RCK1 domain of BK channels." *Proc Natl Acad Sci U S A* 107 (43):18700–5. doi:10.1073/pnas.1010124107.

Zhang, G., X. Xu, Z. Jia, Y. Geng, H. Liang, J. Shi, M. Marras, C. Abella, K. L. Magleby, J. R. Silva, J. Chen, X. Zou, and J. Cui. 2022. "An allosteric modulator activates BK channels by perturbing coupling between Ca^{2+} binding and pore opening." *Nat Commun* 13 (1):6784. doi:10.1038/s41467-022-34359-6.

Zhou, Y., X. M. Xia, and C. J. Lingle. 2020. "The functionally relevant site for Paxilline inhibition of BK channels." *Proc Natl Acad Sci U S A* 117 (2):1021–6. doi:10.1073/pnas.1912623117.

8

Small-Conductance Calcium-Activated Potassium (SK) Channels

Miao Zhang and Heike Wulff

CONTENTS

8.1 Introduction

Calcium-activated K^+ channels with a higher calcium sensitivity and a much smaller conductance than the more easily detectable large-conductance Ca^{2+}-activated K^+ (BK) channels (see Chapter 7) were known to exist in neurons and muscle since the early 1980s. The molecular identity of these small-conductance Ca^{2+}-activated K^+ (SK) channels became clear when the channels were cloned by John Adelman and colleagues in 1996 (Kohler et al. 1996). SK channels are encoded by the *KCNN* genes, including *KCNN1* for SK1 ($K_{Ca}2.1$), *KCNN2* for SK2 ($K_{Ca}2.2$) and *KCNN3* for SK3 ($K_{Ca}2.3$). The Ca^{2+}-binding protein calmodulin (CaM) is constitutively associated with the SK channels and serves as their Ca^{2+}-sensing beta subunit (Xia et al. 1998). Elevated intracellular Ca^{2+} levels in the vicinity of the channels cause conformational changes in the SK–CaM complex and result in channel opening.

8.2 Physiological Function

The three SK channel subtypes are widely expressed in the central and peripheral nervous system and are best known for mediating the medium afterhyperpolarization (mAHP), which regulates neuronal firing frequency (Adelman, Maylie, and Sah 2012). Ca^{2+} release from the endoplasmic reticulum through ryanodine receptors (RyRs) or inositol-1,4,5-trisphosphate receptors (IP$_3$Rs), as well as Ca^{2+} influx through calcium-permeable channels such as voltage-gated Ca^{2+} (Cav) channels, *N*-methyl-*D*-aspartate (NMDA) receptors, nicotinic acetylcholine receptors (nAchRs) and transient receptor potential (TRP) channels, triggers local Ca^{2+} signals to activate SK channels in neurons (Figure 8.1A) (Adelman, Maylie, and Sah 2012). Functional coupling between SK channels and their Ca^{2+} sources contributes to the regulation of neuronal excitability by Ca^{2+} signaling. SK channels have been reported to underlie the mAHP in many types of neurons, including spinal motoneurons, neurosecretory neurons in the supraoptic area of the hypothalamus, vagal motoneurons, pyramidal neurons in the sensory cortex and the lateral and basolateral amygdala, interneurons in the nucleus reticularis of the thalamus, striatal cholinergic interneurons, hippocampal interneurons in the stratum oriens-alveus and the stratum radiatum, cholinergic nucleus basalis neurons, paraventricular neurons, rat subthalamic neurons, cerebellar Purkinje neurons, noradrenergic neurons of the locus coeruleus, serotonergic neurons of the dorsal raphe, midbrain dopamine neurons, circadian clock neurons in the suprachiasmatic nucleus of the hypothalamus, and mitral cells of the olfactory bulb (Adelman, Maylie, and Sah 2012). Inhibition of SK channels by the bee venom peptide apamin significantly enhances long-term potentiation (LTP) in hippocampal CA1 and amygdala pyramidal neurons, whereas the potentiation of SK channels suppresses LTP. Accordingly, blocking of SK channels with apamin enhances learning and memory, while increasing SK channel expression and/or activity impairs learning in animal models, underscoring the importance of SK channels in information processing and storage at the systems level (Adelman, Maylie, and Sah 2012).

All three subtypes of SK channels are also expressed in the heart, with more abundant expression in the atria, pace-making cells and Purkinje fibers than the ventricles under physiological conditions (Zhang et al. 2021). SK channels, L-type Cav channels and RyRs are within hundreds of nanometers from each other in cardiomyocytes. Ca^{2+} influx through L-type Cav channels together with Ca^{2+} release from the sarcoplasmic reticulum through RyRs may be the primary Ca^{2+} sources for cardiac SK channels (Figure 8.1B). Ablation of SK2 in SK2 knockout mice led to significantly prolonged action potential duration (APD), particularly in the late phase of atrial repolarization. Inhibition of SK channels prolongs APD in isolated human atrial myocytes. On the other hand, overexpression of SK3 channels results in shortening of APD in atrial myocytes. These findings support a critical role of SK channels in cardiac repolarization, particularly in atrial myocytes where they have a more prominent expression (Zhang et al. 2021).

The SK3 channel subtype and its cousin, the intermediate-conductance Ca^{2+}-activated K^+ channel (IK, SK4 or K$_{Ca}$3.1 encoded by *KCNN4*), are both expressed in the vascular endothelium (Wulff and Kohler 2013). Unlike IK channels that are activated via calcium release induced by stimulation of GPCRs like muscarinic acetylcholine or bradykinin receptors, endothelial SK3 channels sense local Ca^{2+} increases produced by mechanical deformation during shear stress stimulation. Ca^{2+} release from the endoplasmic reticulum as well as Ca^{2+} entry through TRP channels have been suggested as the Ca^{2+} sources for vascular endothelial SK3 channels (Figure 8.1C). The activation of SK3 and/or IK channels induces hyperpolarization in vascular endothelial cells, which spreads to the underlying vascular smooth muscle cell layer and causes endothelium-derived hyperpolarization (EDH), leading to blood vessel dilation. Genetic deficit of SK3 and/or IK channels in mice compromises the EDH response and causes hypertension, suggesting a role for endothelial SK and IK channels in blood pressure regulation (Wulff and Kohler 2013).

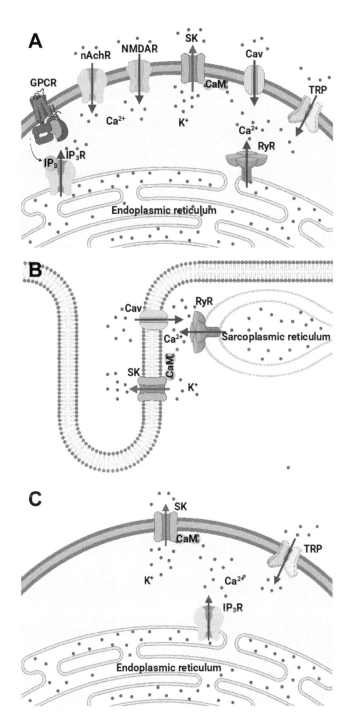

FIGURE 8.1

Ca^{2+} sources of SK channels in (A) neurons, (B) cardiomyocytes and (C) endothelial cells. GPCR, G protein-coupled receptor; IP_3R, inositol-1,4,5-trisphosphate receptor; nAChR, nicotinic acetylcholine receptor; NMDAR, *N*-methyl-*D*-aspartate receptor; Cav, voltage-gated Ca^{2+} channel; RyR, ryanodine receptor; TRP, transient receptor potential channel. (Created with Biorender.com.)

8.3 Subunit Diversity and Structure

The three SK channel subtypes share 80%–90% identity in their six transmembrane domains but vary in amino acid sequence and length at their cytoplasmic N- and C- termini (Kohler et al. 1996). The SK channel subtypes are only ~40% homologous to IK, the fourth member of the *KCNN* gene family. In the IUPHAR nomenclature, IK accordingly has been named $K_{Ca}3.1$, making it a separate family (Kaczmarek et al. 2017). At least in heterologous expression systems, SK and IK channels are able to form functional heteromeric channels suggesting that this might also be possible in neurons (Higham et al. 2019). Unlike BK channels that achieve Ca^{2+} sensitivity through the cytoplasmic Ca^{2+}-binding sites, both SK and IK channels are operated by a Ca^{2+}/CaM gating mechanism. CaM is constitutively associated with SK and IK channels at the calmodulin-binding domain (CaM-BD) and confers Ca^{2+} sensitivity to these channels (Figure 8.2). IK channels exhibit a slightly higher apparent Ca^{2+} sensitivity with reported EC_{50} values for Ca^{2+} ranging from 0.1 to 0.4 μM, than the three subtypes of SK channels with EC_{50} values from 0.3 to 0.75 μM (Brown et al. 2020).

SK channels share the same tetrameric assembly with the voltage-gated K^+ (Kv) channels. Each channel subunit comprises six transmembrane α-helical domains denoted S1–S6. Unlike Kv channels, SK channels have only three positively charged amino acid residues in the S4, and thus are voltage-independent. While full-length structures are not

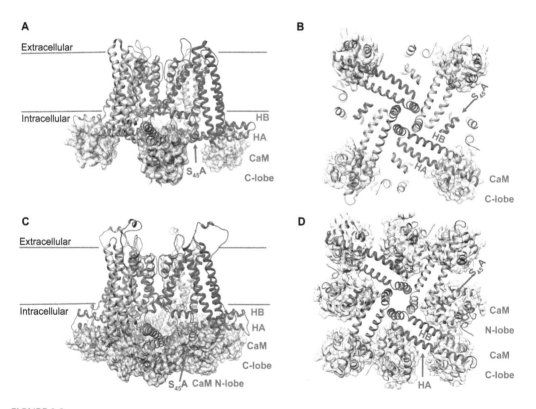

FIGURE 8.2
Gating mechanism of SK2 channels shown by the homology model. Side view (A) and view from the cytoplasmic side (B) of rat SK2 channels in the absence of Ca^{2+}. Side view (C) and view from the cytoplasmic side (D) of rat SK2 channels in the presence of Ca^{2+}. The distal intracellular N and C-termini of SK2 channels are not included in the model. (Created with UCSF Chimera program.)

available for the three SK channel subtypes, Roderick MacKinnon and colleagues (Lee and MacKinnon 2018) determined the full-length cryogenic electron microscopy (cryo-EM) structures of IK ($K_{Ca}3.1$, SK4) channels in the absence and presence of Ca^{2+}, which shed light on the Ca^{2+}/CaM gating mechanism shared by SK and IK channels. Transmembrane helices S1–S4 interact with the pore-forming helices S5 and S6 from the same subunit, which is similar to BK channels (see Chapter 7), without the domain swapping seen in Kv1 to Kv7 channels (see Chapter 4). Homology models of the rat SK2 channels were generated using the IK channel structures as templates (Figure 8.2). There are four CaM molecules in the tetrameric channel structure. The CaM-BD forms two α-helices, HA and HB, while the linker between the S4 and S5 transmembrane domains (S4–S5 linker) consists of two α-helices, $S_{45}A$ and $S_{45}B$. In the absence of Ca^{2+}, the C-lobe of CaM is seen bound to the HA and HB helices, which form the CaM-BD, while the N-lobe of CaM is poorly resolved due to high flexibility (Figure 8.2A, B). The N-lobe of CaM becomes well structured when bound with Ca^{2+} and is seen making substantial contacts with the $S_{45}A$ helix of the S4-S5 linker of a neighboring subunit (Figure 8.2C, D). As such, CaM, the HA and HB helices (= CaM-BD), and the $S_{45}A$ and $S_{45}B$ helices (S4–S5 linker) from a neighboring channel subunit become a stable structural group right at the bottom of the pore-forming S6 transmembrane domain.

The full-length structure of IK (Lee and MacKinnon 2018), which is very likely also representative of SK2 channels in the presence of Ca^{2+}, demonstrated several discrepancies with the crystal structure of the Ca^{2+}-CaM/CaM-BD complex previously reported (Schumacher et al. 2001). Since the S4–S5 linker was not included in the crystal structure, both the N- and C-lobes of CaM grab the two helices formed by CaM-BD and form a dimeric structure. This dimeric Ca^{2+}-CaM/CaM-BD complex structure was also utilized in subsequent crystallographic studies exploring the binding sites of small molecule modulators of SK channels. However, the lesson we learned from the full-length IK structure and the most recent validation of the full-length model of SK2 channels (Shim et al. 2019, Nam et al. 2020) is that the dimeric Ca^{2+}-CaM/CaM-BD complex crystal structure is very likely an artifact, and existing ideas about the binding site of SK channel modulators, therefore, may need to be revised (Brown et al. 2020).

8.4 Gating and Ion Permeability

As described in the previous paragraph, SK channels are operated by a unique Ca^{2+}/CaM gating mechanism (Figure 8.2). In the absence of Ca^{2+}, the CaM C-lobe is associated with the channel at the HA and HB helices formed by CaM-BD, while the N-lobe is flexible and as such invisible in the structure. The channel pore is closed in the absence of Ca^{2+}. In the presence of Ca^{2+}, the two EF hands of the CaM N-lobe become calcified, and the hydrophobicity of the CaM N-lobe and, therefore, its affinity for the $S_{45}A$ helix within the S4-S5 linker increases. The Ca^{2+}-bound N-lobe then pulls the $S_{45}A$ helix toward the cytoplasm, forcing the $S_{45}B$ helix to follow. This movement expands the S6 transmembrane domain and opens the channel pore.

The single-channel conductance of SK channels is ~10 pS in symmetrical K^+ solutions (Figure 8.3A), which is smaller than that of the IK channels (~40 pS) and dramatically smaller than that of BK channels (~200 pS) (Kaczmarek et al. 2017). The inside-out and whole-cell currents of rat SK2 channels are inwardly rectified (Figure 8.3B, C). Similar to Kv channels, the selectivity filter located between the S5 and S6 transmembrane domain is responsible for the selective permeability of K^+ ions. Like other K^+ channels, rubidium

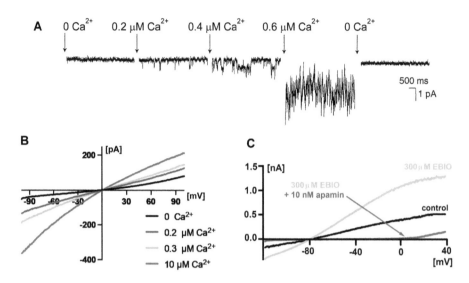

FIGURE 8.3

Examples of rat SK2 channel currents. (A) Single-channel activity of rat SK2 channels in symmetrical K+. (Adopted from Kohler et al. 1996.) (B) Macroscopic recordings of rat SK2 channels with inside-out configuration in symmetrical K+. (Adopted from Nam et al. 2020.) (C) Whole-cell recording of rat SK2 transiently expression in HEK-293 cells following dialysis with 1 μM of free Ca^{2+}. Currents were first activated with EBIO and then blocked by apamin. (Unpublished recording from H. Wulff.)

and thallium ions can flow through the open state of SK channels, making rubidium or thallium-flux assays a useful high-throughput screening method for drug discovery targeting SK channels (Hougaard et al. 2007).

8.5 Pharmacology

Before their cloning, SK channels were often characterized as apamin-sensitive calcium-activated K+ channels. Apamin is an 18-amino acid bee venom toxin that very selectively inhibits SK channel activity. The SK2 channel subtype is the most sensitive to apamin with subnanomolar IC_{50} values, while SK1 and SK3 channels are less sensitive with nanomolar IC_{50} values (Kaczmarek et al. 2017). Apamin has been proposed as a pore blocker based on its interactions with the outer pore, and as a negative allosteric modulator because of its putative binding to the extracellular S3-S4 loop. The larger, 31-amino acid scorpion toxins scyllatoxin (also known as leiurotoxin-I) and tamapin in contrast are bona fide pore blockers of SK channels. A striking feature of apamin is the presence of several positively charged arginine residues. This structural element inspired the design of the two bis-quinolinium cyclophanes UCL1684 and UCL1848, which inhibit SK2 channels with picomolar affinities (Brown et al. 2020).

Similar to other ligand-gated ion channels like $GABA_A$ receptors, SK channels have both positive and negative gating modulators. 1-EBIO was the first positive gating modulator identified more than two decades ago that modulates both SK and IK channel subtypes. SKA-31, which was developed using the neuroprotectant riluzole as a template, has a higher potency than 1-EBIO, but does not differentiate between SK/IK channel subtypes.

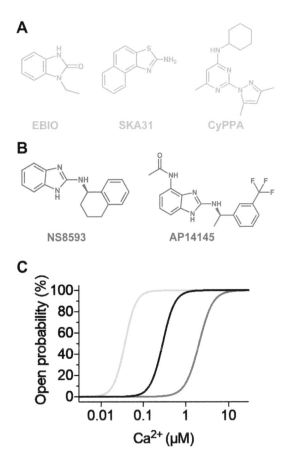

FIGURE 8.4
Pharmacology of SK channels. (A) Positive gating modulators. (B) Negative gating modulators. (C) Positive (green) and negative (red) gating modulators change the channel open probability through shifting the apparent Ca^{2+} sensitivity (black).

The prototypical subtype-selective positive modulator CyPPA potentiates SK2 and SK3 channel activity, but not SK1 or IK channel activity. Negative modulators of SK channels have also been developed, including NS8593, AP14145, and AP30663, which are in clinical trials for atrial fibrillation (Brown et al. 2020). Both the positive and negative SK channel modulators exert their effects through influencing the apparent Ca^{2+} sensitivity of these channels. Positive modulators left-shift, while negative modulators right-shift the Ca^{2+}-dependent activation of SK channels (Figure 8.4).

8.6 Regulation

SK channels are subject to regulation by posttranslational modification (e.g., phosphorylation). SK channels form multiprotein complexes that, in addition to CaM, also contain two regulatory enzymes: casein kinase 2 (CK2) and protein phosphatase 2A (PP2A) (Figure

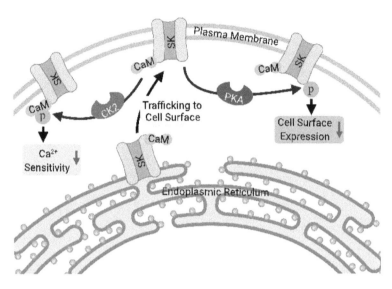

FIGURE 8.5
SK channels are regulated by phosphorylation. CK2, casein kinase 2; PP2A, protein phosphatase 2A; PKA, protein kinase A. (Created with Biorender.com.)

8.5). However, instead of modifying the pore-forming subunit, CK2 and PP2A phosphorylate or dephosphorylate a specific threonine residue (Thr79) in CaM, and thus reduce or increase the apparent Ca^{2+} sensitivity of the SK channels (Adelman, Maylie, and Sah 2012). Most studies on the phosphorylation of the SK–CaM complex by CK2 were performed with the SK2 channel subtype. However, other SK channel subtypes and IK channels may also be regulated by the CK2/PP2A pair at least when expressed heterologously (Nam et al. 2021).

8.7 Cell Biology (Biogenesis, Trafficking, Turnover)

There are mRNA splice variants for all three SK channel subtypes (Girault et al. 2011). The cytoplasmic C-terminus of SK1 channels can be alternatively spliced, which leads to diminished CaM binding. The physiological significance of these SK1 splice variants is unclear. The full-length SK1 channel variant appears to be the predominant form in the human hippocampus, while two truncated channel variants are more abundant in reticulocytes and erythroleukemia cells. Both short and long splice variants of SK2 channels (SK2-S and SK2-L, respectively) have been reported. The SK2-L and SK2-S mRNAs are transcribed from two separate promoters. Compared with the shorter SK2-S, SK2-L has 207 extra amino acid residues at its cytoplasmic N-terminus. SK2-S and SK2-L do not differ in their apparent Ca^{2+} sensitivity when expressed heterologously. The extra 207 amino acid residues at the N-terminus of SK2-L variants may be involved in localization of the channel in the postsynaptic density (PSD) of dendritic spines on mouse CA1 pyramidal neurons. Two splice variants of SK3 channels, SK3-1B (Tomita et al. 2003) and SK3-1C (Kolski-Andreaco et al. 2004), function as dominant-negative suppressors of all other SK channel subtypes and of IK. When expressed alone, SK3-1B and SK3-1C do not produce functional channels. SK3-1B or SK3-1C might co-assemble with and sequester full-length SK channel proteins in the endoplasmic reticulum, and prevent the trafficking of full-length SK channel proteins to the cell membrane. Another SK3 channel splice variant, SK3_ex4, has 15 additional amino acid residues between the S5 transmembrane domain and the pore region. The insertion of these residues does not influence channel activity. Instead, it reduces the sensitivity of SK3_ex4 toward apamin and scyllatoxin (Girault et al. 2011).

The expression of SK channels on the cell surface requires their Ca^{2+}-independent interactions with CaM (Girault et al. 2011). $CaM_{1,2,3,4}$, a CaM in which the binding of all four of the EF hands to Ca^{2+} was abolished by mutations, can still facilitate the trafficking of SK channels to the plasma membrane (Lee et al. 2003). The trafficking of SK2 channels is also regulated by phosphorylation. The distal C-terminus of the SK2 channel can be phosphorylated by the protein kinase A (PKA), which leads to reduced cell surface expression of the channel protein (Figure 8.5). Activation of the β-adrenoceptors can stimulate PKA, which in turn regulates synaptic SK2 channel expression on the dendrites of lateral amygdala pyramidal neurons. In mouse hippocampal CA1 pyramidal neurons, LTP induction activates PKA and downregulates SK2 channel activity through the internalization of the channel protein (Adelman, Maylie, and Sah 2012). Tonic PKA activity is also involved in the enrichment of SK channels on the dendrites compared to the soma of cultured hippocampal neurons (Abiraman et al. 2016).

8.8 Channelopathies and Disease

SK2 channels are expressed in cerebellar Purkinje neurons and serve as one of the principal ion channels involved in pace-making by these important neurons. Loss of firing precision in Purkinje neurons underlies the symptoms of movement disorders such as ataxia. This was first demonstrated in transgenic mice where dominant-negative suppression of SK2 channels with SK3-1B produced a phenotype resembling cerebellar ataxia (Shakkottai et al. 2004). Mutations that diminish SK2 channel activity have been linked with tremor in rodents (Kuramoto et al. 2017). In humans, loss-of-function mutations in the *KCNN2* gene also lead to neurodevelopmental movement disorders including cerebellar ataxia and tremor (Balint et al. 2020; Mochel et al. 2020). Subtype-selective SK2/SK3 positive modulators such as CyPPA and NS13001 have been shown to improve ataxia in mouse models and are proposed as potential therapeutic agents.

Genome-wide association studies identified multiple single-nucleotide polymorphisms in *KCNN3* genes linked with lone atrial fibrillation (Zhang et al. 2021). Gain-of-function mutations increase the apparent Ca^{2+} sensitivity of SK3 channels and may subsequently lead to Zimmermann–Laband syndrome (ZLS) (Bauer et al. 2019). Based on the contribution of SK3 channels to the aforementioned EDH mediated vasodilation, it has been speculated that gain-of-function mutant SK3 channels expressed in vascular endothelium may be related to vascular damage during limb development of ZLS patients.

SK2 and SK3 channels are also expressed in several tumors. SK2 channels may enhance cancer cell proliferation, while SK3 channels are involved in cancer cell migration. SK3 channels are not expressed in normal breast cells, but they are expressed in breast cancer cell lines. The synthetic alkyl-lipids edelfosine and ohmline are SK3 channel inhibitors and have been described to reduce cancer cell migration (Girault et al. 2011).

In the normal heart, SK channels are expressed mostly in the atrial cardiomyocytes rather than the ventricles. However, in failing ventricular cardiomyocytes, SK channels may be upregulated, along with elevated channel protein expression and increased apparent Ca^{2+} sensitivity, suggesting their pathophysiological role in heart failure (Zhang et al. 2021).

Considering both their physiological and pathophysiological roles, SK channels have been proposed as drug targets for ataxia, tremor, alcohol use disorders, atrial fibrillation and several other diseases. A positive modulator of SK2/SK3 channels, CAD-1883 from Cadent Therapeutics, has been evaluated for essential tremor (ClinicalTrials.gov Identifier: NCT03688685). A clinical trial of the same drug was also initiated for spinocerebellar ataxia (ClinicalTrials.gov Identifier: NCT04301284), but was later withdrawn after Cadent Therapeutics was acquired by Novartis. Chlorzoxazone, a drug approved by the US Food and Drug Administration, which has a positive modulatory effect on SK channels, was evaluated in a clinical trial for alcohol use disorder (ClinicalTrials.gov Identifier: NCT01342341). A clinical trial of the SK channel negative modulator AP30663 is currently ongoing for atrial fibrillation (ClinicalTrials.gov Identifier: NCT04571385).

8.9 Conclusions

SK channels are voltage-independent and operated by a Ca^{2+}/CaM gating mechanism. CaM is constitutively associated with SK channels on their cytoplasmic C-termini. Upon binding with Ca^{2+}, CaM induces conformational changes that lead to channel opening. SK channels mediate the mAHP and thus modulate the excitability of neurons and cardiomyocytes. Positive and negative gating modulators exert their effects through influencing the apparent Ca^{2+} sensitivity of SK channels and constitute promising therapeutic agents for the treatment of neurological and cardiovascular diseases.

Suggested Reading

Abiraman, K., M. Sah, R. S. Walikonis, G. Lykotrafitis, and A. V. Tzingounis. 2016. Tonic PKA activity regulates SK channel nanoclustering and somatodendritic distribution. *J Mol Biol* 428 (11):2521–37.

Adelman, J. P., J. Maylie, and P. Sah. 2012. Small-conductance Ca^{2+}-activated K^+ channels: form and function. *Annu Rev Physiol* 74:245–69.

Balint, B., R. Guerreiro, S. Carmona, N. Dehghani, A. Latorre, C. Cordivari, K. P. Bhatia, and J. Bras. 2020. KCNN2 mutation in autosomal-dominant tremulous myoclonus-dystonia. *Eur J Neurol* 27 (8):1471–7.

Bauer, C. K., P. E. Schneeberger, F. Kortum, J. Altmuller, F. Santos-Simarro, L. Baker, J. Keller-Ramey, S. M. White, P. M. Campeau, K. W. Gripp, and K. Kutsche. 2019. Gain-of-function mutations in KCNN3 encoding the small-conductance Ca^{2+}-activated K^+ channel SK3 cause Zimmermann-Laband syndrome. *Am J Hum Genet* 104 (6):1139–57. doi:10.1016/j.ajhg.2019.04.012.

Brown, B. M., H. Shim, P. Christophersen, and H. Wulff. 2020. Pharmacology of small- and intermediate-conductance calcium-activated potassium channels. *Annu Rev Pharmacol Toxicol* 60:219–40.

Girault, A., J. P. Haelters, M. Potier-Cartereau, A. Chantome, P. A. Jaffres, P. Bougnoux, J. Joulin, and C. Vandier. 2011. Targeting SKCa channels in cancer: potential new therapeutic approaches. *Curr Med Chem* 19:697–713.

Higham, J., G. Sahu, R. M. Wazen, P. Colarusso, A. Gregorie, B. S. J. Harvey, L. Goudswaard, G. Varley, D. N. Sheppard, R. W. Turner, and N. V. Marrion. 2019. Preferred formation of hetero-meric channels between coexpressed SK1 and IKCa channel subunits provides a unique phar-macological profile of Ca^{2+}-activated potassium channels. *Mol Pharmacol* 96 (1):115–26.

Hougaard, C., B. L. Eriksen, S. Jorgensen, T. H. Johansen, T. Dyhring, L. S. Madsen, D. Strobaek, and P. Christophersen. 2007. Selective positive modulation of the SK3 and SK2 subtypes of small conductance Ca^{2+}-activated K^+ channels. *Br J Pharmacol* 151 (5):655–65.

Kaczmarek, L. K., Aldrich, R. W., Chandy, K. G., Grissmer, S., Wei, A. D., Wulff, H. 2017. International Union of Basic and Clinical Pharmacology. C. Nomenclature and Properties of Calcium-Activated and Sodium-Activated Potassium Channels. *Pharmacol Rev* 69(1): 1–11. doi: 10.1124/pr.116.012864. Epub 2016 Nov 15.

Kohler, M., B. Hirschberg, C. T. Bond, J. M. Kinzie, N. V. Marrion, J. Maylie, and J. P. Adelman. 1996. Small-conductance, calcium-activated potassium channels from mammalian brain. *Science* 273 (5282):1709–14.

Kolski-Andreaco, Aaron, Hiroaki Tomita, Vikram G. Shakkottai, George A. Gutman, Michael D. Cahalan, J. Jay Gargus, and K. George Chandy. 2004. SK3-1C, a dominant-negative suppressor of SKCa and IKCa channels. *J Biol Chem* 279 (8):6893–904.

Kuramoto, T., M. Yokoe, N. Kunisawa, K. Ohashi, T. Miyake, Y. Higuchi, K. Yoshimi, T. Mashimo, M. Tanaka, M. Kuwamura, S. Kaneko, S. Shimizu, T. Serikawa, and Y. Ohno. 2017. Tremor dominant Kyoto (Trdk) rats carry a missense mutation in the gene encoding the SK2 subunit of small-conductance Ca^{2+}-activated K^+ channel. *Brain Res* 1676:38–45.

Lee, C. H., and R. MacKinnon. 2018. Activation mechanism of a human SK-calmodulin channel com-plex elucidated by cryo-EM structures. *Science* 360 (6388):508–13.

Lee, Wei-Sheng, Thu Jennifer Ngo-Anh, Andrew Bruening-Wright, James Maylie, and John P. Adelman. 2003. Small conductance Ca^{2+}-activated K^+ channels and calmodulin: cell surface expression and gating. *J Biol Chem* 278 (28):25940–6.

Mochel, F., A. Rastetter, B. Ceulemans, K. Platzer, S. Yang, D. N. Shinde, K. L. Helbig, D. Lopergolo, F. Mari, A. Renieri, E. Benetti, R. Canitano, Q. Waisfisz, A. S. Plomp, S. A. Huisman, G. N. Wilson, S. S. Cathey, R. J. Louie, D. Del Gaudio, D. Waggoner, S. Kacker, K. M. Nugent, E. R. Roeder, A. L. Bruel, J. Thevenon, N. Ehmke, D. Horn, M. Holtgrewe, F. J. Kaiser, S. B. Kamphausen, R. Abou Jamra, S. Weckhuysen, C. Dalle, and C. Depienne. 2020. Variants in the SK2 channel gene (KCNN2) lead to dominant neurodevelopmental movement disorders. *Brain* 143 (12):3564–73. doi:10.1093/brain/awaa346.

Nam, Y. W., M. Cui, R. Orfali, A. Viegas, M. Nguyen, E. H. M. Mohammed, K. A. Zoghebi, S. Rahighi, K. Parang, and M. Zhang. 2020. Hydrophobic interactions between the HA helix and S4-S5 linker modulate apparent Ca(2+) sensitivity of SK2 channels. *Acta Physiol (Oxf)* 231 (1):e13552.

Nam, Y. W., D. Kong, D. Wang, R. Orfali, R. T. Sherpa, J. Totonchy, S. M. Nauli, and M. Zhang. 2021. Differential modulation of SK channel subtypes by phosphorylation. *Cell Calcium* 94:102346.

Schumacher, M. A., A. F. Rivard, H. P. Bachinger, and J. P. Adelman. 2001. Structure of the gat-ing domain of a Ca^{2+}-activated K^+ channel complexed with Ca^{2+}/calmodulin. *Nature* 410 (6832):1120–4.

Shakkottai, V. G., C. H. Chou, S. Oddo, C. A. Sailer, H. G. Knaus, G. A. Gutman, M. E. Barish, F. M. LaFerla, and K. G. Chandy. 2004. Enhanced neuronal excitability in the absence of neurodegen-eration induces cerebellar ataxia. *J Clin Invest* 113 (4):582–90.

Shim, H., B. M. Brown, L. Singh, V. Singh, J. C. Fettinger, V. Yarov-Yarovoy, and H. Wulff. 2019. The trials and tribulations of structure assisted design of KCa channel activators. *Front Pharmacol* 10:972. doi:10.3389/fphar.2019.00972.

Tomita, H., V. G. Shakkottai, G. A. Gutman, G. Sun, W. E. Bunney, M. D. Cahalan, K. G. Chandy, and J. J. Gargus. 2003. Novel truncated isoform of SK3 potassium channel is a potent dominant-negative regulator of SK currents: implications in schizophrenia. *Mol Psychiatry* 8 (5):524–35.

Wulff, H., and R. Kohler. 2013. Endothelial small-conductance and intermediate-conductance KCa channels: an update on their pharmacology and usefulness as cardiovascular targets. *J Cardiovasc Pharmacol* 61 (2):102–2.

Xia, X. M., B. Fakler, A. Rivard, G. Wayman, T. Johnson-Pais, J. E. Keen, T. Ishii, B. Hirschberg, C. T. Bond, S. Lutsenko, J. Maylie, and J. P. Adelman. 1998. Mechanism of calcium gating in small-conductance calcium-activated potassium channels. *Nature* 395 (6701):503–7.

Zhang, X. D., P. N. Thai, D. K. Lieu, and N. Chiamvimonvat. 2021. Cardiac small-conductance calcium-activated potassium channels in health and disease. *Pflugers Arch* 473 (3):477–89.

9

Inward Rectifier Potassium Channels

Camden Driggers, Min-Woo Sung, and Show-Ling Shyng

CONTENTS

9.1 Introduction

Inward rectification refers to the tendency of an ion channel to allow greater inward than outward currents at any given driving force of opposite directions. Potassium currents showing strong inward rectification were first described in K^+-depolarized muscle by Bernard (Katz 1949), who used the term "anomalous rectification" to contrast their properties from those of "normal" delayed rectification observed in the squid axon. We now know that the channel underlying the anomalous currents is a member of the inward rectifier potassium (Kir) channel family, and the inward rectification is caused by voltage-dependent block of the channel pore by intracellular Mg^{2+} and polyamines. Kir channels exhibit a range of inward rectification properties and are regulated by diverse signals including ions, nucleotides, lipids and proteins. The ability of Kir channels to regulate membrane potential in response to diverse intra and extracellular signals grants them the ability to govern a wide range of vital physiological processes. Recent high-resolution structures of Kir channels obtained by X-ray crystallography and single-particle cryo-electron microscopy (cryo-EM) have offered insights into the structural bases that underlie ion selectivity, ligand-induced gating and inward rectification. Genetic studies in animals and linkage of Kir channel dysfunction to a range of human disorders continue to inform the importance of this class of potassium channels in physiology and pathophysiology.

DOI: 10.1201/9781003096276-9

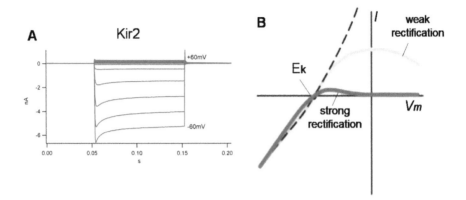

FIGURE 9.1

Inward rectification. (A) Whole-cell current traces of Kir2 channels expressed in HEK 293 cells showing strong inward rectification. The voltage steps are from −60 mV to +60 mV in 10 mV increments. (Graph in the public domain, https://en.wikipedia.org/wiki/File:Inward-rectification.png.) (B) Current–voltage relationships of strong and weak inward rectifiers. (From Nichols and Lee, JBC, 2018).

9.2 Physiological Roles

Found in many cell types, Kir channels play central roles in controlling membrane excitability and K$^+$ homeostasis. They can be classified into strong (Kir2.x, Kir3.x), intermediate (Kir4.x) and weak (Kir1.1 and Kir6.x) inward rectifiers based on the extent of inward rectification (Figure 9.1). The biological significance of strong inward rectification is best seen in excitable cells, such as cardiac and neuronal cells. The resting membrane potential of these cells is typically slightly positive to the equilibrium potential for potassium (E$_K$), enabling Kir channels to sustain a small amount of outward currents to keep the membrane potential near E$_K$. However, when the cell becomes excited and other voltage-gated Na$^+$ or Ca^{2+} channels open to trigger an action potential, the strong voltage-dependent block or ligand-induced closure of strong inward rectifiers ensures the action potential does not become short-circuited. In contrast, weak inward rectifiers will both stabilize the resting potential and shorten the action potential as their activity remains high at depolarizing membrane potentials. An example is seen in the ATP-sensitive potassium (K$_{ATP}$) channels formed by Kir6.2 and SUR2A. When activated in the heart under ischemic conditions, K$_{ATP}$ channels reduce the duration of cardiac action potentials, protecting the heart against ischemic injuries.

Kir channels have also been classified into four different categories based on their functional and physiological properties: classical Kir channels, G protein-coupled Kir channels, K$_{ATP}$ channels and K$^+$ transport channels (Hibino et al. 2010) (Figure 9.2). Classical Kir channels are formed by members from the Kir2 subfamily. They exhibit strong inward rectification with little current flow at potentials positive to −40 mV. Kir2.1 is the predominant isoform in skeletal muscle. In atrial and ventricular myocytes, Kir2.1 associates with Kir2.2 or Kir2.3 to form strong inwardly rectifying I$_{K1}$ currents. In the heart, the high K$^+$ conductance at negative voltages ensures a stable resting potential, but at depolarized potentials inward rectification results in suppression of conductance, which allows for the normally long plateau potentials that ensure a long refractory period to avoid cardiac arrhythmias (Anumonwo and Lopatin 2010). In vascular endothelial cells, classical Kir channels are mostly formed by Kir2.2 subunits and have a prominent role in modulating vascular tone.

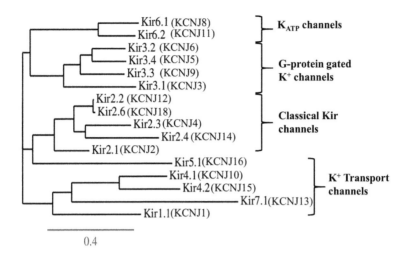

FIGURE 9.2

Phylogenetic tree of 16 known human Kir channels based on amino acid sequence alignment and structural similarity. Sequence alignments were generated using PROSMALS3D, a multiple sequence and structure alignment server, using select structures of Kir channels as a guide (Protein Data Bank IDs: 4KFM, 3SYO, 3SPI, 3JYC and 6BAA). The phylogenetic tree that reflects Kir evolutionary relations and distances was calculated using ConSurf and rendered using TreeDyn on the phylogeny.fr server. Functional classification of the channels is shown on the right.

Kir2.x channels are also expressed in skeletal muscle and neurons in the brain, where they regulate cell excitability.

G protein-gated Kir channels, or GIRK channels (Kir3.x), underlie G protein-coupled receptor (GPCR)-activated currents in the heart, brain, and endocrine tissues. Upon $G\alpha i$ or $G\alpha o$-coupled GPCR activation, $G\beta\gamma$ dissociates from the heterotrimeric G protein and binds directly to Kir3.x to activate the channel. In atrial cardiac myocytes, acetylcholine-activated K^+ channels (I_{KACh}), generated by heteromeric assemblies of Kir3.1/3.4, are activated by the GPCR M2 muscarinic receptor in response to acetylcholine released by the vagus nerve to slow the heart rate (Anumonwo and Lopatin 2010). Kir3.x channels composed of homomeric Kir3.2 or heteromers of Kir3.1, Kir3.2 and Kir3.3 are also prominent in the brain, where they have been implicated in synaptic plasticity and in mechanisms underlying drug addiction (Luscher and Slesinger 2010). In pancreatic islets, activation of Kir3-mediated currents in β-cells in response to catecholamines and somatostatin suppresses insulin secretion (Jacobson and Shyng 2020).

K_{ATP} channels are hetero-octameric complexes of pore-forming Kir6.x subunits and regulatory sulfonylurea receptor subunits (SURx) (Nichols 2006). They are expressed in most excitable cells and serve to couple cell energetics to membrane potential. Two Kir6.x isoforms, (Kir6.1 and Kir6.2) and three major SUR isoforms (SUR1 and two splice variants of SUR2, SUR2A and SUR2B) co-assemble to form K_{ATP} channel subtypes with distinct tissue distribution, gating characteristics and function. The best studied is the Kir6.2/SUR1 subtype found in pancreatic endocrine cells and neurons. In pancreatic β-cells, K_{ATP} channels couple glucose metabolism with insulin secretion and are vital to glucose homeostasis. In cardiac myocytes, Kir6.2/SUR2A channels are the most prominent and are important for protecting the heart against ischemic insults. In vascular smooth muscle and endothelial cells, K_{ATP} channels formed by Kir6.1/SUR2B have a key role in regulating vascular tone. K_{ATP} channels are also present in skeletal muscles where they are involved in glycogen storage and muscle atrophy sensing.

K+ transport channels (Kir1.x, Kir4.x, Kir5.x and Kir7.x) are intermediate or weak recti-fiers predominantly expressed in epithelial cells and glial cells. Renal epithelial Kir chan-nels play an essential role in maintaining homeostasis of the urine and blood, not only by regulating K+ transport but also the concentrations of other ions such as Na+ and Cl– through functional coupling with various transporters. Kir1.1 channels are expressed in the apical surface facing the lumen. K+ efflux through Kir1.1 channels provides the K+ con-centration gradient across the apical membrane needed to support the function of the Na+/K+/2Cl– cotransporter critical for reabsorption of Na+ (Manis et al. 2020). In the basolateral membrane facing the capillary, Kir channels formed by heteromeric assembly of Kir4.1 and Kir5.1 are expressed. K+ transport through these channels provides K+ needed to drive the activity of Na-K ATPase for unidirectional Na+ reabsorption back to the plasma. Thus, Kir4.1/Kir5.1 channels are significant determinants of the transepithelial voltage, which maintains electrolyte homeostasis, regulates blood volume and ultimately contributes to long-term blood pressure control. In astrocytes and glial cells of the nervous system, homomeric Kir4.1 channels mediate uptake of excess extracellular K+ into astroglial cells during high neuronal activity, a process referred to as "K+ buffering" (Beckner 2020). This occurs as a result of a shift in E_K toward a more positive value than V_m as K+ accumulates in the extracellular space, which causes K+ influx into astroglial cells to restore normal extracellular K+ concentrations. Finally, Kir7.1 is one of the most recently cloned members of the Kir channel family and the least understood. The channel is broadly expressed as a homotetramer in epithelial cells of the eye, gut, kidney, choroid plexus, respiratory tract and certain neurons in the brain. In neurons from the paraventricular nucleus of the hypothalamus, activation of the melanocortin-4 receptor (MC4R) depolarizes neurons via G protein-independent closure of Kir7.1 channels (Ghamari-Langroudi et al. 2015), and deletion of Kir7.1 in these neurons results in late-onset obesity (Anderson et al. 2019). These studies point to a role of Kir7.1 in energy homeostasis.

9.3 Subunit Diversity and Basic Structural Organization

The first Kir genes, encoding Kir1.1 and Kir2.1, were isolated in 1993 via expression clon-ing (Ho et al. 1993; Kubo et al. 1993). Since then, 16 Kir channel genes have been identified forming seven subfamilies (Figure 9.2). Most Kir channels are tetrameric assemblies of identical or similar Kir proteins each containing two transmembrane helices M1 and M2, and cytosolic N- and C-termini. The lone exception is the ATP-sensitive potassium (K_{ATP}) channel, which requires co-assembly of the Kir6 tetramer with four additional sulfonyl-urea receptors (SURs), members of the ABC transporter family. Prokaryotes also possess Kir channels and so far nine genes have been identified. They share the same architecture as mammalian Kir channels and have been widely used for high-resolution structural studies. However, they have different gating properties and will not be considered further in this chapter.

Various subunit combinations give rise to functional diversity in Kir channels. This is most clearly demonstrated in recombinant expression systems using defined mRNAs or cDNAs. For example, homomeric Kir4.1 channels have lower single-channel conductance and reduced pH sensitivity compared with heteromeric Kir4.1/Kir5.1 channels. K_{ATP} chan-nels formed by different Kir6.x and SURx combinations have distinct unitary conductance, open probability, and sensitivities to intracellular ATP and ADP. In native tissues, different

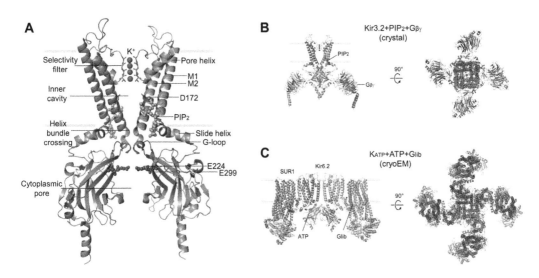

FIGURE 9.3

Structures of Kir channels. (A) Crystal structure of chicken Kir2.2 bound to PIP_2 (PDB ID3SPI). Major domains and structural features are labeled. D172 is the rectification controller that binds polyamines with high affinity. E224 and E299 in the cytoplasmic domain form weaker interactions with polyamines. Only two of four subunits are shown for clarity. (B) Crystal structure of mouse Kir3.2 in complex with PIP_2 and Gβ γ viewed from the side and the top (PDB ID4KFM). (C) Cryo-EM structure of a rodent Kir6.2/SUR1 K_{ATP} channel bound to ATP and glibenclamide (Glib) (PDB ID6BAA).

Kir proteins have different but overlapping tissue distributions. Precise identification of Kir channel compositions that underlie native currents can be challenging, and largely relies on evidence from mRNA and protein expression, electrophysiology, and genetic deletion experiments. In addition to subunit compositions, alternative splicing serves as another mechanism to increase diversity. Multiple splice variants have been reported for Kir1.1, Kir2, Kir3.2, Kir3.3, Kir4.1 and Kir5.1. Among mammalian Kir members, Kir7.1 is the least conserved in amino acid sequence and has some unusual properties including an extremely small unitary conductance at ~50 fS, a larger Rb^+-to-K^+ conduction ratio, and an approximately tenfold lower sensitivity to Ba^{2+} and Cs^+ block.

Each Kir subunit has two transmembrane helices (M1 and M2), and cytosolic N- and C-termini (Figure 9.3). The two transmembrane helices are connected by an extracellular turret region, a short pore helix and the selectivity filter. This topology is the simplest of all potassium channels. High-resolution structures have now been solved for a number of prokaryotic and mammalian Kir channels using either X-ray crystallography or more recently single-particle cryo-EM. The transmembrane domain formed by the two TM helices and the pore helix bear structural similarity to the bacterial K_{CSA} channels. The central cavity is lined by the four M2 helices, which converge at the cytoplasmic end to form a constriction known as the helix bundle crossing. The outer vestibule including the turret loop varies in size and shape, which contributes to differences in unitary conductance and sensitivity to various blockers including ions and peptide toxins (Whorton and MacKinnon 2011). The cytoplasmic domain is formed by association of the N-terminus and C-terminus of neighboring subunits. The N-terminus of Kir protein contains an amphipathic helix named the slide helix or interfacial helix that runs parallel to the inner membrane, and has been proposed to move during gating. Notably, to obtain crystal structures of Kir2 and Kir3 channels, the distal N-terminus of the proteins had to be removed (Hansen, Tao, and

MacKinnon 2011; Whorton and MacKinnon 2011). In cryo-EM structures of K_{ATP} channels and Kir3 channels, the distal N-terminus of the Kir proteins is also poorly resolved (Martin, Yoshioka, et al. 2017; Niu et al. 2020). Thus, the distal N-terminus is likely disordered and relatively flexible. Interestingly, in cryo-EM structures of K_{ATP} channels bound to glibenclamide, a channel inhibitor, at the SUR subunit, the Kir6.2 N-terminus is located in the cavity formed by the two halves of the ABC core of SUR, adjacent the bound inhibitor, suggesting drug binding stabilizes the intrinsically disordered Kir6 N-terminus (Martin et al. 2019). The cytoplasmic C-terminal domain, linked to M2 via a C-linker, extends the transmembrane channel pore into the cytoplasm, nearly doubling the conduction pathway to ~60 Å. Within the cytoplasmic pore, an intrinsically flexible loop forms a girdle around the central pore axis (Pegan et al. 2005). This constriction point, called the G-loop gate, sits at the apex of the cytoplasmic domain below the helix bundle crossing gate, and it is thought that both constriction points need to open in order to pass K^+. The interfaces between subunits formed by N- and C-termini of adjacent subunits are important for ligand sensing and gating. A prominent example is seen in K_{ATP} channels where ATP inhibits channel activity by binding at the interface between two subunits, being stabilized by residues from the N-terminus of one subunit and the C-terminus of a neighboring subunit (Martin, Yoshioka, et al. 2017).

9.4 Gating of Kir Channels

9.4.1 Regulators of Kir Channel Gating

Gating of Kir channels at the helix bundle crossing and G-loop gates is under the regulation of diverse signals, including lipids, proteins, nucleotides and ions. Common to eukaryotic Kir channels, however, is their activation by membrane phosphatidylinositol 4,5-bisphosphate (PIP_2). Crystal structures of Kir2 and Kir3 channels in complex with PIP_2 show that PIP_2 binds within a pocket near the cytoplasmic end of the TM helices. Positively charged and hydrophobic residues from M1 and M2 as well as the cytoplasmic domain interact with the negatively charged phosphate head groups and lipid tails, respectively. Mutation of these residues reduce channel open probability, which can be reversed by addition of exogenous PIP_2 (Shyng and Nichols 1998; Baukrowitz et al. 1998; Huang, Feng, and Hilgemann 1998). In excised membrane patches, Kir channel activity gradually declines, a phenomenon known as rundown, which is attributed to the breakdown of PIP_2 by Mg^{2+}-dependent lipid phosphatases and phospholipases. Rundown can be reduced or prevented by inclusion of EDTA to chelate Mg^{2+} or MgATP to replenish PIP_2 via lipid kinases (Lin et al. 2003); conversely, it can be artificially induced by application of PIP_2 antibodies or polycations such as polylysine or neomycin (Rapedius et al. 2007) that compete with Kir channels for PIP_2 binding.

Kir channels display a wide range of spontaneous activity. This activity, referred to as intrinsic P_o, has been postulated to reflect apparent affinity of the channel for PIP_2. Thus, Kir1 and Kir2 channels, which show high spontaneous activity, are thought to have stronger interactions with PIP_2; while Kir3 channels having low spontaneous activity are thought to have weaker interactions with PIP_2 (Zhang et al. 1999). Other factors can modulate channel activity in an additive, synergistic or competitive manner with PIP_2.

For example, binding of both Gβγ and Na$^+$ markedly increase the activity of neuronal Kir3.2 and cardiac Kir3.1/Kir3.4 (K$_{ACh}$) channels by stabilizing PIP$_2$ interactions (Whorton and MacKinnon 2011; Rosenhouse-Dantsker et al. 2008). A similar effect of Na$^+$ is seen in Kir4.1/Kir5.1 channels (Rosenhouse-Dantsker et al. 2008). In Kir2 channels, recent studies show that a second phospholipid-binding site is critical for PIP$_2$ sensitivity and has a synergistic effect with PIP$_2$ on channel opening (Lee et al. 2013). K$_{ATP}$ channels containing Kir6.2 pore subunits show relatively high intrinsic P_o. However, when artificially expressed in the absence of SUR, Kir6.2 has very low open probability similar to Kir3 channels without Gβγ and Na$^+$ (Tucker et al. 1997). SUR thus appears to enhance or stabilize the interaction between Kir6.2 and PIP$_2$, akin to the effect of Gβγ on Kir3 channels. A cardinal feature of K$_{ATP}$ channels is their inhibition by intracellular ATP. ATP competes with PIP$_2$ to allosterically close or open the channel, respectively. ATP binds at the interface between two adjacent Kir6 subunits and may engage residues that are also involved in PIP$_2$ binding (Driggers and Shyng, 2023). Finally, intracellular pH affects the activity of a number of Kir channels (Hibino et al. 2010). Intracellular acidification reduces the activity of Kir1.1 and Kir2.3, and channels containing Kir4.1. K$_{ATP}$ channels exhibit a bell-shaped response to pH and are activated by mild acidic pH, which has been proposed to result from protonation of a histidine next to the PIP$_2$-binding site for increased PIP$_2$ binding.

9.4.2 Conformational Changes during Kir Channel Gating

In single-channel recordings, several Kir channels show bursts of openings interspersed with long closures. Within a burst, brief openings or flickerings are observed. These events are referred to as fast gating. The transitions between bursts of activity are referred to as slow gating. Mutations within the pore loop of Kir channels affect the intraburst kinetics, indicating that fast gating occurs at the selectivity filter. There is evidence that the selectivity filter and the outer pore undergo conformational changes, and that constriction in this region may trap permeant ions and cause brief channel closures. In contrast, the frequency and duration of bursts are affected by mutations at the intracellular end of M2, indicating that slow gating involves the helix bundle crossing gate formed by bulky hydrophobic residues. Movements of the M2 helices leading to widening of the bundle crossing gate have been proposed to occur during gating. In particular, a highly conserved glycine residue in the middle of M2 has been proposed to serve as a hinge to allow bending of M2 necessary to open the cytoplasmic gate; however, this notion has recently been challenged and remains unresolved (Black et al. 2021). In addition to M2, M1 and the N-terminal slide helix preceding M1 are also implicated in gating. M1 hydrogen bonds with M2 near the helix bundle crossing and this interaction is altered during gating by protons and PIP$_2$ (Rapedius et al. 2007), while the slide helix is thought to couple ligand binding to channel gating. It has been proposed that gating at the selectivity filter and the cytoplasmic end may be coupled; however, whether and how this occurs remains unresolved (Black et al. 2021).

Comparison of the crystal structures of Kir2.1 in the absence and presence of PIP$_2$ revealed a PIP$_2$-induced translocation of the cytoplasmic domain toward the transmembrane domain by ~6 Å (Hansen, Tao, and MacKinnon 2011). The C-linker, which connects M2 to the cytoplasmic C-terminal domain, transforms from an unstructured loop in the apo structure to a helix (also called the tethering helix) in the PIP$_2$-bound structure, bringing positively charged residues in the N- and the C-terminal domains close to membrane PIP$_2$ for interactions. A similar PIP$_2$-dependent "docking" of the cytoplasmic domain to

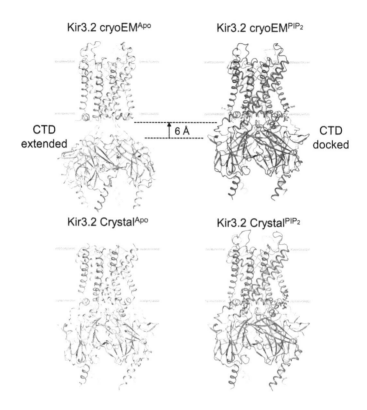

Kir3.2 cryoEM^Apo   Kir3.2 cryoEM^PIP2

CTD extended   ↑6 Å   CTD docked

Kir3.2 Crystal^Apo   Kir3.2 Crystal^PIP2

FIGURE 9.4
Comparison of apo and PIP$_2$-bound Kir3.2 structures resolved using cryo-EM and X-ray crystallography. The top two structures were resolved by single-particle cryo-EM. In the predominant apo cryo-EM structure (Protein Data Bank ID: 6XIS), the cytoplasmic domain (CTD) is extended away from the membrane by ~6 Å compared to the docked CTD in the predominant PIP$_2$-bound structure (Protein Data Bank ID: 6XIT). The bottom two structures were solved by X-ray crystallography. Note in both apo (Protein Data Bank ID: 3SYO) and PIP$_2$-bound (Protein Data Bank ID: 3SYA) structures, the CTD is closely docked to the membrane. The docked CTD in the apo structure is likely caused by crystal contacts (Niu et al. 2020).

the transmembrane domain is also observed in cryo-EM structures of Kir3.2 (Niu et al. 2020). Note, earlier crystal structures of Kir3.2 channels with and without PIP$_2$ do not show a difference in the location of the cytoplasmic domain relative to the transmembrane domain; in light of the recent Kir3.2 cryo-EM structures, the docked cytoplasmic domain seen in the apo structure has now been attributed to artifacts caused by crystal contacts (Figure 9.4). In addition to movement of the cytoplasmic domain perpendicular to the membrane, rigid body rotations of the cytoplasmic domain have been observed. For example, in Kir3 channels, a rotation of the cytoplasmic domain (clockwise when viewed from the extracellular side) is observed in PIP$_2$-bound structure relative to the apo state structure. Interestingly, the upward movement and rotation of the cytoplasmic domain in Kir3.2 channels render the cytoplasmic domain permissive to Gβγ binding, which is consistent with functional studies showing that Gβγ activation of the channel requires PIP$_2$. Similar translational and rotational movements of the cytoplasmic domain have also been observed in Kir6.2 of the Kir6.2/SUR1 K$_{ATP}$ channel complex, and have been correlated with ligand binding and functional state of the channel (Driggers and Shyng, 2023).

If PIP_2 and Kir channel agonists stimulate channel activity by opening the cytoplasmic gates, one would expect ligand-dependent widening of these gates. In Kir2.2, which is principally gated by PIP_2, the helix bundle crossing gate has been shown to widen in the presence of PIP_2, with little change in the G-loop gate, which is already open even without PIP_2 (Hansen, Tao, and MacKinnon 2011). In the case of Kir3.2, the helix bundle crossing and G-loop gates are both closed in the unliganded state. $G\beta\gamma$ binding widens the G-loop gate but not the helix bundle crossing gate, while PIP_2 widens the helix bundle crossing gate but not the G-loop gate. Only when both PIP_2 and $G\beta\gamma$ are present do both the helix bundle crossing and G-loop gates widen (Whorton and MacKinnon, 2011; Whorton and MacKinnon, 2013). It is worth noting that even when the gates widen in the mammalian Kir channel structures solved to date, the openings are not enough to allow fully hydrated K^+ to pass through, including recent open structures of K_{ATP} channels harboring a Kir6.2 mutation near the helix bundle crossing that renders the channel constitutively active (Zhao and MacKinnon, 2021). There is evidence from MD simulation studies that K^+ conduction may not require wide-open lower gates that accommodate fully hydrated K^+. Rather, local partial dehydration of K^+ near the gate and interactions of partially dehydrated K^+ with flexible side-chains of gate residues may be sufficient to permit K^+ conduction (Black et al. 2020; Bernsteiner et al. 2019).

9.5 Ion Selectivity and Conduction

Kir channels are highly selective for K^+. This high K^+ selectivity is conferred by the selectivity filter, the narrowest part of the conduction pathway with the signature sequence motifs TXGYG or TXGFG. High-resolution structures of Kir channels show that like other K^+ channels, the hydroxyl oxygens of threonine and main chain carbonyl oxygens of the conserved K^+ channel signature sequence form a square antiprism at each of the K^+-binding sites (Figure 9.3).

In vivo, conduction of all Kir channels is voltage-dependent due to inward rectification. The idea that inward rectification might result from a positively charged substance blocking the channel from the internal vestibule was proposed by Armstrong in 1969, who noted a resemblance between a voltage-dependent block of Kv channels treated with internal TEA and the voltage-dependent currents of Kir channels exhibiting strong inward rectification. Indeed, inward rectification disappeared when channels were isolated from the intracellular environment. Application of Mg^{2+} to the cytoplasmic side of the patch reconstituted most of the inward rectification; however, a very strong voltage-dependent rectification component was still missing. This component could be recovered by moving the patch close to the cell, suggesting substance released by the cell produced rectification. Biochemical analysis revealed that intracellular polyamines are responsible for strongly voltage-dependent inward rectification (Lopatin, Makhina, and Nichols 1994). Polyamines are positively charged rod-shaped molecules present in cells at submillimolar concentrations. It is now well documented that strong inward rectification is caused by binding of polyamines to the channel, through initial interactions that are weakly voltage-dependent involving acidic residues near the cytoplasmic end of the pore (e.g., Glu224 and Glu229 in Kir2.1; Figure 9.3A), followed by strongly voltage-dependent interactions involving a negatively charged residue in M2 lining the transmembrane pore (D172 in Kir2.1, known as the rectification controller). In Kir channel subtypes that show weak inward rectification, such as Kir1.1 and Kir6.x-containing K_{ATP} channels, the rectification controller position in M2 is occupied by an uncharged Asn residue. Of note, in all Kir channels except for Kir7.1, the voltage dependence of inward rectification is modulated by extracellular K^+ concentrations. Increasing external K^+ relieves the rectification by K^+ binding at external sites and knocking off Mg^{2+} and polyamines from sites deeper inside the pore. As a result, the voltage range of channel opening shifts to more positive voltages.

9.6 Pharmacology

Among Kir channels, K_{ATP} channels have the richest pharmacology. Sulfonylureas were discovered to have hypoglycemic effects in 1942 by Mercel Janbon while studying sulfonamide antibiotics. The serendipitous finding led to further development of sulfonylureas and other related compounds such as glinides for the treatment of type 2 diabetes. These drugs inhibit pancreatic β-cell K_{ATP} channels to stimulate insulin secretion and lower blood glucose. Recent cryo-EM structures of K_{ATP} channels in complex with the high-affinity sulfonylurea glibenclamide show that the drug binds to a pocket formed by the transmembrane helix bundle consisting of TMs 6, 7, 8, and 11 from TMD1, and TMs 15 and 16 from TMD2 (Martin, Kandasamy, et al. 2017). In the structure, the two nucleotide-binding domains (NBDs) of SUR1 are separate and the distal N-terminus of Kir6.2 is wedged in a cleft formed by the two transmembrane bundles of the SUR1 ABC core structure, next to the bound drug. The structure suggests glibenclamide inhibits channel activity by preventing dimerization of NBDs in SUR1 and trapping the Kir6.2-distal N-terminus to prevent Kir6.2 from conformational changes required to open the channel. Additional structurally distinct K_{ATP} inhibitors, including carbamazepine and repaglinide, a prototypical glinide, share similar structural mechanisms. Sulfonylureas and glinides also inhibit other K_{ATP} channel isoforms including the cardiac Kir6.2/SUR2A channel and vascular Kir6.1/SUR2B channel, but with lower affinities. Drugs that open K_{ATP} channels such as diazoxide, pinacidil, and cromakalim are also used clinically to treat congenital hyperinsulinism, hypertension, and hair loss. They too target the SUR subunits by binding to a pocket formed by transmembrane helices of the SUR ABC core when the two SUR nucleotide binding domains dimerize (Wu, Ding, and Chen, 2022). A challenge facing clinical use of existing K_{ATP} channel drugs is lack of desired isoform specificity. While current drugs all exhibit preferential effects on certain K_{ATP} channel isoforms, they have significant off-target effects on other K_{ATP} channel isoforms. For example, diazoxide, used to treat congenital hyperinsulinism by activating pancreatic K_{ATP} channels also stimulates vascular K_{ATP} channels, which results in drug-induced Cantú syndrome normally associated with gain-of-function mutations in vascular K_{ATP} channels. In addition to drugs targeting the SUR subunits, those targeting Kir6.x subunits have also been reported, including the inhibitors cyanoguanidines (e.g., PNU-37883) and rosiglitazone. However, they also lack high isoform specificity and are less well characterized. High-throughput screening (HTS) has identified newer inhibitors and activators, and continuing efforts on this front are needed to overcome current limitations.

By comparison, pharmacological compounds have been historically underdeveloped for other Kir channels. The mainstay inhibitors have been Ba^{2+} or Cs^+, which inhibit all Kir channels. Recent molecular target-based approaches using HTS of small-molecule libraries and medicinal chemistry have led to Kir channel modulators with improved potency and specificity. For example, Bhave and colleagues employed a thallium flux assay in a HTS to identify VU591 as the first potent (IC_{50} of 240 nM) and selective small-molecule pore blocker of the Kir1.1 channel. The compound shares structural similarity to a Kir1.1 inhibitor developed by Merck & Co. called compound A, which has been shown to induce natriuresis and diuresis without causing potassium wasting, lower blood pressure, and protect the kidney from hypertension-induced kidney injury. Derivatives of these compounds are also being explored as insecticides against *Aedes aegypti* mosquitoes to prevent diseases spread by mosquitoes such as Zika virus and dengue fever, as inhibition of mosquito Kir1 channels disrupts diuresis and hemolymph K^+ homeostasis (Weaver and Denton 2021). Another example involves Kir3 channels, which are increasingly recognized as important targets for diseases such as epilepsy, and drug and alcohol addiction. Recent function and structure-based screening methods have led to the identification of ML297 as an activator of Kir3.1/Kir3.2 channels (EC_{50} of 160 nM). Interestingly, the effect of ML297 is dependent on PIP_2 but not G$\beta\gamma$, and thus is GPCR-independent. Some ML297 derivatives have been shown to have analgesic effects in animal studies and improved potency and/or selectivity over cardiac Kir3.1/3.4 channels. Photo-switchable versions of ML297-based compounds called light-operated GIRK channel openers (or LOGOs) have also been developed for research, although their use is limited due to lack of isoform specificity. Finally, using a structural model of Kir3.2 bound to alcohol, an activator called GiGA1 has been developed using virtual screenings. The compound activates Kir3.1/3.2 channels in a manner that is also independent of G protein but is mechanistically different from ML297. Significantly, GiGA1 has been shown to reduce seizure durations and severity in an acute epilepsy mouse model (Zhao et al. 2021). These studies highlight the potential of structure-based drug development.

9.6.1 Peptide Toxins

In addition to small molecules, several peptide toxins that target Kir channels have been identified. Tertiapin, isolated from honeybee venom, is a potent inhibitor of both Kir1.1 and Kir3.1/3.4 channels, and binds in the outer vestibule of the channel pore with nanomolar affinities. More recently, the spider toxin SpTx-1 has been shown to inhibit human K_{ATP} channels composed of Kir6.2 and SUR1 by binding to the extracellular side of Kir6.2 with a K_d of 15 nM. SpTx-1 was shown to inhibit several gain-of-function mutations associated with permanent neonatal diabetes not responsive to sulfonylureas, raising the possibility that the toxin may be of use in designing Kir6.2-targeting inhibitors to treat neonatal diabetes or DEND (developmental delay, epilepsy, and neonatal diabetes) syndrome caused by gain-of-function K_{ATP} channel mutations.

9.7 Regulation

Posttranslational modifications affecting channel expression, trafficking and/or function have been reported in many Kir channels. The most common posttranslational modification is phosphorylation. Phosphorylation by protein kinases such as protein kinases A and C (PKA and PKC, respectively) has been shown to modulate surface expression or gating of various Kir channels, including Kir1.1, Kir2.x, K_{ATP} and Kir7.1, either positively or negatively in site-specific manners. In addition to phosphorylation, palmitoylation of a cysteine residue (C166) in M2 of Kir6.2 has been recently reported in association with enhanced channel sensitivity to PIP_2 and increased K_{ATP} channel opening (Yang et al. 2020).

Membrane PIP_2 increases the activity of all Kir channels in isolated membranes. However, whether PIP_2 is absolutely necessary for channel activity and whether physiological changes of PIP_2 levels regulate Kir channel activity is less well resolved. A recent study shows that purified K_{ATP} channels reconstituted in lipid bilayers lacking PIP_2 can still open (Zhao and MacKinnon, 2021). While there is evidence that hydrolysis of PIP_2 by phospholipase C following activation of Gq protein-coupled receptors (GqPCRs) inhibits the activity of some Kir channels including Kir3.x, Kir2.1 and K_{ATP} channels, there is also evidence that channel inhibition is due to PKC-mediated channel phosphorylation following GqPCR activation (Harraz, Hill-Eubanks, and Nelson 2020). In addition to PIP_2, cholesterol is another lipid that has been implicated in Kir channel regulation. Cholesterol suppresses the activity of Kir2.1, while increasing the activity of Kir3.2 (Levitan 2009). Although cholesterol is initially postulated to impact channel function by affecting membrane properties, recent structural and functional studies provide evidence that cholesterol modulates these channels by direct binding to the channels, which may alter channel interactions with PIP_2 (Rosenhouse-Dantsker 2019).

Like cholesterol, *n*-alcohols such as methanol, ethanol, and 1-propanol activate Kir3.x channels in the presence of PIP_2 (Glaaser and Slesinger 2015). An alcohol-binding pocket in Kir3.2 has been identified in the subunit interface in the cytoplasmic domain, overlapping with the region involved in channel activation by Gβγ. The alcohol-binding site does not overlap with the proposed cholesterol binding sites near the membrane, and the activation effects of alcohol and cholesterol are additive, suggesting the two regulators activate the channel via distinct mechanisms.

Lastly, physiological signals affecting the levels of Kir channel ligands can regulate Kir channel activity. Kir3.x channels are activated by neurotransmitters and hormones acting on GPCRs that result in the release of Gβγ from trimeric G proteins. Conversely, regulator of G protein signaling (RGS) proteins accelerate GTP hydrolysis in the Gα subunit and reduce the amount of free Gβγ, resulting in reduced activity of Kir3.x channels (Hibino et al. 2010). For K_{ATP} channels, a key physiological regulator is glucose. High glucose increases intracellular ATP to ADP ratios to inhibit K_{ATP} channel activity, whereas low glucose decreases ATP to ADP ratios to open the channel.

9.8 Cell Biology: Biogenesis, Trafficking, Subcellular Targeting

Surface expression and localization of Kir channels are regulated by diverse mechanisms, including endoplasmic reticulum (ER) export signals, ER retention signals, Golgi exit signals, endocytosis signals and signals targeting specific membrane compartments. When expressed in heterologous systems, different Kir channels show different surface expression levels, with Kir1.1 and Kir2.x having much stronger expression than Kir3.x and Kir6.x. This is largely due to the presence of ER export motifs present in Kir1.1 and Kir2.1. In Kir1.1, VLS and EXD motifs in the C-terminal domain promote ER exit and forward trafficking to the plasma membrane. In Kir2.x, a FCYENE motif in the C-terminal domain serves as an ER export signal, while additional motifs in the cytosolic N-terminal (R44, R46) and C-terminal (YXXΦ) domains are required for trafficking from Golgi to plasma membrane. For Kir3.x channels, surface expression is dependent on subunit compositions. Homomeric Kir3.1 channels are retained in the ER; by contrast, heteromeric Kir3.1/3.2 or Kir3.1/3.4 channels are efficiently expressed at the cell surface. Furthermore, Kir3.3 suppresses surface expression of channels containing Kir3.2 or Kir3.4. The complex expression patterns are due to the interplay of ER exit, post-ER to plasma membrane trafficking, and ER to lysosome trafficking signals, which are differentially present in different subunits. Kir6.x channels are unique in requiring SURx for surface expression. A tripeptide -RKR- motif present in the C-terminus of Kir6.x, and also in the cytoplasmic domain of SURx between the first transmembrane domain and the first nucleotide-binding domain, acts as an ER retention signal to prevent unassembled or incompletely assembled channel proteins from leaving the ER (Zerangue et al. 1999). Upon octameric assembly, the -RKR- motifs are thought to be shielded, likely involving binding of 14-3-3 proteins (Heusser et al. 2006), to allow the complex to leave the ER. The ER retention signals in both subunits thus provide a quality control mechanism to ensure only correctly assembled functional channels traffic to the cell surface (Figure 9.5). A similar ER retention signal, RAR, is present in Kir1.1, but phosphorylation of an N-terminal serine by the serum-glucocorticoid-regulated kinase SGK-1 overrides this signal to promote channel surface expression.

Many Kir channels are expressed in polarized cells and their functions in these cells are dependent on correct targeting to specific membrane regions. For example, in the postsynaptic membrane, Kir3.1/3.2 is associated with Gi/o-coupled receptors, such as γ-aminobutyric acid type B (GABA$_B$), to produce ligand-induced inhibitory postsynaptic currents. In the apical membrane of renal epithelia, Kir1.1 is functionally associated with the Na$^+$-K$^+$-2Cl$^-$ cotransporter to balance salt and water homeostasis. Polarized expression can be achieved by channel interactions with scaffold proteins. In renal epithelial cells binding of Kir1.1 via its PDZ domain binding motif to the PDZ domains of NHERF-1 and NHERF-2 results in targeted expression of the channel to apical membranes. In neurons and epithelial cells, Kir2 channels are targeted to the postsynaptic dendrites and

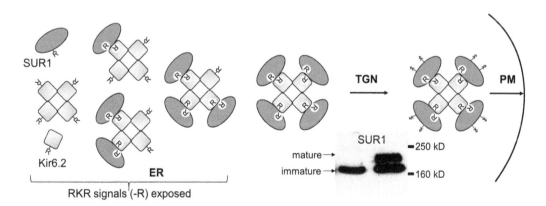

FIGURE 9.5

K$_{ATP}$ channel biogenesis and trafficking regulation. Hetero-octameric assemblies of K$_{ATP}$ channels occur in the endoplasmic reticulum (ER). Both Kir6.2 and SUR1 contain an -RKR- ER retention motif. In unassembled or partially assembled subunits, the RKR signal is exposed, causing ER retention. Upon complete assembly, the RKR signal becomes shielded allowing ER exit of the complex. As the assembled channels travel through the trans Golgi network (TGN), the two N-linked glycosylation sites in SUR1 are further modified to become mature, complex glycosylated SUR1 that is distinguishable on Western blot from the immature, core-glycosylated SUR1 found in the ER. The Western blot shown was from COSm6 cells transfected with SUR1 alone (left lane) or SUR1 together with Kir6.2 (right lane) and probed with an anti-SUR1 antibody.

basolateral domains, respectively, via interactions between the PDZ binding motifs in the C-terminal domain of Kir2.x channels and the PDZ domain containing scaffold proteins. Similar mechanisms account for expression of Kir4.1/Kir5.1 channels in the basolateral membrane of renal epithelial cells. Interestingly, a PDZ domain-mediated interaction has also been reported to promote endocytic trafficking of Kir3 channels via their binding to PDZ domain containing sorting nexin 27 (Hibino et al. 2010).

In addition to intrinsic signals present in Kir channels, extrinsic signals also regulate Kir trafficking to modulate the number of surface channels. The rapid and dynamic regulation of trafficking affords adaptability to changing physiological demands. In pancreatic β-cells, in response to both glucose starvation and the adipocyte-derived hormone leptin, surface expression of K_{ATP} channels increases via increased channel trafficking to the plasma membrane, which results in suppression of insulin secretion (Cochrane and Shyng 2019). In renal epithelial cells, elaborate signaling mechanisms involving several protein kinases regulate surface expression of Kir1.1 channels. In addition to the SGK-1 mediated phosphorylation that promotes Kir1.1 forward trafficking in response to high K^+ or vasopressin, protein tyrosine kinase Src and WNK kinases (with no lysine kinases) can downregulate surface Kir1.1 via endocytic trafficking (Welling and Ho 2009). These checks and balances ensure proper responses to physiological demands to maintain K^+ homeostasis.

9.9 Kir Channelopathies

Multiple human diseases have now been linked to mutations in Kir genes. These mutations can reduce or enhance channel function by affecting Kir gene transcription, Kir protein translation, folding, assembly, trafficking, turnover and gating. As Kir channels are widely expressed throughout the body, these mutations often cause complex pleiotropic pathologies. A prominent example is loss-of-function mutations in the classical inward rectifier Kir2.1. These mutations cause Andersen–Tawil syndrome (ATS), a rare multisystem disorder characterized by a long QT interval and ventricular arrhythmias, periodic paralysis, and dysmorphic features. In contrast, gain-of-function Kir2.1 mutations cause short QT, which can also lead to cardiac arrhythmias. Loss-of-function mutations in another classical inward rectifier Kir2.6 are associated with thyrotoxic hypokalemic periodic paralysis, which affects skeletal muscle excitability under thyrotoxic conditions.

As important inhibitory downstream effectors of the G protein-mediated signaling pathway, Kir3.x channels are implicated in a number of human diseases, including epilepsy, cardiac arrhythmias, addiction and alcohol abuse. Kir3.2 gain-of-function mutations cause Keppen–Lubinsky syndrome, a rare condition that encompasses lipodystrophy, hypertonia, hyperreflexia, developmental delay and intellectual disability. Familial hyperaldosteronism type III, characterized by early onset of severe hypertension and hypokalemia, is associated with mutations in or near the selectivity filter of Kir3.4. Mutant channels lack K^+ specificity and are permeant to Na^+. In zona glomerulosa cells, where the mutant channels are expressed, the inflow of Na^+ causes cell depolarization and increased intracellular Ca^{2+} concentrations, which activate transcription pathways that increase aldosterone production. Loss-of-function mutations in Kir3.4, a component of cardiac K_{ACh} channels, are associated with long QT syndrome, which indicates that these acetylcholine-activated channels, also play a role in ventricular repolarization. Finally, a recent genome-wide association study of alcoholics found single-nucleotide polymorphisms (SNPs) in Kir3.2 (KCNJ6), though it remains to be determined what impact the SNPs have on Kir3.2 function. Nonetheless, these studies suggest Kir3.2 as a potential target for the treatment of alcohol abuse.

K_{ATP} channels have critical roles in sensing cellular energetics. Loss-of-function mutations in Kir6.2 cause congenital hyperinsulinism. Conversely, activating mutations in Kir6.2 cause transient or permanent neonatal diabetes and DEND syndrome characterized by developmental delay, epilepsy, and neonatal diabetes (Gloyn et al. 2004). Additionally, a polymorphism in Kir6.2 that slightly increases channel activity is associated with increased risk for type 2 diabetes. Gain-of-function mutations in Kir6.1 have now been associated with Cantú syndrome, a rare disorder with clinical symptoms including hypertrichosis, heart valve abnormalities and congenital hypertrophic cardiomyopathy (Grange, Nichols, and Singh 1993).

K^+ transport Kir channels are essential for K^+ homeostasis and overall electrolyte balance. Loss-of-function mutations in Kir1.1 underlie Bartter syndrome type II, a salt-losing nephropathy resulting in hypokalemia and alkalosis. The pleiotropic SeSAME (seizures, sensorineural deafness, ataxia, mental retardation and electrolyte imbalance) or EAST (epilepsy, ataxia, sensorineural deafness and tubulopathy) syndrome is caused by loss-of-function mutations in Kir4.1. Impaired function of Kir4.1 in the kidney leads to hypokalemic metabolic acidosis, while impaired Kir4.1 function in glial cells increases neural tissue potassium levels giving rise to neuron depolarization. Loss of Kir4.1 function also abolishes endocochlear potential and causes deafness. Recently, antibodies against Kir4.1 have been detected in some multiple sclerosis (MS) patients, suggesting that some MS cases may belong to a group of autoimmune disorders caused by ion channel dysfunction. Lastly, Kir7.1 is critical for regulating the direction of fluid transport across the retinal pigment epithelium and subretinal space volume, and helps to maintain K^+ homeostasis around the photoreceptor outer segment. Loss-of-function mutations in Kir7.1 are linked to snowflake vitreoretinal degeneration (SVD), a developmental and progressive eye disorder, and Leber congenital amaurosis (LCA), which results in visual impairment from the first year of life.

9.10 Conclusion

Since the discovery of Kir currents more than half a century ago, much progress has been made in understanding the channels underlying these currents including their mechanisms of inward rectification, gating, trafficking, and their role in physiology and disease. However, many questions remain. We still do not have a complete picture of how Kir channel gating occurs at the structural level, the mechanisms by which Kir channel functions are regulated temporally and spatially, and how genetic variations in Kir genes contribute to human physiology and pathophysiology. As the tools available to study Kir channels continue to advance, it is anticipated that our knowledge in various aspects of Kir channel biology will continue to grow and refine. Particularly exciting is the prospect of harnessing the growing knowledge toward advancing health by targeted manipulations of faulty channels.

Acknowledgments

We thank Dr. Bruce Patton for comments on the manuscript. The work is supported by R01DK057699 and R01DK066485 to Show-Ling Shyng and an OHSU N. L. Tartar Trust Research Fellowship to Camden M. Driggers.

Suggested Reading

This chapter includes additional bibliographical references hosted only online as indicated by citations in blue color font in the text. Please visit https://www.routledge.com/9780367538163 to access the additional references for this chapter, found under "Support Material" at the bottom of the page.

Armstrong, C. M. 1969. "Inactivation of the potassium conductance and related phenomena caused by quaternary ammonium ion injection in squid axons." *J Gen Physiol* 54 (5):553–75. doi:10.1085/jgp.54.5.553.

Baukrowitz, T., U. Schulte, D. Oliver, S. Herlitze, T. Krauter, S. J. Tucker, J. P. Ruppersberg, and B. Fakler. 1998. "PIP2 and PIP as determinants for ATP inhibition of KATP channels." *Science* 282 (5391):1141–4. doi:10.1126/science.282.5391.1141.

Beckner, M. E. 2020. "A roadmap for potassium buffering/dispersion via the glial network of the CNS." *Neurochem Int* 136:104727. doi:10.1016/j.neuint.2020.104727.

Bernsteiner, H., E. M. Zangerl-Plessl, X. Chen, and A. Stary-Weinzinger. 2019. "Conduction through a narrow inward-rectifier K(+) channel pore." *J Gen Physiol* 151 (10):1231–46. doi:10.1085/jgp.201912359.

Black, K. A., R. Jin, S. He, and J. M. Gulbis. 2021. "Changing perspectives on how the permeation pathway through potassium channels is regulated." *J Physiol* 599 (7):1961–76. doi:10.1113/JP278682.

Cochrane, V., and S. L. Shyng. 2019. "Leptin-induced trafficking of KATP channels: a mechanism to regulate pancreatic beta-cell excitability and insulin secretion." *Int J Mol Sci* 20 (11). doi:10.3390/ijms20112660.

Driggers, C. M., S.L. Shyng. 2023. Mechanistic insights on KATP channel regulation from cryo-EM structures. *J Gen Physiol* 2023 Jan 2; 155(1): e202113046. doi: 10.1085/jgp.202113046. Epub 2022 Nov 28. PMID: 36441147; PMCID: PMC9700523.

Ghamari-Langroudi, M., G. J. Digby, J. A. Sebag, G. L. Millhauser, R. Palomino, R. Matthews, T. Gillyard, B. L. Panaro, I. R. Tough, H. M. Cox, J. S. Denton, and R. D. Cone. 2015. "G-protein-independent coupling of MC4R to Kir7.1 in hypothalamic neurons." *Nature* 520 (7545):94–8. doi:10.1038/nature14051.

Gloyn, A.L., E.R. Pearson, J.F. Antcliff, P. Proks, G.J. Bruining, A.S. Slingerland, N. Howard, S. Srinivasan, J.M. Silva, J. Molnes, E.L. Edghill, T.M. Frayling, I.K. Temple, D. Mackay, J.P. Shield, Z. Sumnik, A. van Rhijn, J.K. Wales, P. Clark, S. Gorman, J. Aisenberg, S. Ellard, P.R. Njølstad, F.M. Ashcroft, A.T. Hattersley. 2004 Activating mutations in the gene encoding the ATP-sensitive potassium-channel subunit Kir6.2 and permanent neonatal diabetes. *N Engl J Med* 350(18): 1838–49. doi: 10.1056/NEJMoa032922. Erratum in: N Engl J Med. 2004 Sep 30;351(14):1470. PMID: 15115830.

Grange, D. K., C. G. Nichols, and G. K. Singh. 2014Oct 2 [updated 2020 Oct 1]. "Cantú Syndrome." In *GeneReviews((R))*, [Internet]. edited by M.P. Adam, D.B. Everman, G.M. Mirzaa, R.A. Pagon, S.E. Wallace, L.J.H. Bean, K.W. Gripp, A. Amemiya. Seattle, WA: University of Washington, Seattle; 1993–2022. PMID: 25275207..

Hansen, S. B., X. Tao, and R. MacKinnon. 2011. "Structural basis of PIP2 activation of the classical inward rectifier K+ channel Kir2.2." *Nature* 477 (7365):495–8. doi:10.1038/nature10370.

Harraz, O. F., D. Hill-Eubanks, and M. T. Nelson. 2020. "PIP2: a critical regulator of vascular ion channels hiding in plain sight." *Proc Natl Acad Sci U S A* 117 (34):20378–89. doi:10.1073/pnas.2006737117.

Hibino, H., A. Inanobe, K. Furutani, S. Murakami, I. Findlay, and Y. Kurachi. 2010. "Inwardly rectifying potassium channels: their structure, function, and physiological roles." *Physiol Rev* 90 (1):291–366. doi:10.1152/physrev.00021.2009.

Ho, K., C. G. Nichols, W. J. Lederer, J. Lytton, P. M. Vassilev, M. V. Kanazirska, and S. C. Hebert. 1993. "Cloning and expression of an inwardly rectifying ATP-regulated potassium channel." *Nature* 362 (6415):31–8. doi:10.1038/362031a0.

Huang, C. L., S. Feng, and D. W. Hilgemann. 1998. "Direct activation of inward rectifier potassium channels by PIP2 and its stabilization by Gbetagamma." *Nature* 391 (6669):803–6. doi:10.1038/35882.

Kubo, Y., T. J. Baldwin, Y. N. Jan, and L. Y. Jan. 1993. "Primary structure and functional expression of a mouse inward rectifier potassium channel." *Nature* 362 (6416):127–33. doi:10.1038/362127a0.

Lee, S. J., S. Wang, W. Borschel, S. Heyman, J. Gyore, and C. G. Nichols. 2013. "Secondary anionic phospholipid binding site and gating mechanism in Kir2.1 inward rectifier channels." *Nat Commun* 4:2786. doi:10.1038/ncomms3786.

Levitan, I. 2009. "Cholesterol and Kir channels." *IUBMB Life* 61 (8):781–90. doi:10.1002/iub.192.

Lopatin, A. N., E. N. Makhina, and C. G. Nichols. 1994. "Potassium channel block by cytoplasmic polyamines as the mechanism of intrinsic rectification." *Nature* 372 (6504):366–9. doi:10.1038/372366a0.

Manis, A. D., M. R. Hodges, A. Staruschenko, and O. Palygin. 2020. "Expression, localization, and functional properties of inwardly rectifying K(+) channels in the kidney." *Am J Physiol Ren Physiol* 318 (2):F332–F7. doi:10.1152/ajprenal.00523.2019.

Martin, G. M., C. Yoshioka, E. A. Rex, J. F. Fay, Q. Xie, M. R. Whorton, J. Z. Chen, and S. L. Shyng. 2017. "Cryo-EM structure of the ATP-sensitive potassium channel illuminates mechanisms of assembly and gating." *eLife* 6. doi:10.7554/eLife.24149.

Nichols, C. G. 2006. "KATP channels as molecular sensors of cellular metabolism." *Nature* 440 (7083):470–6. doi:10.1038/nature04711.

Nichols, C.G., S.J. Lee. 2018. Polyamines and potassium channels: A 25-year romance. *J Biol Chem* 293(48): 18779–18788. doi: 10.1074/jbc.TM118.003344. Epub 2018 Oct 17. PMID: 30333230; PMCID: PMC6290165.

Niu, Y., X. Tao, K. K. Touhara, and R. MacKinnon. 2020. "Cryo-EM analysis of PIP2 regulation in mammalian GIRK channels." *eLife* 9. doi:10.7554/eLife.60552.

Pegan, S., C. Arrabit, W. Zhou, W. Kwiatkowski, A. Collins, P. A. Slesinger, and S. Choe. 2005. "Cytoplasmic domain structures of Kir2.1 and Kir3.1 show sites for modulating gating and rectification." *Nat Neurosci* 8 (3):279–87. doi:10.1038/nn1411.

Shyng, S. L., and C. G. Nichols. 1998. "Membrane phospholipid control of nucleotide sensitivity of KATP channels." *Science* 282 (5391):1138–41. doi:10.1126/science.282.5391.1138.

Tucker, S. J., F. M. Gribble, C. Zhao, S. Trapp, and F. M. Ashcroft. 1997. "Truncation of Kir6.2 produces ATP-sensitive K+ channels in the absence of the sulphonylurea receptor." *Nature* 387 (6629):179–83. doi:10.1038/387179a0.

Weaver, C. D., and J. S. Denton. 2021. "Next-generation inward rectifier potassium channel modulators: discovery and molecular pharmacology." *Am J Physiol Cell Physiol.* 320(6):C1125-C1140. doi:10.1152/ajpcell.00548.2020. Epub 2021 Apr 7. PMID: 33826405; PMCID: PMC8285633.

Welling, P. A., and K. Ho. 2009. "A comprehensive guide to the ROMK potassium channel: form and function in health and disease." *Am J Physiol Ren Physiol* 297 (4):F849–63. doi:10.1152/ajprenal.00181.2009.

Whorton, M. R., and R. MacKinnon. 2011. "Crystal structure of the mammalian GIRK2 K+ channel and gating regulation by G proteins, PIP2, and sodium." *Cell* 147 (1):199–208. doi:10.1016/j.cell.2011.07.046.

Whorton, M.R., R. MacKinnon. 2011. Crystal structure of the mammalian GIRK2 K+ channel and gating regulation by G proteins, PIP2, and sodium. *Cell* 147(1): 199–208. doi: 10.1016/j.cell.2011.07.046. PMID: 21962516; PMCID: PMC3243363.

Wu, J. X., D. Ding, M. Wang, Y. Kang, X. Zeng, and L. Chen. 2018. "Ligand binding and conformational changes of SUR1 subunit in pancreatic ATP-sensitive potassium channels." *Protein Cell* 9 (6):553–67. doi:10.1007/s13238-018-0530-y.

Wu, J. X., D. Ding, L. Chen. 2022. The Emerging Structural Pharmacology of ATP-Sensitive Potassium Channels. *Mol Pharmacol* 102(5): 234–239. doi: 10.1124/molpharm.122.000570. Epub 2022 Sep 2. PMID: 36253099.

Yang, H. Q., W. Martinez-Ortiz, J. Hwang, X. Fan, T. J. Cardozo, and W. A. Coetzee. 2020. "Palmitoylation of the KATP channel Kir6.2 subunit promotes channel opening by regulating PIP2 sensitivity." *Proc Natl Acad Sci U S A* 117 (19):10593–602. doi:10.1073/pnas.1918088117.

Zerangue, N., B. Schwappach, Y. N. Jan, and L. Y. Jan. 1999. "A new ER trafficking signal regulates the subunit stoichiometry of plasma membrane K(ATP) channels." *Neuron* 22 (3):537–48. doi:10.1016/s0896-6273(00)80708-4.

C. Zhao, R. MacKinnon. 2021. Molecular structure of an open human KATP channel. *Proc Natl Acad Sci U S A* 118(48):e2112267118. doi: 10.1073/pnas.2112267118. PMID: 34815345; PMCID: PMC8640745.

10

Two-Pore-Domain Potassium Channels

Leigh D. Plant and Steve A.N. Goldstein

CONTENTS

10.1 Introduction to the Physiology of K2P Channels

The ability of cells to receive, integrate and process information from their environment depends on the operation of a finely tuned orchestra of receptors and ion channels that reside at the plasma membrane. Among the ensemble on the plasma membrane are many types of K^+ channels that act together to control the membrane potential (V_M) of a broad range of tissues and the electrical activity of excitable cells in the central nervous system, heart and muscles of the body. It is valuable, therefore, to understand the structural basis for K^+ channel function, the role of K^+ channels in physiology, and the factors that regulate K^+ channel activity in health and disease. Here, we provide a brief review of the two-pore-domain K^+ (K2P) channels. We discuss the structure, function and pathophysiology of K2P channels, as well as their emerging pharmacology. This perspective is perforce brief, so the curious reader is directed to excellent papers that that explore the specific aspects of structure, function, pathophysiology and pharmacology of individual K2P channels.

Encoded by 15 *KCNK* genes in humans (Figure 10.1A), K2P channels open and close (gate) in response to a variety of neurotransmitters, physicochemical stimuli (such as changes in pH) and drugs. K2P channels have a distinct structure of four transmembrane domains (TMDs) and two-pore (P)-forming loops in each subunit (Figure 10.1B). The channels operate across the physiological voltage range (Figure 10.1C) and therefore they influence the

DOI: 10.1201/9781003096276-10

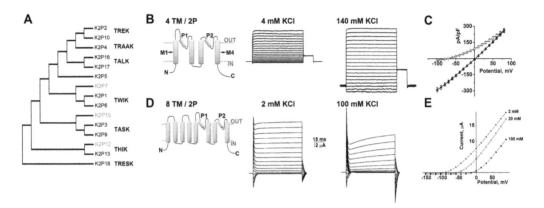

FIGURE 10.1

Classification and operation of two-pore domain channels. (A) A phylogenetic tree calculated to show the related-ness of the 15 K2P subunits found in humans based on ClustalW alignments of the IUPHAR accession numbers for each clone (see http://www.iuphar-db.org/DATABASE/VoltageGatedSubunitListForward). K2P8, K2P11 and K2P14 were found to be homologs of previously described subunits and are not recognized by IUPHAR. K2P8 is equivalent to K2P7. To date, functional expression has not been observed for K2P7, K2P12 and K2P15 (gray text). Abbreviations: TWIK, tandem of P-domains in a weak inward-rectifying K+ channel; TREK, TWIK- related K+ channel; TRAAK, TWIK-related arachidonic acid-stimulated K+ channel; TALK, TWIK-related alkaline-activated K+ channel; TASK, TWIK-related, acid-sensitive K+ channel; THIK, TWIK-related halothane-inhibited K+ channel; TRESK, TWIK-related spinal cord K+ channel. (B) Left: K2P subunits are integral membrane proteins with inter-nal amino (N) and carboxy (C) termini, four transmembrane domains (M1–M4), and two-pore forming (P)-loops. Middle: Current recorded from a Chinese hamster ovary cell heterologously expressing active K2P1 channels (in whole cell mode with deSUMOylating enzyme in the pipette). (Adapted from Plant et al. 2010.) The inside of the cell contains 140 mM K+ and the external solution contains 4 mM K+. Right: The same cell recorded with 140 mM K+ on both sides of the membrane. (C) Mean current–voltage relationships for cells studied as in panel B. Active K2P1 channels show open (GHK) rectification passing more outward current under quasi-physiologic conditions (Δ, 4 mM external K+) and a linear current–voltage relationship with symmetrical 140 mM K+ ($\blacktriangle$). (D) Left: Fungal two-P domain subunits have eight transmembrane domains. Middle and right: Current recorded from *Xenopus laevis* oocytes expressing TOK1 channels studied in 2 mM (middle) or 100 mM external K+ (right). (Adapted from Ketchum et al. 1995.) (E) Current–voltage relationships for the cell studied in panel D shows outward rectification, that is, a shift of the potential where outward current is measured with changes in the external concentrations of K+ that accord with the Nernst potential for K+ (E_K).

stability of V_M at hyperpolarized voltages below the firing threshold for action potentials and the shape of excitable events at depolarized potentials, including the rise to activation threshold, the shape of action potentials, the rate of recovery from firing (Plant et al. 2012) and afterhyperpolarizations.

Since the *kcnk* genes were identified in the late-1990s, a wealth of new research and dis-covery on K2P channels have propelled background K+ currents from the insightful pre-dictions by Hodgkin, Huxley and Katz that underpinned models of neuronal biophysics to descriptions of the molecular bases for establishment and regulation of the resting state from which action potentials rise and return. K2P channels are also widely expressed in non-excitable cells including epithelia, fibroblasts and astroglia where they regulate K+ homeostasis and endocrine function.

Here, we offer a brief overview of K2P channels and the regulatory pathways that mod-ulate their roles in physiology. By necessity this is also a snapshot in time because our understanding of the mechanistic basis for their operation and the extent of their respon-sibilities in vivo is still rapidly expanding. Indeed, some K2P channels show little to no activity when they reach the membrane surface (or pass current only if they carry muta-tions), so their roles remain mysterious.

10.2 Structural Organization and Subunit Diversity

K2P subunits were first identified in the genomes of the budding yeast *Saccharomyces cerevisiae* and nematode *Caenorhabditis elegans* (Ketchum et al. 1995). Each subunit of the yeast channel (called TOK1 for two-pore-domain outwardly rectifying K^+ channel 1) has two reentrant P-loops and eight TMDs, whereas subunits in the roundworm have two P-loops and four TMDs like their congeners in higher organisms (Figure 10.1). Subunits with two P-loops and four TMDs were described shortly thereafter in *Drosophila melanogaster* (Goldstein et al. 1996) and in mammals (Lesage et al. 1996).

To date, TOK-like channels with eight TMDs have been identified only in fungi, such as *Saccharomyces cerevisiae* and *Neurospora crassa*, and opportunistic pathogens like *Candida albicans*, *Aspergillus fumigatus* and *Cryptococcus neoformans* (Lewis et al. 2020). In addition to their unique subunit architecture, TOK channels are known to pass large outward K^+ currents when the membrane is depolarized above E_K (Nernst potential for K^+), but little or no inward currents when the V_M is below E_K (Figure 10.1D).

Two-P domain K^+ subunits with four TMDs are now called K2P channel subunits. K2PØ, encoded by *kcnkØ*, was cloned from a *Drosophila melanogaster* neuromuscular gene library, and shown to exhibit the functional properties expected for background K^+ channels (Goldstein et al. 1996). At first called dORK channels, to reflect their identification in *Drosophila melanogaster* and operation as openly rectifying and K^+ selective pores, both the cloned channels and native background currents show Goldman–Hodgkin–Katz (GHK or open) rectification – that is, the simple property of an ion selective pore to pass permeant ions more readily across the membrane from the side of higher ion concentration to the side of lower concentration (Figure 10.1). Thus, intracellular K^+ in mammals is higher than extracellular and this favors K^+ efflux at most V_M. Also, like native background currents, cloned K2PØ channels expressed in heterologous cells open and close with little or no dependence on voltage and time.

In the decade that followed the identification of two-P domain channels in yeast, worms and fruit flies, 15 *KCNK* genes for K2P channel subunits were identified in the genomes of humans, rats and mice (Figure 10.1A). The mammalian channels were initially named according to the biophysical, physiological or pharmacological attributes of the currents that they passed. However, discrepant observations in different species and experimental systems encouraged adoption of a formal nomenclature in 2005 (Plant et al. 2005). Although proteins encoded by *kcnk* genes are now called K2P subunits (and are numbered to match their encoding gene, thus *kcnk1* and K2P1), the original descriptive names remain in use to correlate cloned channels with the background K^+ currents they mediate in native cells.

Mammalian K2P channels have the membrane topology and subunit stoichiometry demonstrated for K2PØ channels: each subunit has two P-loops and four TMDs, the amino- and carboxy-termini reside in the cytoplasm, and channels operate as homodimers or heterodimers. The first X-ray structures of K2P channels were achieved for human K2P1 (Miller and Long 2012) and K2P4 (Brohawn, Campbell, and Mackinnon 2013), and appear to show the open state of the channels validating prior homology models derived from other classes of K^+ channels and electrophysiology studies of K2PØ channel point mutants (Kollewe et al. 2009). Thus, both crystal structures reveal a four fold symmetric ion conduction pore within a bilaterally symmetric channel corpus. Unexpectedly, the first external loop of each subunit was observed to extend ~35 Å beyond the outer leaflet of the plasma membrane to form a cap domain above the outer mouth of the pore, bifurcating the entry to the K^+ conduction pathway (Figure 10.2). The cap domain has been observed on all K2P channel structures solved to date, including K2P1, K2P2, K2P3, K2P4 and K2P10, and has

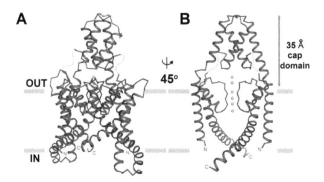

FIGURE 10.2

A 3D structure of the K2P4 channel. (A) A ribbon representation of the X-ray structure of human K2P4 resolved at 2.75 Å. (Adapted from Brohawn, Campbell, and Mackinnon 2013.) Fab fragments used to stabilize the protein have been removed for clarity. The channel is viewed from the membrane plane with one subunit in red and the other in blue, K^+ ions are shown in green, and the boundary of the membrane is in gray. Loops for which structure was not resolved are suggested by dashed lines. (B) A view of the channel rotated by ~45° demonstrating the unique domain swap wherein the outer pore helix interacts with the inner helix from the other subunit rather than its own. Note also the ~35 Å cap domain above the outer mouth of the pore that bifurcates the entrance to the K^+ conduction pathway.

not been described in other potassium channels; the cap may explain the failure of pore-blocking peptide neurotoxins to act on K2P channels due to limited access. Structural studies have also revealed that the two K2P subunits that comprise a holochannel adopt a domain-swapped configuration. That is, the outer pore helices of each subunit appear to interact with the inner helix from the other subunit, an unexpected means for intersubunit communication (Brohawn, Campbell, and Mackinnon 2013). In addition, several K2P structures contain lateral portals that expose the ion conduction pore to the hydrophobic interior of the lipid bilayer.

The crystal structure of K2P1 predicts a C-helix that follows directly from the final (M4) TMD and runs close to the intracellular lipid interface in K2P1. In K2P4, an analogous helix is modeled as part of the second (M2) TMD. A role for these segments in gating is proposed based on their location below the conduction pore in the structure and previous reports that mutation of residues in the domains modify K2P channel activity.

10.3 Gating K2P Channels

Although K2P channels can open across the physiological voltage range this does not mean that they continuously pass K^+ ions. Indeed, diligent biophysical analysis shows that the open probability of most K2P channels is low and subject to exquisite control by a plethora of regulators and second messengers (Plant et al. 2005). Regulatory pathways that decrease K2P channel activity serve to increase cellular excitability and are now recognized to act in several ways: by decreasing the number of channels at the cell surface, by blocking the conduction pore, by decreasing single-channel conductance, by reducing open probability or by altering ion selectivity. Reciprocally, those regulators that increase the activity of K2P channels increase K^+ efflux and dampen excitability. The regulatory mechanisms vary with the K2P channel subtype, and many are tissue dependent. Biophysical and structure–function studies show that the closure of K2P channels is analogous to the C-type inactivation that occurs in voltage-gated K^+ channels, that is, constriction of the K^+ permeation pathway at the selectivity filter (Zilberberg, Ilan, and Goldstein 2001; Lolicato et al. 2020). Indeed, allosteric coupling through the corpus of the channel integrates physicochemical and second messenger signals at the cytoplasmic C-terminus to affect C-type closure of the channel (Zilberberg, Ilan, and Goldstein 2001).

Although the gating of most K2P channels is voltage-independent, some members of the superfamily can show voltage-dependent gating under the influence of specific regulatory pathways. For example, protein kinase A (PKA)-mediated phosphorylation of K2P2 on Ser^{348} rapidly converts the channel from an open rectifier to a voltage-dependent channel in hippocampal neurons and on heterologous expression (Bockenhauer, Zilberberg, and Goldstein 2001).

10.4 The Regulation and Cell Biology of the K2P Channels

Here, we summarize the regulation and cell biology of the six subfamilies of K2P channels identified by protein sequence similarity and that often share regulatory and biophysical attributes.

10.4.1 K2P1, K2P6 and K2P7 Channels

Studies of K2P1 channels have been exciting, at times confusing and ultimately very informative. Encoded by the *kcnk1* gene, and initially named TWIK1, for tandem of P-domains in a weak inward rectifying K^+ channel (Lesage et al. 1996), the functional attributes of K2P1 proved to be enigmatic in heterologous expression systems. However, researchers were motivated by the observation that transcripts were detected at high levels in specific tissues, including the placenta, lungs, kidneys, heart and central nervous system (Talley et al. 2001, Lesage et al. 1996,) and once validated antibodies became available, K2P1 protein was shown in cerebellar granule neurons (Plant et al. 2012). Despite a broad expression profile in native cells, K2P1 channels showed little, or no, electrical activity precluding reproducible characterization of the channels or the development of pharmacological tools. Several groups have suggested findings to explain the low activity of K2P1 channels: silencing by SUMOylation of the channels despite expression at the plasma membrane, rapid endocytosis of the channels removing them from the surface and hydrophobic dewetting of the pore.

SUMOylation is an enzyme-mediated, posttranslational modification that takes place in all cells to regulate transcription factor activity, which links one of three ~100 amino acid *s*mall *u*biquitin-like *mo*difier (SUMO) proteins to the ε-amino group of lysine residues within specific recognition motifs. Although SUMOylation was not thought to occur away from the nucleus, K^+ selective currents were observed when SUMO was cleaved from membrane-localized K2P1 channels by SUMO-specific proteases (SENPs) or when the SUMOylation site (K2P1-Lys274) was mutated to prevent covalent linkage of SUMO to the channel (Rajan et al. 2005). SUMOylation is now known to regulate the activity of an array of ion channels by enzymes at the membrane surface, including voltage-gated potassium and sodium channels. SUMOylation is rapid, induced by environmental stimuli such as hypoxia and reversed by a large family of deSUMOylases (Plant et al. 2020). In keeping with many other SUMO substrates, purification of SUMOylated K2P1 is difficult to demonstrate after detergent purification due to rapid deSUMOylation (Feliciangeli et al. 2007); in contrast, studies in live cells by electrophysiology, FRET spectroscopy and single-molecule fluorescent microscopy show the majority of channels are linked to SUMO (Plant et al. 2010).

Rapid endocytic recycling of K2P1 from the plasma membrane in studies with MDCK and HEK293 cells has also been implicated in low channel activity. The process depends on dynamin under control of a K2P1 di-isoleucine motif such that mutation of Ile293 and Ile294 results in measurable currents upon heterologous expression. Further, K2P1 was found to associate with ARF6, a small G protein that modulates endocytosis at the apical surface of epithelial cells (Decressac et al. 2004).

Following the solution of the crystal structure of human K2P1, molecular dynamic simulations (MDS) of ion permeation identified a "hydrophobic cuff" in the inner vestibule of the channel below the selectivity filter composed of four residues: Leu146 on M2 and Leu261 on M4, from each subunit (Aryal et al. 2014; Miller and Long 2012). MDS revealed that stochastic motion of the cuff restricted the access of water molecules to the internal entrance of the pore, creating an energetic barrier to the permeation of K^+ ions. Based on this model, substitution of Leu146 with hydrophilic residues resulted in a K2P1 channel variant that passed robust currents in *Xenopus* oocytes (Aryal et al. 2014).

Determining how SUMOylation, rapid endocytosis and the hydrophobic gating barrier contribute individually or together to the regulation of K2P1 in native cells remains an area of active study that is spurred by observations that K2P1 knockout mice exhibit altered physiology in several tissues, including pancreatic β cells and the kidney. Similarly, K2P1 has been shown to play a role in the physiology of the heart, mediating paradoxical depolarization of cardiomyocytes under hypokalemic conditions. The enigmatic character of K2P1 can be attributed in part to its heterodimerization with K2P3 and K2P9 subunits, which has been demonstrated in rat neurons and on heterologous expression. Thus, the various homomeric and heteromeric channels have distinct properties, for example, in response to volatile, halogenated ether-based anesthetics and in their regulation by SUMOylation, the latter conferred to SUMO-insensitive K2P3 and K2P9 by incorporation of K2P1 into mixed complexes (Plant et al. 2012).

The TWIK subfamily also includes K2P6 (TWIK2, *KCNK6*) and K2P7 (Kcnk8, *KCNK7*) subunits. K2P6 subunits share 34% sequence identity with K2P1 and 94% homology with K2P7 subunits. Despite broad transcript and protein expression, both have remained electrically silent thus far.

10.4.2 K2P2, K2P4 and K2P10 Channels: Polymodal Rheostats

K2P2 (or TREK1, for TWIK-related K^+ channel 1) and K2P4 (or TRAAK, for TWIK-related arachidonic acid-activated K^+ channel) are expressed throughout the central and peripheral nervous systems (Lesage et al. 1996, Fink et al. 1998) as well as in atrial and ventricular myocytes. The functional properties of neuronal K2P2 channels are subject to a novel type of regulation that leads to Na^+ permeation through the channels and thus altered cellular excitability. Because the *kcnk2* gene has a weak Kozak, initiation sequence translation can also start at a second codon to generate K2P2Δ, a channel variant lacking the first 56 residues (Thomas et al. 2008). This alternative translation initiation (ATI) of *kcnk2* mRNA is regulated across brain regions and during development. K2P2Δ channels have a truncated intracellular N-terminus that modifies operation of the K^+ conduction pathway so that Na^+ ions pass under physiological conditions with a relative permeability of 0.18 compared to K^+, nearly an order of magnitude greater than for full-length channels (0.02) (Thomas et al. 2008).

The activities of K2P2, K2P4 and K2P10 channels are also regulated by a diverse array of lipids, including arachidonic acid, neurotransmitter-activated G protein-coupled receptor pathways, anesthetics and drugs. Several other studies have reported on modulation of this family of K2P channels by changes in temperature and membrane stretch (Maingret et al. 2000). Because these stimuli can occur simultaneously, K2P2 channels have been posited to act as a polymodal signal integrators.

The temperature-sensitivity of K2P2 and K2P4 channels has been reported to contribute to thermal regulation of nociceptive neurons. The activity of these channels increases as body temperature rises above normal (37°C) and falls again at the threshold for noxious heat (>~43°C). Thus, increased activity is expected to dampen the excitability of nociceptors up to the threshold for the detection of noxious heat, beyond which decreased background K^+ currents would facilitate excitability. The current amplitude of K2P2 and K2P4 channels increases by about sixfold for a 10°C rise, e.g., $Q_{10} = 6$. In contrast, the TRPV family of channels has a Q_{10} close to 20. TRPV1 channels open in response to noxious heat while TRPV3 and TRPV4 channels are activated by more moderate temperatures above 30°C. It remains unclear if temperature modulation of K2P2 and K2P4 is a direct effect on the channels or because temperature changes alter second messenger pathways, phosphorylation cascades or free fatty acid concentrations that then act on the channels to modify activity. Whatever the mechanism, the findings suggest that K2P2 and K2P4 channels contribute as thermal rheostats, tuning the responsiveness of neurons to changes in temperature.

The operation of K2P2 channels is subject to regulation by protein kinase A-dependent phosphorylation of Ser^{348}. This regulatory change is rapid and notable because it transforms the gating of K2P2 channels in hippocampal cells from openly-rectifying to voltage-dependent (Bockenhauer, Zilberberg, and Goldstein 2001). Furthermore, it appears that increased intracellular concentrations of phosphatidylinositol-4,5-bisphosphate (PIP_2) shift the voltage-dependence of K2P2 channels to hyperpolarized potentials, augmenting current magnitude across the physiological voltage range (Lopes et al. 2005).

10.4.3 The Acid-Sensitive Channels: K2P3 and K2P9

Originally called TASK1 and TASK3, for TWIK-related acid-sensitive K^+ channel 1 and 3, K2P3 (Duprat et al. 1997) and K2P9 (Rajan et al. 2001) pass K^+ selective currents that are blocked by protonation of a histidine residue in the outer mouth of the pore in the P1 loop (Gly-Tyr-Gly-His) (Lopes et al. 2000). This mechanism of pH regulation is shared with K2P1. K2P3 and K2P9 channels are expressed throughout the central and peripheral nervous systems where they contribute to maintaining V_M. Thus, acidification depolarizes neurons by reducing K^+ currents through K2P3 and K2P9 channels (Talley et al. 2001). K2P3 is also expressed in the kidney, in adrenal glomerulosa cells, in the cardiac conduction pathway and in the carotid body where they are posited to depolarize type 1 glomus cells in response to acidosis and hypoxia. While the TASKs are not the only channels modulated by pH, changes in the magnitude of background K^+ currents correlate well with the slow kinetics of depolarization in somatosensory neurons in response to acidification.

K2P3 and K2P9 subunits are also notable because they form heterodimeric TASK channels (Czirjak and Enyedi 2002) with distinct sensitivities to acidification and pungent stimuli such as hydroxy-α-sanshool, the active ingredient in *Xanthoxylum* Szechuan peppercorns. As noted earlier, K2P3 and K2P9 co-assemble with K2P1 subunits to form heterodimeric, SUMO- and pH-regulated channels in central neurons (Plant et al. 2012).

Surface levels of K2P3 and K2P9 are tightly regulated by the opposing action of signaling motifs that determine retention of the channels in the endoplasmic reticulum (ER)

(O'Kelly et al. 2002). Thus, forward trafficking of K2P3 to the cell membrane requires the phosphorylation-dependent binding of the ubiquitous, soluble adapter protein 14-3-3β. Binding of 14-3-3 suppresses the interaction between K2P3 and βCOP, a vesicular transport protein that promotes retention of the channel in the ER (O'Kelly et al. 2002). βCOP binding is mediated by separate, basic motifs on the N- and C-termini of K2P3 subunits, and disruption of either site promotes forward trafficking of the channel to the cell surface.

K2P15 shows sequence similarity to K2P3 and K2P9, and *kcnk15* transcripts are identified in the pancreas, liver, lung, ovary, testis and heart, and are reported to be elevated in ovarian cancer cells. Despite this expression profile, the biophysical attributes of the channel and its role in physiology remains unclear because it has thus far failed to pass currents in heterologous expression systems.

10.4.4 Alkaline-Activated Channels: K2P5, K2P16 and K2P17

K2P5 (previously TASK2) was initially considered to be a member of the TASK subfamily because it passed background K+ currents that are regulated by changes in extracellular pH (Fink et al. 1998). However, further characterization demonstrated that the channel passes currents only when the extracellular solution exceeds pH 7.5. K2P16 and K2P17 (also known as TALK1 and TALK2, respectively, for TWIK-related alkaline-activated K+ channel) are also activated by extracellular alkalization above normal levels, and like K2P5 they are expressed at high levels in the pancreas. All three channels have been implicated in mediating the apical K+ conductance that facilitates the secretion of bicarbonate from the epithelial cells in the tubular lumen of the exocrine pancreas. K2P5 has been associated with bicarbonate handling in proximal tubule cells and the papillary collecting ducts of the kidney. Thus, K2P5 knockout mice show impaired bicarbonate reabsorption, metabolic acidosis, hyponatremia and hypotension. Based on these findings, a role is suggested for K2P5 in renal acidosis syndromes.

Gating of K2P5, K2P16 and K2P17 by alkaline extracellular pH is proposed to be due to the neutralization of basic residues (arginine in K2P5 and K2P16, lysine in K2P17) that reside near the second P-loop and influence properties of the K+ selectivity filter. One study employing concatemers of cloned K2P5 subunits showed that both putative pH sensors must be neutralized for dimeric channels to conduct.

10.4.5 K2P12 and K2P13 Channels

K2P12 and K2P13 (or THIK2 and THIK1, respectively, for TWIK-related halothane-inhibited K+ channel 2 and 1) are highly expressed in the heart, skeletal muscle and pancreas (Rajan et al. 2001). In situ hybridization suggests that both subunits are also expressed in proximal tubules, thick ascending limbs, cortical collecting ducts of human kidney, and the central nervous system.

In common with other K2P clones, K2P13 channels pass K+ selective leak currents on heterologous expression. The currents are insensitive to physiological changes in pH, temperature and free fatty acids. In contrast to many other K2P channels, K2P13 currents are inhibited by the volatile halogenated anesthetics, such as halothane with a K_i of 2.8 mM (Rajan et al. 2001). Inhibition of a K2P13-like conductance has been proposed to activate central respiratory chemoreceptor neurons in the retrotrapezoid nucleus, and this mechanism is posited to preserve adequate respiratory motor activity and ventilation during anesthesia.

K2P12 clones do not pass currents in experimental systems despite being trafficked to the plasma membrane. Nor does coexpression of K2P12 appear to impact the magnitude of currents passed by K2P13, suggesting that these channel subunits do not form heterodimers (Rajan et al. 2001).

10.4.6 K2P18 Channels and the Pathophysiology of Pain

Transcripts for the last K2P channel to be identified in the human genome, K2P18 (also called TRESK, for TWIK-related spinal cord K^+ channel), has been found in human spinal cord (Sano et al. 2003). In common with other K2P channels, K2P18 channels pass K^+ selective currents and function as open rectifiers.

The activity of K2P18 channels is regulated by the intracellular concentration of Ca^{2+}. An increase in cytosolic Ca^{2+} activates the calmodulin-dependent protein phosphatase calcineurin, and this leads to dephosphorylation of an intracellular serine (Ser^{264}) precluding the binding of 14-3-3 proteins that impact the surface expression and activity of the channels. K2P18 is notable for its expression in the spinal cord, as well as the trigeminal and dorsal root ganglia, where it has been suggested to pass a significant portion of the background K^+ current that determines the excitability of somatosensory nociceptive fibers.

Formal correlation of K2P18 channels and pain was provided by linkage of a frame-shift mutation (F139WfsX24) in the *kcnk*18 gene associated with familial migraine with aura (Lafreniere et al. 2010). The F139WfsX24 variant of K2P18 is truncated at 162 residues and appears to act as a dominant-negative subunit to suppress the function of wild-type K2P18 channels.

10.5 Pharmacology and Emerging Perspectives on the Role of K2P Channels in Pathophysiology

Despite accumulating evidence to implicate K2P channels in disease, our understanding of their role in pathophysiology is incomplete. Slow progress in this direction can be explained, in part, by the challenge that some of the channels do not pass current on heterologous expression and variation in the attributes of K2P subunits in vivo when they operate as heterodimers (Plant et al. 2012; Lengyel et al. 2020). The dearth of specific, high affinity ligands for K2P channels has also been a limiting factor in identifying and characterizing the roles of these proteins in disease states. However, recent advances in single-cell, high-capacity genomics are revealing how expression profiles of KCNK genes change during different pathophysiological processes. A prominent example is the increase in transcript levels for several K2P channels in various tumors and cancer cell lines (Zuniga et al. 2022; Arevalo et al. 2022). While knockout strategies suggest that K2P channels play a role in the proliferation and migration of cancer cells, changes in K2P channel expression may prove to be secondary to other oncogenic factors that drive metastatic potential. The search for selective K2P channel pharmacophores to characterize the roles of the channels is advancing following the elucidation of K2P channel structures and application of high-throughput techniques. While the unique cap domain present in the K2P channel structures solved to date protects the external pore from infiltration by larger classical K^+ channel blockers such as protein neurotoxins, computational chemistry and molecular dynamic simulations have begun to identify drug-binding sites within the channel corpus including a binding pocket behind the pore in K2P2 that coordinates small-molecule channel activators (Lolicato et al. 2017). Other efforts have focused on promising small-molecules, for example derivatization of 3-benzamidobenzoic acid produced a pore blocker for K2P3 that was used to demonstrate a role for the channel in the proliferation of the breast cancer cell line, MCF-7 (Arevalo et al. 2022).

10.6 Conclusions

Over the last two decades, identification and study of channels passing background K^+ currents have advanced our understanding of their biophysical operation and begun to reveal their roles in various tissues. The K2P channels are active in all phases of excitable membrane function, influencing the stability of the resting V_M, the shape of excitable events at depolarized voltages (Plant et al. 2012), and recovery from excitation. The channels change their level of activity in response to influences as varied as anesthetics, pH, temperature and membrane stretch, and are subject to regulation by SUMOylation–deSUMOylation, kinases, phosphatases, GPCRs, lipids and trafficking pathways (Plant et al. 2005). K2P channels have roles in clinical disorders, including cardiac arrhythmia associated with hypokalemia (K2P1), oncogenesis (K2P9) and pain (K2P18) (Lafreniere et al. 2010). Although, we are learning where K2P channels operate at different times in different tissues, in many cases we have yet to discover what they do, most notably, isolates that have thus far failed to demonstrate ion conduction despite expression at the plasma membrane. The breadth of their emerging roles suggests that despite their recent addition to the family of K^+ channels of known molecular identity, K2P channels are attractive, tissue-specific targets for therapy and that development of a pharmacopoeia will speed delineation of their roles of in physiology and disease.

Suggested Reading

Arevalo, B. et al. 2022. "Selective TASK-1 inhibitor with a defined structure–activity relationship reduces cancer cell proliferation and viability." https://doi.org/10.1021/acs.jmedchem.1c00378

Aryal, P., F. Abd-Wahab, G. Bucci, M. S. P. Sansom, and S. J. Tucker. 2014. "A hydrophobic barrier deep within the inner pore of the TWIK-1 K2P potassium channel." *Nat Commun* 5 (1):1–9.

Bockenhauer, D., N. Zilberberg, and S. A. Goldstein. 2001. "KCNK2: reversible conversion of a hippocampal potassium leak into a voltage-dependent channel." *Nat Neurosci* 4 (5):486–91. doi:10.1038/87434.

Brohawn, S. G., E. B. Campbell, and R. Mackinnon. 2013. "Domain-swapped chain connectivity and gated membrane access in a Fab-mediated crystal of the human TRAAK K+ channel." *Proc Natl Acad Sci U S A* 110 (6):2129–34. doi:10.1073/pnas.1218950110.

Czirjak, G., and P. Enyedi. 2002. "Formation of functional heterodimers between the TASK-1 and TASK-3 two-pore domain potassium channel subunits." *J Biol Chem* 277 (7):5426–32. doi:10.1074/jbc.M107138200.

Decressac, S., M. Franco, S. Bendahhou, R. Warth, S. Knauer, J. Barhanin, M. Lazdunski, and F. Lesage. 2004. "ARF6-dependent interaction of the TWIK1 K+ channel with EFA6, a GDP/GTP exchange factor for ARF6" *EMBO Rep* 5 (12):1171–5.

Duprat, F., F. Lesage, M. Fink, R. Reyes, C. Heurteaux, and M. Lazdunski. 1997. "TASK, a human background K+ channel to sense external pH variations near physiological pH." *EMBO J* 16 (17):5464–71. doi:10.1093/emboj/16.17.5464.

Feliciangeli, S., S. Bendahhou, G. Sandoz, P. Gounon, M. Reichold, R. Warth, M. Lazdunski, J. Barhanin, and F. Lesage. 2007. "Does SUMOylation control K2P1/TWIK1 background K+ channels?" *Cell* 130 (3):563–9.

Fink, M., F. Lesage, F. Duprat, C. Heurteaux, R. Reyes, M. Fosset, and M. Lazdunski. 1998. "A neuronal two P domain K+ channel stimulated by arachidonic acid and polyunsaturated fatty acids." *EMBO J* 17 (12):3297–308. doi:10.1093/emboj/17.12.3297.

Goldstein, S. A., L. A. Price, D. N. Rosenthal, and M. H. Pausch. 1996. "ORK1, a potassium-selective leak channel with two pore domains cloned from *Drosophila melanogaster* by expression in *Saccharomyces cerevisiae*." *Proc Natl Acad Sci U S A* 93 (23):13256–61.

Ketchum, K. A., W. J. Joiner, A. J. Sellers, L. K. Kaczmarek, and S. A. Goldstein. 1995. "A new family of outwardly rectifying potassium channel proteins with two pore domains in tandem." *Nature* 376 (6542):690–5. doi:10.1038/376690a0.

Kollewe, A., A. Y. Lau, A. Sullivan, B. Roux, and S. A. Goldstein. 2009. "A structural model for K2P potassium channels based on 23 pairs of interacting sites and continuum electrostatics." *J Gen Physiol* 134 (1):53–68. doi:10.1085/jgp.200910235.

Lafreniere, R. G., M. Z. Cader, J. F. Poulin, I. Andres-Enguix, M. Simoneau, N. Gupta, K. Boisvert, F. Lafreniere, S. McLaughlan, M. P. Dube, M. M. Marcinkiewicz, S. Ramagopalan, O. Ansorge, B. Brais, J. Sequeiros, J. M. Pereira-Monteiro, L. R. Griffiths, S. J. Tucker, G. Ebers, and G. A. Rouleau. 2010. "A dominant-negative mutation in the TRESK potassium channel is linked to familial migraine with aura." *Nat Med* 16 (10):1157–U1501. doi:10.1038/nm.2216.

Lengyel et al. 2020. "TRESK and TREK-2 two-pore-domain potassium channel subunits form functional heterodimers in primary somatosensory neurons." doi: 10.1074/jbc.RA120.014125

Lesage, F., E. Guillemare, M. Fink, F. Duprat, M. Lazdunski, G. Romey, and J. Barhanin. 1996. "TWIK-1, a ubiquitous human weakly inward rectifying K+ channel with a novel structure." *EMBO J* 15 (5):1004–11.

Lewis, A., Z. A. McCrossan, R. W. Manville, M. O. Popa, L. G. Cuello, and S. A. N. Goldstein. 2020. "TOK channels use the two gates in classical K(+) channels to achieve outward rectification." *FASEB J* 34 (7):8902–19. doi:10.1096/fj.202000545R.

Lolicato, M., C. Arrigoni, T. Mori, Y. Sekioka, C. Bryant, K. A. Clark, and D. L. Minor Jr. 2017. "K2P2.1 (TREK-1)-activator complexes reveal a cryptic selectivity filter binding site." *Nature* 547 (7663):364–8. doi:10.1038/nature22988.

Lolicato, M., A. M. Natale, F. Abderemane-Ali, D. Crottes, S. Capponi, R. Duman, A. Wagner, J. M. Rosenberg, M. Grabe, and D. L. Minor Jr. 2020. "K2P channel C-type gating involves asymmetric selectivity filter order-disorder transitions." *Sci Adv* 6:44. doi:10.1126/sciadv.abc9174.

Lopes, C. M., P. G. Gallagher, M. E. Buck, M. H. Butler, and S. A. Goldstein. 2000. "Proton block and voltage gating are potassium-dependent in the cardiac leak channel Kcnk3" *J Biol Chem* 275 (22):16969–78. doi:10.1074/jbc.M001948200.

Lopes, C. M., T. Rohacs, G. Czirjak, T. Balla, P. Enyedi, and D. E. Logothetis. 2005. "PIP2 hydrolysis underlies agonist-induced inhibition and regulates voltage gating of two-pore domain K+ channels." *J Physiol* 564 (1):117–29. doi:10.1113/jphysiol.2004.081935.

Maingret, F., I. Lauritzen, A. J. Patel, C. Heurteaux, R. Reyes, F. Lesage, M. Lazdunski, and E. Honore. 2000. "TREK-1 is a heat-activated background K(+) channel." *EMBO J* 19 (11):2483–91. doi:10.1093/emboj/19.11.2483.

Miller, A. N., and S. B. Long. 2012. "Crystal structure of the human two-pore domain potassium channel K2P1." *Science* 335 (6067):432–6. doi:10.1126/science.1213274.

O'Kelly, I., M. H. Butler, N. Zilberberg, and S. A. Goldstein. 2002. "Forward transport. 14-3-3 binding overcomes retention in endoplasmic reticulum by dibasic signals." *Cell* 111 (4):577–88. doi:10.1016/s0092-8674(02)01040-1.

Plant, L. D., D. A. Bayliss, D. Kim, F. Lesage, S. A. Goldstein. 2005. "International Union of Pharmacology. LV. Nomenclature and molecular relationships of two-P potassium channels." *Pharmacol Rev* 57 (4):527–40. doi:10.1124/pr.57.4.12.

Plant, L. D., I. S. Dementieva, A. Kollewe, S. Olikara, J. D. Marks, and S. A. Goldstein. 2010. "One SUMO is sufficient to silence the dimeric potassium channel K2P1." *Proc Natl Acad Sci U S A* 107 (23):10743–8. doi:10.1073/pnas.1004712107.

Plant, L. D., L. Zuniga, D. Araki, J. D. Marks, and S. A. Goldstein. 2012. "SUMOylation silences heterodimeric TASK potassium channels containing K2P1 subunits in cerebellar granule neurons." *Sci Signal* 5 (251):ra84. doi:10.1126/scisignal.2003431.

Plant et al. 2020. "Hypoxia Produces Pro-arrhythmic Late Sodium Current in Cardiac Myocytes by SUMOylation ofNaV1.5 Channels." *Cell Reports*. https://doi.org/10.1016/j.celrep.2020.01.025

Rajan, S., L. D. Plant, M. L. Rabin, M. H. Butler, and S. A. N. Goldstein. 2005. "SUMOylation silences the plasma membrane leak K+ channel K2P1." *Cell* 121 (1):37–47.

Rajan, S., E. Wischmeyer, C. Karschin, R. Preisig-Muller, K. H. Grzeschik, J. Daut, A. Karschin, and C. Derst. 2001. "THIK-1 and THIK-2, a novel subfamily of tandem pore domain K+ channels." *J Biol Chem* 276 (10):7302–11. doi:10.1074/jbc.M008985200.

Sano, Y., K. Inamura, A. Miyake, S. Mochizuki, C. Kitada, H. Yokoi, K. Nozawa, H. Okada, H. Matsushime, and K. Furuichi. 2003. "A novel two-pore domain K(+) channel, TRESK, is localized in the spinal cord." *J Biol Chem* 278 (30):27406–12. doi:10.1074/jbc.M206810200.

Talley, E. M., G. Solorzano, Q. B. Lei, D. Kim, and D. A. Bayliss. 2001. "CNS distribution of members of the two-pore-domain (KCNK) potassium channel family." *J Neurosci* 21 (19):7491–505.

Thomas, D., L. D. Plant, C. M. Wilkens, Z. A. McCrossan, and S. A. Goldstein. 2008. "Alternative translation initiation in rat brain yields K2P2.1 potassium channels permeable to sodium." *Neuron* 58 (6):859–70. doi:10.1016/j.neuron.2008.04.016.

Zilberberg, N., N. Ilan, and S. A. Goldstein. 2001. "KCNKO: opening and closing the 2-P-domain potassium leak channel entails "C-type" gating of the outer pore." *Neuron* 32 (4):635–48. doi:10.1016/s0896-6273(01)00503-7.

Zuniga, L. et al. 2022. "Potassium channels as a target for cancer therapy: Current perspectives." https://doi.org/10.2147/OTT.S326614

11

Cyclic Nucleotide-Gated Channels

Michael D. Varnum and Gucan Dai

CONTENTS

11.1 Introduction

Ion channels that are principally activated by direct binding of intracellular cyclic nucleotides have evolved as cation nonselective transducers for intracellular pathways that regulate synthesis or destruction of cyclic nucleotides. These channels play essential roles in sensory systems, where they interpret the chemical signal produced by external stimulus input – a change in the intracellular concentration of guanosine 3,5-cyclic monophosphate (cGMP) or adenosine 3,5-cyclic monophosphate (cAMP) – into an electrical response via a change in cation conductance through the channel pore. Cyclic nucleotide-gated (CNG) channels are tetrameric complexes of homologous subunits, with each subunit contributing a cyclic nucleotide-binding domain (CNBD), part of the ion conduction pathway, and the fundamental machinery to covert cGMP or cAMP binding into channel openings. Here, we review their functional and structural properties, and their physiological and pathophysiological contributions with an emphasis on new structural and mechanistic information.

DOI: 10.1201/9781003096276-11

11.2 Physiological Roles of CNG Channels

The best characterized role of CNG channels is within sensory receptor neurons where
they help convert sensory information into electrical responses (Fesenko et al., 1985; Kaupp
et al., 1989). In vertebrate rod and cone photoreceptors, these channels conduct a cation current (the "dark current") in the absence of light; photoactivation of the light receptor (opsin)
activates the G_T protein, which in turn activates a cGMP-specific phosphodiesterase (PDE)
leading to the hydrolysis of cGMP and closure of CNG channels (Figure 11.1). Decreased
channel activity produces membrane hyperpolarization and decreased neurotransmitter
release onto second-order cells. In addition, CNG channels located at photoreceptor synapses have been shown to modulate synaptic transmission and mediate the effects of nitric
oxide (Rieke and Schwartz, 1994; Savchenko et al., 1997). CNG channels also are essential
for olfactory transduction. Olfactory sensory neurons detect odorants via diverse G protein-
coupled receptors that drive G_{olf} activation, which stimulates adenylate cyclase (AC), thus
leading to increased production of cAMP. Increased cAMP opens olfactory CNG channels,
which depolarize the olfactory neurons via Na^+/Ca^{2+} entry and amplification by opening
of Ca^{2+}-activated chloride channels. A subset of olfactory receptor neurons utilize cGMP
signaling components, including a specialized CNG channel, rather than the cAMP-based
components found in the principal olfactory neurons (Meyer et al., 2000).

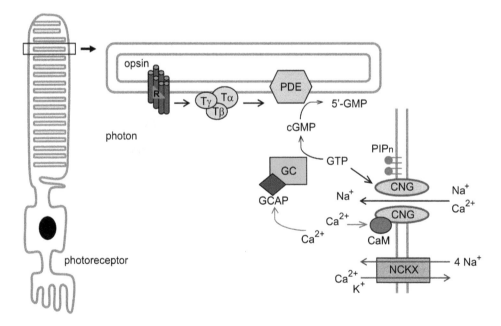

FIGURE 11.1

Phototransduction cascade in vertebrate photoreceptors. Illustration of phototransduction cascade within
single-membrane disc as well as the plasma membrane of a rod photoreceptor outer segment (left). Photons
isomerize 11-cis retinal (R) to all-trans retinal, which in turn activates the 7TM G protein-coupled receptor,
opsin, and trimeric G protein transducin (G_T). The α subunit of G_T activates a cGMP-specific phosphodiesterase (PDE), which hydrolyses cGMP to 5-GMP, closing cation-selective CNG channels. Ca^{2+} entering through
CNG channels regulates guanylyl cyclase (GC) via binding to guanylyl cyclase activating protein (GCAP). CNG
channel activity can be modulated by Ca^{2+}–calmodulin (CaM) or membrane-bound phosphoinositides (PIP_n).
K^+-dependent, Na^+–Ca^{2+} exchangers (NCKX) extrude Ca^{2+} from the photoreceptor, helping balance Ca^{2+} levels
within the outer segment (Szerencsei et al., 2002).

CNG channels are expressed in a wide range of cell types and locations, where they are proposed to have functional roles outside of their canonical contributions to visual and olfactory signaling. CNG channels can contribute to other sensory pathways and systems, including taste submodalities (Misaka et al., 1997), vestibular and cochlear hair cells (Selvakumar et al., 2013), and as cold-sensing channels in hypothalamic neurons (Feketa et al., 2020) and Grueneberg ganglion neurons (Mamasuew et al., 2010). CNG channels also may play a role in neuronal development and remodeling, as they have been shown to contribute to nerve growth cone guidance and morphology (Togashi et al., 2008; Lopez-Jimenez et al., 2012) and may facilitate spinal cord repair following injury (Zhou et al., 2022). Furthermore, CNG channels appear to be important for nociceptive processing, with CNGA3 subunits having a role in inhibitory interneurons and/or glial cells modulating inflammatory pain in rodents (Heine et al., 2011) and CNGB1 subunits contributing to neuropathic pain responses (Kallenborn-Gerhardt et al., 2020). In addition to proposed pain-related functions, CNG channels are expressed in neurons and glia at other CNS locations, where they may regulate neurogenesis (Podda et al., 2013), glial cell functions (Podda et al., 2012), and synaptic plasticity and learning within the hippocampus and amygdala (Kuzmiski and MacVicar, 2001; Michalakis et al., 2011a; Parent et al., 1998). Finally, evidence suggests that CNG channels have functional importance for vascular endothelial cell calcium influx (Shen et al., 2008; Bader et al., 2017).

11.3 CNG Channel Subunit Diversity and Structural Organization

Six paralogous genes in mammals encode the fundamental building blocks (subunits) of CNG channels (Figure 11.2): *CNGA1, CNGA2, CNGA3, CNGA4*; and evolutionarily divergent *CNGB1* and *CNGB3* (Kaupp et al., 1989; Dhallan et al., 1990; Bönigk et al., 1993; Chen et al., 1993; Bradley et al., 1994; Liman and Buck, 1994; Gerstner et al., 2000). Homologous CNG channels exist in more distant parts of the phylogenetic landscape, including TAX2 and TAX4 of *Caenorhabditis elegans* (Hellman and Shen, 2011), plant CNG channels (Kaplan et al., 2007), and bacterial CNG channels (Brams et al., 2014). Furthermore, extended subunit diversity exists with CNG channel ohnologs characterized in lineages (e.g., teleost) having undergone an additional whole-genome duplication event. Mammalian CNG channels are homo- or heteromeric assemblies of some combination of the six possible pore-forming subunits (Figure 11.2). Only the CNGA1–3 subunits can form functional channels when expressed alone, but co-assembly with modulatory subunits tune channel properties and regulation for the specialized roles they play in different cell types. Well-characterized subunit combinations include A1/B1 for rod photoreceptors; A3/B3 for cone photoreceptors; and A2/A4/B1 for olfactory receptors (Figure 11.2). Other possible subunit combinations likely exist in other cell types. CNG channels are part of the pore-loop, cation-selective superfamily of ion channels, and their basic architecture is similar to tetrameric voltage-dependent K^+ channels, with each CNG channel subunit having a transmembrane ion-transport module, functionally divided into a voltage-sensor domain (VSD; S1–S4) and a pore-forming domain (S5–pore loop/helix–S6) contributing to the ion conduction pathway and gate. Each CNG channel subunit also presents cytoplasmic amino (N)-terminal and carboxy (C)-terminal domains, with a CNBD located within the C-terminal region. A C-linker domain (also referred to as the "gating ring") connects the CNBD to the pore-forming domain, serving to couple ligand binding to channel opening.

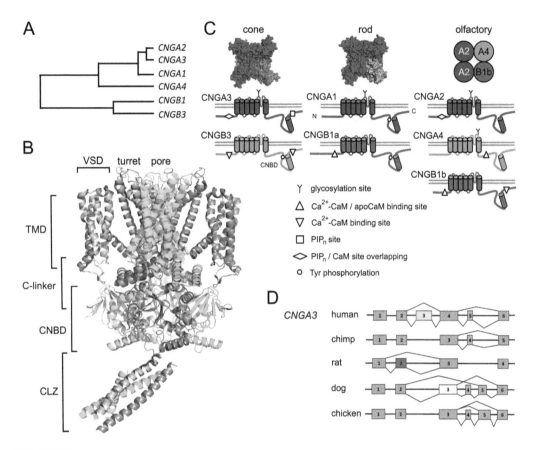

FIGURE 11.2
Subunit diversity and structural organization of CNG channels. (A) Phylogenetic relationship for six genes encoding mammalian CNG channel subunits. (B) Model for CNGA1/CNGB1 channel (PDB 7RHL) showing modular organization of subunits: TMD, transmembrane domain, with pore, and VSD, voltage-sensor domain, modules; C-linker gating ring; CNBD, cyclic nucleotide-binding domain; CLZ, C-terminal leucine zipper domain. (C) Subunit composition for main sensory CNG channels (PDB: cone 7RHS; rod, 7RHH), along with basic topology of individual subunits and sites for regulation and/or modification (see key). (D) Alternative splicing generates CNG channel subunit diversity. *CNGA3* splicing patterns across representative species. Exon (box) but not intron (line) length presented to scale.

The nearest relatives of CNG channels are voltage-dependent: the hyperpolarization-activated cyclic nucleotide-regulated (HCN) channels, and the KCNH family of voltage-gated potassium channels that instead possess a cyclic nucleotide-binding *homology* domain (CNBHD) containing an "intrinsic ligand" (Brelidze et al., 2013). At the level of the transmembrane ion-transport unit, CNG channels and these close relatives all present a non-domain-swapped configuration for the VSD relative to the pore domain (Figure 11.2), such that the VSD is associated with the pore unit within the same subunit (Li et al., 2017). The vestigial VSD present in CNG channels appears to be uncoupled from the gating machinery, with only weak voltage dependence exhibited for cGMP apparent affinity. Consistently, CNG channel structures reveal an S4 helical segment that is divided and less ordered compared to S4 within VSDs of voltage-dependent channels. However, this VSD can support voltage sensing (a) when transplanted into a voltage-dependent channel relative (Tang and Papazian, 1997), or (b) following select mutations at the interface between the pore helix and VSD, which appears to reanimate voltage sensitivity (Martínez-François et al., 2009; Mazzolini et al., 2018). The basic structural organization of CNG channels at the level of the C-linker/gating ring and CNBD shows "domain-swapped" features providing intersubunit contacts in homomeric and heteromeric channels. In addition, intersubunit N–C interactions are a well-characterized feature of CNG channels, having intricate roles in channel regulation (Varnum and Zagotta, 1997; Rosenbaum and Gordon, 2002; Trudeau and Zagotta, 2002b; Zheng et al., 2003; Michalakis et al., 2011b; Dai and Varnum, 2013), although detailed structural information about these N–C interactions is currently missing. Another defined structural element, located in the post-CNBD region of CNGA subunits, is a leucine-zipper (CLZ) domain demonstrated to play an important role for channel assembly and subunit stoichiometry (Trudeau and Zagotta, 2002a; Zhong et al., 2002; Shuart et al., 2011).

In addition to combinatorial assembly of differentially expressed subunits, CNG channel diversity also arises from alternative splicing of precursor mRNA. The generation of CNG channel subunit variants via alternative splicing was first described for the *CNGB1* gene. CNGB1 long and short isoforms (B1a and B1b) are expressed in rod photoreceptors and olfactory receptor neurons, respectively (Körschen et al., 1995; Sautter et al., 1998). In addition, alternative splicing can generate two soluble isoforms representing the N-terminal glutamic acid-rich protein (GARP) domain of B1, without the channel-forming region. The GARP domain of B1a and related soluble forms have been shown to be critical for protein–protein interactions and modulation of gating (Körschen et al., 1999; Ritter et al., 2011; Michalakis et al., 2011b; Pearring et al., 2021). In addition, *CNGA3* transcripts produce subunit isoforms via optional cassette exons that encode regions of the N-terminal cytoplasmic domain (Figure 11.2). While alternative splicing of *CNGA3* is conserved across distantly related species, some incorporated optional exons and the exact splicing patterns produced can differ among orthologs. *CNGA3* alternative splicing appears to primarily control sensitivity to modulation by second messengers and other regulatory events (Bönigk et al., 1996; Dai et al., 2014) rather than fundamental channel properties.

11.4 Gating of CNG Channels

CNG channels are activated by direct binding of cyclic nucleotides to cyclic nucleotide-binding domains, typically without channel desensitization (Figure 11.3). Cyclic nucleotide binding induces conformational rearrangements at the CNBD that are transduced via the C-linker/gating ring to the transmembrane pore-forming module, ultimately opening the channel gate to allow ion conduction.

11.4.1 Ligand Binding at CNBD

For CNG channels, apparent ligand affinity and ligand efficacy vary depending on subunit composition, but for most channels cGMP acts as a better agonist compared to cAMP (Figure 11.3). Biophysical studies and high-resolution structures have illuminated the interactions and conformational changes accompanying ligand binding. The CNBD is comprised of three α-helices (A-, B- and C-helix) and a β-roll domain with eight β strands located between the A-helix and the B-helix (Figure 11.4). Conceptually, two fundamental steps occur during ligand interactions within the CNBD: initial ligand docking followed by conformational changes including C-helix movement that help couple binding

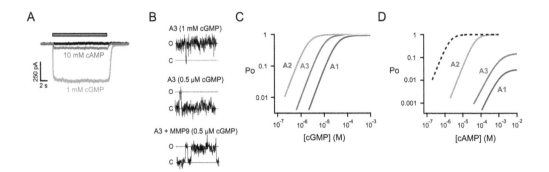

FIGURE 11.3

Diverse ligand efficacy and apparent affinity of CNG channels. (A) Continuous recording at –60 mV for inside-out patch expressing CNGA3 channels during application (bar) of saturating concentrations of cGMP, cAMP, or no cyclic nucleotide (black). (B) Representative CNGA3 single-channel recordings at +80 mV in saturating or low cGMP, with or without extracellular MMP modification. (C) Representative cGMP dose–response relationships for the activation of homomeric CNGA1, CNGA2 and CNGA3 channels. Continuous curves represent fits using the Hill equation, with parameters as follows: for A1, $K_{1/2,cGMP}$ = 33.0 μM, h = 2.0; for A2, $K_{1/2,cGMP}$ = 1.9 μM, h = 2.0; for A3, $K_{1/2,cGMP}$ = 9.8 μM, h = 2.0. (D) Representative cAMP dose–response relationships for the activation of homomeric channels. The Hill parameters are as follows: for A1, $K_{1/2,cAMP}$ = 1.40 mM, h = 1.6; for A2, $K_{1/2,cAMP}$ = 49.2 μM, h = 2.0; for A3, $K_{1/2,cAMP}$ = 1.18 mM, h = 1.5. The dotted curve represents the cGMP dose–response relationships for A2 in panel A, emphasizing agonist selectivity for cGMP.

to channel opening. Several conserved residues within the CNBD β-roll coordinate the ligand-binding step and interactions with the ribose and cyclic phosphate moiety of the cyclic nucleotide. Importantly, a conserved glycine and glutamic acid within the loop between the β-6 and β-7 strands interact with the 2-OH of the ribose, and a conserved β-7 strand arginine interacts with the cyclic phosphate of cGMP or cAMP; mutating this R to E dramatically decreases ligand affinity but does not alter ligand efficacy (Tibbs et al., 1998). Therefore, this R-to-E mutation has been used to selectively "ablate" a ligand-binding site in order to investigate subunit contributions to binding and activation (Liu et al., 1998; Nache et al., 2012; Waldeck et al., 2009). Moreover, the threonine adjacent to this arginine is important for ligand docking, as well as contributing to selectivity for cGMP over cAMP in CNG channels (Altenhofen et al., 1991; Varnum et al., 1995). Together, β-roll interactions with the ligand are thought to primarily determine the initial binding affinity for different cyclic nucleotides.

After the initial ligand-docking event, the CNBD C-helix closes and translates toward the membrane to interact with the purine ring of cyclic nucleotides, and this movement is coupled through other parts of the channel to the opening conformational change (Varnum et al., 1995; Evans et al., 2020; Rheinberger et al., 2018). In addition, stabilization of the helical structure of the C-helix itself has been linked to promoting channel opening after ligand binding (Puljung and Zagotta, 2013). Recent structural studies confirm that cAMP binds to the CNBD in the *anti* configuration (orientation of the purine ring relative to the ribose of cyclic nucleotide), while cGMP binds to the CNBD in the *syn* conformation (Xue et al., 2022). Interactions between the C-helix and ligand are thought to primarily determine ligand efficacy.

For CNG channels, ligand selectivity is highly dependent on ligand discrimination residues located near the C-terminal end of the C-helix (an aspartic acid in A1 and A3; glutamic acid in A2) (Figure 11.4). For these CNGA subunits, the D/E residue is thought to

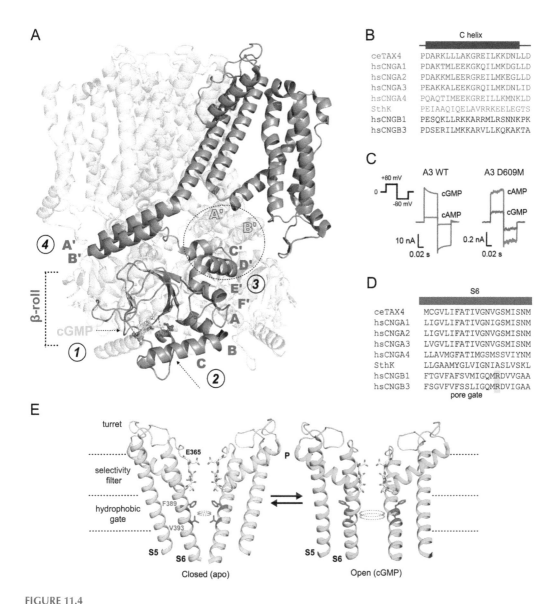

B C helix

ceTAX4	PDARKLLLAKGREILKKDNLLD
hsCNGA1	PDAKTMLEEKGKQILMKDGLLD
hsCNGA2	PDAKKMLEERGREILMKEGLLD
hsCNGA3	PEAKKALEEKGRQILMKDNLID
hsCNGA4	PQAQTIMEEKGREILLKMNKLD
SthK	PEIAAQIQELAVRRKEELEGTS
hsCNGB1	PESQKLLRKKARRMLRSNNKPK
hsCNGB3	PDSERILMKKARVLLKQKAKTA

C

A3 WT   A3 D609M

cGMP cAMP
cAMP cGMP

10 nA   0.2 nA
0.02 s   0.02 s

D S6

ceTAX4	MCGVLIFATIVGNVGSMISNM
hsCNGA1	LIGVLIFATIVGNIGSMISNM
hsCNGA2	LIGVLIFATIVGNVGSMISNM
hsCNGA3	LVGVLIFATIVGNVGSMISNM
hsCNGA4	LLAVMGFATIMGSMSSVIYNM
SthK	LLGAAMYGLVIGNIASLVSKL
hsCNGB1	FTGVFAFSVMIGQMRDVVGAA
hsCNGB3	FSGVFVFSSLIGQMRDVIGAA
	pore gate

E

turret

E365   P

selectivity filter

hydrophobic gate
F389
V393

S5 S6   S5 S6

Closed (apo)   Open (cGMP)

FIGURE 11.4

Model for CNBD and C-linker module coupling ligand binding to pore gating. (A) Single CNGA1 subunit (magenta) depicted in context of background A1/B1 tetramer (gray), with bound cGMP (PDB 7RHH). The A to F helices of the C-linker region, the β-roll region, and A-, B- and C-helices of the CNBD are indicated. Also highlighted are three key residues within the β-roll that are important for initial ligand binding (E544, red; R559, yellow; T560, cyan). Numbers highlight general sequence for ligand binding and gating transitions; dashed circle highlights intersubunit coupling between C/D and A/B helices of adjacent subunits. (B) Sequence alignment for CNBD C-helix in paralogs; key ligand discrimination site highlighted. (C) Traces showing inverted agonist efficacy following D609M mutation in CNGA3. Currents were elicited by saturating cGMP (1 mM) or cAMP (10 mM). (D) Sequence alignment showing conserved hydrophobic gate of channel paralogs (yellow), with additional B1/B3 cytoplasmic constriction (orange). (E) Gating conformational change within pore module of CNGA1 (PDB 7LTF,7LFW); selectivity filter (yellow, E365 labeled); F389/V393 gate (purple); P, pore helix. Central gate constriction: radius ~1Å in closed/apo state, >4Å in open/cGMP-bound state (Xue et al., 2021).

present a favorable electrostatic interaction with cGMP, but unfavorable interactions with cAMP, possibly due to repulsion between the negative charge of the side chain and the unshared pair of electrons in the sp^2 orbital at N1 of the adenine (Varnum et al., 1995). These differences help make cGMP nearly a full agonist and cAMP a partial agonist for A1 and A3 channels. A2 also is selective for cGMP over cAMP, showing an apparent affinity for cGMP ~25-fold higher than that for cAMP (Figure 11.3). Mutating the aspartic acid to hydrophobic methionine (A1-D604M; A3-D609M) reverses ligand selectivity (Figure 11.4) (Varnum et al., 1995). Residues at equivalent positions within B1 and A4 also help explain increased cAMP efficacy of heteromeric channels containing those subunits (Pagès et al., 2000; Shapiro and Zagotta, 2000). B3 subunit orthologs present a lysine at this position. However, B3 subunits do not appear to be specialized to discriminate between cGMP and cAMP; instead, increased cAMP efficacy may reflect more favorable intrinsic, ligand-independent gating energy, as suggested by increased spontaneous open probability observed for A3/B3 channels compared to A3 homomers. At the level of the CNBDs, each subunit of CNG channels contributes one of four sites for binding of cGMP or cAMP; yet, in heteromeric channels, the B1 or B3 subunits may not participate in the same way as CNGA subunits to ligand-dependent transitions associated with channel gating. Intriguingly, cGMP binding to B1 subunits produces a less robust C-helix conformational change in the CNBD compared to cGMP binding to A1 subunits (Barret et al., 2022a 2022; Xue et al., 2022). This asymmetry may reflect a more limited role for B1 subunits in relaying the binding event at their CNBD, thus contributing less energy overall to the ligand-dependent opening process.

11.4.2 Gating Conformational Changes through C-Linker Module to Pore Gate of CNG Channels

Recent work has solidified our understanding of structural mechanisms underlying communication of ligand-binding events to the pore module of CNG channels during channel opening, and these are in agreement with previous biophysical studies of the C-linker/gating ring region connecting S6 and the CNBD (Gordon and Zagotta, 1995; Zong et al., 1998; Zhou et al., 2004). Following ligand docking, C-helix translocation and stabilization, plus other CNBD conformational changes, movement of the CNBD is coupled to E/F and C/D helices of the C-linker module (Figure 11.4). This transition appears to be relayed between subunits via the C/D helix (shoulder) to A/B helix (elbow) of adjacent subunits, and the intersubunit conformational rearrangement promotes state-dependent A/B interactions with the TM/pore module, thereby inducing S6 dilation and rotation, with opening of the gate within the permeation pathway (Evans et al., 2020; Zheng et al., 2020; Rheinberger et al., 2018; Xue et al., 2021). Overall, the structural evidence suggests that the open conformation of CNG channels involves a net upward translation (and twisting) of the cytoplasmic gating machinery. Other evidence supports this compaction and decrement in flexibility at the intracellular/cytoplasmic aspect of the channel in the ligand-bound, open state, with reciprocal decreased structural rigidity observed for the pore turret region located at the extracellular face of the channel (Mazzolini et al., 2018). Intriguingly, within A1/B1 channels there appear to be distinct asymmetries in the way B1 subunits communicate cGMP binding to adjacent A1 subunits, with remarkable evidence for subunit independence and minimal or incomplete B1 coupling to the other regions of the channel, including the pore gate (Barret et al., 2022a; Xue et al., 2022); this gating asymmetry may be influenced by B1-dependent regulation (Barret et al., 2022b).

Unlike voltage-dependent K$^+$ channels, the primary pore gate of CNG channels is not located at the inner helix bundle crossing (Flynn and Zagotta, 2001; Contreras and Holmgren, 2006; Nair et al., 2009). In addition, recent structural information reveals that the main gating conformational changes within the pore (in presence versus absence of bound cGMP at the CNBD) unexpectedly do not occur at the selectivity filter. Instead, the CNG channel pore gate resides at the level of the central cavity and is formed by two bulky hydrophobic side chains (phenylalanine and valine/isoleucine in most CNGA/CNGB subunits) contributed by each of the four subunits (Figure 11.4) (Zheng et al., 2020). B1 and some B3 orthologs also exhibit an additional pore constriction at R880/ R442 in bovine B1/ human B3, which potentially implicates an additional cytoplasmic-entryway gate, at least for some heteromeric CNG channels (Zheng et al., 2022). Furthermore, in A1/B1 channels the open hydrophobic gate within the pore cavity is asymmetrically dilated. This asymmetrical configuration within the pore of A1/B1 (but not A1 homomers) is thought to help configure a unique binding site for L-cis-diltiazem block (Xue et al., 2022).

11.4.3 CNG Channel Activation Schemes and Subunit Contributions

Several different biophysical approaches and activation schemes have been used to interpret the gating behavior of CNG channels. A Monod–Wyman–Changeux (MWC) concerted allosteric model (Monod et al., 1965) has been useful to describe ligand binding and channel activation (Goulding et al., 1994; Varnum and Zagotta, 1996). An advantage of this model is that it can account for observed spontaneous opening events in the absence of ligands, which reflect the ligand-independent intrinsic gating properties of the channels. For homomeric CNG channels, the spontaneous open probability (P_{sp}) in rank order is estimated to be (from low to high): A1, P_{sp} = ~10^{-5}; A3, P_{sp} = ~10^{-4}; and A2, P_{sp} = ~10^{-3} (Ruiz and Karpen, 1997; Tibbs et al., 1997; Gerstner et al., 2000). Furthermore, B3 subunits have been shown to increase (P_{sp}) when expressed with A3 subunits (Gerstner et al., 2000; Meighan et al., 2015). However, an MWC-like model can be less than satisfactory in describing all properties integral to activation of CNG channels, particularly as divergent subunit types produce asymmetrical subunit contributions that are an essential feature of heteromeric channel gating. Sequential models involving two to four ligand-binding steps followed by an allosteric opening conformational change also have been proposed to describe activation of CNG channels (Karpen et al., 1988; Gordon and Zagotta, 1995). Important tools employed to probe ligand binding and gating of CNG channels include (a) photolysis-induced jumps of "caged" cGMP or cAMP (Karpen et al., 1988; Nache et al., 2005); (b) photoaffinity ligands to covalently lock channels in different ligand-bound states (Ruiz and Karpen, 1997); (c) tandem-subunit constructs to control subunit composition (Varnum and Zagotta, 1996); (d) subunit tagging or restriction, e.g., using CNBD ablation mutations in select subunits (Liu et al., 1998; Waldeck et al., 2009; Nache et al., 2012); and (e) the patch-clamp fluorometry technique using fluorescent cGMP analogues to simultaneously monitor ligand binding and channel activation (Biskup et al., 2007). Collectively, these studies reveal several CNG channel gating complexities. First, CNG channel activation is highly cooperative. Second, depending in part on channel subunit composition, it is not necessary for all four CNBDs to be occupied by cGMP in order to produce nearly full channel activation. Third, subunits can be specialized for binding and channel activation by cAMP versus cGMP, e.g., A4 subunits for cAMP, and A2 and A3 subunits for cGMP. Fourth, individual subunits do not contribute uniformly to activation; CNGA subunits appear to contribute more to ligand-dependent stabilization of the open state than CNGB subunits. This aspect agrees with structural information showing asymmetrical conformational changes in A1/B1 channels following ligand binding (Xue et al., 2022). Asymmetrical binding and gating conformational changes also may occur even in the context of homomeric CNG channels (Liu et al., 1998; Nache et al., 2013).

11.5 Pore Properties of CNG Channels

CNG channels are nonselective cation channels, conducting monovalent K^+, Na^+, Li^+, Rb^+ and Cs^+, as well as being highly permeable to Ca^{2+}, yet partially blocked by Ca^{2+} from both intracellular and extracellular access (Zimmerman and Baylor, 1992; Karpen et al., 1993; Root and MacKinnon, 1993). Similar to K^+ channels, the selectivity filter of CNG channels is formed by the reentrant "P-loop" and the supporting pore helix between S5 and S6 of each subunit. The presence of a glutamic acid within the selectivity filter of CNGA subunits (Figure 11.4) is critical for coordination of Ca^{2+} ions for Ca^{2+} permeability and block; E to Q mutation here in A1 channels disrupts one of two possible calcium binding sites and produces voltage-dependent gating (Xue et al., 2021). Ion selectivity and the fraction of current carried by Ca^{2+} differ among homomeric CNGA channels and for heteromeric channels containing CNGB subunits (Frings et al., 1995; Dzeja et al., 1999), with B1 and B3 subunits having a glycine at the corresponding position within the selectivity filter. This G versus E difference in CNGB subunits alters the architecture of the selectivity filter (Xue et al., 2022), producing (1) decreased Ca^{2+} affinity, therefore reducing the dwell time of Ca^{2+}; (2) relief of Ca^{2+} block of monovalent cation permeation through the pore, thus enhancing the permeation of Na^+ and K^+ ions; and (3) flickering single-channel kinetic behavior. For native cone photoreceptor CNG channels, the selectivity for Ca^{2+} over Na^+ was found to increase as the open probability of the channel increased (Hackos and Korenbrot, 1999).

11.6 Pharmacology of CNG Channels

The best-known blocker for CNG channels is L-cis-diltiazem (LCD), which blocks native channels from photoreceptors and olfactory neurons (Koch and Kaupp, 1985; Frings et al., 1992; Haynes, 1992) and recombinant heteromeric channels containing B1 or B3 subunits, while having little effect on homomeric CNG channels (Chen et al., 1993; Gerstner et al., 2000). Recently, the structural basis for LCD block of heteromeric A1/B1 channels was determined, demonstrating LCD trapped deep within the hydrophobic central cavity region above the gate, essentially coordinated between subunits within an asymmetrical, partially open conformation of the pore (Xue et al., 2022).

In general, caution is warranted for interpreting actions of several blockers or inhibitors of other channel types or enzymes, as some also have been shown to block or inhibit CNG channels. These include Na^+- and Ca^{2+}-channel blockers such as pimozide, D-600 and nifedipine, as well as ruthenium red, neomycin and W7 (Brown et al., 2006). The local anesthetic tetracaine blocks CNG channels in a state-dependent manner (Fodor et al., 1997); this profound state dependence for tetracaine block makes it a useful reporter for the equilibrium between closed and open channels. Dequalinium, an extracellular blocker of SK channels, inhibits CNG channel activity at the intracellular side (Rosenbaum et al., 2003). Finally, pseudechetoxin, a peptide toxin isolated from snake venom, inhibits homomeric CNG channels in a voltage-dependent manner, with subunit-specific differences within the turret region influencing blocker affinity (Brown et al., 2003).

11.7 Regulation of CNG Channels

CNG channel activity can be adjusted by second-messenger systems and other regulatory pathways outside of direct cyclic nucleotide-dependent actions on the channels (Figure 11.2). Regulation of CNG channels is important for adaptation in photoreceptors and olfactory receptors, and for autocrine/paracrine and circadian controls within photoreceptors. CNG channels are both highly permeable to calcium ions and subject to calcium-feedback regulation of apparent ligand affinity, most notably via calcium-sensor proteins that interact with channel subunits, such as calmodulin (CaM) and CNG-modulin (Hsu and Molday, 1993; Kurahashi and Menini, 1997; Hackos and Korenbrot, 1997; Chen et al., 2010; Rebrik et al., 2012). Calcium-feedback mechanisms contribute to (a) adjusting ligand sensitivity during adaptation in sensory neurons under continuous stimulus conditions, such that a *change* in sensory input can be successfully reported; and (b) rapid termination of sensory neuron responses (Song et al., 2008). The principal subunits conferring calcium-feedback regulation are B1, B3, A2 and/or A4 subunits, although the precise role of each can depend on context and on interdomain/intersubunit interactions (Varnum and Zagotta, 1997; Zheng et al., 2003; Trudeau and Zagotta, 2002b; Bradley et al., 2004; Peng et al., 2003a).

Another key aspect of CNG channel regulation is sensitivity to membrane phosphoinositides like PIP_2 and PIP_3, ubiquitous ion channel modulators (Hilgemann et al., 2018). CNG channels appear to be primarily inhibited by increased PIP_2/PIP_3 levels (Womack et al., 2000; Spehr et al., 2002; Dai et al., 2013). For cone CNG channels, sensitivity to phosphoinositide regulation depends on defined features within cytoplasmic N- and C-terminal regions, as well as intersubunit and interdomain interactions that can mask or unlock regulation (Dai et al., 2013; Dai and Varnum, 2013). Phosphoinositide regulation can exhibit crosstalk with other regulatory features, such as Ca^{2+}/CaM (Brady et al., 2006). Some regulatory features can be governed by alternative splicing of pre-mRNA (e.g., *CNGA3*), producing variants with enhanced sensitivity to regulation (Dai et al., 2014).

CNG channel ligand affinity and overall activity can be controlled by other diverse mechanisms. These include extracellular proteolysis by matrix metalloproteinases (Meighan et al., 2012), and by intracellular oxidizing conditions that promote disulfide-bond formation or cysteine modifications (Rosenbaum and Gordon, 2002). Other studies offer evidence for CNG channel regulation downstream of PLC activation (Gordon et al., 1995), by S/T- and Y-phosphorylation pathways (Müller et al., 2001; Molokanova et al., 2003), and via various protein–protein interactions. Notably, dephosphorylation of a Y residue within the CNBD of B1, A1 or B3 subunits can produce increased apparent cGMP affinity for photoreceptor channels (Molokanova et al., 2003; Bright et al., 2007). Physiological correlates of CNG channel regulation events are not completely defined but likely relate to adjustments in channel activity following paracrine signals, light input and/or circadian controls (Ko et al., 2003, 2004; Chen et al., 2007; Chae et al., 2007; Gupta et al., 2010); or are related to developmental and tissue-remodeling cues.

11.8 CNG Channel Cell Biology

Compared to biophysical and structural characterizations of CNG channels described earlier, less is known about cell biology aspects of their function, including channel biogenesis and degradation mechanisms. During biogenesis, CNGA (but not CNGB) subunits are modified by N-glycosylation at the extracellular turret (Figure 11.2). The precise site for N-glycosylation can vary, with paralog and species-specific differences. The lack of glycan addition can report defects in channel folding or maturation (Liu and Varnum, 2005; Duricka et al., 2012) but is not essential for channel function. In addition, subunit glycosylation within the turret can protect channels from matrix metalloproteinase (MMP)-dependent modification (Meighan et al., 2013). Furthermore, the N-terminal region of some CNGA subunits undergoes proteolytic processing in photoreceptors (Bönigk et al., 1993). Regarding channel degradation, CNG channels appear to be subject to turnover via the ubiquitin-proteasome system, particularly in terms of quality-control surveillance (Michalakis et al., 2006; Becirovic et al., 2010). The ubiquitin system also may exert a role in mediating normal turnover of CNG channels. The half-life of cone photoreceptor CNG channels appears to be <12 hours (Ko et al., 2001), much shorter than the expected turnover rate via phagocytosis mediated by the retinal pigment epithelium, thought to engulf ~7%–10% of OS macromolecules/day (Kevany and Palczewski, 2010; Imanishi, 2019). These contradictions suggest that new information is needed regarding photoreceptor channel turnover mechanisms.

Another crucial aspect of CNG channel function is evidence for subunit-specific trafficking features. Recombinant CNGA1-3 channels exhibit efficient plasma-membrane localization in the absence of B1 or B3 subunits, but CNGB and A4 subunits are retained within intracellular compartments in the absence of CNGA subunits (Trudeau and Zagotta, 2002a; Zheng and Zagotta, 2004; Nache et al., 2012; Peng et al., 2003b). For native cone CNG channels, B3 knockout in mice decreases but does not eliminate A3 outer-segment localization (Ding et al., 2009). In contrast, CNGB1 has been shown to be essential for ciliary trafficking of CNG channels (Hüttl et al., 2005; Michalakis et al., 2006). Several other protein–protein interactions regulate CNG channel trafficking and localization within cilia. These include motor proteins (Jenkins et al., 2006; Avasthi et al., 2009) in olfactory neurons and cone photoreceptors, respectively; cytoskeletal and adaptor proteins (Kizhatil et al., 2009; Nemet et al., 2014; Ramamurthy et al., 2014); and contacts made with protein components of the outer-segment disc membranes including peripherin-2 (Pearring et al., 2021; Poetsch et al., 2001) for B1a in rod photoreceptors.

11.9 CNG Channel Disease Mechanisms

The most common forms of CNG channelopathies involve disturbances of CNG channel function in the retina (Biel and Michalakis, 2007). Mutations in the genes encoding rod CNG channel subunits (*CNGA1* and *CNGB1*) are associated with autosomal recessive retinitis pigmentosa (arRP) characterized by progressive degeneration of rod photoreceptors followed by loss of cones (Dryja et al., 1995; Paquet-Durand et al., 2011). *CNGA3* and *CNGB3* mutations are linked to achromatopsia, cone dystrophy, oligocone trichromacy and inherited macular degeneration (Sundin et al., 2000; Kohl et al., 1998, 2000; Michaelides et al., 2004; Khan et al., 2007; Vincent et al., 2011). These disorders are typically inherited in

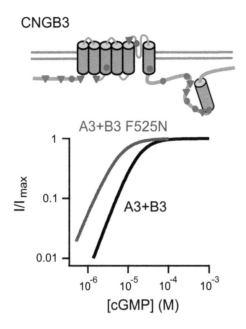

FIGURE 11.5
Disease-associated mutations in CNG channels can produce gain-of-function changes in channel gating. Illustration of mutations in CNGB3 subunit (top), where circles indicate missense mutations; triangles represent truncations produced by frame shifts, defective splice sites and/or premature stop codons. The achromatopsia-associated F525N mutation in CNGB3 enhances apparent cGMP affinity compared to WT channels (bottom). Representative cGMP dose–response relationships with continuous curves representing fits using the Hill equation: for WT channels, $K_{1/2}$ = 17.6 μM, h = 1.8; for A3+B3–F525N, $K_{1/2}$ = 5.9 μM, h = 1.6.

an autosomal recessive manner, and show absent or limited cone function, compromised visual acuity, and in many cases progressive cone degeneration.

Extensive work has focused on determining molecular/cellular mechanisms arising from disease-associated mutations in CNG channels. Predominant features include loss of function at the molecular level – with decreased ligand sensitivity or absence of functional subunits, impaired folding/assembly/localization, and/or increased subunit turnover for cone (Tränkner et al., 2004; Duricka et al., 2012; Patel et al., 2005; Liu and Varnum, 2005; Reuter et al., 2008) and rod (Dryja et al., 1995; Trudeau and Zagotta, 2002a) channel types. ER stress has been implicated in disease progression for channel folding/assembly/localization defects leading to ER accumulation and induction of the unfolded protein response (Duricka et al., 2012; Thapa et al., 2012). CNG channelopathies also are associated with gain-of-function changes in CNG channel activity (Peng et al., 2003b; Bright et al., 2005; Meighan et al., 2015; Zheng et al., 2022) (Figure 11.5). Hyperactive CNG channels are expected to disturb calcium homeostasis and lead to photoreceptor dysfunction and death (Liu et al., 2013; Das et al., 2021). With disease-causing mutations in other critical phototransduction proteins such as PDE, producing elevated cGMP levels, CNG channel knockout or knockdown has a rescue effect, slowing progression of retinal degeneration in mice (Paquet-Durand et al., 2011; Tosi et al., 2011). In other circumstances, however, inducible repair of endogenous B1 expression saved visual function and rod photoreceptors, demonstrating retinal plasticity with channel restoration (Wang et al., 2019). Other functional defects include altered pore properties such as selectivity and/or single-channel conductance (Dryja et al., 1995; Peng et al., 2003b; Tränkner et al., 2004). Finally, disease-associated mutations in CNG channels can interfere with intersubunit interactions critical for channel assembly, gating and/or regulation (Trudeau and Zagotta, 2002a; Michalakis et al., 2011b; Dai and Varnum, 2013b). For example, a CNGA3 mutation was shown to enhance channel sensitivity to regulation by phosphoinositides by altering intersubunit coupling (Dai and Varnum, 2013).

CNG channel gene knockouts in mice represent informative disease models mimicking pathological features arising from CNG channel null mutations in humans. *Cnga3$^{-/-}$* mice demonstrate impaired cone photoresponses and progressive degeneration (Biel et al., 1999; Michalakis et al., 2005, 2011a). *Cngb1* knockout leads to functional and structural defects in both rod photoreceptors and olfactory receptor neurons (Hüttl et al., 2005; Michalakis et al., 2006). In contrast, the effect of *Cngb3$^{-/-}$* appears less severe, with subtle early changes, evidence of residual cone function, and slower progression overall (Ding et al., 2009). The emergence of new CNG channelopathy models in other species, including cone-rich, diurnal organisms such as zebrafish, are expected to provide new insights into pathophysiological mechanisms and rescue strategies.

11.10 Conclusions

Since the discovery of CNG channels over 35 years ago, detailed physiological and biophysical studies have provided many insights into fundamental channel mechanisms. Recent structural studies have amplified and enriched our understanding of CNG channel structure and function, including detailed "visions and revisions" of pore dynamics and block, structural intricacies of gating conformational changes, and mechanisms for asymmetrical contributions of CNGB1 and CNGB3 subunits to gating and pore properties. Issues that require further study include several cell biology aspects of CNG channel function, including protein–protein interactions, regulatory pathways, and biogenesis/turnover in native cells, as well as the intersection of these features with normal physiology and pathophysiology. Also needed is expanded structural information related to less ordered regions or domains missing in recent cryo-EM structures; this information might help account for subunit interactions involving flexible N- and C-terminal cytoplasmic regions thought to underlie channel regulation events, and the evolutionary conservation of alternative splicing features and outcomes tuning those events. Finally, mechanisms and roles for CNG channels in nonsensory neurons, glia, or other diverse cell types and circumstances remain underappreciated and deserve deeper interrogation.

Suggested Reading

This chapter includes additional bibliographical references hosted only online as indicated by citations in blue color font in the text. Please visit https://www.routledge.com/9780367538163 to access the additional references for this chapter, found under "Support Material" at the bottom of the page.

Barret, D. C. A., Schertler, G. F., Benjamin Kaupp, U., Marino, J. 2022a. The structure of the native CNGA1/CNGB1 CNG channel from bovine retinal rods. *Nat. Struct. Mol. Biol.* 29:32–39. doi.org/10.1038/s41594-021-00700-8

Becirovic E., K. Nakova, V. Hammelmann, R. Hennel, M. Biel, and S. Michalakis. 2010. The retinitis pigmentosa mutation c.3444+1G>A in CNGB1 results in skipping of exon 32. *PLOS ONE*;5(1):e8969. doi:10.1371/journal.pone.0008969.

Biel M., M. Seeliger, A. Pfeifer, K. Kohler, A. Gerstner, A. Ludwig, G. Jaissle, S. Fauser, E. Zrenner, and F. Hofmann. 1999. Selective loss of cone function in mice lacking the cyclic nucleotide-gated channel CNG3. *Proc. Natl. Acad. Sci. U. S. A.*;96(13):7553–7557.

Brady J.D., E.D. Rich, J.R. Martens, J.W. Karpen, M.D. Varnum, and R.L. Brown. 2006. Interplay between PIP3 and calmodulin regulation of olfactory cyclic nucleotide-gated channels. *Proc. Natl. Acad. Sci. U. S. A.*;103(42):15635–15640. doi:10.1073/pnas.0603344103.

Brown R.L., T. Strassmaier, J.D. Brady, and J.W. Karpen. 2006. The pharmacology of cyclic nucleotide-gated channels: Emerging from the darkness. *Curr. Pharm. Des.*;12(28):3597–3613. doi:10.2174/138161206778522100.

Dai G., T. Sherpa, and M.D. Varnum. 2014. Alternative splicing governs cone cyclic nucleotide-gated (CNG) channel sensitivity to regulation by phosphoinositides. *J. Biol. Chem.*;289(19):13680–13690. doi:10.1074/jbc.M114.562272.

Dai G., and M.D. Varnum. 2013. CNGA3 achromatopsia-associated mutation potentiates the phosphoinositide sensitivity of cone photoreceptor CNG channels by altering intersubunit interactions. *Am. J. Physiol. Cell Physiol* 305(2):C147–C159. doi:10.1152/ajpcell.00037.2013.

Duricka D.L., R.L. Brown, and M.D. Varnum. 2012. Defective trafficking of cone photoreceptor CNG channels induces the unfolded protein response and ER-stress-associated cell death. *Biochem. J.*;441(2):685–696. doi:10.1042/BJ20111004.

Evans E.G.B., J.L.W. Morgan, F. DiMaio, W.N. Zagotta, and S. Stoll. 2020. Allosteric conformational change of a cyclic nucleotide-gated ion channel revealed by DEER spectroscopy. *Proc. Natl. Acad. Sci. U. S. A.*;117(20):10839–10847. doi:10.1073/pnas.1916375117.

Feketa V.V., Y.A. Nikolaev, D.K. Merriman, S.N. Bagriantsev, and E.O. Gracheva. 2020. Cnga3 acts as a cold sensor in hypothalamic neurons. *eLife*;9:1–22. doi:10.7554/eLife.55370.

Hackos D.H., and J.I. Korenbrot. 1999. Divalent cation selectivity is a function of gating in native and recombinant cyclic nucleotide-gated ion channels from retinal photoreceptors. *J. Gen. Physiol.*;113(6):799–818.

Heine S., S. Michalakis, W. Kallenborn-Gerhardt, R. Lu, H.-Y. Lim, J. Weiland, D. Del Turco, T. Deller, I. Tegeder, M. Biel, G. Geisslinger, and A. Schmidtko. 2011. CNGA3: A target of spinal nitric oxide/cGMP signaling and modulator of inflammatory pain hypersensitivity. *J. Neurosci.*;31(31):11184–11192. doi:10.1523/JNEUROSCI.6159-10.2011.

Hsu Y.T., and R.S. Molday. 1993. Modulation of the cGMP-gated channel of rod photoreceptor cells by calmodulin. *Nature*;361(6407):76–79. doi:10.1038/361076a0.

Hüttl S., S. Michalakis, M. Seeliger, D.-G. Luo, N. Acar, H. Geiger, K. Hudl, R. Mader, S. Haverkamp, M. Moser, A. Pfeifer, A. Gerstner, K.-W. Yau, and M. Biel. 2005. Impaired channel targeting and retinal degeneration in mice lacking the cyclic nucleotide-gated channel subunit CNGB1. *J. Neurosci.*;25(1):130–138. doi:10.1523/JNEUROSCI.3764-04.2005.

Kallenborn-Gerhardt W., K. Metzner, R. Lu, et al. 2020. Neuropathic and cAMP-induced pain behavior is ameliorated in mice lacking CNGB1. *Neuropharmacology*;171(April):108087. doi:10.1016/j.neuropharm.2020.108087.

Ko G.Y.-P., M.L. Ko, and S.E. Dryer. 2003. Circadian phase-dependent modulation of cGMP-gated channels of cone photoreceptors by dopamine and D2 agonist. *J. Neurosci.*;23(8):3145–3153.

Li M., X. Zhou, S. Wang, I. Michailidis, Y. Gong, D. Su, H. Li, X. Li, and J. Yang. 2017. Structure of a eukaryotic cyclic-nucleotide-gated channel. *Nature*;542(7639):60–65. doi:10.1038/nature20819.

Liu D.T., G.R. Tibbs, P. Paoletti, and S.A. Siegelbaum. 1998. Constraining ligand-binding site stoichiometry suggests that a cyclic nucleotide-gated channel is composed of two functional dimers. *Neuron*;21(1):235–248.

Mazzolini M., M. Arcangeletti, A. Marchesi, L.M.R. Napolitano, D. Grosa, S. Maity, C. Anselmi, and V. Torre. 2018. The gating mechanism in cyclic nucleotide-gated ion channels. *Sci. Rep.*;8(1). doi:10.1038/s41598-017-18499-0.

Meighan P.C., S.E. Meighan, E.D. Rich, R.L. Brown, and M.D. Varnum. 2012. Matrix metalloproteinase-9 and -2 enhance the ligand sensitivity of photoreceptor cyclic nucleotide-gated channels. *Channels Austin Tex*;6(3):181–196. doi:10.4161/chan.20904.

Michalakis S., X. Zong, E. Becirovic, V. Hammelmann, T. Wein, K.T. Wanner, and M. Biel. 2011b. The glutamic acid-rich protein is a gating inhibitor of cyclic nucleotide-gated channels. *J. Neurosci.*;31(1):133–141. doi:10.1523/JNEUROSCI.4735-10.2011.

Molokanova E., J.L. Krajewski, D. Satpaev, C.W. Luetje, and R.H. Kramer. 2003. Subunit contributions to phosphorylation-dependent modulation of bovine rod cyclic nucleotide-gated channels. *J. Physiol.*;552(2):345–356. doi:10.1113/jphysiol.2003.047167.

Nache V., T. Zimmer, N. Wongsamitkul, R. Schmauder, J. Kusch, L. Reinhardt, W. Bönigk, R. Seifert, C. Biskup, F. Schwede, and K. Benndorf. 2012. Differential regulation by cyclic nucleotides of the CNGA4 and CNGB1b subunits in olfactory cyclic nucleotide-gated channels. *Sci. Signal.*;5(232):ra48. doi:10.1126/scisignal.2003110.

Nemet I., G. Tian, and Y. Imanishi. 2014. Submembrane assembly and renewal of rod photoreceptor cGMP-gated channel: Insight into the actin-dependent process of outer segment morphogenesis. *J. Neurosci.*;34(24):8164–8174. doi:10.1523/JNEUROSCI.1282-14.2014.

Pearring J.N., J. Martínez-Márquez, J.R. Willer, E.C. Lieu, R.Y. Salinas, and V.Y. Arshavsky. 2021. The GARP domain of the rod CNG channel s β1-subunit contains distinct sites for outer segment targeting and connecting to the photoreceptor disk rim. *J. Neurosci.*;41(14):3094–3104. doi:10.1523/JNEUROSCI.2609-20.2021.

Peng C., E.D. Rich, and M.D. Varnum. 2003b. Achromatopsia-associated mutation in the human cone photoreceptor cyclic nucleotide-gated channel CNGB3 subunit alters the ligand sensitivity and pore properties of heteromeric channels. *J. Biol. Chem.*;278(36):34533–34540. doi:10.1074/jbc.M305102200.

Podda M.V., R. Piacentini, S.A. Barbati, A. Mastrodonato, D. Puzzo, M. D'Ascenzo, L. Leone, and C. Grassi. 2013. Role of cyclic nucleotide-gated channels in the modulation of mouse hippocampal neurogenesis. *PLoS ONE*;8(8):1–14. doi:10.1371/journal.pone.0073246.

Rebrik T.I., I. Botchkina, V.Y. Arshavsky, C.M. Craft, and J.I. Korenbrot. 2012. CNG-modulin: A novel Ca-dependent modulator of ligand sensitivity in cone photoreceptor cGMP-gated ion channels. *J. Neurosci.*;32(9):3142–3153. doi:10.1523/JNEUROSCI.5518-11.2012.

Rheinberger J., X. Gao, P.A.M. Schmidpeter, and C.M. Nimigean. 2018. Ligand discrimination and gating in cyclic nucleotide-gated ion channels from apo and partial agonist-bound cryo-EM structures. *eLife*;7:e39775. doi:10.7554/eLife.39775.

Savchenko A., S. Barnes, and R.H. Kramer. 1997. Cyclic-nucleotide-gated channels mediate synaptic feedback by nitric oxide. *Nature*;390(6661):694–698. doi:10.1038/37803.

Tibbs G.R., D.T. Liu, B.G. Leypold, and S.A. Siegelbaum. 1998. A state-independent interaction between ligand and a conserved arginine residue in cyclic nucleotide-gated channels reveals a functional polarity of the cyclic nucleotide binding site. *J. Biol. Chem.*;273(8):4497–4505.

Togashi K., M.J. von Schimmelmann, M. Nishiyama, C.-S. Lim, N. Yoshida, B. Yun, R.S. Molday, Y. Goshima, and K. Hong. 2008. Cyclic GMP-gated CNG channels function in Sema3A-induced growth cone repulsion. *Neuron*;58(5):694–707. doi:10.1016/j.neuron.2008.03.017.

Tränkner D., H. Jägle, S. Kohl, E. Apfelstedt-Sylla, L.T. Sharpe, U.B. Kaupp, E. Zrenner, R. Seifert, and B. Wissinger. 2004. Molecular basis of an inherited form of incomplete achromatopsia. *J. Neurosci.*;24(1):138–147. doi:10.1523/JNEUROSCI.3883-03.2004.

Trudeau M.C., and W.N. Zagotta. 2002a. An intersubunit interaction regulates trafficking of rod cyclic nucleotide-gated channels and is disrupted in an inherited form of blindness. *Neuron*;34(2):197–207.

Trudeau M.C., and W.N. Zagotta. 2002b. Mechanism of calcium/calmodulin inhibition of rod cyclic nucleotide-gated channels. *Proc. Natl. Acad. Sci. U. S. A.*;99(12):8424–8429. doi:10.1073/pnas.122015999.

Varnum M.D., K.D. Black, and W.N. Zagotta. 1995. Molecular mechanism for ligand discrimination of cyclic nucleotide-gated channels. *Neuron*;15(3):619–625.

Varnum M.D., and W.N. Zagotta. 1997. Interdomain interactions underlying activation of cyclic nucleotide-gated channels. *Science*;278(5335):110–113.

Wang T., J. Pahlberg, J. Cafaro, R. Frederiksen, A.J. Cooper, A.P. Sampath, G.D. Field, and J. Chen. 2019. Activation of rod input in a model of retinal degeneration reverses retinal remodeling and induces formation of functional synapses and recovery of visual signaling in the adult retina. *J. Neurosci.*;39(34):6798–6810. doi:10.1523/JNEUROSCI.2902-18.2019.

Xue J., Y. Han, W. Zeng, Y. Wang, and Y. Jiang. 2021. Structural mechanisms of gating and selectivity of human rod CNGA1 channel. *Neuron*;109(8):1302–1313.e4. doi:10.1016/j.neuron.2021.02.007.

Xue J., Y. Han, W. Zeng, and Y. Jiang. 2022. Structural mechanisms of assembly, permeation, gating, and pharmacology of native human rod CNG channel. *Neuron*;110(1):86-95.e5. doi:10.1016/j. neuron.2021.10.006.

Zheng X., Z. Fu, D. Su, et al. 2020. Mechanism of ligand activation of a eukaryotic cyclic nucleotide–gated channel. *Nat. Struct. Mol. Biol.*;27(7):625–634. doi:10.1038/s41594-020-0433-5.

Zheng X., Z. Hu, H. Li, and J. Yang. 2022. Structure of the human cone photoreceptor cyclic nucleotide-gated channel. *Nat. Struct. Mol. Biol.*;29(1):40–46. doi:10.1038/s41594-021-00699-y.

Zhong H., L.L. Molday, R.S. Molday, and K.-W. Yau. 2002. The heteromeric cyclic nucleotide-gated channel adopts a 3A:1B stoichiometry. *Nature*;420(6912):193–198. doi:10.1038/nature01201.

12

HCN Channels

Colin H. Peters and Catherine Proenza

CONTENTS

12.1 Introduction

Slowly activating currents elicited by membrane hyperpolarization were first characterized in the 1970s and 1980s in rod photoreceptors, cardiac pacemaker and conduction tissue, and hippocampal neurons. These currents in different tissues were originally termed I_f (for funny current), I_h (hyperpolarization-activated current) or I_q (queer current), reflecting their atypical activation by membrane hyperpolarization, slow gating kinetics, sensitivity to cyclic nucleotides and mixed cationic conductance. Nearly two decades later, in the late 1990s, the underlying hyperpolarization-activated cyclic nucleotide-sensitive (HCN) ion channels were cloned. HCN channels are sometimes called pacemaker channels due to their expression in cardiac pacemaker cells and spontaneously active neurons. However, HCN channels are widely expressed throughout the body, where they exert complex effects on excitability depending on the cellular context.

12.2 Basic Structural Organization and Subunit Diversity

Despite their unusual properties, HCN channels have a familiar structure, similar to that of voltage-gated K^+ (K_v) channels. Thus, the channels are tetramers in which each subunit has six transmembrane-spanning domains (S1–S6) with intracellular amino and carboxyl

DOI: 10.1201/9781003096276-12

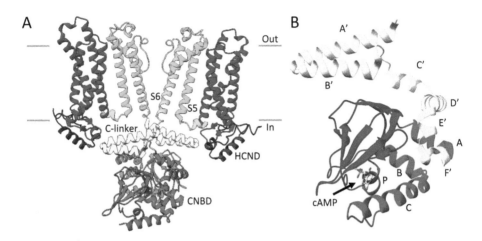

FIGURE 12.1
Structure of HCN channels. (A) Cryo-EM structure of HCN4 (Protein Data Bank ID: 6GYO) illustrating the HCN domain (purple), voltage-sensing domain (blue), S4–S5 linker (cyan), pore domain (green), C-linker (yellow) and CNBD (orange). For clarity, only two subunits of the tetrameric channel are shown. The distal portions of the N- and C-terminals are not resolved in the structure. (B) Structure of the six C-linker helices (A–F; yellow), four CNBD helices (A, P, B and C; red), and CNBD β-roll (blue) coordinating cAMP at the cyclic nucleotide binding site (Protein Data Bank ID: 1Q5O). (From Zagotta et al., 2003.)

terminals (Figure 12.1A). Although HCN channels are activated by membrane hyperpolarization, they share a similar voltage sensor architecture to that of depolarization-activated channels. The S1–S4 segments form the voltage-sensing domain, with a positively charged S4 segment that moves across the electrical field of the cell membrane. The S5–S6 segments form the pore domain, which is similar to that of K_v channels, with the voltage-dependent gate located in the lower part of the S6 segment.

In contrast to the structures of most voltage-gated K^+, Na^+ and Ca^{2+} channels, the voltage-sensor domains in HCN channels are non-domain swapped. That is, the voltage-sensing domain of each subunit in an HCN channel is associated with the pore domain of the same subunit. In domain-swapped channels, a long helical S4–S5 linker couples conformational changes in the voltage-sensor domain to gating of the pore domain. In contrast, HCN channels have only short non-helical S4–S5 linkers, which are not required for channel activation by hyperpolarization, indicating a different mechanism for coupling between the voltage sensor and the gate. However, the non-domain-swapped architecture alone is not sufficient for gating by hyperpolarization; a similar arrangement is found in the KCNH family of depolarization-activated channels including EAG, ELK and ERG channels. Additionally, the S4 transmembrane segment of HCN channels is considerably longer than S4 segments in other voltage-gated channels, extending two additional helical turns on the cytoplasmic side and contacting the C-linker of an adjacent subunit.

The cyclic nucleotide sensitivity of HCN channels is mediated by the proximal portion of the intracellular carboxyl terminus, which consists of the C-linker domain and the cyclic nucleotide-binding domain (CNBD) (Figure 12.1B). The CNBD is highly homologous with the CNBDs of cyclic nucleotide-gated channels (see Chapter 11), the catabolite gene activator protein and the cAMP-activated protein kinase (protein kinase A). It consists of an initial α helix (A-helix), followed by an eight-stranded antiparallel β-roll (β1–β8), a short B-helix and a long C-helix. Cyclic nucleotides bind within the β-roll and also interact with the C-helix. Different cyclic nucleotides bind to the same site with opposite orientations. Conformational changes from binding of cyclic nucleotides are transmitted to the channel pore and voltage-sensing domains via the six α-helices (A to F) of the C-linker, which not only connects to S6 but also contacts the N-terminal HCN domain and the S4–S5 linker.

As outlined next, the first structures of functional HCN channels revealed a number of unique features within this common overall plan that provide insights into the basis for some of the unusual properties of HCN channels (Lee and MacKinnon, 2017). One such case is a unique three-helix structure termed the HCN domain (HCND) formed by the proximal ~45 amino acids of N-terminus preceding S1. In intact channels, the HCND lies between the voltage sensor and the cytoplasmic domains. It contacts the S4 segment of the voltage sensor from the same subunit and the C-linker and CNBD from an adjacent subunit.

HCN channels are present throughout the animal kingdom, in both invertebrates and vertebrates. In mammals, the HCN channel family is composed of four isoforms (HCN1–4). HCN1–3 are primarily expressed in the nervous system, whereas HCN4 is the predominant isoform in pacemaker cells of the sinoatrial node of the heart. Differences between subunits reside mainly in the distal N- and C-terminal regions, which diverge in both length and sequence; whereas, the core transmembrane domains and proximal N- and C-terminals are well conserved between the isoforms. Functionally, HCN1–4 differ in their rates of activation and deactivation and in their response to cyclic nucleotides. In heterologous expression systems, mammalian HCN channel isoforms can combine to form heteromeric channels with slightly altered functional properties. Although multiple HCN channels isoforms are expressed in many tissues, usually a single isoform predominates. In most cases it is not known whether the native channels are heteromers or homomers.

12.3 Gating

HCN channels are gated by hyperpolarization of the plasma membrane, with activation occurring at potentials negative to about –40 mV depending on the tissue and experimental conditions (Figure 12.2A, B). Unlike cyclic nucleotide-gated (CNG) channels, cyclic nucleotide binding to the CNBD does not directly gate HCN channels, but rather shifts

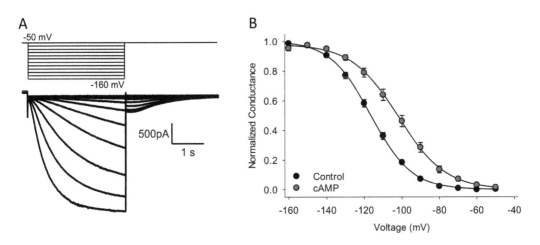

FIGURE 12.2
HCN channel currents. (A) Representative current family recorded from a HEK cell expressing HCN4 in response to hyperpolarizing voltage steps. Note the slow time course of activation. (B) Average conductance–voltage relationships for HCN4 in the presence and absence of cAMP.

the voltage range for hyperpolarization-activated gating. The mammalian HCN channel isoforms do not display overt inactivation of the current during hyperpolarizing voltage steps. However, the sea urchin HCN channel inactivates in absence of cAMP.

The molecular mechanisms responsible for the reversed gating polarity in HCN channels in response to S4 movement have not been fully resolved. Several structural features appear to contribute, including the long S4 segment; the non-domain-swapped arrangement of the pore and voltage sensor domains; and the close packing of S4, S5 and S6 (Lee and MacKinnon, 2017). It is known that the charged helical S4 segment moves downward during hyperpolarization and then breaks into two helices, an upper one that remains fairly vertical relative to the membrane, and a lower helix that assumes a more horizontal conformation, parallel to the membrane. This unique hyperpolarized conformation of S4 is thought to contribute to channel opening because it is accompanied by a large displacement of S5 relative to S6.

Gating of HCN channels is also noteworthy for its very slow rates of activation and deactivation, with time constants in the 100 ms to seconds range, depending on the isoform. Of the mammalian HCN channels, HCN1 activates and deactivates the fastest, HCN2 and HCN3 have intermediate rates, and HCN4 is the slowest. The slow gating of the sinoatrial HCN4 isoform contributes to its role in driving pacemaker activity in sinoatrial node myocytes by allowing it to remain open throughout the entire cardiac cycle (Peters et al., 2021). A similar phenomenon may also occur for the faster HCN channel isoforms during the much shorter neuronal action potential.

12.4 Ion Permeability

HCN channels pass a mixed cationic current with a very low single-channel conductance (~2 pS) (DiFrancesco, 1986). The selectivity filter of HCN channels contains a consensus GYG motif, which is a hallmark of K^+-selective channels. However, HCN channels also conduct Na^+, with a K^+:Na^+ permeability ratio of about 4:1. The ability of HCN channels to conduct Na^+ arises due to differences in the residues surrounding the GYG motif, which cause the conserved Tyr residue to be oriented differently in HCN channels. As a consequence, the selectivity filter of HCN channels has only two K^+ binding sites instead of four as in K^+ selective channels, a feature which is thought to allow Na^+ to pass (Lee and MacKinnon, 2017). HCN channels are blocked by Cs^+ in a voltage-dependent manner and they are not permeable to Ca^{2+} or other divalent ions.

The mixed Na^+/K^+ permeability yields a net reversal potential of approximately −30 mV for HCN channels in physiological solutions. Thus, HCN channels have a largely excitatory effect in many cells. However, the reversal potential along with the slow gating allows HCN channels to conduct both inward and outward currents in response to physiological voltage stimuli such as sinoatrial node action potentials. This allows HCN channels to drive oscillations in membrane potential by contributing driving force in both directions around their intermediate reversal potential.

12.5 Regulation and Cell Biology

Cyclic nucleotides are the best studied regulators of HCN channels. They bind to the CNBD to canonically shift the voltage-dependent activation to more positive potentials, speed the rate of activation, and slow the rate of deactivation (Figure 12.2). While cAMP is a full agonist and has the highest affinity, other cyclic nucleotides, such as cGMP and cCMP, can also bind to and regulate HCN channels. The effects of cyclic nucleotides are isoform-specific. In HCN2 and HCN4, cAMP causes a 10–20 mV depolarizing shift in voltage dependence. In contrast, HCN1 responds to cAMP with a more modest shift (2–6 mV), whereas HCN3 is insensitive to cyclic nucleotides.

The allosteric coupling between cyclic nucleotide binding and gating remains an active research area. Binding of cyclic nucleotides induces conformational changes in the CNBD and C-linker, which influence voltage-dependent gating through a series of intramolecular interactions. These include the direct connection of the C-linker to the S6 transmembrane domain as well as contacts between the C-linker and S4, and possibly the HCND (Figure 12.1). Deletion of the CNBD partially mimics the effects of cAMP binding by causing a positive shift in voltage dependence and speeding channel activation. Hence, it is thought that the unliganded CNBD inhibits channel gating and that cAMP binding relieves this "auto inhibition" (Wainger et al., 2001). Interestingly, deletion of the CNBD does not recapitulate the slowing of deactivation caused by cyclic nucleotide binding, indicating that distinct molecular pathways mediate the effects of cAMP on activation and deactivation (Wicks et al., 2011).

A number of proteins have been shown to associate with HCN channels to regulate their expression and gating. The best studied HCN channel interaction partner is the brain-specific protein TRIP8b (TPR-containing Rab8b interacting protein), which was first discovered in 2004 (Santoro et al., 2004) (Figure 12.3A). TRIP8b binds to and regulates all four mammalian HCN channel isoforms with a 1:1 stoichiometry. There are two distinct interaction sites that have different effects. An upstream site of TRIP8b interacts with the CNBD and regulates gating by antagonizing cAMP binding. And a downstream site, comprised of the TPR domains of TRIP8b, interacts with the three terminal amino acids of the C-terminus to regulate surface expression. TRIP8b is subject to extensive alternative splicing that is predicted to yield at least ten distinct translated proteins. Different variants have been reported to either increase or decrease HCN channel surface expression, depending in part on the cellular context. However, all known splice variants act as competitive antagonists for cAMP. TRIP8b variants are thought to be important determinants of the enriched expression of HCN channels in dendrites. And TRIP8b regulation of HCN channels contributes to learning and memory, major depressive disorder, and epilepsy.

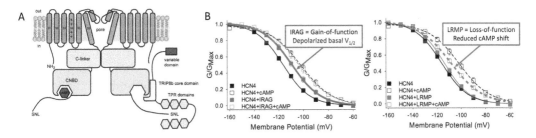

FIGURE 12.3

HCN channel regulation by interacting proteins. (A) TRIP8b interacts with HCN channels at two sites. Interaction of the TRIP8b core domain with the CNBD regulates gating by antagonizing binding of cAMP. Interaction of the TPR domains with the distal C-terminus regulates HCN channel expression. (From DeBerg et al., 2015.) (B) IRAG1 and LRMP (IRAG2) have opposing effects on HCN channel gating. IRAG1 causes a gain of function by depolarizing the voltage dependence in the absence of cAMP. LRMP causes a loss of function by inhibiting the cAMP-induced shift. The binding sites for IRAG1 and LRMP on HCN channels have not yet been resolved. (From Peters et al., 2020.)

Less is known about other HCN channel regulatory proteins. The ER transmembrane proteins inositol 1,4,5-trisphosphate receptor-associated protein (IRAG) and lymphoid-restricted membrane protein (LRMP, also known as IRAG2) are recently identified isoform-specific regulators of HCN4, which exert opposing effects on channel activation (Peters et al., 2020). IRAG potentiates HCN4 by depolarizing the basal voltage dependence, whereas LRMP inhibits HCN4 by reducing the cAMP-dependent shift in activation (Figure 12.3B). IRAG is expressed along with HCN4 at very high levels in the sinoatrial node, and may play a role in heart rate regulation.

MiRP1 (MinK-related peptide) and KCR1 (K+ channel regulator) are transmembrane proteins that regulate a number of different ion channels. Both proteins have been reported to have isoform-specific effects on HCN channel expression and function, including regulation of activation rate and voltage dependence. Other protein interactors such as Filamin A, Mint2 and Tamalin are thought to contribute to expression and subcellular localization of HCN channels.

Like most ion channels, HCN channels are also regulated by membrane phosphoinositides such as PIP_2. PIP_2 can either potentiate or inhibit HCN channels. Direct application of PIP_2 produces a large depolarizing shift in the voltage dependence of HCN channels. However, depletion of PIP_2 by receptor-mediated phospholipase C activation also causes a depolarizing shift, suggesting that PIP_2 regulation of HCN channels in vivo may be mediated by local pools. Although the effects of PIP_2 and cAMP can be similar, they appear to act via different molecular pathways. PIP_2 is thought to bind to both the transmembrane domains and C-linker of HCN channels.

Transcriptional control of HCN channels is exerted by the MiR1 microRNA as well as Tbx3 (T-box transcription factor 3) and NRSF (neuron-restrictive silencer factor) transcription factors. Posttranslational modifications such as phosphorylation likely contribute to regulation of HCN channels in native cells, although they have not been extensively described. Protein kinase A can directly phosphorylate HCN4 to cause a depolarizing shift in voltage dependence. The Src tyrosine kinase speeds the rate of activation of multiple HCN channel isoforms via phosphorylation of a conserved residue in the C-linker.

12.6 Pharmacology

A number of structurally diverse HCN channel blockers have been described, including cilobradine, ZD7288, zatebradine and ivabradine. All are pore blockers with affinities in the low micromolar range, and all equally block the HCN1–4 isoforms. All known HCN channel blockers cause bradycardia by blocking the funny current in sinoatrial myocytes; however, none are specific for HCN channels, leading to a pro-arrhythmic profile at higher concentrations. Of these compounds, ivabradine is the sole drug approved by the US Food and Drug Administration that acts on HCN channels. It is prescribed for treatment of some types of heart failure and angina. Ivabradine blocks HCN channels by entering the pore from the intracellular side, through the voltage-dependent gate. Its therapeutic benefit is conferred by reduction of heart rate and consequent increase in ventricular filling time. While side effects are limited because ivabradine does not cross the blood–brain barrier, off-target block of retinal HCN channels results in visual side effects. Ivabradine also blocks voltage-gated Na+, Ca2+ and K+ channels, including hERG, limiting its value as a research tool.

Several drugs that block other types of ion channels also block HCN channels, such as the general anesthetic propofol. Neuronal blockade of HCN channels by propofol is thought to contribute to its anesthetic properties, whereas block of sinoatrial HCN channels is likely responsible for its bradycardic side effect. Block of HCN channels by the local anesthetic lidocaine may similarly contribute to its anesthetic effects. Analgesics such as dexmedetomidine and clonidine also block HCN channels and have bradycardic side effects.

12.7 Physiological Roles

Different HCN channel isoforms are widely expressed in tissues throughout the body where they regulate cellular excitability and often contribute to the generation of spontaneous action potentials.

12.7.1 Cardiac Pacemaking

HCN channels are perhaps best known for their role in cardiac pacemaking. Indeed, HCN4 is used as a molecular marker of sinoatrial node myocytes (SAMs), which are cardiac pacemaker cells. In SAMs, HCN channels produce the funny current (I_f) that contributes to the generation of spontaneous action potentials (APs; Figure 12.4) and thus helps to determine heart rate. A critical role for HCN4 and I_f in pacemaking is indicated by mutations in HCN4, which cause sinus node arrhythmias (Figure 12.5), and by blocking I_f, which slows heart rate. Global or cardiac-specific knockout of HCN4 is embryonically lethal due to impaired pacemaker potentials, and mice with conditional knockout of HCN4 have sinus pauses, bradycardia and atrioventricular node block.

It has long been assumed that I_f activation during hyperpolarized phases of the sinoatrial AP produces an inward current that contributes to the spontaneous diastolic depolarization. Yet the precise role of I_f in pacemaking remains enigmatic because the voltage range for channel activation is quite negative relative to the sinoatrial AP and the rate of

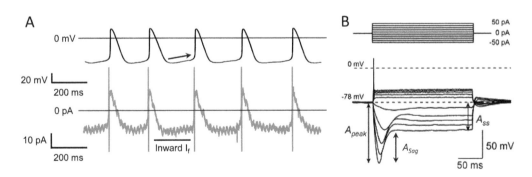

FIGURE 12.4
Examples of HCN channel currents in native cells. (A) I_f (red trace) flowing in response to spontaneous APs (black trace) in a sinoatrial node myocyte. The inward component of I_f is a major determinant of the spontaneous depolarization during diastole that drives pacemaking (arrow). (From Peters et al., 2021.) (B) Characteristic sag (A_{sag}) in membrane potential from the peak hyperpolarization (A_{peak}) to steady-state (A_{ss}) in response to injection of hyperpolarizing current in a cerebellar mossy fiber. (From Byczkowicz et al., 2019.)

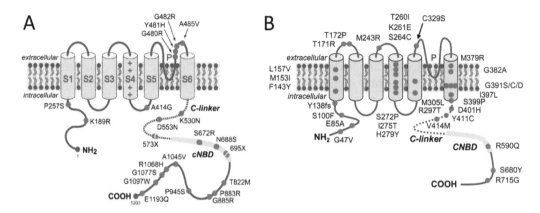

FIGURE 12.5
HCN channelopathy mutations. (A) HCN4 mutations associated with sinus node dysfunction. (Adapted from Verkerk and Wilders, 2015.) (B) HCN1 mutations associated with epilepsy. (Based on Marini et al., 2018.)

activation is quite slow compared to the cardiac cycle. Consequently, only a small fraction of channels can be opened during diastole. However, a recent study shows that despite its low open probability, I_f contributes a substantial fraction of the net charge movement during the pacemaker cycle because the channels do not deactivate during systole (Peters et al., 2021).

The contribution of HCN channels and I_f to the sympathetic fight-or-flight increase in heart rate also remains unresolved. Stimulation of β-adrenergic receptors on SAMs causes a depolarizing shift in voltage dependence of I_f and increases the fraction of net current during the sinoatrial AP. Yet, mice bearing mutations in the CNBD of HCN4 still exhibit some increase in heart rate in response to sympathetic stimulation, indicating that the fight-or-flight response is caused by several cAMP-sensitive currents, including I_f.

12.7.2 HCN Channel Function in the Nervous System

In neurons, HCN channels produce the hyperpolarization-activated current I_h, which is an important determinant of resting membrane potential and excitability. In general, HCN channels tend to have an excitatory effect in neurons. They are active at resting membrane potentials where they produce an inward current that stabilizes the resting potential in the subthreshold range. This behavior is apparent as a characteristic "sag" in membrane potential in response to a hyperpolarizing current step – the injected current causes activation of I_h and the resulting inward current drives the membrane potential back toward its original value (Byczkowicz et al., 2019) (Figure 12.4). Importantly, I_h can also have an inhibitory effect in neurons by acting as a shunt current. In this capacity, I_h activity at resting potentials decreases the input resistance, so that more current from other channels is required to excite the cell. Ultimately, the net excitatory or inhibitory effect of HCN channels depends on their expression level, regulatory pathways and the net activity of other currents in the cell.

The functions of HCN channels in neurons are further refined by their localization to different subcellular compartments. For example, gradients of HCN channel expression along dendrites shape the temporal profile of excitability of neocortical and hippocampal neurons (Magee, 1998). And axon-specific expression of HCN channels ensures fast and reliable AP propagation in GABAergic interneurons (Roth and Hu, 2020). The differential distribution of HCN channels in different subcellular compartments is thought to be controlled in part by interaction with the auxiliary subunit TRIP8b and its many splice variants.

In the central nervous system, HCN channels are widely expressed, including in neurons of the cortex, thalamus, brainstem, basal ganglia and hippocampus. They have been implicated in a plethora of brain functions. In thalamocortical relay neurons, I_h regulates burst firing during periods of inattentiveness and sleep (McCormick and Pape, 1990). HCN channels also play a critical role in motor learning and neuronal integration by cerebellar Purkinje cells (Nolan et al., 2003), and they are important for movement responses in motor cortex. In central auditory neurons, HCN channels play a fundamental role in coding temporal information and sound localization. HCN channels in the retina are similarly important for temporal resolution of fast stimuli. Intriguingly, HCN channels are required for noncanonical phototransduction in some intrinsically photosensitive retinal ganglion cells that mediate non-image-forming vision involved in processes such as circadian rhythms and pupillary light reflexes (Jiang et al., 2018). I_h has also been implicated as a modulator of rhythmic breathing frequency.

In the peripheral nervous system, HCN channel expression in dorsal root ganglion neurons increases in response to nerve injury and HCN channel inhibitors attenuate pain responses. Gastrointestinal neurons expressing HCN4 regulate retrograde peristalsis, and mice lacking HCN2 have abnormal gastrointestinal function that reduces food consumption and gastrointestinal motility.

12.8 Channelopathies, Disease, and Therapeutic Opportunities

The widespread tissue distribution of HCN channels and the phenotypes of knockout mice provide broad insights into the range of pathologies potentially associated with the channels. Knockout of HCN1 increases seizure susceptibility and causes motor abnormalities. HCN2 knockout is more severe, with animals exhibiting ataxia and spontaneous absence seizures as well as gastrointestinal impairment. HCN3 knockout mice have a mild phenotype, with impaired processing of contextual information. Finally, global and cardiac-specific knockout of HCN4 is embryonically lethal, whereas brain-specific knockout of HCN4 slows cortical and thalamic oscillations associated with alert states.

HCN channelopathies have been described in HCN1, HCN2 and HCN4. Gain-of-function mutations in HCN1 and HCN2 are strongly associated with epilepsy. These variants are thought to contribute to seizures by depolarizing the resting potential and increasing excitability. Mutations in HCN4 are closely associated with sinoatrial node dysfunction, although some variants are also linked to other cardiac arrhythmias such as atrial fibrillation. Mutations in the coding region of HCN4 primarily cluster in the pore domain and intracellular C-terminus, while HCN1 mutations associated with epilepsy are distributed throughout the channel (Verkerk and Wilders, 2015; Marini et al., 2018) (Figure 12.5).

HCN channels have been identified as high-priority drug targets for treatment of a number of diseases, the most obvious example of which is heart rate control through the HCN4 isoform. As mentioned earlier, the nonspecific HCN channel blocker ivabradine is approved for treatment of angina and heart failure, where it acts by slowing heart rate. In principle, HCN4-specific activators would also be useful to increase heart rate to treat age-related sinoatrial node dysfunction. Whereas, specific targeting of HCN1 or HCN2 has been identified as a potential treatment for neuropathic and inflammatory pain, as well as depression and anxiety. Unfortunately, development of new drugs is limited by the lack of isoform-specific compounds, HCN channel activators and drugs that cross the blood–brain barrier. Since the existing pharmacological tools all act as HCN channel pore blockers, modification of these compounds is unlikely to yield substantive advances. Therefore, new approaches will need to target non-pore domains of the channels to potentially act as allosteric regulators of channel activity.

12.9 Conclusions

Hyperpolarization-activated cyclic nucleotide-sensitive (HCN) channels are unusual voltage-gated channels. Despite a structure similar to that of voltage-gated K^+ channels, which are selective for K^+ and activated by depolarization, HCN channels are permeable to both Na^+ and K^+ and are activated by hyperpolarization. Voltage-dependent gating of HCN channels is also allosterically regulated by cyclic nucleotides. The four mammalian HCN channel isoforms are widely expressed throughout the body where their unusual properties make them important determinants of cellular excitability. HCN channels are critical for cardiac pacemaking in the sinoatrial node of the heart. In the nervous system, HCN channels contribute to sleep/wake cycles, motor learning and pain. The unique properties and tissue distribution of HCN channels make them alluring drug targets for treatment of conditions ranging from heart failure and epilepsy to neuropathic pain and depression.

Suggested Reading

Byczkowicz, Niklas, Abdelmoneim Eshra, Jacqueline Montanaro, Andrea Trevisiol, Johannes Hirrlinger, Maarten H. P. Kole, Ryuichi Shigemoto, and Stefan Hallermann. 2019. "HCN Channel-Mediated Neuromodulation Can Control Action Potential Velocity and Fidelity in Central Axons." *eLife* 8. https://doi.org/10.7554/eLife.42766.

DeBerg, H. A., J. R. Bankston, J. C. Rosenbaum, P. S. Brzovic, W. N. Zagotta, S. Stoll. 2015. Structural Mechanism for the Regulation of HCN Ion Channels by the Accessory Protein TRIP8b. *Structure* 23(4):734–744. https://doi.org/10.1016/j.str.2015.02.007.

DiFrancesco, D. 1986. "Characterization of Single Pacemaker Channels in Cardiac Sino-Atrial Node Cells." *Nature* 324(6096): 470–73. https://doi.org/10.1038/324470a0.

Jiang, Zheng, Wendy W. S. Yue, Lujing Chen, Yanghui Sheng, and King-Wai Yau. 2018. "Cyclic-Nucleotide- and HCN-Channel-Mediated Phototransduction in Intrinsically Photosensitive Retinal Ganglion Cells." *Cell* 175(3): 652–664.e12. https://doi.org/10.1016/j.cell.2018.08.055.

Lee, Chia-Hsueh, and Roderick MacKinnon. 2017. "Structures of the Human HCN1 Hyperpolarization-Activated Channel." *Cell* 168(1–2): 111–120.e11. https://doi.org/10.1016/j.cell.2016.12.023.

Magee, Jeffrey C. 1998. "Dendritic Hyperpolarization-Activated Currents Modify the Integrative Properties of Hippocampal CA1 Pyramidal Neurons." *The Journal of Neuroscience* 18(19): 7613–24.

Marini, Carla, Alessandro Porro, Agnès Rastetter, Carine Dalle, Ilaria Rivolta, Daniel Bauer, Renske Oegema, et al. 2018. "HCN1 Mutation Spectrum: From Neonatal Epileptic Encephalopathy to Benign Generalized Epilepsy and Beyond." *Brain: A Journal of Neurology* 141(11): 3160–78. https://doi.org/10.1093/brain/awy263.

McCormick, D. A., and H. C. Pape. 1990. "Properties of a Hyperpolarization-Activated Cation Current and Its Role in Rhythmic Oscillation in Thalamic Relay Neurones." *The Journal of Physiology* 431(December): 291–318.

Nolan, Matthew F., Gaël Malleret, Ka Hung Lee, Emma Gibbs, Joshua T. Dudman, Bina Santoro, Deqi Yin, et al. 2003. "The Hyperpolarization-Activated HCN1 Channel Is Important for Motor Learning and Neuronal Integration by Cerebellar Purkinje Cells." *Cell* 115(5): 551–64. https://doi.org/10.1016/s0092-8674(03)00884-5.

Peters, Colin H., Pin W. Liu, Stefano Morotti, Stephanie C. Gantz, Eleonora Grandi, Bruce P. Bean, and Catherine Proenza. 2021. "Bidirectional Flow of the Funny Current (If) during the Pacemaking Cycle in Murine Sinoatrial Node Myocytes." *Proceedings of the National Academy of Sciences of the United States of America* 118(28). https://doi.org/10.1073/pnas.2104668118.

Peters, Colin H., Mallory E. Myers, Julie Juchno, Charlie Haimbaugh, Hicham Bichraoui, Yanmei Du, John R. Bankston, Lori A. Walker, and Catherine Proenza. 2020. "Isoform-Specific Regulation of HCN4 Channels by a Family of Endoplasmic Reticulum Proteins." *Proceedings of the National Academy of Sciences of the United States of America* 117(30): 18079–90. https://doi.org/10.1073/pnas.2006238117.

Roth, Fabian C., and Hua Hu. 2020. "An Axon-Specific Expression of HCN Channels Catalyzes Fast Action Potential Signaling in GABAergic Interneurons." *Nature Communications* 11 (May). https://doi.org/10.1038/s41467-020-15791-y.

Santoro, B., B. J. Wainger, and S. A. Siegelbaum. 2004. "Regulation of HCN Channel Surface Expression by a Novel C-Terminal Protein-Protein Interaction." *The Journal of Neuroscience* 24(47): 10750–62. https://doi.org/10.1523/JNEUROSCI.3300-04.2004.

Verkerk, Arie O., and Ronald Wilders. 2015. "Pacemaker Activity of the Human Sinoatrial Node: An Update on the Effects of Mutations in HCN4 on the Hyperpolarization-Activated Current." *International Journal of Molecular Sciences* 16(2): 3071–94. https://doi.org/10.3390/ijms16023071.

Wainger, B. J., M. DeGennaro, B. Santoro, S. A. Siegelbaum, and G. R. Tibbs. 2001. "Molecular Mechanism of CAMP Modulation of HCN Pacemaker Channels." *Nature* 411(6839): 805–10.

Wicks, N. L., T. Wong, J. Sun, Z. Madden, and E. C. Young. 2011. "Cytoplasmic CAMP-Sensing Domain of Hyperpolarization-Activated Cation (HCN) Channels Uses Two Structurally Distinct Mechanisms to Regulate Voltage Gating." *Proceedings of the National Academy of Sciences of the United States of America* 108(2): 609–14. https://doi.org/10.1073/pnas.1012750108.

Zagotta, W. N., N. B. Olivier, K. D. Black, E. C. Young, R. Olson, E. Gouaux. 2003. "Structural Basis for Modulation and Agonist Specificity of HCN Pacemaker Channels." *Nature* 425:200–205. https://doi.org/10.1038/nature01922

13

CLC Chloride Channels and Transporters

Anna K. Koster and Merritt Maduke

CONTENTS

13.1 Introduction

Members of the ChLoride Channel (CLC) family are united by an evolutionarily conserved homodimeric architecture and selectivity for anions. Despite the name, many CLCs are not actually channels but rather are secondary active transporters, catalyzing stoichiometric exchange of two Cl^- for one H^+ in opposite directions across cell membranes. That is, they use the energy of one ion flowing down its electrochemical gradient to pump another ion electrochemically uphill. Aside from the loss of stoichiometric substrate coupling that separates transporters from channels, individual CLC homologs display a surprising diversity of biophysical properties, including differences in voltage dependence, auxiliary subunits, and sensitivity to H^+, Cl^- and Ca^{2+}. The molecular precepts underlying these differing characteristics are only beginning to be elucidated and will likely provide further insight into how CLCs are functionally regulated in vivo.

CLCs are found in all phyla. In humans (and most other mammals), there are nine homologs (Figure 13.1). The CLC channels are expressed at the plasma membrane, while the CLC transporters reside intracellularly in endosomes, lysosomes, synaptic vesicles and a variety of specialized locations, such as the osteoclast ruffled border. In addition to differences

DOI: 10.1201/9781003096276-13

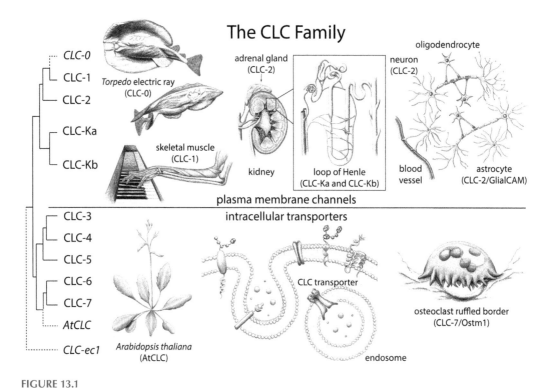

FIGURE 13.1

CLC family tree. The nine human homologs, indicated with solid lines in the diagram at left, carry out wide-spread physiological functions. The founding CLC family member, CLC-0 from the *Torpedo* ray, is most closely related to the mammalian skeletal muscle homolog, CLC-1. CLC-ec1, from *E. coli*, was the first CLC discovered to be an antiporter (Accardi and Miller, 2004) and has served as a paradigm for the family. (Artwork copyright 2021 by Anna Koster; used with permission.)

in subcellular localization, CLC channels and transporters are differentially expressed in specific cell types and tissues, leading to a varied physiological repertoire.

13.2 Physiological Roles

The physiology of each mammalian CLC subtype will be briefly discussed in this chapter, serving as a prelude to Bretag et al. in Volume III, Chapter 11. Select nonmammalian CLCs will also be discussed. A comprehensive review of CLC physiology can be found in Jentsch and Pusch (2018).

13.2.1 CLC Channels

CLC-0, the founding member of the CLC family, is highly expressed in the *Torpedo* electric organ, a stack of cells that together form a high-current battery used for electrocuting prey. Here, CLC-0 provides a low-resistance pathway for anion current flow (Miller, 2014). CLC-0 is also expressed in the *Torpedo* skeletal muscle, where presumably it plays a role analogous to mammalian CLC-1, as discussed later.

CLC-1 channels are mainly found in skeletal muscle. In contrast to other excitable tissues, in which action potential repolarization relies solely on opening of K^+ channels, skeletal muscle additionally employs voltage-dependent opening of CLC-1 channels. This mechanism prevents hyperexcitability of muscle fibers that would occur upon overaccumulation of K^+ in the narrow lumen of t-tubules. CLC-1 is dynamically regulated in response to cellular metabolism during exercise. The channel is inhibited by intracellular ATP, an effect that is magnified at low intracellular pH. Based on these in vitro observations, it has been proposed that buildup of acid in muscle during intense exercise decreases Cl^- currents and maintains action potential firing in the short term as oxygen-deprived muscles engage in anaerobic respiration. Conversely, depletion of ATP during sustained exercise may reduce CLC-1 inhibition, leading to decreased action potential firing and correlating with muscle fatigue (Pedersen et al., 2016; Altamura et al., 2020). Transcript levels of CLC-1 exhibit differential expression patterns during development and aging, sharply increasing after birth and decreasing during the aging process. Transcript levels also vary between fast- and slow-twitch muscle phenotypes, with higher expression seen in fast-twitch muscle fibers.

CLC-2 is broadly expressed in a variety of tissues. In epithelial membranes of the lung and intestine, CLC-2 is localized to the basolateral membrane, where it participates in Cl^- resorption. In the adrenal glands, CLC-2 currents have been recorded from glomerulosa cells. Although its physiological function here is not fully understood, human gain-of-function mutations in CLC-2 cause hyperaldosteronism (see Volume III, Chapter 11), likely by contributing to cellular depolarization that leads to downstream release of aldosterone. In the central nervous system (CNS), CLC-2 is expressed in neurons and glia. In neurons, CLC-2 plays a role at fast-spiking inhibitory synapses, facilitating Cl^- efflux and thereby preventing Cl^- overload during prolonged synchronized network oscillations. It has also been proposed that CLC-2 has nonsynaptic functions in neurons. More thorough characterization of the subcellular localization(s) of CLC-2 within different neuronal populations will be critical to elucidating specific neuronal functions. In glia, CLC-2 associates with an adhesion molecule called GlialCAM (Jeworutzki et al., 2014). GlialCAM abolishes channel voltage dependence and is required for localization of CLC-2 at cell–cell contacts and astrocyte endfeet surrounding blood vessels. The association of CLC-2 with GlialCAM could facilitate significant voltage-independent Cl^- flux at these cellular junctions to assist with rapid K^+ clearance following neuronal activity.

CLC-Ka and CLC-Kb are expressed in the kidney and inner ear (Fahlke and Fischer, 2010). Both proteins associate with a β-subunit called barttin (*BSND*) that is necessary for trafficking and insertion into the plasma membrane. Within the kidney, CLC-Ka and CLC-Kb are expressed in the thin (t) and thick (T) ascending limbs (AL) of the loop of Henle, respectively. Localization to the tAL suggests that Cl^- flux through CLC-Ka helps to establish the steep solute concentration gradient in the renal medulla that drives water absorption from the collecting duct. In support of this mechanism, loss of CLC-Ka in mice increases urine volume under conditions of water restriction. In the TAL, CLC-Kb contributes to bulk NaCl reabsorption. In the inner ear, CLC-Ka and CLC-Kb are expressed in the marginal epithelial cells of the cochlear *stria vascularis* and dark cells of the vestibular organ. Here, they have mutually redundant functions to enable K^+ secretion (through parallel movement of Cl^-), which is necessary to drive the electrochemical gradient required for hair cell mechanotransduction.

13.2.2 CLC Transporters

Relatively little is known about the physiological roles of CLC-3, CLC-4 and CLC-6, which have broad expression in intracellular membranes (Jentsch and Pusch, 2018). CLC-3 resides in late endosomes, synaptic vesicles and synaptic-like microvesicles. Knockout mouse models have shown reduced acidification of these compartments, but the specific physiological function of CLC-3 is not yet certain. Nonetheless, the phenotype of the knockout mouse, which includes severe neurodegeneration after birth, implies a core CNS function. CLC-4 is closely related to CLC-3 and is also widely expressed, including in the brain. The physiological roles of CLC-4 are almost entirely unknown, but rare loss-of-function CLC-4 mutations in humans are associated with learning and behavioral deficits, as well as seizure disorders. In contrast, CLC-4 knockout mice do not exhibit the same phenotypes and have no apparent morphological changes in the brain. CLC-6 is more closely related to CLC-7 than to CLC-3 and CLC-4. CLC-6 transcripts are found in a wide variety of tissues, including epithelial cells of the intestines, pancreas, lungs and trachea, as well as the testis, brain, spinal cord, eyes and trigeminal/dorsal root ganglia. Despite widespread expression at the transcript level, detection of the CLC-6 protein is primarily restricted to the brain and seems to be localized to late endosomes. CLC-6 knockout animals display lysosomal storage abnormalities with axonal swelling in hippocampal and dorsal root ganglia neurons.

CLC-5 is primarily expressed within endosomes of the kidney, where it plays a critical role in endocytosis (Zifarelli, 2015; Gianesello et al., 2020). It is highly expressed in proximal tubule cells, which is where the majority of small-protein resorption occurs. Correspondingly, CLC-5 knockout mouse and human disease mutations (see Volume III, Chapter 11) are associated with low-molecular-weight proteinuria. Systematic analysis of CLC-5 knockout mice revealed the physiological importance of CLC-5 in influencing levels of peptide hormones, which can lead to widespread physiological effects, most notably on phosphate and calcium regulation. Before the discovery that CLC-5 is a transporter and not a channel, it was hypothesized that the protein serves as a Cl⁻ shunt, neutralizing positive charge buildup from H⁺ movement through V-type H⁺-ATPase and thus facilitating acidification of the endosomal compartment. A CLC-5 knock-in mutant designed to remove H⁺ transport (while retaining movement of Cl⁻ across the endosomal membrane) showed that the antiport function is essential for the role of CLC-5 in endocytosis but not for normal endosomal acidification. The detailed roles of CLC-5 in regulating Cl⁻ and/or H⁺ levels remain to be elucidated.

CLC-7 is a ubiquitous transporter that is localized to lysosomes (Zifarelli, 2015). In osteoclasts, CLC-7 resides in both lysosomes and the ruffled border of the plasma membrane, a highly acidic region that interfaces with bone and is critical for bone resorption. Given the ubiquitous expression of CLC-7, constitutive knockout animals unsurprisingly show many deficits (including severe retinal and CNS degeneration), and they die soon after birth. Whether the function of CLC-7 in lysosomes is to regulate Cl⁻ concentrations or pH has been a matter of debate. A role in maintaining lysosomal pH is supported by a human gain-of-function mutant that is associated with lysosomal hyperacidification (Nicoli et al., 2019).

In plants, CLCs have been most extensively studied in the model organism *Arabidopsis thaliana* (At). The seven AtCLC homologs are expressed in intracellular membranes and presumably act as transporters. AtCLC-a has a mutation in the anion permeation pathway that renders it selective for NO₃⁻ over Cl⁻, and this homolog is responsible for NO₃⁻ transport into vacuoles. Because nitrogen is often a limiting nutrient, such transport and storage are critical to plant physiology. Other AtCLC homologs are involved in salt tolerance and stomatal opening.

In *E. coli*, there are two CLC homologs that appear to play mutually redundant roles. A double knockout of both homologs was needed to reveal their role in allowing *E. coli* to survive extreme acid shock. CLCs are also proposed to be important for the daily life of obligate microbial acidophiles.

13.3 Subunit Diversity and Basic Structural Organization

All CLCs have the same basic homodimeric structural organization (Jentsch and Pusch, 2018), and it is not generally feasible to distinguish CLC channels from transporters based on structure alone. Unlike many ion channels, in which multiple subunits come together to form a symmetric pore, the CLC Cl$^-$ and H$^+$ permeation pathways are contained within a single subunit (Figure 13.2A). A highly conserved glutamate residue (Glu$_{ex}$) is positioned at the intersection of the Cl$^-$ and H$^+$ permeation pathways. Glu$_{ex}$ is essential for H$^+$ movement in transporters and acts as a gate for Cl$^-$ transport in both transporters and channels.

CLC subunits are functionally independent and can operate as monomers when mutated to disrupt the dimer interface. Nevertheless, interactions between subunits

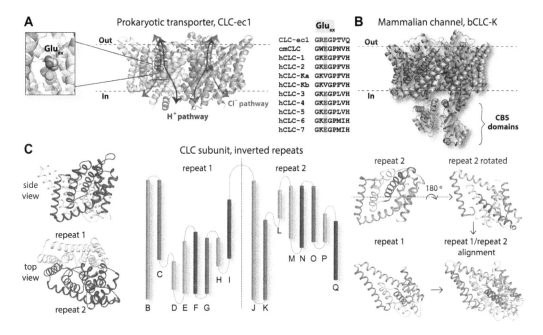

FIGURE 13.2

CLC structural organization. (A) CLC-ec1 structure, highlighting the Cl$^-$ and H$^+$ permeation pathways within each subunit of the homodimer. These pathways are shared along the extracellular entryway and diverge at Glu$_{ex}$, a key residue that is essential for H$^+$ transport. Glu$_{ex}$ gates the anion pathway in CLC transporters and in some CLC channels (see alignment at right). (B) Bovine CLC-K (bCLC-K) structure. CBS domains are present in all eukaryotic and some prokaryotic homologs. (C) CLC inverted repeat structural domains. Left: Side and top views of a single subunit, showing the inverted repeat domains in different shades of gray. Middle: Topology diagram of the repeat domains, colored to indicate the homologous helices. Helices D–H (repeat 1) are localized toward the intracellular side of the membrane, while corresponding helices L–P (repeat 2) are localized toward the extracellular side of the membrane. Right: Structural homology of the repeats. Structures were aligned following rotation of repeat 2 around an axis parallel the membrane. Helix colors match those illustrated in the topology diagram.

appear critical for proper biophysical and physiological function. Mutations at the CLC dimer interface generally influence common gating (described later). The ability of CLCs to heterodimerize has been demonstrated for CLC-1/CLC-2 and for CLC-4 with CLC-3 and CLC-5. The physiological consequences of CLC heterodimerization have not been fully explored.

All eukaryotic and some prokaryotic CLCs possess a pair of intracellular cystathionine β-synthase (CBS) domains. This region of protein, which is seen at high resolution in several CLC structures, forms a dimeric complex that abuts the intracellular surface of the transmembrane domain (Figure 13.2B). The precise function of the CLC CBS domains is not fully understood, but the physiological importance of these domains is evidenced by numerous disease-causing mutations in this region (CLC-Kb, CLC-1, CLC-5 and CLC-7). Many of these mutations alter gating kinetics and voltage dependence, which suggests a role for these domains in channel/transporter regulation. Within the membrane, each subunit of the CLC homodimer contains 16 helices, many dipping only part way through the membrane. These helices comprise two homologous structural domains, arranged with inverted topology (Figure 13.2C). Such inverted repeat symmetry is characteristic of many classes of transporters.

13.4 Gating

13.4.1 CLC Channels

Chris Miller's pioneering studies on the CLC-0 chloride channel from the *Torpedo* electric ray (Miller, 2014) provided the foundation for what we know about CLC channel gating. CLC-0 has two main gating modes. The first mode, known as "fast" or "protopore" gating, occurs on the milliseconds time scale and operates on each pore of the homodimer independently. The second mode, known as "slow" or "common" gating, occurs orders of magnitude more slowly and involves intersubunit cooperation that closes both pores simultaneously. When the common gate is open, the protopores open and close independently. In single-channel recordings, this can be visualized as bursts of activity with three distinct and equidistant conductance levels corresponding to a closed channel, one pore open, or both pores open, interspersed by long silent periods when the slow gate is closed (Figure 13.3).

CLC-0 is voltage-gated. The protopore gate is opened by depolarization, whereas the common gate is opened by hyperpolarization. Both processes have a relatively shallow voltage dependence, with a gating charge of ~1. CLC-0 protopore gating can be studied in isolation from common gating due to the large difference in time scale between the two processes. Following cloning of CLC-0 by Thomas Jentsch, Pusch et al. studied CLC-0 gating in a heterologous expression system. By studying how voltage dependence is influenced by changes in the concentration of Cl^- and other anions, they came to the remarkable conclusion that the CLC-0 gating charge likely arises not from the protein itself, but rather from movement of the permeant ion (Cl^-) through the protein (Jentsch and Pusch, 2018). Thus, voltage-dependent gating in CLC-0 is fundamentally different from that of voltage-gated cation channels.

Among the mammalian CLC channel homologs, CLC-1 and CLC-2 share many characteristics with CLC-0, including voltage dependence that arises from movement of the permeant ion (either Cl^- or H^+, as discussed later). In contrast, CLC-K has distinct gating characteristics and lacks voltage dependence. Next, we compare and contrast the gating characteristics of CLC-0, CLC-1 and CLC-2 before describing the unique gating of the CLC-K homologs.

The central player in CLC-0, CLC-1 and CLC-2 protopore gating is Glu_{ex} (Figure 13.2A). When deprotonated, this residue occupies a Cl⁻ binding site, thus blocking the ion permeation pathway. Neutralization of Glu_{ex} by protonation decreases its affinity for the electropositive pore, making way for the entry of Cl⁻. Mutation of Glu_{ex} to a neutral residue generates constitutively open channels lacking voltage dependence. Glu_{ex} protonation under conditions of low extracellular pH produces a similar gating phenotype. These results suggest that the Glu_{ex} side-chain movement may be fully responsible for protopore gating. However, studies using small-molecule inhibitor probes indicate that protopore gating could involve additional protein conformational change (Accardi and Pusch, 2003). Although the details are not fully elucidated, this proposal is consistent with the observation of global conformational change during outer-gate opening of CLC transporters (see Section 13.4.2).

The voltage dependence of protopore gating is modulated by H⁺ (Miller, 2006). Following the discovery that some CLCs act as Cl⁻/H⁺ exchangers, the mechanism by which H⁺ modulates channel gating drew closer scrutiny. Insightful evaluation of the literature and clever experiments generated compelling proposals that H⁺ – moving from the intracellular side to the Glu_{ex} site – could be the gating charge in CLC-0 (Miller, 2006; Traverso et al., 2006). In this scenario, extracellular Cl⁻ would induce a conformational change to grant intracellular protons access to Glu_{ex} – exactly as it does in the transporters (see Figure 13.2 and following discussion). The finding that transmembrane H⁺ gradients drive CLC-0 gating asymmetry (indicating that H⁺ transport occurs during each gating cycle) provides strong support for this "degraded transporter" model of CLC-0 gating (Lisal and Maduke, 2008).

In contrast to the protopore gate, the common gate involves coordinated opening and closing of pores that are separated by tens of angstroms. The molecular details of CLC common gating are not well understood but are postulated to involve a large-scale conformational change. Consistent with this idea, mutations dispersed throughout the protein influence common gating. Many of these mutations cluster at the dimer interface. This observation, together with biophysical studies on Cd^{2+}-mediated inhibition (Yu et al., 2015), strongly suggest a conformational rearrangement at the subunit interface during common gating. Further evidence for large conformational change comes from fluorescence resonance energy transfer studies focusing on movement in the C-terminal region during the common gating process.

To a first approximation, CLC-1 and CLC-2 gating processes are parallel to that of CLC-0, exhibiting a voltage dependence that is strongly modulated by H⁺ and Cl⁻ and a gating charge of ~1. However, these processes also have striking differences. In CLC-1, common gating is much faster than in CLC-0, such that common and protopore gating are not kinetically separable. Moreover, the common gate of CLC-1 is activated by depolarization, not hyperpolarization as in CLC-0. In CLC-2, common and protopore gating are also not kinetically separable, but both are substantially slower than the equivalent processes in CLC-1, and both are activated by hyperpolarization (Zuniga et al., 2004). The net effect is that electrophysiological current traces of CLC-1 and CLC-2 appear remarkably dissimilar (Figure 13.3). These differences arise, at least in part, from differences in interactions between the channel and ions (Cl⁻ and H⁺) and how these ions modulate voltage dependence (Traverso et al., 2006; De Jesus-Perez et al., 2016).

Unlike CLC-1 and CLC-2, the CLC-K channels have essentially no voltage dependence, a characteristic that can be explained by the existence of a valine in place of Glu_{ex}. Gating of CLC-K channels by protons occurs via protonation/deprotonation of extracellular His and Lys residues, standing in contrast to the Glu_{ex}-mediated pH gating in CLC-1 and CLC-2.

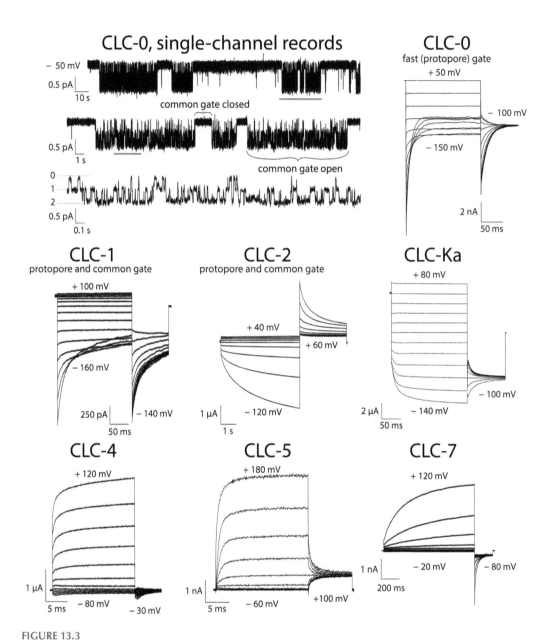

FIGURE 13.3

Electrophysiological overview of CLC channels and transporters. CLC-0 data were obtained in the Maduke laboratory, using inside-out patch-clamp recording on *Xenopus* oocytes. All other traces are courtesy of Michael Pusch. CLC-2, CLC-4 and CLC-Ka traces were recorded using two-electrode voltage clamp on *Xenopus* oocytes. CLC-1, CLC-5 and CLC-7 traces are from whole-cell patch-clamp recordings on HEK293 cells. In the single-channel traces, regions indicated by red bars under the top and middle trace are expanded and shown in the middle and bottom traces, respectively.

Among CLCs, the CLC-K channels are uniquely dependent on extracellular Ca^{2+} (Jentsch and Pusch, 2018). These channels contain extracellular Ca^{2+} binding sites formed by sets of acidic residues interacting across the dimer interface. The modularity of these sites was demonstrated by the conferral of Ca^{2+} sensitivity onto CLC-0.

13.4.2 CLC Transporters

Unlike channels, in which gating leads to an open pore, gating of secondary active transporters must avoid such opening, which would counteract coupling. While it is well known that transporter gating must involve some type of alternating access to either side of the membrane, understanding how this happens with CLCs requires a novel framework. All other known transporters exchange like-charged substrates that share pathways. In contrast, CLC transporters exchange an anion for a cation, using divergent pathways.

The stoichiometry of anion/proton exchange for most CLCs, including prokaryotic, mammalian and plant homologs, is 2:1. A working model for understanding the 2:1 CLC exchange mechanism is shown in Figure 13.4. This model is based on a wealth of functional data, together with high-resolution structures that reveal four distinct positions for the Glu_{ex} side chain: "middle," "up," "down" and "out." The up and out conformations seen in mutants of the CLC-ec1 transporter are also observed in wild-type structures of CLC-1 and CLC-7, respectively, lending credence to the idea that these are universal CLC conformations. Mutation of Glu_{ex} to a neutral residue abolishes H^+ transport in all CLC transporters, illustrating the central role of Glu_{ex} in the transport cycle. As an alternative to the type of model shown in Figure 13.4, multiscale kinetic modeling has suggested that multiple conformational pathways, each with different stoichiometries, may combine to yield the functionally measured 2:1 exchange (Mayes et al., 2018).

Electrophysiological recordings of CLC transporters reveal varied characteristics. The mammalian CLCs (representative traces shown in Figure 13.3) show strong outward rectification. In contrast, CLC-ec1 shows only slight rectification. The slow gating kinetics of CLC-7 allow tail-current analysis, which reveals that the rectification is a gating phenomenon; that is, CLC-7 is a gated transporter. Mutagenesis results indirectly suggest that this phenomenon may be analogous to the common gating that occurs in CLC channel homologs (Ludwig et al., 2013). The rectification of CLC-3 to CLC-6 may reflect a gating process, voltage-dependent H^+ block or some combination of these factors (Picollo et al., 2010; De Stefano et al., 2013).

13.5 Ion Permeability

CLCs are impermeable to cations, except H^+, as discussed in the gating section. Among the halides, CLCs exhibit $Cl^- > Br^- > I^-$ selectivity (Jentsch and Pusch, 2018). This selectivity is opposite that of other anion-channel families, including the Ca^{2+}-activated Cl^- channels (Chapter 14); cystic fibrosis transmembrane conductance regulators (CFTR; Volume III, Chapter 10); and volume-regulated anion channels, which exhibit $I^- > Br^- > Cl^-$ selectivity. The selectivity of CLC channels for Cl^- over the larger Br^- may reflect that these channels possess a narrow permeation pathway, an idea consistent with the relatively low single-channel conductance of CLC channels (~1 pS for CLC-1 ranging to ~20 pS for CLC-K) and the narrow pores seen in structures. That said, the most thorough study of CLC permeability (on CLC-1) revealed that the large organic cation benzoate is substantially permeable (0.15 relative to Cl^-) (Rychkov et al., 1998). This result suggests that there may be conformational changes accompanying channel opening that have not yet been observed in CLC channel structures.

The CLC transporters generally exhibit the same anion-selectivity properties as the channels. As discussed in Section 13.2, certain plant homologs evolved into NO_3^-/H^+ antiporters via a mutation at the intracellular side of the selectivity filter. Another selectivity

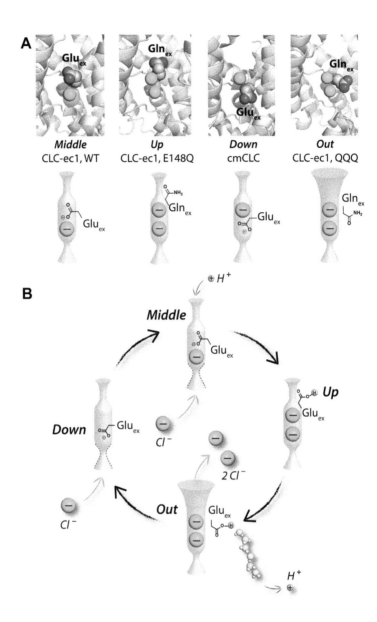

FIGURE 13.4

Glu$_{ex}$ conformations and the CLC Cl$^-$/H$^+$ exchange cycle. (A) Illustration of the four Glu$_{ex}$ conformations seen in different crystal structures, with cartoon representations below each structural depiction (Protein Data Bank: 1OTS, 1OTU, 3ORG and 6V2J). The E148Q and "QQQ" CLC-ec1 mutants mimic different protonation states of the transporter (Dutzler et al., 2003; Chavan et al., 2020). cmCLC is a transporter from red algae (Feng et al., 2010). The QQQ structure (Glu$_{ex}$ "out") exhibits backbone changes that open the extracellular anion permeation pathway, while all other Glu$_{ex}$ variants have backbone structures similar to WT CLC-ec1. (B) Working model for the CLC transport cycle, showing how transitions through the four Glu$_{ex}$ conformations, coordinated with Cl$^-$/H$^+$ movements, can achieve 2:1 Cl$^-$/H$^+$ antiport. This mechanism yields 2:1 antiport in both directions; for clarity, only one direction is depicted. Dashed lines along the inner gate in the depictions of the "down" and "middle" states indicate uncertainty as to the mechanism of inner-gate opening, which may involve as-yet unseen conformational change (Basilio et al., 2014; Accardi, 2015) or a static kinetic barrier (Feng et al., 2010). Water pathways to Glu$_{ex}$ (shown as light blue molecules in the "out" diagram) are not seen in structures but have been observed in molecular dynamics simulations by several groups. CLC-ec1's "Glu$_{in}$" residue, which strongly influences the formation of water pathways in CLC-ec1, is not universally required among CLC transporters and therefore not depicted here.

variant is seen in a distantly related subclass of prokaryotic CLCs that lack conservation in the intracellular region of the selectivity filter. These CLCs share the same structural architecture as other CLC transporters but act as F^-/H^+ antiporters, with a 1:1 stoichiometry (Last et al., 2018).

13.6 Pharmacology

Many classic anion-channel inhibitors, which are notoriously nonselective, are effective against CLCs at high micromolar concentrations. These inhibitors fit into two broad categories: derivatives of stilbene disulfonates and a variety of organic carboxylates (Figure 13.5). These compounds are active in their negatively charged form, which predominates at physiological pH.

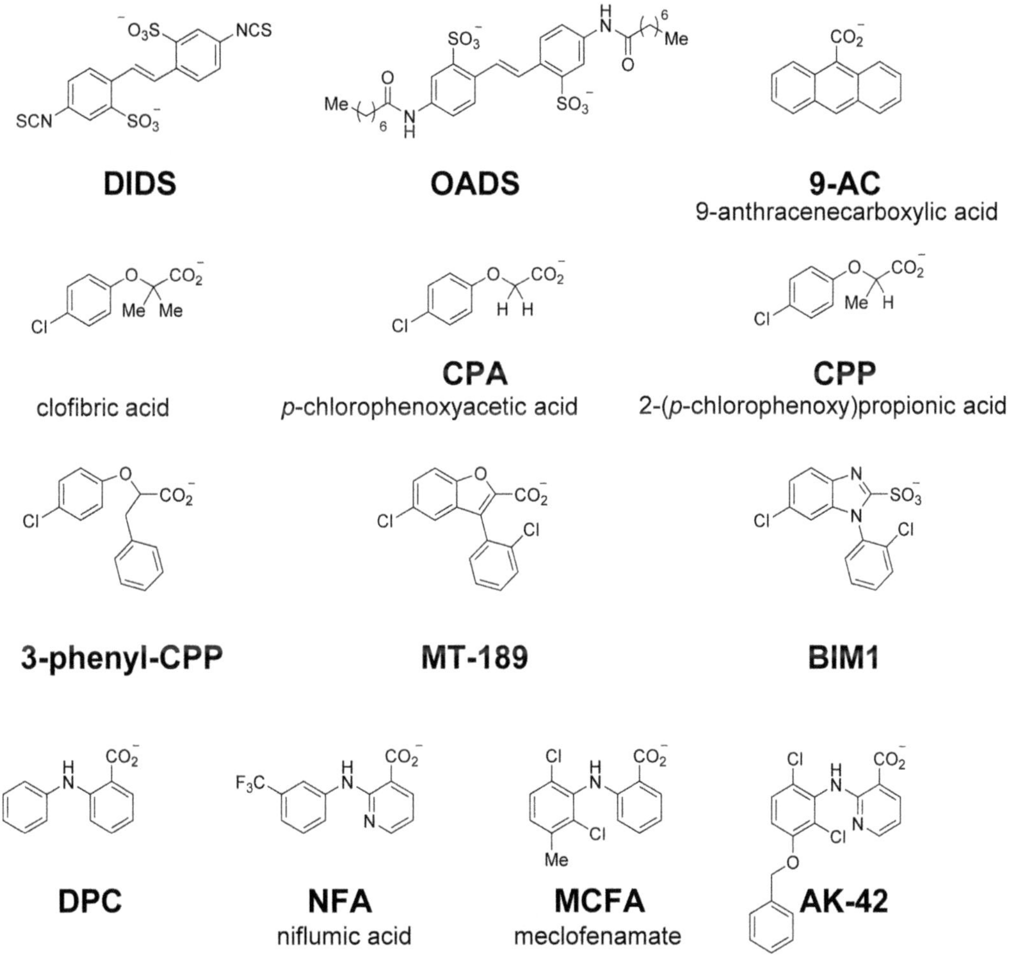

FIGURE 13.5
CLC pharmacology. Structures of inhibitors discussed in the text. All compounds are shown in their active (negatively charged) form, which predominates at physiological pH.

Stilbene disulfonates, such as DIDS (4,4-diisothiocyanatostilbene-2,2-disulfonate), have been used for decades as generic blockers of anion currents and are active against CLC-0, CLC-Ka, CLC-7 and CLC-ec1. The instability of the DIDS scaffold makes it rather intractable for further structure-activity relationship studies; however, it was used as the basis for the development of OADS, a stable molecule that inhibits the CLC-ec1 transporter.

The effects of organic carboxylates have been most extensively studied for the CLC-0, CLC-1 and CLC-Ka channels. CLC-1 is inhibited from the intracellular side by 9-anthracenecarboxylic acid (9-AC) and by clofibric acid derivatives, such as CPA and CPP. CLC-0 is also inhibited by clofibric acid derivatives, and clever mechanistic studies demonstrated that these compounds are state-dependent. That is, they have different apparent affinities for the closed and open states of the channel (Accardi and Pusch, 2003), making them useful for biophysical studies of CLC-0 gating. While CPA and CPP are not potent against the CLC-K channels, modifications to the CPP scaffold led to the development of two CLC-Ka inhibitors: 3-phenyl-CPP and a benzofuran heterocycle called MT-189. MT-189 has improved potency against CLC-Ka (IC_{50} = 7 μM) relative to 3-phenyl-CPP and impressive threefold selectivity for CLC-Ka over CLC-Kb despite the 90% sequence identity between these homologs (Liantonio et al., 2008; Gradogna and Pusch, 2010). In contrast to the intracellularly mediated inhibition of CLC-0 and CLC-1 by clofibric acid derivatives, the benzofuran carboxylates inhibit CLC-K channels from the extracellular side. Using the molecular geometry of MT-189 as a model, a novel class of benzimidazole sulfonates (BIMs) was developed. The BIMs retain the low micromolar potency of MT-189 (IC_{50} = 9 μM), but with selectivity improved to >20-fold for CLC-Ka over CLC-Kb (Koster et al., 2018).

Another class of organic carboxylates that act on CLC channels belong to a family of nonsteroidal anti-inflammatory drugs called fenamates. The unelaborated fenamate scaffold, diphenylamine-2-carboxylate (DPC) has long been used as a generic blocker of anion currents. For the CLC-K channels, fenamates are thought to bind in the same extracellular vestibule as the benzofurans, BIMS and clofibrates. Another nonspecific blocker within this class is niflumic acid (NFA). NFA inhibits CLC-1 currents in the mid-micromolar range and has no effect on CLC-2 below at least 120 μM. Interestingly, NFA potentiates instead of inhibits CLC-Ka currents at concentrations below 1 mM and acts via an unknown mechanism that is distinct from that of the inhibitory binding site (Gradogna and Pusch, 2010). No other direct small-molecule activators of CLCs are currently known. Starting with a fenamate hit compound called meclofenamate (MCFA), the first low-nanomolar potency small-molecule inhibitor of the CLC family was recently developed. This compound, AK-42, has an IC_{50} of 17 nM for CLC-2, is ~10,000 times more potent against CLC-2 compared to the next closest homolog (CLC-1), and has been validated against endogenous CLC-2 currents in CA1 hippocampal neurons (Koster et al., 2020).

In contrast to the voltage-gated cation channels, for which various prokaryotes and venomous animals have provided a rich source of specific toxin-based inhibitors and activators, biologically derived modulators of CLC function are still generally lacking. The one exception is a peptide identified from scorpion venom, GaTx2, which inhibits CLC-2 with ~20 pM potency (Thompson et al., 2009). Inhibition by GaTx2, however, saturates at ~50%.

Many CLCs are sensitive to Zn^{2+} or Cd^{2+} in the low to mid-micromolar range. Zn^{2+} and Cd^{2+} act by allosteric closure of the slow gate in the CLC channels (-0, -1, -2, -Ka/Kb). The CLC-4 transporter is also inhibited by Zn^{2+}, likely through extracellular histidine residues that are not conserved in the channels. Small-molecule modulators for the mammalian CLC transporters are currently lacking. CLC-7 – the most pharmacologically characterized of the mammalian transporters – was inhibited by 4 of 16 classic (nonselective) Cl^- channel inhibitors that were tested (Schulz et al., 2010).

13.7 Regulation

CLCs are variously regulated by phosphorylation, palmitoylation, nucleotide (ATP) binding to the CBS domains, and by auxiliary subunits. In addition, CLC-1 is regulated by the cellular redox environment and NAD^+. Lipid regulation has not yet been demonstrated for CLCs, but the structural observation of phosphatidylinositol 3-phosphate (PI3P) bound to CLC-7 (Schrecker et al., 2020) suggests that such regulation may exist.

Auxiliary subunits are known for CLC-2, CLC-K and CLC-7. CLC-2 is expressed in many cell types without an auxiliary subunit but associates with an adhesion molecule called GlialCAM in glia, as described in Section 13.2. In contrast, the CLC-K auxiliary subunit barttin is always found coexpressed with the CLC-K channel subunits. Barttin, a two-span membrane protein, is essential for CLC-K trafficking and has relatively modest effects on the channels' functional properties. The CLC-7 auxiliary subunit Ostm1 is not required for CLC-7 localization, but its heavy glycosylation appears to protect CLC-7 from lysosomal proteases.

13.8 Cell Biology

Subcellular sorting is crucial for most, if not all, CLCs. In epithelial cells, CLC-2 is sorted to the basolateral membrane via interactions between a dileucine motif in the first CBS domain and the clathrin adaptor AP-1. In the nephron and inner ear, barttin is essential for trafficking CLC-K channels to the proper membrane. In skeletal muscle, the localization of CLC-1 within the t-tubules or sarcolemma is controversial, but its localization is likely important given the unique architecture of these regions. For the intracellular CLCs, various sorting motifs that interact with adaptor proteins and direct subcellular localization have been identified (Stauber and Jentsch, 2010). In turn, the intracellular CLC transporters influence vesicular trafficking and function (Stauber and Jentsch, 2013).

13.9 Channelopathies and Disease

Loss-of-function and dominant negative mutations in CLCs are responsible for a variety of human diseases, including myotonia congenita (CLC-1), Bartter's syndrome (CLC-Kb), neurological and endocrine disorders (CLC-2, CLC-3, CLC-4, CLC-6, CLC-7), and Dent disease (CLC-5). These are detailed in Volume III, Chapter 11.

13.10 Conclusions

CLCs are a fascinating family of double-barreled anion channels and transporters. Studies of CLC proteins have revealed novel gating mechanisms and illuminated how variants of a single structural scaffold can carry out thermodynamically different functions. Moreover, CLCs are central to physiology – from endowing foodborne pathogenic microorganisms with extreme acid tolerance to driving complex physiological pathways in higher organisms. Mechanistic studies to uncover the details of these functions will continue, providing critical life insights into the foreseeable future.

Acknowledgments

We thank Alessio Accardi, Allan Bretag, and Michael Pusch for comments on the manuscript. We thank the National Institutes of Health for supporting our CLC research (R01 GM113195, R01 NS113611, and R01 DK128881).

Suggested Reading

Accardi, A. 2015. "Structure and gating of CLC channels and exchangers." *J Physiol* 593(18):4129–38. doi: 10.1113/JP270575.

Accardi, A., and C. Miller. 2004. "Secondary active transport mediated by a prokaryotic homologue of ClC Cl– channels." *Nature* 427(6977):803–7. doi: 10.1038/nature02314.

Accardi, A., and M. Pusch. 2003. "Conformational changes in the pore of CLC-0." *J Gen Physiol* 122(3):277–93. doi: 10.1085/jgp.200308834.

Altamura, C., J. F. Desaphy, D. Conte, A. De Luca, and P. Imbrici. 2020. "Skeletal muscle ClC-1 chloride channels in health and diseases." *Pflugers Arch* 472(7):961–75. doi: 10.1007/s00424-020-02376-3.

Basilio, D., K. Noack, A. Picollo, and A. Accardi. 2014. "Conformational changes required for H(+)/ Cl(−) exchange mediated by a CLC transporter." *Nat Struct Mol Biol* 21(5):456–63. doi: 10.1038/nsmb.2814.

Chavan, T. S., R. C. Cheng, T. Jiang, I. I. Mathews, R. A. Stein, A. Koehl, H. S. McHaourab, E. Tajkhorshid, and M. Maduke. 2020. "A CLC-ec1 mutant reveals global conformational change and suggests a unifying mechanism for the CLC Cl(−)/H(+) transport cycle." *eLife* 9. doi: 10.7554/eLife.53479.

De Jesus-Perez, J. J., A. Castro-Chong, R. C. Shieh, C. Y. Hernandez-Carballo, J. A. De Santiago-Castillo, and J. Arreola. 2016. "Gating the glutamate gate of CLC-2 chloride channel by pore occupancy." *J Gen Physiol* 147(1):25–37. doi: 10.1085/jgp.201511424.

De Stefano, S., M. Pusch, and G. Zifarelli. 2013. "A single point mutation reveals gating of the human ClC-5 Cl−/H+ antiporter." *J Physiol* 591(23):5879–93. doi: 10.1113/jphysiol.2013.260240.

Dutzler, R., E. B. Campbell, and R. MacKinnon. 2003. "Gating the selectivity filter in ClC chloride channels." *Science* 300(5616):108–12. doi: 10.1126/science.1082708.

Fahlke, C., and M. Fischer. 2010. "Physiology and pathophysiology of ClC-K/barttin channels." *Front Physiol* 1:155. doi: 10.3389/fphys.2010.00155.

Feng, L., E. B. Campbell, Y. Hsiung, and R. MacKinnon. 2010. "Structure of a eukaryotic CLC transporter defines an intermediate state in the transport cycle." *Science* 330(6004):635–41. doi: 10.1126/science.1195230.

Gianesello, L., D. Del Prete, M. Ceol, G. Priante, L. A. Calo, and F. Anglani. 2020. "From protein uptake to Dent disease: An overview of the CLCN5 gene." *Gene* 747:144662. doi: 10.1016/j.gene.2020.144662.

Gradogna, A., and M. Pusch. 2010. "Molecular pharmacology of kidney and inner ear CLC-K chloride channels." *Front Pharmacol* 1:130. doi: 10.3389/fphar.2010.00130.

Jentsch, T. J., and M. Pusch. 2018. "CLC chloride channels and transporters: Structure, function, physiology, and disease." *Physiol Rev* 98(3):1493–590. doi: 10.1152/physrev.00047.2017.

Jeworutzki, E., L. Lagostena, X. Elorza-Vidal, T. Lopez-Hernandez, R. Estevez, and M. Pusch. 2014. "GlialCAM, a CLC-2 Cl(−) channel subunit, activates the slow gate of CLC chloride channels." *Biophys J* 107(5):1105–16. doi: 10.1016/j.bpj.2014.07.040.

Koster, A. K., A. L. Reese, Y. Kuryshev, X. Wen, K. A. McKiernan, E. E. Gray, C. Wu, J. R. Huguenard, M. Maduke, and J. Du Bois. 2020. "Development and validation of a potent and specific inhibitor for the CLC-2 chloride channel." *Proc Natl Acad Sci U S A* 117(51):32711–21. doi: 10.1073/pnas.2009977117.

Koster, A. K., C. A. P. Wood, R. Thomas-Tran, T. S. Chavan, J. Almqvist, K. H. Choi, J. Du Bois, and M. Maduke. 2018. "A selective class of inhibitors for the CLC-Ka chloride ion channel." *Proc Natl Acad Sci U S A* 115(21):E4900–E9. doi: 10.1073/pnas.1720584115.

Last, N. B., R. B. Stockbridge, A. E. Wilson, T. Shane, L. Kolmakova-Partensky, A. Koide, S. Koide, and C. Miller. 2018. "A CLC-type F(-)/H(+) antiporter in ion-swapped conformations." *Nat Struct Mol Biol* 25(7):601–6. doi: 10.1038/s41594-018-0082-0.

Liantonio, A., A. Picollo, G. Carbonara, G. Fracchiolla, P. Tortorella, F. Loiodice, A. Laghezza, E. Babini, G. Zifarelli, M. Pusch, and D. C. Camerino. 2008. "Molecular switch for CLC-K Cl− channel block/activation: Optimal pharmacophoric requirements towards high-affinity ligands." *Proc Natl Acad Sci U S A* 105(4):1369–73. doi: 10.1073/pnas.0708977105.

Lisal, J., and M. Maduke. 2008. "The ClC-0 chloride channel is a "broken" Cl−/H+ antiporter." *Nat Struct Mol Biol* 15(8):805–10.

Ludwig, C. F., F. Ullrich, L. Leisle, T. Stauber, and T. J. Jentsch. 2013. "Common gating of both CLC transporter subunits underlies voltage-dependent activation of the 2Cl−/1H+ exchanger ClC-7/Ostm1." *J Biol Chem* 288(40):28611–9. doi: 10.1074/jbc.M113.509364.

Mayes, H. B., S. Lee, A. D. White, G. A. Voth, and J. M. J. Swanson. 2018. "Multiscale kinetic modeling reveals an ensemble of Cl(−)/H(+) exchange pathways in ClC-ec1 antiporter." *J Am Chem Soc* 140(5):1793–804. doi: 10.1021/jacs.7b11463.

Miller, C. 2006. "ClC chloride channels viewed through a transporter lens." *Nature* 440(7083):484–9. doi: 10.1038/nature04713.

Miller, C. 2014. "In the beginning: A personal reminiscence on the origin and legacy of ClC-0, the 'Torpedo Cl− channel'." *J Physiol* 593(Pt 18):4085–90. doi: 10.1113/jphysiol.2014.286260.

Nicoli, E. R., M. R. Weston, M. Hackbarth, A. Becerril, A. Larson, W. M. Zein, P. R. Baker 2nd, J. D. Burke, H. Dorward, M. Davids, Y. Huang, D. R. Adams, P. M. Zerfas, D. Chen, T. C. Markello, C. Toro, T. Wood, G. Elliott, M. Vu, Network Undiagnosed Diseases, W. Zheng, L. J. Garrett, C. J. Tifft, W. A. Gahl, D. L. Day-Salvatore, J. A. Mindell, and M. C. V. Malicdan. 2019. "Lysosomal storage and albinism due to effects of a de novo CLCN7 variant on lysosomal acidification." *Am J Hum Genet* 104 (6):1127–38. doi: 10.1016/j.ajhg.2019.04.008.

Pedersen, T. H., A. Riisager, F. V. de Paoli, T. Y. Chen, and O. B. Nielsen. 2016. "Role of physiological ClC-1 Cl− ion channel regulation for the excitability and function of working skeletal muscle." *J Gen Physiol* 147(4):291–308. doi: 10.1085/jgp.201611582.

Picollo, A., M. Malvezzi, and A. Accardi. 2010. "Proton block of the CLC-5 Cl−/H+ exchanger." *J Gen Physiol* 135(6):653–9. doi: 10.1085/jgp.201010428.

Rychkov, G. Y., M. Pusch, M. L. Roberts, T. J. Jentsch, and A. H. Bretag. 1998. "Permeation and block of the skeletal muscle chloride channel, ClC-1, by foreign anions." *J Gen Physiol* 111(5):653–65.

Schrecker, M., J. Korobenko, and R. K. Hite. 2020. "Cryo-EM structure of the lysosomal chloride-proton exchanger CLC-7 in complex with OSTM1." *eLife* 9:e59555. doi: 10.7554/eLife.59555.

Schulz, P., J. Werner, T. Stauber, K. Henriksen, and K. Fendler. 2010. "The G215R mutation in the Cl−/H+-antiporter ClC-7 found in ADO II osteopetrosis does not abolish function but causes a severe trafficking defect." *PLOS ONE* 5(9):e12585. doi: 10.1371/journal.pone.0012585.

Stauber, T., and T. J. Jentsch. 2010. "Sorting motifs of the endosomal/lysosomal CLC chloride transporters." *J Biol Chem* 285(45):34537–48. doi: 10.1074/jbc.M110.162545.

Stauber, T., and T. J. Jentsch. 2013. "Chloride in vesicular trafficking and function." *Annu Rev Physiol* 75:453–77. doi: 10.1146/annurev-physiol-030212-183702.

Thompson, C. H., P. R. Olivetti, M. D. Fuller, C. S. Freeman, D. McMaster, R. J. French, J. Pohl, J. Kubanek, and N. A. McCarty. 2009. "Isolation and characterization of a high affinity peptide inhibitor of ClC-2 chloride channels." *J Biol Chem* 284(38):26051–62.

Traverso, S., G. Zifarelli, R. Aiello, and M. Pusch. 2006. "Proton sensing of CLC-0 mutant E166D." *J Gen Physiol* 127(1):51–66.

Yu, Y., M. F. Tsai, W. P. Yu, and T. Y. Chen. 2015. "Modulation of the slow/common gating of CLC channels by intracellular cadmium." *J Gen Physiol* 146(6):495–508. doi: 10.1085/jgp.201511413.

Zifarelli, G. 2015. "A tale of two CLCs: Biophysical insights toward understanding ClC-5 and ClC-7 function in endosomes and lysosomes." *J Physiol* 593(18):4139–50. doi: 10.1113/JP270604.

Zuniga, L., M. I. Niemeyer, D. Varela, M. Catalan, L. P. Cid, and F. V. Sepulveda. 2004. "The voltage-dependent ClC-2 chloride channel has a dual gating mechanism." *J Physiol* 555(3):671–82.

14

Calcium-Activated Cl⁻ Channels

H. Criss Hartzell

CONTENTS

14.1 Introduction

Many cells express anion-selective ion channels that are activated by increases in cytosolic Ca^{2+} concentration ($[Ca^{2+}]_i$). These ligand-gated Ca^{2+}-activated Cl⁻ channels (CaCCs) support a wealth of diverse, crucial physiological functions (Hartzell et al., 2005). CaCCs are encoded by six genes. "Classical" CaCCs are encoded by two members of the Transmembrane Protein-16/*Anoctamin* gene family: *TMEM16A/ANO1* and *TMEM16B/ANO2*. The official HUGO name, Anoctamin, is a misnomer based on incorrect predictions that all Tmem16s are *an*ion channels (most are phospholipid scramblases) and have

DOI: 10.1201/9781003096276-14

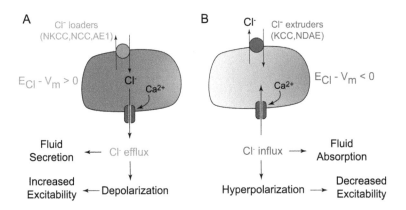

FIGURE 14.1
CaCC function depends on the Cl⁻ electrochemical gradient. (A) Some cells, like epithelial cells and sensory neurons, accumulate Cl⁻ via Cl⁻-loading transporters. CaCC opening triggered by increases in cytosolic Ca^{2+} results in Cl⁻ efflux and depolarization. Fluid secretion occurs because Cl⁻ efflux is followed by cations and water to maintain charge balance and osmolarity. (B) Other cells, such as many mature neurons, extrude Cl⁻ so that CaCC opening leads to Cl⁻ influx and hyperpolarization. Typically, CaCCs are located on the apical membrane facing the lumen (for example, Tmem16a is apical in salivary gland acinar cells), but Best1 is located on the basolateral membrane of retinal pigment epithelial (RPE) cells and Best2 is on the basolateral membrane of colonic goblet cells. Abbreviations: NKCC, sodium potassium chloride cotransporter; NCC, sodium chloride cotransporter; AE1, anion exchanger 1; KCC, potassium chloride cotransporter; NDAE, sodium-dependent anion exchanger.

eight (*oct*) transmembrane helices (they have ten), so the TMEM16 nomenclature is used here. The other four CaCC genes are bestrophins (*BEST1–BEST4*). Even though they are both CaCCs, Tmem16s and bestrophins have completely different structures, mechanisms of gating, tissue expression patterns and physiological functions. Here, we describe and compare these two physiologically important channels and their roles in human disease.

14.2 Physiological Roles

CaCCs transduce an elevation of cytosolic Ca^{2+} into Cl⁻ flux across the plasma membrane. Under resting conditions CaCCs are closed, but they open in response to stimuli, such as autonomic neurotransmitters, that elevate cytosolic Ca^{2+}. Cl⁻ then may enter or leave the cell and produce either hyperpolarization or depolarization, depending on the Cl⁻ electrochemical gradient, which varies greatly in different cell types and physiological conditions (Figure 14.1). Tmem16 and Best functions are discussed specifically in the following sections.

14.3 Tmem16/Anoctamin Channels

14.3.1 Physiological Roles

"Classical" CaCCs, now known to be encoded by *TMEM16A* and *TMEM16B*, were first described in the 1980s (Hartzell et al., 2005). One of their first described functions was fast block to polyspermy (Volume III, Chapter 6), but CaCCs are best known for their role in epithelial fluid secretion. In secretory epithelia, intracellular Cl⁻ is accumulated above electrochemical equilibrium by basolateral transporters like NKCC1, a Na⁺–K⁺–2Cl⁻ transporter. Increases in $[Ca^{2+}]_i$ gate open apical CaCCs that allow Cl⁻ efflux, which is followed by Na⁺ and water to maintain electroneutrality and osmolarity. This saline secretion forms the liquid component of bodily fluids like saliva, tears, milk, semen and pancreatic juice, and is crucial for hydrating proteins and other components secreted by these tissues. For example, antibacterial proteins and enzymes secreted by salivary acinar cells are solubilized and swept into the salivary ducts by CaCC-driven fluid secretion.

Some functions of Tmem16a are linked to regulation of membrane potential (V_m) (Hartzell et al., 2005; Pedemonte and Galietta, 2014). In certain smooth muscles, such as arterial smooth muscle, Tmem16a regulates contractility by directly controlling smooth muscle V_m (Leblanc et al., 2015). Because arterial smooth muscle $[Cl^-]_i$ is above electrochemical equilibrium, CaCC activation leads to Cl⁻ efflux. The resulting depolarization opens voltage-gated Ca^{2+} channels that trigger vasoconstriction. In other smooth muscles like gut, Tmem16a controls muscle tone indirectly by regulating V_m of interstitial cells of Cajal (Sanders, 2019), which are electrically coupled to smooth muscle cells. Interstitial cells of Cajal are pacemaker cells responsible for gut peristalsis.

By regulating V_m, Tmem16a and Tmem16b also control excitability. For example, Tmem16a is a nociceptive ion channel that operates in conjunction with Trpv1 (Chapter 24) in mediating pain sensation via dorsal root ganglion somatosensory neurons (Shah et al., 2020). Tmem16b regulates excitability in some central nervous system neurons including cerebellar Purkinje cells, and neurons in the hippocampus, thalamus, lateral septum, amygdala and the olivocerebellar system, but depending on the Cl⁻ electrochemical potential, Tmem16b can mediate increased or decreased excitability. In olfactory neurons, Ca^{2+} influx mediated by odorant-activated cyclic nucleotide-gated channels (CNG channels; Chapter 11) activates Tmem16b, which shapes the odor response.

Tmem16a plays an important role in iodide homeostasis in the thyroid gland, which accumulates iodide for the synthesis of the hormone thyroxine.

14.3.2 Subunit Diversity and Basic Structural Organization

Tmem16s are expressed in most eukaryotes. Whereas green plants and fungi including yeast have only one or two *TMEM16* genes, other phyla have multiple *TMEM16* genes that arose during evolution by gene duplication. Of the ten *TMEM16* genes in vertebrates, only two, *TMEM16A* and *TMEM16B*, encode CaCCs (Whitlock and Hartzell, 2017). Most other *TMEM16* genes encode Ca^{2+}-activated phospholipid scramblases that facilitate the passive transport of phospholipids between the inner and outer leaflets of the membrane bilayer. Scramblases also conduct ions, but with low ionic selectivity. Although Tmem16 scramblase currents can be as large as Tmem16a currents (nanoamps) and can conduct Ca^{2+}, the physiological role of these currents remains an open question.

In addition to diversity endowed by multiple CaCC genes, alternative splicing adds another level of complexity (Pedemonte and Galietta, 2014) (Figure14.2A). Alternative exon 6b (segment "b," 22 amino acids) regulates Ca^{2+} sensitivity, exon 13 (segment "c," 4 amino acids) controls voltage dependence and regulation by PI(4,5)P₂, and exon 15 (segment "d," 26 amino acids) affects activation/deactivation kinetics. Segment "a" is encoded by an alternative start site. Splice variants also affect the sensitivity to pharmacological blockers. Alternative exons can exist in various combinations and they are expressed in a tissue-dependent fashion, which suggests that functional differences among these isoforms subserve tissue-specific roles.

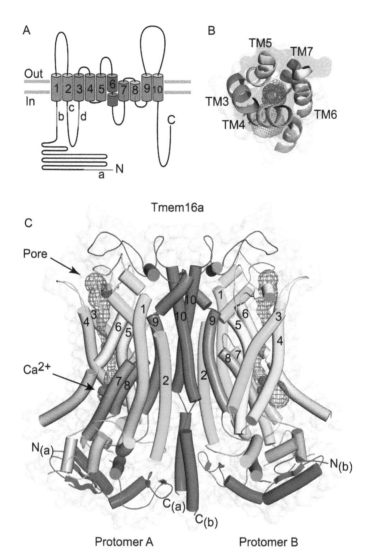

FIGURE 14.2
Structure of Tmem16a/Ano1. (A) Illustration of single protomer. Splice variants are indicated by orange segments labeled a (alternative start), b (exon 6b), c (exon 13) and d (exon 15). (B) The protomer pore viewed from the extracellular space. (C) Dimeric Tmem16a viewed from the plane of the membrane. Pore is shown as blue mesh. Ca^{2+} ions are magenta spheres. Protein Data Bank: 5OYB.

All Tmem16 Cl⁻ channels and scramblases have a similar overall architecture (Kalienkova et al., 2021; Falzone et al., 2018). The mature protein is a homodimer with back-to-back protomers forming a double-barreled channel (Figure 14.2). Each protomer functions independently and has its own gate controlled by Ca^{2+} binding. While CLC channels are also double-barreled homodimers (Chapter 13), unlike Tmem16a, CLC channels have a common gate for both protomers in addition to individual gates.

The Tmem16a conductance pathway is lined by transmembrane helices 3–7 that form a dumbbell-shaped hydrophilic pore with large vestibules on the cytoplasmic and extracellular sides connected by a narrow neck. The gate is composed of hydrophobic amino acids located in the narrowest region near the center of the membrane. However, the exact nature of the permeation pathway remains in question because cryo-EM structures have not yet captured a fully open state. The diameter of the neck in the available structures is ~2.5 Å, significantly less than the 3.6 Å Cl– ion diameter. A wealth of data shows that Tmem16a has multiple open and closed conformations. Because these channels spontaneously inactivate in the presence of Ca^{2+} (see Section 14.3.6), it might be expected that protein isolation for cryo-EM studies in the presence of Ca^{2+} would preferentially capture nonconducting states.

14.3.3 Gating

Each protomer of the dimeric protein binds two Ca^{2+} ions (Kalienkova et al., 2021). The Ca^{2+} binding site is located adjacent to the Cl– permeation pathway a short distance into the membrane. Ca^{2+} ions are coordinated in their binding site by side-chain oxygen atoms of seven hydrophilic residues located on TMs 6–8 of each protomer (Figures 14.2 and 14.3). Four of these residues (E654, E705, E734 and D738) are conserved in all Tmem16 proteins.

Comparison of Tmem16a with and without bound Ca^{2+} reveals conformational changes associated with channel gating (Figure 14.3). In the Ca^{2+}-free state, the lower half of TM6 is unfolded and bends toward TM4 around a glycine hinge. This happens because the negatively charged amino acid side chains in the vacant Ca^{2+} binding site electrostatically repel. Ca^{2+} binding neutralizes the negative charge and pulls TM6 closer to TM7 and TM8, and away from TM4. These conformational changes around the Ca^{2+} binding site allosterically open the gate in the neck.

An intriguing feature of Tmem16a currents is their dramatically different gating behavior with different intracellular Ca^{2+} concentrations (Figure 14.3) (Hartzell et al., 2005). At $[Ca^{2+}]_i$ <1 μM, the currents are voltage-dependent: they outwardly rectify and activate slowly with time upon depolarization. At higher $[Ca^{2+}]_i$, the currents are much less voltage-dependent. The voltage dependence at low $[Ca^{2+}]_i$ is explained by the location of the Ca^{2+} binding site within the voltage field of the membrane. Positive V_m increases the apparent Ca^{2+} binding affinity by driving Ca^{2+} into its binding pocket. The location of the Ca^{2+} binding site also affects current rectification because neutralization of negative charges adjacent to the inner vestibule reduces an electrostatic barrier to anion permeation.

A full understanding of Tmem16a channel gating is hampered by its small (~3 pS) single-channel conductance. However, structurally guided mutagenesis and macroscopic current modeling have provided robust kinetic gating schemes that support the existence of multiple closed and open states (Kalienkova et al., 2021).

14.3.4 Ion Permeability

Tmem16a exhibits a lyotropic permeability sequence in which larger, more easily dehydrated anions, such as Br–, I– and SCN–, are more permeable than Cl–. Molecular dynamics simulations suggest that conducted Cl– ions are partially hydrated, which is consistent with biophysical measurements that indicate the pore has a dielectric constant near ten, about half that of water (Hartzell et al., 2005). Because Tmem16a can conduct ions with diameters up to 8–10 Å, at least one of the open conformations of the channel is certainly quite different than the presently available structures.

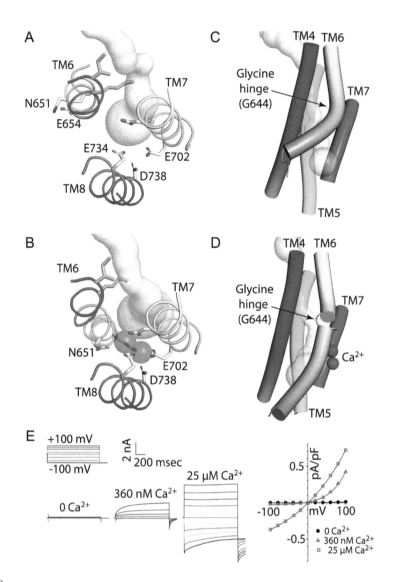

FIGURE 14.3
Gating of Tmem16a. (A and B) Ca^{2+}-binding site in the Ca^{2+}-free (A) and Ca^{2+}-bound (B) states. Ca^{2+} ions are shown as magenta spheres. Surface shows the pore calculated by Caver 3.0. (C and D) Side view of pore-forming helices in Ca^{2+}-free (C) and Ca^{2+}-bound (D) states. (E) Currents and I–V plots of TMEM16A activated by different intracellular $[Ca^{2+}]$.

14.3.5 Pharmacology

The pharmacology of anion channels has lagged that of cation channels (Chapter 13). Niflumic acid (NFA) is a traditional CaCC blocker, but it is nonspecific and low affinity (IC_{50} ~10 μM range). Recently, high-throughput screens have identified several compounds (CACCinh-A01, T16AinhA01, MONNA, Ani-9) that block Tmem16a with micromolar or submicromolar affinities, but in general, the mechanisms of action and specificity have not been completely evaluated. Considering its important physiological functions, Tmem16a is an obvious therapeutic target for disorders like hypertension, but its nearly ubiquitous tissue expression raises questions about off-target effects that may require clever drug delivery approaches to avoid. Recently, it has been shown that many Tmem16a inhibitors act indirectly by altering intracellular Ca^{2+} signaling (Genovese et al., 2022).

14.3.6 Regulation

Tmem16a current amplitude decreases with time of recording ("rundown"), especially when exposed to high concentrations of cytosolic Ca^{2+}. As shown for many K^+ channels, Tmem16a inactivation is explained by depletion of $PI(4,5)P_2$ (Arreola and Hartzell, 2019). Rundown in excised patches can be slowed or abrogated by addition of excess $PI(4,5)P_2$ to the cytoplasmic face of the membrane. Mutagenesis and molecular dynamics simulations have identified several potential $PI(4,5)P_2$ binding sites. Mutations that affect rundown or sensitivity to $PI(4,5)P_2$ cluster around the cytoplasmic ends of TM2-6, but details of the mechanisms of regulation by $PI(4,5)P_2$ remain unclear partly because Tmem16a is highly allosteric. Understanding how $PI(4,5)P_2$ binding to one region affects gating or conduction mediated by another discontinuous region is challenging. In addition to regulation by $PI(4,5)P_2$, Tmem16a is also regulated by cholesterol, calmodulin (although this is controversial) and auxiliary subunits.

14.3.7 Cell Biology (Biogenesis, Trafficking, Turnover)

Tmem16a expression is regulated at the transcriptional, translational and posttranslational levels (Dulin, 2020). Several interleukins, including the proinflammatory cytokine IL-4 that is involved in the pathogenesis of asthma, regulate the production of airway surface liquid (ASL) by controlling transcription of TMEM16A (and other channels) in the airway epithelium. ASL plays a critical role in maintaining mucociliary clearance, an important innate defense mechanism that clears respiratory pathogens from the airway and lung. Tmem16a translation is regulated by several microRNAs, (miRNA-9, -29, -132, -144, -381 and -422), which are small single-stranded noncoding RNAs that induce mRNA degradation or translational repression. Trafficking and turnover of Tmem16a is regulated by ubiquitination, extended synaptotagmins, Clca1 (chloride channel accessory 1) and 14-3-3 proteins.

14.4 Bestrophins

14.4.1 Physiological Roles

Whereas the Tmem16s are widely expressed and have diverse functions, bestrophin is more restricted. Best1 is highly expressed in the retinal pigment epithelium (RPE), a layer of cells in contact with the photoreceptors that is crucial in retinal homeostasis. RPE is essential for regeneration of visual pigment and daily renewal of photoreceptor discs. In addition, the fluid and ionic composition surrounding photoreceptors is regulated by the RPE in a light-dependent manner. RPE cells accumulate Cl⁻ (Figure 14.1A). Illumination activates Best1, which results in Cl⁻ efflux and depolarization of the RPE (Hartzell et al., 2008; Johnson et al., 2017), but the end function of the depolarization remains an open question. Cl⁻ flux may be involved in cell volume regulation by a noncanonical pathway (Milenkovic et al., 2015). Canonically, when cells swell, they regulate their volume by activating volume-regulated anion channels (VRAC) that expedite the efflux of Cl⁻, organic solutes and water. In vertebrates, VRAC is encoded by *LRRC8* genes and Lrrc8a is required for cell volume regulation in many mammalian cells. However, Bestrophins and Tmem16s, are also sensitive to cell volume. In *Drosophila*, Best1 is essential for cell volume regulation. It has been suggested that Best1 might play a noncanonical role in cell volume regulation in mammalian RPE independent of Lrrc8 (Milenkovic et al., 2015).

Best1 has been reported to be permeable to neurotransmitters glutamate and GABA, and may play a role in release of these molecules by astrocytes in the brain to modulate synaptic activity (Elorza-Vidal et al., 2019).

Best2 is expressed in the basolateral plasma membrane of nonpigmented epithelial (NPE) cells of the ciliary body and regulates intraocular pressure (IOP). Best2 may regulate IOP directly by Cl^--dependent fluid secretion (Figure 14.1A), a key determinant of aqueous humor formation. However, because Best2 is also highly permeable to HCO_3^-, Best2 may regulate IOP indirectly via HCO_3^- transport, which controls intracellular and intraocular pH. Best2 plays a role in HCO_3^- secretion by goblet cells of the colon (Yu et al., 2010).

14.4.2 Subunit Diversity and Basic Structural Organization

Bestrophin genes and homologs are found in eukaryotes and prokaryotes. Most vertebrates have 3 or 4 *BEST* genes, but *Caenorhabditis elegans* boasts 26. The only bacterial *BEST* homolog studied functionally is not a Cl^- channel, but a cation-selective channel that lacks the Ca^{2+} sensor of vertebrate Bests (Yang et al., 2014). The structures of chicken Best1, bovine Best2 and a bacterial homolog have been solved. Three splice variants have been reported, but their functional significance remains unclear.

Tmem16s and Bests have completely different architectures. Bests are homopentamers with five protomers symmetrically arranged around a 95 Å-long central pore (Figure 14.4) (Miller et al., 2019). Each protomer has four transmembrane helices, three (S2a, S3a and S4a) of which are connected to helical segments (S2b, S3b and S4b) that extend into the cytoplasm ~55 Å to form a large cytosolic domain. The ion-conducting pore is formed by both the transmembrane and cytosolic helices, and extends the entire length of the protein (95 Å).

The long cylindrical pore has two constrictions. Just inside the outer vestibule, a ~15 Å-long narrow "neck" forms the gate. This neck is formed by three hydrophobic and aromatic residues (I76, F80 and F84) that provide a significant barrier to ion permeation. In the open conformation, this narrow neck widens from 3.5 Å to 13 Å in diameter, which is sufficiently large for partially hydrated anions to pass. The neck opens into a large hydrophilic cavity in the center of the protein. At the cytosolic end there is a second constriction, called the "aperture." The aperture plays a role in determining the anion selectivity of the channel and may also act as a gate for certain anions in some Best channels (Owji et al., 2022).

14.4.3 Gating

Although Best channels are bona fide CaCCs, they are not responsible for classical CaCC currents because they do not exhibit the characteristic Ca^{2+}-dependent voltage dependence shown in Figure 14.3. Best1 and Best2 typically exhibit linear I–V relationships and are activated by lower intracellular Ca^{2+} concentrations than Tmem16a (Figure 14.5 D). Best3 and Best4 are less well studied.

Best channel opening is controlled by Ca^{2+} binding to five Ca^{2+} clasps located at the interface between the cytosol and the membrane (Figure 14.5). In each clasp, Ca^{2+} is coordinated by oxygen atoms of side-chain residues in S4a and S4b (Figure 14.5D) and backbone carbonyls from S1 of the adjacent subunit. Binding of Ca^{2+} results in the gating residues I76, F80 and F84 in S2 to rotate outward to allow for ion permeation. The coupling between

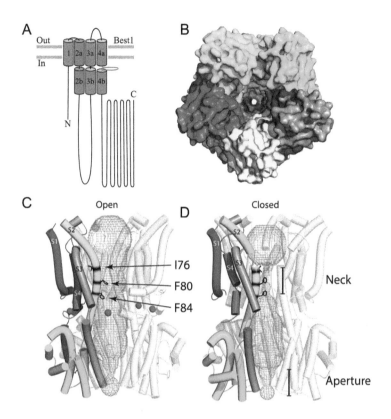

FIGURE 14.4
Structure of Best1. (A) Illustration of a single protomer. (B) The pore viewed from the extracellular surface. Each protomer has a different color. (C and D) Open (C) and closed (D) states of the channel. Only four protomers are shown; one in front has been removed for clarity. One protomer is colored. Mesh shows the pore. I76, F80 and F84 gating residues are shown in stick representation and are colored black. Magenta spheres are Ca^{2+}. Protein Data Bank: 6N28.

S4 and S2 is mediated by W287 located between the Ca^{2+} clasp and the neck. In the open conformation, W287 packs with F80 and F84 in the adjacent subunit, but in the closed conformation it adopts a different rotamer and moves to the space between adjacent subunits.

14.4.4 Ion Permeability

Best1, like Tmem16a, exhibits a lyotropic permeability sequence that suggests the permeant ions are partially dehydrated as they pass through the channel. Because most of the pore in the open conformation is sufficiently wide to allow hydrated ions to pass, dehydration is thought to occur at the aperture. Mutation of the aperture residue V205 in chicken Best1 to smaller residues abolishes the lyotropic permeability sequence, whereas mutation to bulkier isoleucine exaggerates the differences in ion permeability in the lyotropic sequence.

It has been proposed that Best-1 and Best-2 can conduct glutamate and play a role in release of the neurotransmitter glutamate from astrocytes in the brain (Lee et al., 2022). Recently, it has been shown that glutamine synthase, an enzyme that converts glutamate to glutamine, can bind to Best-2. It has been suggested that glutamine synthase can sensitize Best-2 to glutamate (Owji et al. 2022).

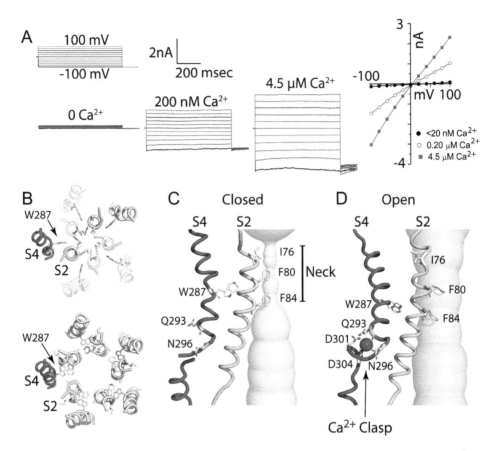

FIGURE 14.5
Bestrophin gating. (A) Whole-cell patch-clamp traces of mouse Best2 currents expressed in HEK cells. (B) Closed (top) and open (bottom) pore of chicken Best1 viewed from the extracellular face. S2 (cyan) and S4 (red) of one subunit are colored. Gating residues I76, F80 and F84 are in yellow, and W287 is in tan. (C and D) Pore viewed from plane of membrane in closed (C) and open (D) conformation. Only one subunit is shown. Ca^{2+} ions are magenta spheres. For clarity, the backbone carbonyls from S1 are not shown. D301 and D304 are not shown in (C) because they are unstructured when Ca^{2+} is not bound. Protein Data Bank: 6N26 and 6N28.

14.4.5 Pharmacology

The pharmacology of bestrophin channels is poorly developed. Although they are blocked by nonspecific anion channel blockers such as 4,4-diisothiocyano-2,2-stilbenedisulfonic acid (DIDS) and niflumic acid (NFA), there are no high-affinity specific blockers.

14.4.6 Regulation

Like Tmem16a, Best1 currents run down with time of recording in a Ca^{2+}-dependent manner, but the mechanisms are completely different. Best1 inactivation is caused by the binding of a C-terminal "inactivation peptide" (residues 346–367) to a "receptor" on the channel's cytosolic surface located near the Ca^{2+} clasp. Inactivation is decreased by mutation of the receptor or the inactivation peptide, and can be eliminated by removal of the inactivation peptide. The inactivation peptide is not shown in Figure 14.4 because the C-terminus was intentionally truncated to simplify analysis of gating mediated by the neck.

14.4.7 Cell Biology

Bestrophins clearly traffic to the plasma membrane as shown by the presence of bestrophin currents. However, some investigators have suggested Best1 regulates Ca^{2+} signaling both at the plasma membrane and in the endoplasmic reticulum. Best1 plasma membrane localization in astrocytes is dependent on the 14-3-3γ protein, which binds to the C-terminus in a phosphorylation-dependent manner.

14.5 Channelopathies and Disease

14.5.1 *TMEM16A* and Cancer

Before Tmem16a was identified as a CaCC, it had attracted the interest of cancer biologists as a biomarker for gastrointestinal squamous tumors and other tumors. This is reflected in its original names: DOG1 (Discovered On GIST-1 tumor), ORAOV2 (oral cancer overexpressed) and TAOS-2 (tumor amplified and overexpressed). The chromosomal region (11q13) that includes *TMEM16A* is amplified and Tmem16a expression is upregulated at both the transcriptional and translational level in certain, but not all, cancers (Bill and Alex Gaither, 2017; Crottes and Jan, 2019). Amplification of the *TMEM16A* locus is correlated with poor prognosis for survival, tumor size and metastasis.

The mechanisms by which Tmem16a promotes sustained proliferation and metastasis is an open question. Expression levels of various cell cycle proteins, transcription factors and apoptosis-related proteins correlate with Tmem16a expression in both overexpression systems and tumors. The epidermal growth factor receptor (EGFR) pathway, and its downstream signaling via Mapk, CamkII and protein kinase B have been implicated as important Tmem16a targets. Several studies suggest that Tmem16a Cl⁻ channel activity is not necessary for its pro-oncogenic activity and that Tmem16a protein may have additional functions, possibly by acting as a scaffold or by organizing membrane subdomains (Bill and Alex Gaither, 2017; He et al., 2017).

14.5.2 *TMEM16A* Genetic Diseases

Until recently, no genetic diseases have been unambiguously linked to *TMEM16A*. However, one study describes a severe phenotype characterized by impaired intestinal peristalsis, intestinal pneumatosis, and dysmorphic features that is linked to a truncating mutation in *TMEM16A*. Genome-wide association studies show that single-nucleotide polymorphisms in *TMEM16A* correlate with measures of obstructive and restrictive lung disease, diastolic blood pressure, and dysmorphic features. These correlations are consistent with studies in mice that show Tmem16a is important in blood pressure regulation and that knockout of *TMEM16A* affects tracheal morphogenesis. Due to its involvement in airway Cl⁻ secretion, Tmem16a has been suggested as a potential target for therapy of cystic fibrosis.

14.5.3 *BEST1*-Linked Diseases

Best vitelliform macular dystrophy (BVMD; or "Best disease") is a genetic form of macular degeneration associated with loss-of-function mutations in the *BEST1* gene that are generally inherited in an autosomal dominant manner. Other mutations in *BEST1* produce related retinal phenotypes that collectively have been termed bestrophinopathies. BVMD is characterized by accumulation of lipofuscin pigment in the RPE, fluid accumulation between the photoreceptors and RPE, and subsequent degeneration of the RPE. The vast majority of mutations are missense mutations located in the pore, the Ca^{2+} clasp and aperture regions of the protein. Mutations are associated with loss of Best1 CaCC current in the RPE. In vitro, it has been possible to restore normal CaCC currents to RPE cells derived from patients with BVMD by infection with adeno-associated virus expressing wild-type Best1. This provides proof-of-principle that genetic therapy for this disease is feasible.

14.6 Conclusions

Tmem16 and Best CaCCs differ markedly in their molecular architecture, mechanisms of gating and expression in various tissues. Yet, products of both gene families play essential physiological roles and are associated with debilitating human diseases. Our understanding of these channels has expanded dramatically in the last ten years and we are now on the verge of exploiting our knowledge of these proteins to treat disease.

Suggested Reading

Arreola, J., and H. C. Hartzell. 2019. "Wasted TMEM16A channels are rescued by phosphatidylinositol 4,5-bisphosphate." *Cell Calcium* 84:102103. doi: 10.1016/j.ceca.2019.102103.

Bill, A., and L. Alex Gaither. 2017. "The mechanistic role of the calcium-activated chloride channel ANO1 in tumor growth and signaling." *Adv Exp Med Biol* 966:1–14. doi: 10.1007/5584_2016_201.

Crottes, D., and L. Y. Jan. 2019. "The multifaceted role of TMEM16A in cancer." *Cell Calcium* 82:102050. doi: 10.1016/j.ceca.2019.06.004.

Dulin, N. O. 2020. "Calcium-activated chloride channel ANO1/TMEM16A: Regulation of expression and signaling." *Front Physiol* 11:590262. doi: 10.3389/fphys.2020.590262.

Elorza-Vidal, X., H. Gaitan-Penas, and R. Estevez. 2019. "Chloride channels in astrocytes: Structure, roles in brain homeostasis and implications in disease." *Int J Mol Sci* 20(5):1034. doi: 10.3390/ijms20051034.

Falzone, M. E., M. Malvezzi, B. C. Lee, and A. Accardi. 2018. "Known structures and unknown mechanisms of TMEM16 scramblases and channels." *J Gen Physiol* 150(7):933–947. doi: 10.1085/jgp.201711957.

Genovese, M., M. Buccirossi, D. Guidone, R. De Cegli, S. Sarnataro, D. di Bernardo, L. J. V. Galietta. 2022. "Analysis of inhibitors of the anoctamin-1 chloride channel (transmembrane member 16A, TMEM16A) reveals indirect mechanisms involving alterations in calcium signaling." *Br J Pharmacol* doi: 10.1111/bph.15995. Online ahead of print.PMID: 36444690

Hartzell, C., I. Putzier, and J. Arreola. 2005. "Calcium-activated chloride channels." *Annu Rev Physiol* 67:719–758.

Hartzell, H. C., Z. Qu, K. Yu, Q. Xiao, and L. T. Chien. 2008. "Molecular physiology of bestrophins: Multifunctional membrane proteins linked to best disease and other retinopathies." *Physiol Rev* 88(2):639–672.

He, M., W. Ye, W. J. Wang, E. S. Sison, Y. N. Jan, and L. Y. Jan. 2017. "Cytoplasmic Cl(–) couples membrane remodeling to epithelial morphogenesis." *Proc Natl Acad Sci U S A* 114(52):E11161–E11169. doi: 10.1073/pnas.1714448115.

Johnson, A. A., K. E. Guziewicz, C. J. Lee, R. C. Kalathur, J. S. Pulido, L. Y. Marmorstein, and A. D. Marmorstein. 2017. "Bestrophin 1 and retinal disease." *Prog Retin Eye Res* 58:45–69. doi: 10.1016/j.preteyeres.2017.01.006.

Kalienkova, V., V. Clerico Mosina, and C. Paulino. 2021. "The groovy TMEM16 family: Molecular mechanisms of lipid scrambling and ion conduction." *J Mol Biol* 433(16):166941. doi: 10.1016/j.jmb.2021.166941.

Lee, J. M., C. G. Gadhe, H. Kang, A. N. Pae, C. J. Lee. 2022. "Glutamate Permeability of Chicken Best1." *Exp Neurobiol* 31(5): 277-288. doi: 10.5607/en22038.PMID: 36351838

Leblanc, N., A. S. Forrest, R. J. Ayon, M. Wiwchar, J. E. Angermann, H. A. Pritchard, C. A. Singer, M. L. Valencik, F. Britton, and I. A. Greenwood. 2015. "Molecular and functional significance of Ca(2+)-activated Cl(–) channels in pulmonary arterial smooth muscle." *Pulm Circ* 5(2):244–268. doi: 10.1086/680189.

Milenkovic, A., C. Brandl, V. M. Milenkovic, T. Jendryke, L. Sirianant, P. Wanitchakool, S. Zimmermann, C. M. Reiff, F. Horling, H. Schrewe, R. Schreiber, K. Kunzelmann, C. H. Wetzel, and B. H. Weber. 2015. "Bestrophin 1 is indispensable for volume regulation in human retinal pigment epithelium cells." *Proc Natl Acad Sci U S A* 112(20):E2630–E2639. doi: 10.1073/pnas.1418840112.

Miller, A. N., G. Vaisey, and S. B. Long. 2019. "Molecular mechanisms of gating in the calcium-activated chloride channel bestrophin." *eLife* 8:e43231. doi: 10.7554/eLife.43231.

Owji, A. P., J. Wang, A. Kittredge, Z. Clark, Y. Zhang, W.A. Hendrickson, T. Yang. 2022. "Structures and gating mechanisms of human bestrophin anion channels." *Nat Commun* 13(1): 3836. doi: 10.1038/s41467-022-31437-7.PMID: 35789156

Owji, A. P., K. Yu, A. Kittredge, J. Wang, Y. Zhang, T. Yang. 2022. "Bestrophin-2 and glutamine synthetase form a complex for glutamate release." *Nature* 611(7934):180–187. doi: 10.1038/s41586-022-05373-x. Epub 2022 Oct 26.PMID: 36289327

Owji, A. P., Q. Zhao, C. Ji, A. Kittredge, A. Hopiavuori, Z. Fu, N. Ward, O. B. Clarke, Y. Shen, Y. Zhang, W. A. Hendrickson, and T. Yang. 2020. "Structural and functional characterization of the bestrophin-2 anion channel." *Nat Struct Mol Biol* 27(4):382–391. doi: 10.1038/s41594-020-0402-z.

Pedemonte, N., and L. J. Galietta. 2014. "Structure and function of TMEM16 proteins (anoctamins)." *Physiol Rev* 94(2):419–459. doi: 10.1152/physrev.00039.2011.

Sanders, K. M. 2019. "Spontaneous electrical activity and rhythmicity in gastrointestinal smooth muscles." *Adv Exp Med Biol* 1124:3–46. doi: 10.1007/978-981-13-5895-1_1.

Shah, S., C. M. Carver, P. Mullen, S. Milne, V. Lukacs, M. S. Shapiro, and N. Gamper. 2020. "Local Ca(2+) signals couple activation of TRPV1 and ANO1 sensory ion channels." *Sci Signal* 13(629):eaaw7963. doi: 10.1126/scisignal.aaw7963.

Whitlock, J. M., and H. C. Hartzell. 2017. "Anoctamins/TMEM16 proteins: Chloride channels flirting with lipids and extracellular vesicles." *Annu Rev Physiol* 79:119–143. doi: 10.1146/annurev-physiol-022516-034031.

Yang, T., Q. Liu, B. Kloss, R. Bruni, R. C. Kalathur, Y. Guo, E. Kloppmann, B. Rost, H. M. Colecraft, and W. A. Hendrickson. 2014. "Structure and selectivity in bestrophin ion channels." *Science* 346(6207):355–359. doi: 10.1126/science.1259723.

Yu, K., R. Lujan, A. Marmorstein, S. Gabriel, and H. C. Hartzell. 2010. "Bestrophin-2 mediates bicarbonate transport by goblet cells in mouse colon." *J Clin Invest* 120(5):1722–1735. doi: 10.1172/JCI41129.

15

Acetylcholine Receptors

Cecilia Bouzat and Juan Facundo Chrestia

CONTENTS

15.1 Introduction

Acetylcholine (ACh) is synthesized, stored and released by cholinergic neurons as neurotransmitter and by several types of nonneuronal cells. It is involved in synaptic transmission in the central nervous system (CNS) and peripheral nervous system (PNS), and in the nonneuronal cholinergic system that modulates a great variety of cell activities. ACh acts through two distinct types of receptors: the metabotropic muscarinic receptor and the ionotropic nicotinic receptor (nAChR).

The nAChR has been a key protagonist in establishing foundations of receptor function since Langley in 1905 detected a receptive substance on the striated muscle surface after the application of nicotine. It has also been an object of attention since Claude Bernard investigated the action of the Central American arrow poison, curare. The muscle nAChR was the first neurotransmitter-gated ion channel to be identified and purified, and the first to be biochemically and electrophysiologically characterized.

DOI: 10.1201/9781003096276-15

The nAChR belongs to the Cys-loop receptor family of pentameric ligand-gated ion channels. It operates as a converter of the chemical signal, given by ACh, into an electrical signal, given by the flux of ions through the pore. Thus, nAChRs are involved in chemical synapses and are key actors in vital physiological processes.

In vertebrates, nAChRs comprise a family of ACh-activated cation channels expressed in neuronal and nonneuronal cells. nAChRs are implicated in diverse pathological conditions, including neuromuscular, neurological, neurodegenerative and inflammatory disorders. The wide structural and functional diversity of nAChRs as well as their main modulatory roles in the nervous system have kindled interest as therapeutic targets for a large spectrum of medical conditions and have inspired the development of selective ligands and discovery drug programs.

15.2 Subunit Diversity and Cell Distribution

In vertebrates, there are 17 different nAChR subunits that combine to yield a variety of pentameric receptors with different pharmacological and biophysical properties, physiological roles, and localization (Figure 15.1) (LeNovere and Changeux, 2001; Zoli et al., 2015).

nAChR subunits are of two types: α-type, which contains a conserved disulfide bridge in the agonist-binding site; and non-α-type. Pentameric arrangements can be homomeric (all α subunits) or heteromeric that require an α subunit to form the ACh binding site. Subunits have been divided into muscle type ($\alpha1$, $\beta1$, δ, ε and γ) and neuronal type ($\alpha2$–$\alpha8$, $\alpha9$, $\alpha10$ and $\beta2$–$\beta4$), although neuronal-type subunits are also expressed in other cells.

The rules that govern the combinatorial assembly of functional nAChRs are for the most part unknown. Whereas some subunits can combine with other numerous, albeit specific, subunits, others can only form functional receptors with a limited subset.

The mammalian muscle nAChR exists in two developmentally regulated isoforms. Embryonic muscle nAChRs are composed of two $\alpha1$, one β, one δ and one γ subunit [$(\alpha1)_2\beta\gamma\delta$]. In adult muscle, the γ subunit is replaced by the ε subunit to yield adult nAChRs with the composition $(\alpha1)_2\beta\varepsilon\delta$ that show different conductance and channel kinetics. Receptors carry two agonist-binding sites located at α/γ or ε and α/δ interfaces. Muscle-type nAChRs present in high density in the electric organs of *Torpedo* and *Electrophorus* electric fishes and have been extremely valuable in the field.

In autonomic ganglia, the $\alpha3$ subunit is the predominant α subunit, and $\alpha3\beta4^*$ are the major, ganglia-type, nAChRs. They show different stoichiometries, $(\alpha3\beta4)_2\alpha3$ and $(\alpha3\beta4)_2\beta4$, or combine with other subunits forming $\alpha3\beta4\alpha5$ and $\alpha3\beta4\beta2$ receptors. Several nAChRs are present in dorsal root ganglia.

The $\alpha4$ and $\beta2$ subunits can assemble in two different stoichiometries $(\alpha4\beta2)_2\alpha4$, carrying two ACh binding sites at $\alpha4/\beta2$ interface and a third at $\alpha4/\alpha4$ that reduces agonist sensitivity, or $(\alpha4\beta2)_2\beta2$, with two ACh binding sites. The latter stoichiometry has an approximate hundredfold higher affinity for ACh and nicotine, lower single-channel conductance and calcium permeability, and its expression is selectively upregulated by nicotine (Moroni et al., 2006). Both subunits can also assemble with other neuronal subunits.

$\alpha7$ is the homomeric member of the family and has five identical binding sites, although occupation of only one is enough for activation (Andersen et al., 2013). It has high calcium permeability and extremely fast desensitization. $\alpha7$ can also assemble with $\beta2$, and heteromeric $\alpha7\beta2$ nAChRs are present in specific areas in the human brain. Functional

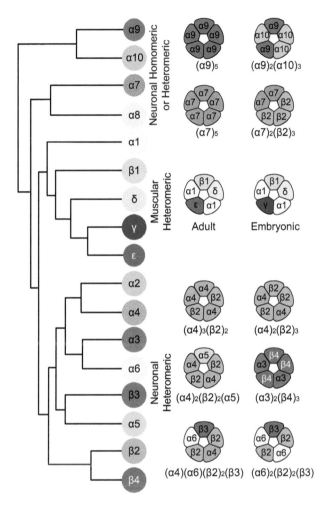

FIGURE 15.1
Cladogram of vertebrate nAChR subunits and models of nAChR subtypes.

α7β2 nAChRs can be formed by one, two or three β2 subunits, with the requirement of a functional binding site at the α7/α7 interface.

α8 is unique to avian tissues and can form homomeric receptors when expressed in *Xenopus* oocytes.

α9 can form homomeric nAChRs *in vitro*, and chick α10, but not mammalian, can form homomeric receptors. The hair cell nAChR has the stoichiometry $(\alpha9)_2(\alpha10)_3$ and an atypical pharmacological profile.

In addition to these major nAChRs, there are other native subtypes with more complex compositions. The α5 subunit forms functional channels when assembled with α4 and β2 or α3 and β4 subunits. The α6 subunit is included in different arrangements combined with β2 and β3, or α4, β2 and β3 subunits.

The identification of novel nAChR subtypes, their functional roles and functional stoichiometries in native tissues, and the discovery of selective ligands are of intense interest.

15.3 Physiological Roles

Due to their ubiquitous distribution and to the broad repertoire of subtypes, nAChRs are involved in many physiological processes in the PNS, CNS and in nonneuronal cells (Taly et al., 2009).

The muscle nAChR plays a key role in neuromuscular transmission. ACh released from the motor neuron terminal interacts with nAChRs at the postsynaptic muscle membrane, leading to the opening of the ion channel that allows the influx of cations (mainly Na^+). This excitatory postsynaptic current causes a local depolarization (the endplate potential) that triggers the opening of voltage-dependent sodium channels, giving rise to a propagated action potential that spreads throughout the muscle cell and initiates muscle contraction. The response is limited because ACh is rapidly removed from the synaptic cleft by acetylcholinesterase (AChE) and diffusion.

Ganglionic nAChRs have a pivotal role in the peripheral autonomic nervous system, including sympathetic, parasympathetic and enteric systems, and are involved in involuntary physiological processes. They are also found in dorsal root ganglia where they are of interest as therapeutic targets for modulating nociceptive signals.

Neuronal nAChRs are found at postsynaptic locations, where they mediate fast responses; presynaptic locations, where they modulate release of different neurotransmitters; and extrasynaptic locations, where they participate in volume transmission by interacting with ACh released from distant sites. They play mainly modulatory roles in the CNS by changing neuronal excitability, altering neurotransmitter release and coordinating the firing of groups of neurons. Neuronal nAChRs mediate synaptic plasticity; contribute to cognitive processes and memory; regulate behaviors; and are involved in neuroprotection, neurodegeneration, and in rewarding motivational effects of nicotine (Picciotto et al., 2012).

There is a vast repertoire of neuronal nAChR subtypes. Among them, the $\alpha4\beta2$ constitutes one of the most important modulatory receptors for a wide range of brain functions, including cognition, mood, consciousness, reward and nociception. Together with $\alpha4\beta2$, $\alpha7$ is one of the most abundant nAChRs in the CNS. It is highly expressed in the hippocampus, cortex and subcortical limbic regions, and is involved in cognition, sensory processing information, attention, working memory and reward pathways. The $\alpha9\alpha10$ nAChR plays a fundamental role in inner ear physiology and mediates synaptic transmission between efferent olivocochlear fibers that descend from the brain stem and hair cells of the auditory sensory epithelium (Elgoyhen et al., 1994).

Neuronal nAChRs are also expressed in nonneuronal cells, such as glia, immune cells, keratinocytes, and endothelial and epithelial cells (Wessler and Kirkpatrick, 2008). Nonneuronal nAChRs respond to ACh released by vagal nerve endings or cells, and control processes such as cell proliferation, adhesion, migration, secretion, survival and apoptosis, in an autocrine, paracrine or juxtacrine manner. nAChRs are present in the immune system where they regulate inflammatory processes; in particular, $\alpha7$ is a key player of the cholinergic anti-inflammatory pathway, which is a link between vagal efferent fibers and the innate immune system.

nAChRs also play key roles in invertebrates. Insects possess several nAChR subtypes involved in excitatory transmission in the nervous system required for physiological processes, including olfactory learning and memory formation. They are targets for the agrochemical industry and have led to the development of novel classes of insecticides such as neonicotinoids.

Nematode nAChRs are of clinical importance as targets of antiparasitic drugs; they have been extensively studied in the free-living nematode *Caenorhabditis elegans*. Curiously, *C.*

elegans possess the largest nAChR subunit family found in a single species and a diverse variety of receptors, including anion-selective nAChRs. nAChRs are involved in many physiological processes. In nematode muscles, there is an heteromeric levamisole-sensitive nAChR involved in neuromuscular transmission and worm locomotion that is a main target of antiparasitic drugs.

15.4 Structural Organization

All nAChR subunits share a basic scaffold composed of a large N-terminal extracellular domain of ~200 amino acids composed of ten β-strands; three transmembrane domains with an α-helix structure separated by short loops (M1–M3); a cytoplasmic loop of variable size and an amino acid sequence carrying two helical regions (MX and MA); and a fourth transmembrane domain (M4) with a relatively short extracellular C-terminal.

High-resolution 3D structures for nAChRs, including muscle, α3β4, α4β2 and α7, have been solved (Unwin, 2005; Morales-Perez et al., 2016; Gharpure et al., 2019; Noviello et al., 2021; Zhao et al., 2021). The five subunits are arranged pseudosymmetrically around a central axis forming a receptor with clear structural modules: (i) the N-terminal extracellular domain (ECD), which carries the orthosteric binding sites; (ii) the transmembrane domain (TMD), composed of four α-helices from each subunit (M1–M4) and contains the ion pore; and (iii) the intracellular domain (ICD). Between the ECD and the TMD there is a structural transition zone, named the coupling region, which is essential for the functional link between agonist binding and channel opening (Bouzat et al., 2004) (Figure 15.2).

15.4.1 Extracellular Domain (ECD) and Neurotransmitter Binding Sites

The ECD consists of a core of ten β-strands (β1–β10) from each subunit forming a β-sandwich structure. Located at the bottom in close contact with the TMD is the linker between β6 and β7 strands, which forms the highly conserved Cys-loop that gives the name to the family.

The ECD carries the orthosteric ligand binding sites at ~30 Å from the ECD-TMD interface. Each binding site is located at interfaces between adjacent subunits; with one face, called the principal face (+), provided by an α-type subunit, and the other, the complementary face (–), provided by either an α or non-α subunit. The principal face is formed by loop A, loop B and loop C, which contribute to a nest of aromatic residues (tryptophane and tyrosine residues) that can make cation-π interactions with the positively charged ammonium group of ACh or other ligands. A cation-π interaction involving a tryptophane residue in loop B (W149 in the muscle nAChR) and the positively charged group of the ligand is a key feature in most nAChRs. The tip of loop C contains the disulfide bond between adjacent cysteines, a feature that defines α-type subunits. The complementary face is composed of loops D–F from separate sections of three β-strands; it is correlated with drug selectivity due to divergent interactions with different ligands (Figure 15.2). In addition to cation-π interactions, stabilization forces of the ligands at the binding site include dipole–cation, hydrogen and van der Waals interactions.

15.4.2 Transmembrane Domain (TMD) and Ion Permeation Pathway

The TMD is composed of four α-helices of each subunit (M1–M4). This region carries the ion channel pore located at the axis of symmetry at the center of the helix bundle and bordered

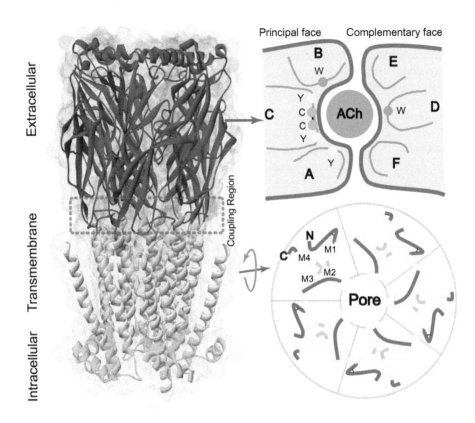

FIGURE 15.2
Structure organization of nAChR. Side view of the α4β2 receptor (Protein Data Bank code: 5KXI) showing the functional domains (Morales-Perez et al., 2016). The schematic representation of the ACh binding site at the interface between two adjacent subunits in the ECD, with key aromatic residues and the conserved disulfide bridge, and of the top view of the TMD are shown.

by the M2 α-helices. The outer ring of 15 α-helices (M1, M3 and M4) shields the channel from the lipids. Lipids influence nAChR function by both conformational selection and kinetic mechanisms. M4 is the most external in contact with the lipid domain and acts as a lipid sensor; mutations in lipids exposed amino acids at this domain alter channel function.

The ion pore bordered by the M2 α-helices shows a stratified organization in which rings of generally homologous residues align along the z axis. In resting/closed state, the pore is maximally constricted in the middle of the membrane by rings of hydrophobic amino acids (positions 16, 9 and 13 from the intracellular end of M2). The narrowest point in the pore cannot allow the passage of a hydrated cation; it corresponds to a conserved ring of leucine residues (L9) that constitutes the channel gate. Three rings of negatively charged residues, located at the extracellular and the intracellular ends of M2, provide a favorable electrostatic environment to allow cation translocation through the pore. At the intracellular end, it is the selectivity filter which has the most significant influence on conductance and ion selectivity (Taly et al., 2009).

Thus, in the permeation pathway through the nAChR, hydrated cations enter the extracellular vestibule facilitated by the negative electrostatic surface potential and encounter a hydrophobic constriction in the closed-pore state in the upper half of the pore. Upon activation, the hydrophobic gate opens and becomes hydrated, allowing passage through the pore. Hydrated cations then exit through the portals, formed by MX and MA α-helices rich in polar and acidic amino acids. Both the extracellular vestibule and the intracellular portals provide determinants outside of the pore for fine-tuning of conductance and cation selectivity (Rahman et al., 2020).

The nAChR is a non-selective cation channel, being permeable to Na^+, K^+, Cs^+ and to bivalent cations Ca^{2+} and Mg^{2+}. Neuronal-type receptors have higher Ca^{2+} permeability (P) than the muscle type. Among neuronal nAChRs, the $P_{Ca}2+/P_{Na}+$ ratio ranges from ~1.1 to >10 in α7.

15.4.3 Intracellular Domain (ICD)

The large M3–M4 loop is the least conserved region among nAChRs and highly variable in sequence and length (from about 80 to 260 amino acids). It consists of a short post-M3 loop, an amphipathic helix (MX) that lies parallel to the plane of the membrane, a poorly conserved and disordered cytoplasmic loop, and the MA helix. The ICD plays a role in functional expression, cellular trafficking and membrane localization, and influences gating and conductance. The ICD contains sites for serine/threonine and tyrosine phosphorylation, and for interaction with cytoskeletal proteins and signaling pathway modulators, such as kinases, G proteins and Ca^{2+} sensors. Therefore, it is a key domain for regulation, localization, upregulation and triggering metabotropic responses.

15.4.4 Structural Changes for Activation

Communication over the 50 Å separating the binding site from the channel gate is essential to nAChR function. Extensive mutagenesis combined with electrophysiological studies have revealed key amino acids and interactions required for receptor function, whereas comparison among 3D structures in different conformational states has provided insights into the underlying structural changes. The biological response begins with agonist binding to the orthosteric site, which generates the transition of loop C from an opened to a closed conformation that caps the agonist. Agonist binding results in an overall more compact ECD. The movements at ECD induce reorganization of electrostatic and hydrophobic interactions in the coupling region, which involves the β1–β2, Cys and β8–β9 loops from the ECD, and the M2–M3 loop, the pre-M1 region and the peripheral C-terminal from the TMD. The changes at the coupling region are translated into movements of M2, causing the residues in the upper half of M2 to move away from the channel axis, opening the hydrophobic gate and allowing the passage of cations through the pore. During desensitization, the TMD adopts again a closed-pore conformation, although different to that of the resting state (Noviello et al., 2021; Zhao et al., 2021).

15.5 Molecular Function and Channel Gating

The nAChR operates as a molecular machine whose design has been tuned to function as a near-perfect on-off switch that responds to ACh. Based on the Monod–Wyman–Changeux model, the functional response can be interpreted as a selection from a few discrete conformational states: closed, open and desensitized (Figure 15.3a) (see Volume I, Chapter 10). In the absence of agonists, the receptor is found mainly in the closed resting state. Agonist binding triggers channel opening resulting in a transient

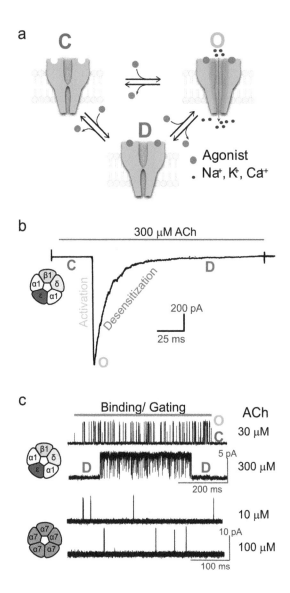

FIGURE 15.3

nAChR activation. (a) Minimal model of nAChR activation showing the three main classes of conformational states: C, closed; O, open; D, desensitized. (b) Typical macroscopic current elicited by 300 μM ACh from whole-cell recordings of adult muscle nAChR. (c) Single-channel activity from cells expressing adult muscle nAChR (top) and α7 (bottom) at two different ACh concentrations. For the muscle nAChR, each cluster includes the binding and gating activity of a single nAChR that adopts closed (C) or open (O) states. The silent periods between clusters of openings are periods when nAChRs in the patch are desensitized (D). As ACh concentration increases, the probability of channel opening increases, which is evidenced by decreased duration of the closings within clusters. In contrast, the temporal pattern of α7 channel currents does not show any concentration dependence with channel activity appearing mainly as brief (~0.1–0.3 ms), isolated events. Channel openings are shown as upward deflections. Membrane potential: –70 mV.

conductive state, a process called gating. A net influx of cations through the pore depolarizes the cell membrane, which increases neuronal excitability and increases calcium entrance. The channel stays open for a few milliseconds and can produce repeated cycles of closing–opening episodes. Prolonged exposure or high neurotransmitter concentrations lead to a desensitized state, a more stable nonconducting state with high agonist affinity.

Although nAChRs exist in at least three main classes of conformational states, the activation mechanism is complex since there are multiple states of each main class and intermediate states between closed and open, and open and desensitized states. In particular, intermediate preactivated closed states of ligand-bound receptors have been described (called flip or primed states), and the ability of an agonist to change the receptor conformation from the closed state to these preactivated states relates to its efficacy (Lape et al., 2008; Mukhtasimova et al., 2009). The number of conformational states in each main class and the transition rates between states define channel kinetics, which is unique for each nAChR subtype. Changes in kinetics originated by mutations lead to channelopathies.

The potency and efficacy of an agonist is classically determined from the relationships between macroscopic peak currents, obtained from whole-cell recordings, and agonist concentration. From macroscopic currents, the decay rate in the presence of agonist has been used as an estimate of the desensitization rate, and the current increase rate as an estimate of the activation rate, although depending on the system and on receptor kinetics these correlations may not be accurate (Figure 15.3b).

Agonist-bound open nAChR channels can transition to nonconducting states by dissociation of the agonist, a process known as deactivation. Most nAChRs show fast activation and deactivation kinetics, which allows them to stay tuned for activation after a minimal time lapse and to follow high-frequency trains of action potentials. Active receptors can also transition to desensitized states; onset and recovery rates of desensitization are particular for each nAChR subtype and can be modulated by cell events, such as phosphorylation. Desensitization may be relevant in controlling synaptic efficacy for specific receptor subtypes and situations. In $\alpha7$, which displays the fastest desensitization among nAChRs, it determines the termination of the response and may act as a filter to avoid cell toxicity due to calcium influx; in the muscle nAChR, desensitization may be important under pathological conditions. Desensitization is also relevant in the CNS since prolonged application of low concentrations of agonist may promote desensitization without significant activation (C to D transition in Figure 15.3a).

Kinetic analysis using high-resolution single-channel data has been key to define mechanistic basis of nAChR activation in normal and pathological states (Mukhtasimova et al., 2009; Sine et al., 1995; Grosman et al., 2000). It allows the generation of gating models that describe the activation mechanism; it can dissect the binding from the gating steps; and it provides information about agonist affinity, the number of bound agonist molecules required for activation, the number of conformational changes that separate the binding events from channel opening, and rate constants for each activation step (see Volume I, Chapter 10). Single-channel recordings from mutant nAChRs have provided detailed information about how amino acids all along the receptor contribute to each step of the activation process and the underlying structural basis.

The nAChR from skeletal muscle holds a special place in the history of application of the patch clamp, for it was the first ion channel from which unitary currents were registered with submillisecond resolution (Colquhoun and Sakmann, 1981) and the first nAChR for which detailed kinetic analysis was performed.

Activation of muscle nAChR occurs in trains of several openings and closings from the same receptor molecule because the time for desensitization is long compared to the time it takes to close and reopen. These activation episodes (clusters) involve closings and openings corresponding to binding/gating transitions and are separated by prolonged closings corresponding to desensitized states. With the increase of ACh concentration, the closed durations within clusters decrease due to an increase of the probability of being open, resulting in a concentration-dependent activation pattern (Figure 15.3c). A different behavior is observed for α7, whose activation pattern appears mainly as isolated brief openings with little agonist concentration dependence. This is because the time for desensitization onset approaches the open-channel lifetime, and desensitization governs the rate of channel closing. Thus, each nAChR subtype has its own kinetic signature, probably tuned to respond to its physiological role.

The classic view of the nAChR as a mere ionotropic receptor has been challenged in the past decades. nAChRs also activate cell signaling pathways and affect more widespread processes. The transient increase in intracellular Ca^{2+} through the channel can be converted into a sustained, wide-ranging phenomenon by calcium release from intracellular stores through a calcium-induced calcium release mechanism, a process involving IP_3 and ryanodine receptors. This process has been extensively studied for α7 and it is associated with physiological roles, including neurite growth (Dajas-Bailador and Wonnacott, 2004). Different intracellular signaling pathways, such as JAK2/STAT3, PI3K/Akt and MAPK/ERK, can be triggered upon receptor activation depending on cell-specific protein–receptor interactions mainly mediated through the ICD. Thus, the ability of nAChRs to operate as dual metabotropic/ionotropic receptors prolongs the transient electrical response in a more sustained response involving multiple intracellular events.

15.6 Pharmacology and Drug Modulation

The nicotinic pharmacopoeia is continuously increasing in response to novel potential therapeutic uses particularly for a variety of neurological and neurodegenerative conditions (Wonnacott, 2014).

15.6.1 Orthosteric Ligands

nAChRs are activated by a large number of compounds, from natural products to synthetic compounds that act as full or partial agonists. The central element of a nAChR agonist is a charged or chargeable nitrogen atom; the simplest quaternary ammonium cation, tetramethylammonium, can activate nAChRs.

In addition to linear chain ACh-derivative agonists, such as choline and succinylcholine, a variety of heterocyclic alkaloids are agonists. The alkaloid (-)-nicotine, extracted from leaves of the tobacco plant *Nicotiana tabacum*, is the prototype agonist, emblematic for determining the pharmacological classification of the family. It activates all nAChR subtypes except α9* and shows remarkably high affinity and efficacy for $(α4β2)_2β2$. Epibatidine, a bicyclic alkaloid from skin extracts of the Amazonian frog, is one of the most potent non-selective agonists; the tobacco alkaloid Anabasine hydrochloride shows selectivity for α7 and α4β2, and even caffeine is a natural agonist. (-)-Cytisine, a tricycle quinolizidine alkaloid found in plants of the *Leguminosae* family, shows high affinity for α4β2 and has become a lead compound for smoking cessation and depression. Varenicline, a clinical drug used for smoking cessation, is a cytisine congener acting as a partial agonist of α4β2 and α6α2β3* receptors. There is a long list of synthetic compounds with different nAChR selectivity; 5-Iodo-A-85380 and PNU-282987 have been extensively used for selective activation of β2* and α7, respectively.

Antagonists have been used as tools for determining nAChR responses. A few of them are used in clinic as muscle blockers. *d*-Tubocurarine binds with high affinity to muscle nAChR and has been used for skeletal muscle relaxation during surgery or mechanical ventilation.

There are a series of toxins that specifically interact with nAChRs and show subtype selectivity. α-Bungarotoxin (α-BTX), isolated from the venom of the Taiwanese banded krait, *Bungarus multicinctus*, shows pseudo-irreversible binding for the muscle and α7 nAChRs. nAChRs are classified as α-BTX-sensitive types, which include α7 and muscle nAChRs, and also α9 receptors that show reversible binding, and α-BTX-insensitive types, which include all neuronal nAChRs. α-Conotoxins are a series of peptides (12–19 amino acids) produced by marine cone snails that act as potent antagonists of nAChRs. Due to their selectivity for nAChR subtypes, they are useful tools for the identification of sub-types as well as leads for therapeutic drugs for clinical conditions (Wonnacott, 2014).

15.6.2 Allosteric Ligands

nAChR responses can be modulated by diverse molecules that bind to distinct sites from the orthosteric ones, called allosteric sites. These allosteric ligands alter energy barriers for transitions between conformational states, and therefore, potentiate or inhibit receptor function. Allosteric ligands are promising therapeutic tools because they maintain the temporal spatial characteristics of the endogenous activation, they show high selectivity since the orthosteric sites are more conserved than allosteric sites, and they allow a high structural diversity and final effects.

From a functional perspective, allosteric ligands are classified as (i) negative allosteric modulators (NAMs); (ii) positive allosteric modulators (PAMs); (iii) silent allosteric modulators (SAMs), which have no effect on orthosteric agonist responses but block allosteric modulation; and (iv) allosteric agonists, which activate receptors from nonorthosteric sites (Figure 15.4). Single-channel recordings have been pivotal for establishing the mechanisms of drug modulation (Figure 15.5).

NAMs inhibit receptor function via different molecular mechanisms and sites. They may stabilize the nAChR in a nonconducting state or increase the rate and extent of desensitization. Inhibition may also occur through open-channel block in which the drug physically blocks ion permeation. NAMs acting at different nAChR subtypes include endogenous compounds, such as steroids, and many others that have other primary targets, such as anesthetics, calcium blockers, antipsychotics and antidepressants. Inhibitors can be of therapeutic benefit. For example, quinidine and fluoxetine are therapeutically effective in shortening prolonged channel openings of the muscle nAChR in slow-channel congenital myasthenic syndromes; hexamethonium bromide, an open-channel blocker, has been used as a ganglionic blocker to treat hypertension and other conditions; and NAMs of α4β2 show antidepressant effects.

The annular (surrounding the perimeter of the receptor) and nonannular lipid domains (between transmembrane helices and subunits) are also sites for a great variety of hydrophobic, endogenous and exogenous compounds that by affecting the lipid environment inhibit function.

PAMs potentiate agonist-evoked responses by altering the energy barriers between resting/open and open/desensitized states and can enhance potency and/or efficacy of the agonist. There is not a common chemical scaffold for a PAM; PAMs embrace a wide variety of structures, from calcium and zinc to complex structures, and include endogenous compounds (such as estradiol for α4β2), natural compounds (such as flavonoids for α7) and synthetic compounds (Chatzidaki and Millar, 2015).

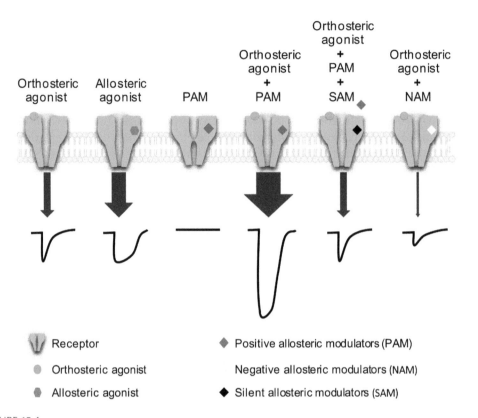

FIGURE 15.4
Scheme of modulation by allosteric ligands. Allosteric ligands can activate nAChRs or can increase (PAMs) or decrease (NAMs) responses elicited by ACh.

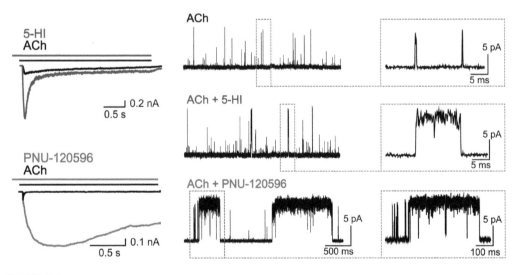

FIGURE 15.5
Actions of PAMs at the single-channel and macroscopic current level of human α7. Macroscopic currents of human α7 elicited by 50 μM ACh in the absence (black trace) or presence of a type I PAM (2 mM 5-HI, blue trace), which only increases the peak current, or a type II PAM (1 μM PNU-120596, red trace), which also decreases desensitization. At the right, traces of single-channel recordings of human α7 activated by 100 μM ACh in the absence and in the presence of 2 mM 5-HI or 1 μM PNU-120596 are shown. Openings are shown as upward deflections. Membrane potential: –70 mV.

α7 PAMs are emerging as novel therapeutics for neurological disorders (such as schizophrenia), neurodegenerative disorders (such as Alzheimer's and Parkinson's diseases) and inflammatory disorders. By enhancing α7 activity, PAMs exert procognitive, neuroprotective and anti-inflammatory actions. They have been classified as type I or II on the basis of their effects on macroscopic currents. Type I PAMs enhance agonist-induced peak currents without significantly affecting current decay, and type II PAMs slow the onset of desensitization and reactivate desensitized receptors. Analysis at the single-channel level have shown that PAMs enhance the brief α7 open-channel lifetime and induce activation in episodes composed of successive openings that can last from a few milliseconds to several seconds, this duration being related to the efficacy of the PAM (Figure 15.5).

α4β2 holds promise as a target for cognitive disorders. At the macroscopic current level, α4β2 PAMs can produce left-shifting of the agonist concentration–response relationships without altering the maximal response, whereas others increase efficacy. Some PAMs can show selectivity for the two different stoichiometries. For instance, desformylflustrabromine (a natural product of marine organism), NS 9283 (that shows in vivo efficacy in models of pain) and Zn^{2+} show higher selectivity for $(\alpha4\beta2)_2\alpha4$, whereas the endogenous steroid 17β-estradiol shows higher selectivity for $(\alpha4\beta2)_2\beta2$.

15.7 Channelopathies and Disease

15.7.1 Muscle and Ganglionic Diseases

Congenital myasthenic syndromes (CMSs) represent a heterogeneous group of disorders in which neuromuscular transmission is compromised by genetic defects in presynaptic, synaptic or postsynaptic proteins, including nAChRs (Engel et al., 2015). Mutations in nAChRs that decrease expression or alter channel kinetics generate this channelopathy. CMS mediated by changes in kinetics divide into slow-channel CMSs that show prolonged ACh-mediated postsynaptic responses due to gain-of-function mutations and fast-channel CMSs that show decreased responses due to loss-of-function mutations. Mutations in slow-channel CMS increase agonist affinity, the probability of opening in the absence of agonist, the opening rate, and/or decrease the closing rate; whereas mutations in fast-channel CMS impair opening, increase the closing rate, and/or decrease agonist affinity. Kinetic analyses of single-channel currents have been especially powerful in giving insight into the mechanistic consequences of the mutations as well as guiding rational therapy (Sine et al., 1995).

nAChRs are the focus of autoimmune diseases. In myasthenia gravis, autoantibodies against the main immunogenic region at the ECD of the muscle nAChR reduce the number of functional receptors, which correlates with disease severity. Autoantibodies specific for ganglionic α3* nAChRs disrupt cholinergic synaptic transmission and lead to autonomic failure.

15.7.2 Neurological and Neurodegenerative Disorders

Autosomal dominant nocturnal frontal lobe epilepsy (ADNFLE) is a form of focal epilepsy. Mutations in α4, β2 and α2 subunits have been found in ADNFLE patients.

Schizophrenia is accompanied by dysfunction in hippocampal nAChRs. It is associated with polymorphism at the α7 gene (CHRNA7) and to a partial duplication of this gene, which results in the chimeric gene CHRFAM7A. Its gene product, dupα7, lacks part of the binding site and acts as a negative modulator of α7 function. α7 agonists and PAMs are promising drugs for schizophrenia.

nAChRs have been associated to other neurological disorders, such as depression, autism spectrum disorders, attention deficit hyperactivity disorder, and pain, and, therefore, several nAChRs are considered as potential druggable sites.

There is a close link between Alzheimer's disease (AD) and nAChRs. AD is accompanied by an early cholinergic deficit; changes in nAChR levels, particularly reduced expression of α7 and α4β2; reduced ACh; and attenuated activity of the cholinergic synthetic enzyme (CHAT). β-amyloid peptides (Aβ) interact with α7 with high affinity; they activate α7 at picomolar concentrations but inhibit at nanomolar concentrations found in patients. Oligomeric Aβ also act as NAMs of α7β2 and α4β2 nAChRs. Agonists and partial agonists of α7 and α4β2 as well as PAMs of α7 have been included in clinical assays for early AD stages due to their procognitive and neuroprotective roles (Dineley et al., 2015).

Dopaminergic neurons in the substantia nigra, that are lost selectively in Parkinson's disease (PD), express multiple subtypes of nAChR. Loss of nicotinic binding sites has been observed in the basal forebrain and cortex of PD patients, and α7- and α6-containing nAChRs emerge as targets for drug development.

15.7.3 Nicotine Dependence

Several nAChR subtypes are strongly associated with nicotine addiction and nicotine-induced behaviors. α4β2 nAChRs have the highest sensitivity to nicotine. Functional upregulation of α4β2 nAChRs, observed in the brains of smokers and animals chronically exposed to nicotine, is combined with a sensitization of the mesolimbic dopamine response to nicotine associated with its overall addictive properties. Other nAChR subunits expressed in brain areas that regulate nicotine/tobacco reinforcement include α7, α6, α5, α4 and β3, and variations of α4, α6 and α5 genes are linked to tobacco dependence (Taly et al., 2009).

15.7.4 Cancer and Inflammatory Disorders

Nonneuronal nAChRs activate cell signals that stimulate cell growth and angiogenesis, and nAChR dysregulation accompanies many cancers. Genome-wide association studies indicate that genetic variants, particularly in CHRNA3, CHRNA5 and CHRNB4, are associated with increased risks of lung cancer. α7 is involved in most of the oncogenic responses, including lung, oral, esophageal, colon and pancreatic cancers, and α9 stimulates the initiation and progression of breast cancer.

nAChRs regulate inflammatory processes. α7 is present in all types of immune cells and plays an important anti-inflammatory role. It also regulates microglial activation inducing neuroprotection, relevant in Parkinson's disease, oxygen and glucose deprivation, and global ischemia. Nonneuronal nAChRs participate in the pathogenesis of several skin disorders and rheumatic diseases.

15.8 Cell Biology and Regulation

Synthesis, folding and assembly of nAChRs take place inside the endoplasmic reticulum (ER), and require the cleavage of the signal peptide, the oxidation of disulfide bonds and the N-glycosylation of some residues. An important role is played by chaperone proteins, which guarantee tight quality control by assisting the subunits to assume the correct folded conformation and eliminating misfolded or unassembled ones. Only correctly assembled pentamers can leave the ER through vesicles that first reach the Golgi apparatus and then the plasma membrane. nAChR chaperones include RIC-3 (resistance to inhibitors of cholinesterase 3), NACHO, small proteins of the lymphocyte antigen-6 (LY6) family and UBXN2A. Nicotine also acts at several steps of nAChR biogenesis and trafficking. Consequently, chronic nicotine exposure upregulates nAChRs in primary neuronal cultures and brains of smokers. By increasing surface receptors, the chaperone activity of nicotine may contribute to its addictive properties and therapeutic effects, as well as to its proliferative capacity of cancer cells (Crespi et al., 2018).

At the neuromuscular junction, nAChRs are clustered at high density at the crests of junctional folds in the postsynaptic membrane of skeletal muscle. Before innervation, nAChRs are evenly distributed in the membrane. The nerve induces clustering by combining signals that trigger a series of transcriptional and posttranslational regulatory mechanisms within the muscle. Agrin, a protein released by the nerve terminal; rapsyn, a protein that interacts with nAChRs; and Musk, a muscle transmembrane tyrosine kinase, are key in the maturation of the neuromuscular junction. Maturation is accompanied by a transcriptional switch of the γ to ε subunit. In the adult muscle, γ-containing nAChRs are found only after denervation and in some CMS and myogenic disorders.

15.9 Conclusions

The nAChR operates as a molecular machine whose design has been tuned to function as a near-perfect on-off switch that responds to ACh. Electrophysiological recordings, particularly single-channel recordings, combined with mutagenesis and cell expression have made invaluable contributions to the understanding of how the nAChR responds to an agonist, the molecular bases of human diseases and the mechanisms of drug modulation. High-resolution 3D structures at defined conformational states have provided structural bases underlying receptor activation and desensitization, and emerged as a means toward defining sites and mechanisms of drug action essential to rational drug design.

Suggested Reading

Andersen, N., Corradi, J., Sine, S.M., and Bouzat, C. 2013. Stoichiometry for activation of neuronal α7 nicotinic receptors. *Proceedings of the National Academy of Sciences of the United States of America* 110(51):20819–20824. doi:10.1073/pnas.1315775110

Bouzat, C., Gumilar, F., Spitzmaul, G., Wang, H.-L., Rayes, D., Hansen, S.B., Taylor, P., and Sine, S.M. 2004. Coupling of agonist binding to channel gating in an ACh-binding protein linked to an ion channel. *Nature* 430(7002):896–900. doi:10.1038/nature02753

Chatzidaki, A. and Millar, N.S. 2015. Allosteric modulation of nicotinic acetylcholine receptors. *Biochemical Pharmacology* 97(4):408–417. doi:10.1016/j.bcp.2015.07.028

Colquhoun, D. and Sakmann, B. 1981. Fluctuations in the microsecond time range of the current through single acetylcholine receptor ion channels. *Nature* 294(5840):464–466.

Crespi, A., Colombo, S.F., and Gotti, C. 2018. Proteins and chemical chaperones involved in neuronal nicotinic receptor expression and function: An update. *British Journal of Pharmacology* 175:1869–1879. doi:10.1111/bph.v175.11/issuetoc

Dajas-Bailador, F. and Wonnacott, S. 2004. Nicotinic acetylcholine receptors and the regulation of neuronal signalling. *Trends in Pharmacological Sciences* 25(6):317–324. doi:10.1016/j.tips.2004.04.006

Dineley, K.T., Pandya, A.A., and Yakel, J.L. 2015. Nicotinic ACh receptors as therapeutic targets in CNS disorders. *Trends in Pharmacological Sciences* 36(2):96–108. doi:10.1016/j.tips.2014.12.002

Elgoyhen, A.B., Johnson, D.S., Boulter, J., Vetter, D.E., and Heinemann, S. 1994. Alpha 9: An acetylcholine receptor with novel pharmacological properties expressed in rat cochlear hair cells. *Cell* 79(4):705–715.

Engel, A.G., Shen, X.M., Selcen, D., and Sine, S.M. 2015. Congenital myasthenic syndromes: Pathogenesis, diagnosis, and treatment. *The Lancet. Neurology* 14(4):420–434. doi:10.1016/S1474-4422(14)70201-7

Gharpure, A., Teng, J., Zhuang, Y., Noviello, C.M., Walsh Jr., R.M., Cabuco, R., Howard, R.J., Zaveri, N.T., Lindahl, E., and Hibbs, R.E. 2019. Agonist selectivity and ion permeation in the α3β4 ganglionic nicotinic receptor. *Neuron* 104(3):501–511. doi:10.1016/j.neuron.2019.07.030

Grosman, C., Zhou, M., and Auerbach, A. 2000. Mapping the conformational wave of acetylcholine receptor channel gating. *Nature* 403(6771):773–776. doi:10.1038/35001586

Lape, R.D., Colquhoun, D., and Sivilotti, L.G. 2008. On the nature of partial agonism in the nicotinic receptor superfamily. *Nature* 454(7205):722–727. doi:10.1038/nature07139

Le Novere, N. and Changeux, J.P. 2001. LGICdb: The ligand-gated ion channel database. *Nucleic Acids Research* 29(1):294–295.

Morales-Perez, C.L., Noviello, C.M., and Hibbs, R.E. 2016. X-ray structure of the human α4β2 nicotinic receptor. *Nature* 538(7625):411–415. doi:10.1038/nature19785

Moroni, M., Zwart, R., Sher, E., Cassels, B.K., and Bermudez, I. 2006. alpha4beta2 nicotinic receptors with high and low acetylcholine sensitivity: Pharmacology, stoichiometry, and sensitivity to long-term exposure to nicotine. *Molecular Pharmacology* 70(2):755–768.

Mukhtasimova, N., Lee, W.Y., Wang, H.L., and Sine, S.M. 2009. Detection and trapping of intermediate states priming nicotinic receptor channel opening. *Nature* 459(7245):451–454. doi:10.1038/nature07923

Noviello, C.M., Gharpure, A., Mukhtasimova, N., Cabuco, R., Baxter, L., Borek, D., Sine, S.M., and Hibbs, R.E. 2021. Structure and gating mechanism of the α7 nicotinic acetylcholine receptor. *Cell* 184(8):2121–2134. doi:10.1016/j.cell.2021.02.049

Picciotto, M.R., Higley, M.J., and Mineur, Y.S. 2012. Acetylcholine as a neuromodulator: Cholinergic signaling shapes nervous system function and behavior. *Neuron* 76(1):116–129. doi:10.1016/j.neuron.2012.08.036

Rahman, M., Teng, J., Worrell, B.T., Noviello, C.M., Lee, M., Karlin, A., Stowell, M.H.B., and Hibbs, R.E. 2020. Structure of the native muscle-type nicotinic receptor and inhibition by snake venom toxins. *Neuron* 106(6):952–962. doi:10.1016/j.neuron.2020.03.012

Sine, S.M., Ohno, K., Bouzat, C., Auerbach, A., Milone, M., Pruitt, J.N., and Engel, A.G. 1995. Mutation of the acetylcholine receptor α subunit causes a slow-channel myasthenic syndrome by enhancing agonist binding affinity. *Neuron* 15(1):229–239.

Taly, A., Corringer, P.J., Guedin, D., Lestage, P., and Changeux, J.P. 2009. Nicotinic receptors: Allosteric transitions and therapeutic targets in the nervous system. *Nature Reviews. Drug Discovery* 8(9):733–750. doi:10.1038/nrd2927

Unwin, N. 2005. Refined structure of the nicotinic acetylcholine receptor at 4A resolution. *Journal of Molecular Biology* 346(4):967–989. doi:10.1016/j.jmb.2004.12.031

Wessler, I. and Kirkpatrick, C.J. 2008. Acetylcholine beyond neurons: The non-neuronal cholinergic system in humans. *British Journal of Pharmacology* 154(8):1558–1571. doi:10.1038/bjp.2008.185

Wonnacott, S. 2014. Nicotinic ACh receptors. *Tocris Bioscience Scientific Reviews Series* 1–31.

Zhao, Y., Liu, S., Zhou, Y., Zhang, M., Chen, H., Xu, H.E., Sun, D., Liu, L., and Tian, C. 2021. Structural basis of human α7 nicotinic acetylcholine receptor activation. *Cell Research* 31(6):713–716. doi:10.1038/s41422-021-00509-6

Zoli, M., Pistillo, F., and Gotti, C. 2015. Diversity of native nicotinic receptor subtypes in mammalian brain. *Neuropharmacology* 96(B):302–311. doi:10.1016/j.neuropharm.2014.11.003

16

Ionotropic Glutamate Receptors

Andrew Plested

CONTENTS

DOI: 10.1201/9781003096276-16

16.1 Introduction

Nine out of every ten synapses in the brain are excitatory (Braitenberg and Schuz, 1998). The glutamate receptor ion channels (iGluRs) that reside in these synapses are cornerstones of fast signaling in the brain. They convert the glutamate released by synaptic terminals into depolarization of their host neurons. Certain iGluRs participate in activity-dependent plasticity, thought to underlie learning and memory. The deployment of iGluRs throughout the nervous system implicates the glutamatergic system in a wide range of neurological and developmental disorders.

16.2 Subunit Diversity; Genes/Paralogs/Orthologs/Subtypes; Alternative Splicing; Evolutionary Relationships

The varied responses of neurons to excitatory amino acids and their derivatives suggested that multiple receptors for these agents should be found in the brain (Watkins and Evans, 1981). This hypothesis was confirmed by molecular cloning (Hollmann and Heinemann, 1994), which delineated the AMPA, NMDA, kainate and delta subtypes of iGluRs. A separate family of metabotropic glutamate receptors act on downstream targets via second messengers. The receptor ion channels are conserved down to worms, and glutamate-receptor-like genes are found in bacteria, plants and algae. There are 18 mammalian genes for iGluRs. Some can form functional homomeric channels, but in vivo, glutamate receptors are predominantly heteromeric complexes. Receptors from each subfamily do not intermix. AMPA receptors, named for their semiselective agonist, alpha-amino-3-hydroxy-y-5-methylisoxazole-4-proprionic-acid (Honoré et al., 1981), are formed from the subunits designated GluA1-4 (with gene names GRIA1–4). N-methyl-D-aspartate (NMDA) receptors are obligate heterotetramers, formed from GluN1 (GRIN1) and any selection from the GluN2A-D (GRIN2A-D) or GluN3A-B (GRIN3A-B) subunits. The kainate receptor subunits GluK1–3 can form channels, but in the brain are usually accompanied by GluK4 or GluK5 subunits (GRIK1–5). In addition to their sparse expression as ion channels, the GluD1 and GluD2 (GRID1,2) subunits also act as scaffolds for synaptogenesis.

In insects, the glutamate receptor template has wide usage. Not only are iGluRs involved in central synaptic transmission but also transmission at the neuromuscular junction. Further, a highly-divergent family of about 60 subunits with the same domain architecture appears to be odorant receptors (Benton et al., 2009). At least some of these genes can generate agonist-activated ion channels with similar pores to AMPA and KA receptors (Abuin et al., 2011). However, it is conceivable that some are not channels, given that delta, NMDA and kainate subtypes can relay signals without ion transport (see Section 16.8.3).

16.2.1 Splice Variation and RNA Editing

Alternative splicing generates diversity across the entire iGluR superfamily (for a comprehensive analysis, see Herbrechter et al., 2021). In AMPA receptors, an alternatively spliced cassette called flip-flop forms part of the active interface between binding domains and thus controls receptor kinetics and modulation. In NMDA receptors, the fifth exon of GluN1 also controls allosteric modulation. In the GluN1 subunits, the cytoplasmic domains are assembled from two alternatively spliced exons. Within these, stretches C0 and C1 (exons 21 and 22, respectively) contain motifs that associate with calmodulin causing strong apparent inhibition of open probability. The GluK1c isoform of kainate receptors has an RXR motif and is retained in ER. The GluN3A and GluD1 receptors have C-terminal splice variants, but their functional significance remains unknown.

AMPA and kainate receptors are subject to broad and essential RNA editing (Herbrechter et al., 2021). As many as one-third of the RNA-editing events that are conserved in mammals occur in the iGluR superfamily. The most prominent example is the Q-R switch in the pore loop of the GluA2 subunit, which occurs through adenine-to-imidine base editing by the ADAR2 enzyme (Rueter et al., 1995). This editing is about 93% complete in the brain. Cognate sites in GluK1–3 are also edited in the same way. Absence of editing by ADAR2 heightens neurodegeneration, perhaps because all-Gln-containing AMPA receptors allow uncontrolled calcium entry. Deficits in ADAR2 lead to activity-related motor neuron loss that may be responsible for amyotrophic lateral sclerosis (ALS) in some patients (Hideyama et al., 2010).

16.3 Structure/Organization

Most tetrameric ion channels are gated by intracellular stimuli or voltage sensors through linkages at their intracellular faces, whereas iGluRs are gated from an external site (Figure 16.1). This inverted topology places the reentrant loop that constrains ion flow at the intracellular mouth of the channel. The extracellular domains are a pair of ancient "clamshell" binding domains appropriated from prokaryotes (Stern-Bach et al., 1994). The layer of clamshell domains closest to the membrane binds agonists and is formed from two discontinuous polypeptide sequences. This observation informed the pioneering work to crystallize a soluble form of the ligand-binding domain (LBD) from rat GluA2 (Armstrong et al., 1998), revealing the chemistry of neurotransmitter binding (Figure 16.2). Subsequent crystallographic analysis of the amino-terminal domains (ATDs) clarified the subunit-specific assembly directed by this domain, and in NMDA receptors, allosteric modulation.

Full-length iGluRs proved more difficult to crystallize, and at first, only an inactive form of the AMPAR GluA2 was available (Sobolevsky et al., 2009). In common with all membrane proteins, the rise of cryo-EM has led to a torrent of iGluR structures for representatives of all four subfamilies (Figure 16.1C). Heteromeric combinations, active (open) forms, complexes with auxiliary subunits and samples from native rodent brain have all been revealed at 3–4 Å resolution.

These structures include the two extracellular domains (ATD and LBD) of each subunit, and the pore domain. In many cryo-EM reconstructions, variability between particles means that the extracellular domains are masked. AMPA and kainate receptors have a similar extended form (Meyerson et al., 2014). NMDA receptors are more compact (Karakas and Furukawa, 2014), with the ATD and LBD of each subunit forming an almost continuous structure. The delta subunits have an extended structure, but do not swap their dimers between layers (Burada et al., 2020a, 2020b). All iGluRs switch from a twofold symmetry in the extracellular domains to approximate fourfold symmetry in the membrane domain, although the alternating arrangement of subunits in heteromeric complexes imposes twofold symmetry throughout the vertical axis of the receptor. In most structures, the individual amino-terminal and ligand-binding domains are largely identical to their isolated

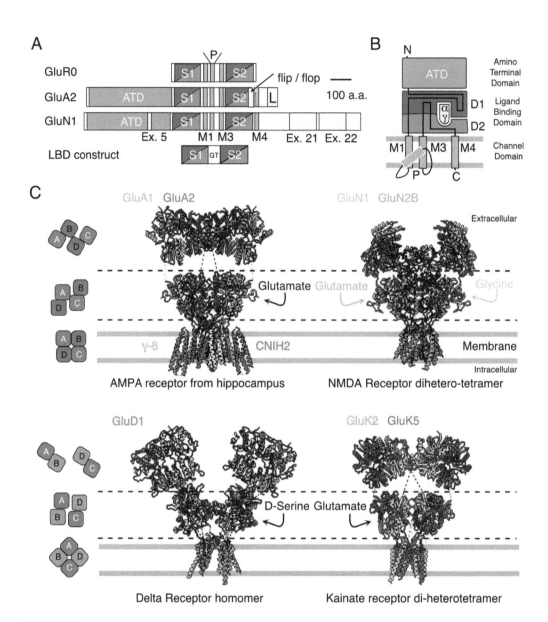

FIGURE 16.1
Receptor structure. (A) Primary structure of iGluRs, showing the increase in complexity from prokaryotic GluR0 to the mammalian NMDA receptor. Alternatively spliced exons are shaded in yellow. Membrane helical segments (M1, M3, M4) and the pore loop (P, or M2) are gray. The S1 and S2 segments are linked in the artificial isolated ligand-binding domain (LBD) construct (used for crystallography and biophysics) by a Gly-Thr linker. ATD, amino-terminal domain; L, long form of GluA2 C-terminus. (B) Topology of a single subunit in the membrane (orange bands). Coloring as in panel A, with alternate exons omitted. D1 and D2, upper and lower lobes of the LBD, respectively. Glutamate (with orientation of α- and γ-carboxyl groups indicated) is in yellow. (C) AMPA, NMDA and kainate heteromeric receptor structures and the GluD1 homomer from cryo-EM. The AMPA receptor from hippocampus is in complex with auxiliary proteins gamma-8 (yellow) and CNIH2 (pink). The common interfaces between the four subunits in the ATD, LBD, and channel layers for AMPA, NMDA, and KAR are indicated in red and cyan squares, and the unique domain architecture of GluD1 is shown below in coral and brown. The Protein Data Bank accession numbers are as follows: AMPAR, 7LDD; KAR, 7KS0; NMDAR, 6WHT; Delta, 7KSP.

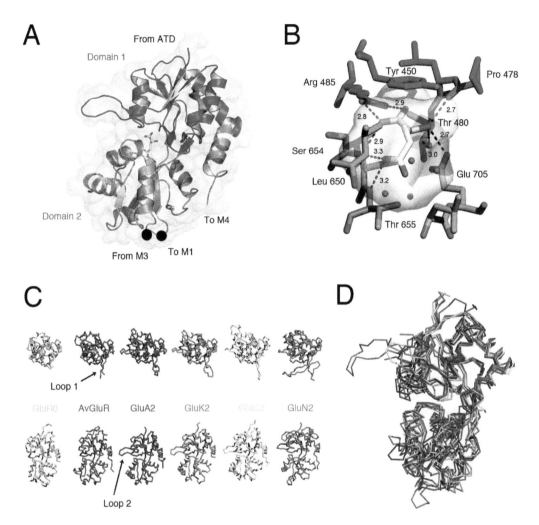

FIGURE 16.2

The glutamate-binding domain. (A) Glutamate (yellow) bound LBD of GluA2 with upper lobe (purple) and lower lobe (green). Surface of the LBD shown in gray. GT linker site is shown as black spheres. (B) Close-up view of glutamate in the binding cavity between lobes. Distances are in angstroms. Water molecules are shown as red balls. Dashed lines indicate hydrogen bonds and the salt bridge between the universally conserved arginine (485 in A2) and the alpha carboxylate of bound glutamate. (C) Top (upper row) and side views (lower row) of LBDs from prokaryotic (GluR0), plant (AvGluR) and mammalian LBDs (A2, K2, D2, N2). Loop regions (Loop 1 and Loop 2) become increasingly ramified with evolutionary distance. All LBDs are glutamate bound, except for GluD2, which does not bind glutamate and is shown in the D-serine bound configuration. The Protein Data Bank accession numbers are as follows: GluA2, 1FTJ; GluR0, 1LL5; GluN2, 2A5S; GluD2, 2V3U; GluK2, 3G3F; AvGluR, 4I02. (D) Overlay of the six LBDs shown in panel C reveals the remarkable conservation of LBD geometry from bacteria to mammalian receptors.

colleagues and form similar interfaces. These similarities bolster the value of biophysical work on isolated domains to understand the full- length receptor.

The intracellular domains remain unresolved; they are largely unstructured. The AMPA, kainate and delta subtypes lack extensive intracellular C-termini, but NMDA receptors have substantial intracellular domains, including sites for binding of calmodulin. This intracellular region shows the greatest variation between NMDAR subtypes.

Bacterial orthologs outline the core of the receptor required for gating. They lack the amino-terminal domain and the final transmembrane domain, and thus have an extracellular C-terminal.

16.4 Physiological Roles; Expression Pattern

The predominant form of AMPA receptor in the brain contains GluA1 and GluA2 subunits. Triheteromeric forms composed of GluA1, GluA2 and GluA3 are also present (Zhao et al., 2019). As well as GluA1 homomers, which may reach the membrane transiently during plasticity (Sanderson et al., 2016), GluA3 may associate with GluA1 to form calcium-permeable AMPARs, at least in the absence of GluA2 (Sans et al., 2003). However, the historical lack of a selective antibody between GluA2 and GluA3 made unequivocal statements about GluA3 difficult. GluA4 is a "fast" AMPA subunit expressed strongly in cerebellum that appears in auditory neurons after the onset of hearing (Taschenberger and von Gersdorff, 2000). It is also strongly expressed in interneurons, some hippocampal principal cells, Bergman glia and astrocytes.

Kainate receptor subunits are somewhat sparsely distributed. GluK1 mRNA is found in various nuclei, and the hippocampus and cortex. GluK2 was cloned from cerebellum and is expressed there, in the granule cell layer (Egebjerg et al., 1991). It is also found in the hippocampus. The GluK5 subunit, which only forms heteromeric receptors with GluK1–3, has wide expression in the brain (Herb et al., 1992), surprising given the limited extent to which kainate receptor currents are detected. On the other hand, GluK4 seems to be exclusively present in the CA3 region of the hippocampus (Werner et al., 1991).

Whilst the GluN1 subunit is expressed throughout the brain, the four GluN2 subunits show distinct spatial and developmental profiles. GluN2B and GluN2D subunits are present at embryonic stages, whereas GluN2A and GluN2C appear later, the latter having very strong expression in the cerebellum. Antibody staining of N3A shows a broadly similar distribution to N1, N2A and N2B (Wong et al., 2002). Evidence for functional receptors incorporating GluN3 subunits has proven hard to obtain but "excitatory glycine receptors" formed from GluN1 and GluN3 subunits are reported in juvenile hippocampus (Grand et al., 2018) and habenula (Otsu et al., 2019).

The GluD1 transcript peaks during development and remains widely expressed in the adult brain, with reports of knockout-controlled functional responses mediated by metabotropic receptor activation from midbrain dopamine neurons (Benamer et al., 2018) and dorsal raphe (Gantz et al., 2020). Delta 2 has strong expression in the Purkinje cells of the cerebellum, where it plays a key role in parallel fiber synapse formation, and can also generate slow mGluR-triggered responses (Ady et al., 2014).

16.4.1 Basal Synaptic Transmission

Glutamate receptors participate in synaptic currents that last from milliseconds to seconds (Logan et al., 2007; Silver et al., 1992), and efficient neural computation might depend on this dynamic range. The classical excitatory current has a fast AMPAR current and a slower NMDAR response (Hestrin et al., 1990). In response to repetitive stimulation, most AMPA and kainate receptors profoundly desensitize, but certain forms can sustain a slow, accumulating response (Figure 16.3). The best studied example is the slow kainate

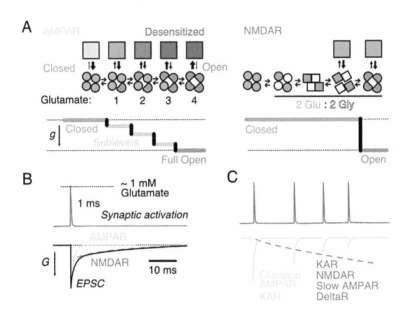

FIGURE 16.3
Kinetics of receptor activation. (A) Left: Illustration of AMPA receptor gating. In the resting state (orange) the pore is closed and the LBDs are vacant. Binding of glutamate activates each subunit independently (green). Subconductance levels (gray) occur in partially bound receptors (openings downward, *g*: single-channel conductance). The fully bound state has a maximally open pore (green). Desensitization (red) is more stable as more glutamate binds (hypothetical transition rates indicated by the thickness of arrows). Right: An analogous scheme for NMDA receptors, which traverse several intermediate states and only open after they are fully bound by glutamate and glycine. NMDA receptors do not desensitize strongly. (B) Synaptic activation and the classic two-component excitatory postsynaptic current (EPSC). The fast (AMPA) and slow (NMDA) components of the conductance change (G) are shown with dashed lines. The glutamate transient due to one vesicle (gray) is about 1 mM for 1 ms. (C) In the case of repetitive stimulation, classical fast AMPA and kainate receptor (KAR) responses (cyan) follow the rapid glutamate transient. All members of the superfamily can also generate slow accumulating currents (dashed curve) operating over hundreds of milliseconds, but these are expressed more selectively. Slow currents endow their host synapses with special properties including short-term potentiation and plateau potentials, and allow them to detonate neuronal spiking and long-term plasticity.

receptor response at the mossy fiber terminal (Castillo et al., 1997). On the subcellular level, mosaic expression of AMPA receptors with slow kinetics has a flat distribution and is not a simple compensation for electrotonic distance (Pampaloni et al., 2021). The higher affinity of NMDA receptors for agonist might support larger responses to spillover of glutamate. NMDA spikes and regenerative events occur in basal dendrites (Nevian et al., 2007). There is some evidence for functional segregation of NMDA receptors formed from different subunits to synaptic and perisynaptic sites. However, most of the functional data involves antagonists that are not particularly selective (such as NVP-AAM077; Liu et al., 2004). Given the abundance of triheteromeric NMDA receptors with distinct properties at hippocampal synapses (Tovar et al., 2013), these data should be interpreted with caution.

Cells spike in the range 0.1 Hz to 1000 Hz, and in this context, tuning of biophysical properties among the iGluR family permits dendritic computation (Branco et al., 2010) and interplay between the subtypes (Geiger et al., 1995). Sharpening of signaling with development has been linked to changes in subunit composition of AMPA receptors (Joshi et al., 2004) and NMDA receptors, where the GluN2A subunit replaces GluN2B (Flint et al., 1997).

16.4.2 Control of Vesicle Release

Although iGluRs are most commonly associated to the postsynaptic membrane, presynaptic localization of "autoreceptors," whether it is in the presynaptic terminal, in the axon itself or even at a somatic site in a presynaptic neuron, influences synaptic function. NMDA and kainate receptors have been identified at certain presynaptic terminals, using knockout mice and precise antagonists like ACET for kainate receptors. NMDA receptors are proposed to enhance GABA release in concert with calcium from intracellular stores in the cerebellum, but evidence of calcium influx at (for example) basket cell axons, where such NMDA receptors should be located, is lacking. The calcium required to enhance release may come from calcium channels activated by passive spread of depolarization from somatodendritic NMDA receptors.

Calcium entry through iGluRs can also trigger neurotransmitter release directly. At reciprocal synapses in the retina, glutamate release precipitates GABAergic feedback inhibition, with distinct time courses. NMDA receptors produce slow feedback, whereas Ca-permeable AMPA receptors provide much faster inhibition (Chávez et al., 2006).

16.4.3 Synaptogenesis

The original observation of "silent synapses" lacking AMPA receptors stimulated the idea that glutamate receptors are central to synapse maturation (Isaac et al., 1995). Roles in synaptogenesis center on ATD interactions with trans-synaptic signaling systems. The interaction of Purkinje cell GluD2 with a protein secreted from cerebellar granule cells, Cbln1, is essential for mature synapse formation (Ito-Ishida et al., 2012). Neurexins in presynaptic membranes form trans-synaptic adhesion complexes with secreted Clbn2 and postsynaptic GluD1 at hippocampal synapses, which regulates AMPA and NMDA receptor accumulation. For GluA4, an interaction with Pentraxins encourages synaptogenesis (Sia et al., 2007). Strikingly, an exogenous combination of Cbln and pentraxin rescues failing synaptic transmission widely across the nervous system (Suzuki et al., 2020).

16.4.4 Synaptic Plasticity: LTP and LTD

The plasticity of glutamatergic synapses is widely proposed to be essential for learning and memory-related behaviors. The cornerstones in this edifice include the absolute NMDA receptor dependence of long-term potentiation (LTP) in some pathways of the hippocampus (Collingridge et al., 1983) and the relatively mild impairments of certain forms of learning by the NMDA receptor antagonist, APV (Morris, 1989). The attractive hypothesis that NMDARs are the molecular embodiment of Hebb's principle of detection of coincident activity in memory (Hebb, 1949) relies on the activity-dependent release of magnesium block and consequent selective local calcium influx (MacDermott et al., 1986). In turn, AMPA receptors should accumulate at synapses. Paradoxically, increases in receptor number (for example, following LTP) measured across many studies (for example, with fluorescence imaging) seem too small to account entirely for the doubling or tripling of the amplitude of population responses (see later). The umbrella concept of "structural plasticity," induced by repetitive glutamate uncaging, offers an indirect route for receptors to be involved in LTP and LTD (long-term depression). The resulting alterations of dendritic spine size or shape correlate to some extent with the magnitude of the synaptic potential.

Some of the first region-specific knockout mice, in which the intention was to delete GluN1 from the hippocampus, showed memory defects, adding emphasis to the

NMDAR–hippocampus–LTP–spatial memory connection (Tsien et al., 1996). Unfortunately, Cre expression in these mice was not restricted to the hippocampus, but extended to the cortex in older mice, and probably led to complete loss of NMDA receptors also in cortex. Cre-CA3 mice (using the CaMKII promoter) showed no defects in spatial memory storage (Bannerman et al., 2012), but did have problems processing spatial information.

The kinases CaMKII and PKC are needed to induce LTP, and probably for its maintenance too (Tao et al., 2021). CaMKII is activated at the level of individual dendritic spines (Lee et al., 2009), by binding to NMDA receptors, and remains active for minutes. The specific activation of CaMKII by the NMDA receptor seems sufficient to stabilize recently formed dendritic spines (Hill and Zito, 2013) – a form of Hebbian plasticity at the microanatomical level. LTD in the cerebellum requires the GluA2 subunit and phosphorylation of Ser880 on this subunit by PKC (Chung et al., 2003). This form of LTD involves the GluD2 subunit in a signaling role that is not related to ion flux, but that may be controlled by D-serine. Tyrosine phosphorylation at a neighboring site (Y876 in GluA2) is able to block LTD, and for this reason, PTPMEG, a phosphatase trapped by the C-terminus of GluD2, is needed to allow LTD (Kohda et al., 2013).

Dialyzing postsynaptic cells with calcium chelators having different kinetics led to the suggestion that a short, sharp Ca influx produces LTP, whereas longer, weaker signals drive LTD. However, uncaging calcium (Neveu and Zucker, 1996) produced potentiation or depression, independent of both the calcium concentration and other parameters. Further, despite the central role of Ca^{2+}, unblock of the NMDA receptor and subsequent calcium influx is not the only possible plasticity signal. Voltage-gated calcium channel activation is probably part of spike-timing-dependent plasticity, among the most physiologically relevant induction protocols for LTP, and store release of calcium is observed in LTD. At least for hippocampal CA1 pyramidal cells, optical quantal analysis suggests that changes in release probability can account for most changes in response amplitude during LTP and LTD (Enoki et al., 2009). In this context, the affinity of classical AMPA receptors for glutamate is low and the peak of neurotransmitter concentration in the cleft probably hits the middle range of the AMPA receptor concentration–response curve (around 1 mM glutamate; Clements et al., 1992). Thus, AMPA receptors might be tuned to respond in a linear way to presynaptic plasticity. Following this logic, synapses whose output depresses are expected to have a high initial release probability (Pr) and those that potentiate should have a low release probability. This powerful concept has one weakness: it requires a stationary postsynaptic receptor response. This assumption is unreliable since certain slow AMPA receptors were demonstrated to produce augmenting responses with repetitive stimulation (Pampaloni et al., 2021).

16.4.5 Short-Term Plasticity

Short-term plasticity refers to intrinsically reversible changes that revert to the basal condition as soon as there is a lull in activity. Potentiation can be accounted for by accumulation of Ca2+ in terminals with low release probability. Vesicle depletion is responsible for depression at some synapses, but others seem indefatigable (Saviane and Silver, 2006). Here receptor desensitization underlies synaptic depression, allowing gain control and hence computation (Rothman et al., 2009). AMPA receptor subtypes with stable desensitization are enough to bring about short-term depression (Rozov et al., 2001). In contrast, AMPA receptors in complex with auxiliary subunits, as well as kainate receptors, can demonstrate indefatigable activity and thus enable a purely postsynaptic mechanism of short-term potentiation (Pampaloni and Plested, 2021).

16.4.6 Homeostasis

Regulation of synaptic strength from hours to days is achieved via changes in AMPA receptor abundance, perhaps triggered by various neurotrophic factors, (reviewed by Turrigiano, 2008). Given that LTP may purely depend on AMPA receptor abundance in extrasynaptic membranes (Granger et al., 2012), this process could be executed by the synthesis and export of more or less receptors to the surface, which would tend to equilibrate by diffusion to distributed synapses.

16.4.7 Expression Outside the Brain

Expression of glutamate receptors outside the nervous system has not been a focus of intense investigation. AMPA receptor subtypes are expressed in beta and delta cells in the islets of Langerhans and can regulate somatostatin release (Muroyama et al., 2004). AMPA receptors are targets of CUX1 in pancreatic cancer, and tumor cell survival in vitro can be altered by high concentrations of AMPAR antagonists (Ripka et al., 2010). There is evidence that NMDA receptors are expressed in the stomach, and gastric epithelial cells may regulate expression (Watanabe et al., 2008) in response to tumors or infection by *Helicobacter pylori* (Seo et al., 2011; Sachs et al., 2011). Details of any agonist-inducted activation of these NMDA receptors are lacking. Kainate receptors are expressed in taste buds (Chaudhari et al., 1996; Chung et al., 2005).

16.5 Pore Properties (Selectivity, Permeation, Gate)

All subtypes of glutamate receptors depolarize cells by allowing sodium entry. AMPA and kainate receptors are nonselective cation channels. Monovalent cations (Na^+, K^+, Cs^+) permeate at similar rates, and are likely hydrated even as they pass the selectivity filter (Biedermann et al., 2021). Calcium permeation is controlled by mRNA editing of GluA2 in AMPA receptors and GluK1–3 in kainate receptors (Burnashev et al., 1995). Deamination of adenosine to imidine places a positively charged arginine at the tip of the pore loop, instead of the genomically encoded glutamine. In the unedited channels, calcium is about equally as permeant as monovalent ions ($P_{Ca}:P_{Na}$ ~1) (Burnashev et al., 1992), and polyamine block is substantial at positive potentials (Bowie and Mayer, 1995). Edited receptors are no longer blocked by polyamines and are barely permeable to calcium. Receptors that contain a full complement of pore arginines are permeant to fluoride but their conductance is very small. The extent to which RNA editing fixes the conductance of channels by regulating stoichiometry remains open. Native and recombinant AMPA receptors display a 2Q:2R ratio, due to the dominant presence of two copies of the GluA2 subunit (Yu et al., 2021). The situation for kainate receptors seems more fluid (Selvakumar et al., 2021).

NMDA receptors are highly permeable to calcium ($P_{Ca}:P_{Na}$ ~10) (MacDermott et al., 1986), and can thus activate adjacent calcium-activated channels, for example, BK potassium channels (Gómez et al., 2021). NMDA receptors also have subunit dependent Ca2+ permeation and conductance properties, with a single residue (S632 in GluN2A and L657 in GluN2D) exerting dominant behavior (Siegler Retchless et al., 2012).

In all members of the superfamily, the highly conserved SYTANLAAF motif forms a gate at the bundle crossing (Figure 16.4A) (Section 16.7.3). In the pore region, structures

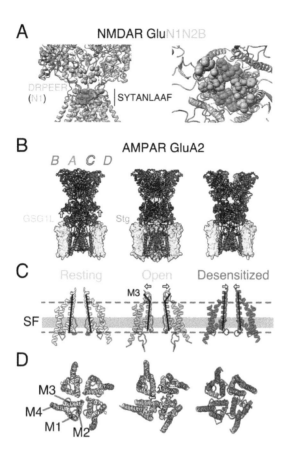

FIGURE 16.4

Gating of glutamate receptors. (A) The universally conserved SYTANLAAF motif that spans the gate region is indicated in spheres on the GluN1N2B structure (Protein Data Bank: 6WHT; GluN1 in blue, GluN2B in green). Extracellular to this, the GluN1 subunit harbors the DRPEER sequence (rose), implicated in binding calcium. (B) Resting and desensitized structures are with GSG1L (Protein Data Bank: 5WEK and 5VHZ, purple translucent surface) and the active, open structure with Stargazin (Protein Data Bank: 5WEO, brown). Arrows show approximate domain motions. (C) Ion channel domains with the B and D subunits of each channel in panel B. The M3 helices are traces with bars to indicate the kink in the open state. The approximate M3 motions are indicated with arrows. The thin selectivity filter region is indicated by a semitransparent gray bar. Dashed lines show the approximate extents of the membrane. (D) As panel C but with a view toward the membrane from the extracellular side. The membrane is represented by dashed lines. Arrows show the approximate movements of the upper M3 helices to open the pore.

of heteromeric receptors reveal subtle deviations from fourfold symmetry. A key feature of glutamate receptor subunits is their propensity to independently adopt different structures to accommodate symmetry mismatches. As more detailed views of the transmembrane domain of native receptors become available (Yu et al., 2021), the way in which higher-order combinations of subunits and auxiliary subunits combine to alter the pore is becoming clear (Zhang et al., 2021a). A key remaining question is the extent to which the desensitized state pore differs from the resting state. Structural evidence suggests that any differences are minor (Twomey et al., 2017b), whereas functional data identify strong coupling of the selectivity filter to desensitization, in certain conditions (Coombs et al., 2019; Poulsen et al., 2019).

16.6 Pharmacology/Blockers

16.6.1 Competitive Antagonists

The first separation between NMDA and non-NMDA currents was made possible by the NMDA "glutamate site" antagonist D-APV (Evans et al., 1982) that binds to GluN2. Later, antagonists of the "non-NMDA" families (effectively, AMPA and kainate receptors), including CNQX and NBQX (Honoré et al., 1988; Sheardown et al., 1990), emerged as complementary tools. Further derivatization of quinoxalinedione has not been successful in distinguishing between AMPA and kainate receptors. Likewise, the binding sites of GluN1 and GluN3 receptors are too similar for strongly selective antagonists. In contrast, highly-selective kainate receptor antagonists including UBP-302 and UBP-310 have been built on the scaffolds of known Willardiine ligands. This work was driven by the structures of the kainate receptor LBDs, which allowed rational exploration of voids flanking the glutamate-binding site (Mayer et al., 2006). Depending on the presence or absence of auxiliary subunits, competitive antagonists of AMPA receptors can be converted into weak partial agonists (Menuz et al., 2007) or become ineffective (reviewed in Pampaloni and Plested, 2021).

16.6.2 Noncompetitive Antagonists and Allosteric Modulators

GYKI 53666 and related drugs bind outside the LBD in the linker region (see Figure 16.5) (Yelshanskaya et al., 2016; Balannik et al., 2005) and so achieve selective block of the AMPA receptor. Derivatives of these noncompetitive antagonists led to the drug perampanel (see Section 16.10). Noncompetitive NMDA receptor antagonists selective for the GluN2C and GluN2D subunits like QZN46 (Hansen and Traynelis, 2011) can also bind at the linkers between LBDs and the TMD.

The NMDA receptor amino-terminal domains contain several modulatory binding sites. The amino-terminal domains cannot interact directly with the channel, but exert their effects by changing either the quaternary arrangement of the LBDs or agonist affinity. Zinc, present in synaptic vesicles at a subset of glutamatergic synapses, inhibits GluN2A subunits with high affinity (Paoletti et al., 2000) by stabilizing a closed form of the GluN2A ATD. Ifenprodil very strongly promotes heterodimerization of isolated GluN1 and GluN2B ATDs (Karakas et al., 2011), binding at an interfacial site (Figure 16.5) nearby to that at which polyamines bind and potentiate the receptor (Mony et al., 2011).

AMPA and kainate receptors are relatively immune to pH changes, but protons strongly inhibit NMDA receptor gating in the physiological range (Traynelis and Cull-Candy, 1991). This inhibition may occur spontaneously in vivo upon glutamate release because of the acidity of vesicles. The principal proton-binding site in NMDARs seems likely to sit between the channel and the ligand-binding domain (Low et al., 2000) (Figure 16.5).

Toxins from conus snails modulate NMDA and AMPA receptor gating. In particular, the con-ikot-ikot toxin binds stably to AMPA receptors and by blocking desensitization is highly neurotoxic (Walker et al., 2009), even though it does not itself promote high activity (Baranovic et al., 2022). These observations mirror the effects of a desensitization-blocking mutation in knock-in mice (Christie et al., 2010). Cyclothiazide and related compounds block desensitization by binding at the dimer interface between AMPA receptor subunits, if position 754 is a serine (as it is in flip isoforms; Partin et al., 1995). Desensitization-blocking drugs based on the same principles that target the kainate receptor have also been developed (Larsen et al., 2017).

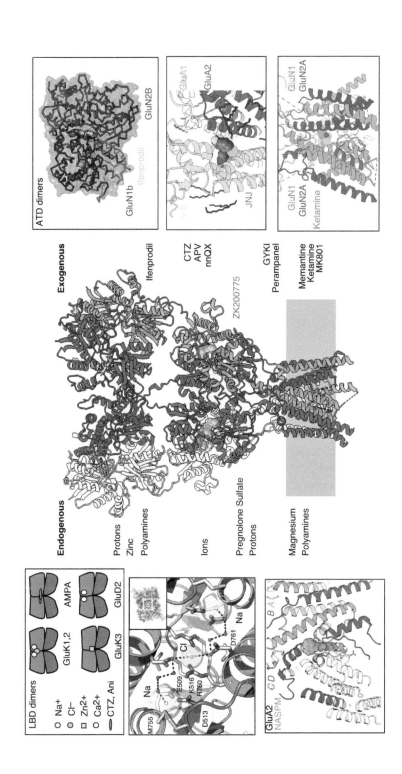

FIGURE 16.5

Modulator binding sites. Endogenous (left) and exogenous (right) modulators and antagonists, with their approximate binding sites respect to the long axis of the AMPA receptor GluA2 with the competitive antagonist ZK200775 bound (Protein Data Bank: 3KG2). nnQX, quinoxaline dione family of competitive antagonists. CTZ, cyclothiazide. (Left upper) Schematic of LBD dimer modulators in the glutamate receptor superfamily. Upper lobes, blue; lower lobes, purple. (Left middle) LBD dimer of GluK2 stabilized by adventitious binding of sodium (gold spheres) and chloride (green sphere) between subunits (orange and blue). Distances of hydrogen bonds in the water network (red spheres) are shown in angstroms. Inset shows the interdimer site in the context of one glutamate-bound LBD dimer (Protein Data Bank: 3C32). (Left bottom) Channel blocker 1-naphthyl acetyl spermine bound in the selectivity filter of GluA2 (Protein Data Bank: 6DM1). (Right top) Ifenprodil (green) binds in the NMDA receptor ATD heterodimer (GluN1, blue; GluN2, red). The upper lobes from both subunits and the lower lobe of GluN2 are involved in the binding site (Protein Data Bank: 3QEL). (Right middle) The gamma-8 specific inhibitor JNJ bound in the cleft between GluA2 and gamma-8 in the AMPAR purified from hippocampus (Protein Data Bank: 7LDD). (Right bottom) The anesthetic ketamine bound to the vestibule of the NMDA receptor (Protein Data Bank: 7EU7).

Several general anesthetics inhibit glutamate receptors, but at clinically relevant doses, only nitrous oxide and xenon inhibit the NMDA receptor. AMPA receptors are inhibited by volatile anesthetics, whereas kainate receptors are potentiated, but these effects all occur at concentrations too high to have clinical relevance.

16.6.3 Pore Blockers

Most subtypes of NMDA receptors are tonically blocked (and thus silent) at the resting membrane potential. Channel block by Mg^{2+} is relieved at around –40 mV, leading to a region of negative conductance (Mayer et al., 1984; Nowak et al., 1984). Unblock by rolling depolarization along dendrites allows nonlinear addition (Branco et al., 2010). The N2C, N2D and N3 subunits reduce the strength of magnesium block (Kuner and Schoepfer, 1996; Chatterton et al., 2002).

Several clinical drugs and drugs of abuse, including phencyclidine and ketamine, are NMDA receptor pore blockers. MK801 is a very stable open-channel blocker (Huettner and Bean, 1988) that traps glutamate and glycine in their binding sites in a voltage-dependent manner. Although they bind at the same site as MK801 in the central channel vestibule (Song et al., 2018), immediately extracellular to the selectivity filter (Figure 16.5), ketamine and memantine allow the channel to close (Zhang et al., 2021b).

Variants of AMPA and kainate receptors without pore-edited subunits are blocked by endogenous cytosolic polyamines at positive potentials (Bowie and Mayer, 1995; Kamboj et al., 1995), which occlude the selectivity filter (Figure 16.5) (Twomey et al., 2018). Generally, activity relieves channel block by endogenous factors, be it depolarization by AMPAR relieving the block of NMDARs by Mg^{2+}, or inward ion flow competing with endogenous polyamines in AMPA or kainate receptors.

A toxin from the Joro spider was identified as a blocker of crustacean glutamatergic neuromuscular transmission, and soon thereafter shown to inhibit hippocampal glutamatergic synapses (Abe et al., 1983; Saito et al., 1985). Similar polyamine toxins from other spiders and wasp venoms, and their derivatives (including 1-napthyl-acetyl-spermine, NASPM and IEM1460) depend on the editing of the pore Q/R site, with Arg-containing pores being spared. They were generally presumed to preferentially isolate receptors containing the GluA2 subunit, consistent with being most effective on the interneuron (calcium-permeable) AMPA receptor population. This simple picture has been confused somewhat by the relief of channel block and changes in polyamine affinity induced by auxiliary proteins (see Section 16.9).

16.6.4 Optical Methods

Various optical approaches for activation and inactivation of glutamate receptors are reported. Photochromic ligands, those tethered to iGluRs by exogenous photoswitches (Volgraf et al., 2006) or unnatural amino acids incorporated directly to the receptor sequence, can convert receptors into light-activated or inhibitable forms (Klippenstein et al., 2014; Zhu et al., 2014). Particularly, unnatural amino acids introduced into the channel domain can identify helical movements during gating (Klippenstein et al., 2017) and link them to activation and desensitization, including an unexpectedly wide role for the selectivity filter (Poulsen et al., 2019). A derivative of the quinoxaline diones containing a photoreactive azido group (ANQX) was used to reveal the time scale of AMPA receptor trafficking in brain slices (Adesnik et al., 2005). For uncaging experiments, MNI-glutamate (Canepari et al., 2001) remains the most used tool. Unfortunately, all caged glutamate variants have some antagonistic activity on the GABA-A receptor and generally have a small two-photon cross-section, mandating their use at high concentrations.

16.7 Gating Mechanisms, Agonist Selectivity, Subunit Contributions and Interactions

16.7.1 LBD Closure as the Driving Force of Channel Activation

Remarkably, the observation from infrared spectroscopy (Cheng et al., 2005) that glutamate binds first to the universally conserved arginine residue in domain 1 of the LBD (Figure 16.2), and then later engages domain 2 via waters and backbone oxygens (Armstrong and Gouaux, 2000) has been reproduced in long unbiased molecular dynamics simulations of glutamate binding (Yu et al., 2018). In such simulations of AMPA and GluN2 (glutamate-binding) NMDA receptor subunits (Yu and Lau, 2018), glutamate enters its binding site along positively charged pathways, and triggered domain closure to adopt the crystallo-graphically observed conformation.

Glutamate analogues show weak selectivity between receptor subtypes because the binding site is so well conserved. Somewhat confusingly, AMPA receptors are activated by the agonist kainate, although the activation is weak in the absence of auxiliary proteins because kainate promotes inactive states of the LBD layer (Salazar et al., 2017). Heteromeric kain-ate receptors and GluK1 are activated by AMPA (Egebjerg et al., 1991). AMPA cannot bind in GluK2 subunit binding sites because of steric clashes (Mayer, 2005), and so presumably activates heteromers by binding to GluK5. In GluN2 subunits, the glutamate residue from domain 2 (e.g., Glu705 in GluA2) that coordinates the amino group in AMPA and kainate-binding pockets is instead an aspartate, which allows NMDA to bind (Furukawa et al., 2005).

The GluN1 subunits bind glycine and D-serine but not glutamate (Furukawa and Gouaux, 2003), explaining the obligate coactivation of GluN1–GluN2 heteromers with glycine and glutamate. GluN3A and GluN3B subunits bind glycine and D-serine with a strong preference over glutamate. Ambient D-serine, largely released by astrocytes, rather than glycine, may be the physiological activator of native NMDA receptors. Likewise, the delta subunits GluD1 and GluD2 are activated by glycine or D-serine, and not glutamate.

The delta subtype of glutamate receptors has long been assumed to lack ion channel activity, even if a mutant form with defective gate (Lurcher) clearly supports cation perme-ation (Kohda et al., 2000). The paradox was solved by the observation that GluD2 receptors can be activated by glycine and D-serine only if Cbln1 is presented to the ATD layer by neurexin 1 from an opposing cell (Carrillo et al., 2021). In the absence of a trans-synaptic partner, the receptor, and in particular the LBD layer, is relaxed in an impotent state and cannot open the membrane ion channel. Intriguingly, similar complexes are involved in control of AMPA and NMDA receptor number at hippocampal synapses, independent from ion channel activity.

The now-canonical view that channel opening is driven by a simple closure of the clam-shell binding domains was developed extensively in AMPA receptors (Armstrong and Gouaux, 2000). The back-to-back active dimer arrangement forces the linkers to the channel domain to separate. The appealing idea that graded closure of individual binding domains corresponded to agonist efficacy was suggested by a series of co-crystal structures of the GluA2 LBD with Willardiine agonists (Jin et al., 2003). However, subsequent work has revealed additional complexity. The LBDs are highly mobile, particularly in the unliganded form, and twisting of the LBD around axes approximately normal to the membrane may also contribute to efficacy. Graded domain closure of GluA2 LBDs was observed in crystals when AMPA was replaced with 5-Bromo-Willardiine. Average occupancy of the binding site could be tracked by the anomalous diffraction from the Willardiine bromine atom and correlated with geometry. However, an individual binding domain can either be occupied by Br-Will or AMPA at a given point in time. Most Willardiine agonists cause similar domain closure in solution NMR experiments and crystal structures, and also do not differ that much in efficacy. Other manipulations, like cross-linking, that reduce linker separation by altering the geometry of the LBD dimers, also reduce efficacy. Indeed, apart from small molecules that stabilize the active dimer arrangement, manipulations that reduce LBD dynamics are in general inhibitory (that is, they make glutamate a partial agonist). Taken together it appears that, except for very bulky antagonists, each ligand can induce a range of LBD conformations, and thus all agonist action is a time-average over occupancy of states of the LBD tetramer (Salazar et al., 2017). In this interpretation, larger partial agonists induce less domain closure and are less efficacious, because they promote more conformations of the LBD layer that do not support channel activation. The structures of NMDA receptor-binding domains in complex with partial and full agonists differ much less for GluN1, GluN2 or GluN3, than for AMPA receptors. Finally, separate observations from molecular dynamics simulations serve as a reminder that there is a much better correlation between efficacy and the energy of binding, than with the simple parameter of the angle of binding domain closure.

16.7.2 Kinetics of Activation

The glutamate transient is very brief (~1 ms) at most synapses (Clements et al., 1992) due to uptake, but the clearance is slowed in the case of terminals with specialized morphologies. Although glutamate is removed to a level of about 10 μM within 5 ms, this concentration is still high enough to preferentially desensitize most forms of AMPA receptors. However, for the fastest AMPA receptors, a time constant of recovery from desensitization in the range of 5 ms allows prolonged high-frequency signaling, where required. Accumulating evidence suggests that slow AMPA receptors, resistant to desensitization and with activation and deactivation kinetics in the 100 ms range, are found throughout the nervous system (Pampaloni and Plested, 2021), operating in parallel with the dominant fast variants.

Selective antagonists revealed that the slow component of an excitatory postsynaptic current normally derives from separate receptor class (the NMDA receptor), but the mechanism of the slow current was only attributed to slow-activating receptors after fast perfusion (Clements et al., 1992) discriminated the ligand concentration profile in time. NMDA receptors are activated slowly (over 100 ms) even if the glutamate pulse is very short, consistent with multiple steps between ligand binding and opening of the pore (Figure 16.3) (Amin et al., 2021). After numerous missteps, the delta-subtype glutamate receptors are now confirmed to be bona fide ion channels, also with a slow activation profile over seconds (Benamer et al., 2018; Carrillo et al., 2021). The mechanisms of this major difference between members of the superfamily remain unknown. Modal gating (a switch between brief and extended activations of individual channels) has been proposed to contribute to the slower, multicomponent decays of NMDA receptors (in the >100 ms range) (Popescu and Auerbach, 2003; Zhang et al., 2008) and slow AMPA receptors probably also rely on this mechanism too. Fast perfusion also allowed the proper discrimination between the stabilities of the desensitized states of different receptor types. The kainate receptor GluK2 is much more stable than any AMPA receptor in the desensitized state, recovering to the resting state over seconds (Bowie and Lange, 2002). The lower lobe of the ligand-binding domain is responsible for this difference (Carbone and Plested, 2012).

16.7.3 Channel Gating

Only AMPA receptors have so far been observed in a bona fide open state (Twomey et al., 2017a; Chen et al., 2017; Zhang et al., 2021a). These open pore conformations were confirmed by molecular dynamics simulations of ion permeation to be approximately equally permeable to Na, K and Cs ions, as AMPA receptors themselves are (Biedermann et al., 2021). The gate opens by an outward movement and kink of the M3 helix. The SYTANLAAF motif is universally conserved (Figure 16.4A), and mutations in this gate region give rise to the spontaneously active Lurcher mutant in GluD2 (see also Section 16.10). This channel gating mechanism is probably not strictly conserved across kainate and NMDA receptors, because the AMPA receptors open in a staircase fashion through several subconductance levels, whereas NMDA receptors open in an all-or-nothing fashion.

In NMDA receptors, the A7 position (Ala 652 in GluN1 and Ala 651 in GluN2B) was convincingly demonstrated to be the narrowest part of the channel in the closed state (Chang and Kuo, 2008), later confirmed by structural studies. Accessibility of this site tracks channel activation, and mutants at this site cannot trap MK-801, presumably due to constitutive activation. The motif DRPEER in GluN1 (Figure 16.4A), extracellular to the A7 gate and also to the membrane, regulates calcium conductance.

16.7.4 Intersubunit Interactions

There is strong evidence that physical stabilization of the back-to-back interaction between the upper lobes of ligand-binding domains blocks desensitization in AMPA and kainate receptors (Figure 16.5 C, D). The Leu to Tyr mutation that was identified from nondesensitizing chimeras (Stern-Bach et al., 1998) maps to this interface. The nearby R/G editing site has some control on AMPAR kinetics. The same residue is an arginine in kainate receptors and participates in a dimer interfacial chloride ion-binding site, which is flanked by two sodium ions (Plested et al., 2008) (Figure 16.5). Cross-linking of active dimers in AMPA and kainate receptors illustrates that D1 breaking is the first trigger that ends an activation before the receptor desensitizes (Weston et al., 2006). Likewise, Ca^{2+} binds at the active dimer interface between LBDs to promote GluD2 activation (Hansen et al., 2009).

Cross-linking dimers in the LBD layer has an inhibitory effect, consistent with lateral movements being required for full activity, at least in AMPARs (Baranovic et al., 2022). As well as breaking the active dimer interfaces, new interfaces between the lower lobes form in desensitized states, most strikingly in kainate receptors where a four-square arrangement is adopted (Meyerson et al., 2014). In NMDA receptors, the LBD interactions are more complicated, partly because the wild-type heterodimers are much more stable, but also because the ATDs control the LBD layer conformation through extensive interactions.

Isolated wild-type AMPA and kainate receptor LBDs do not form dimers in solution (Sun et al., 2002; Weston et al., 2006), but the dimers of ATDs (formed from different subunit pairs) associate tightly, suggesting that they do not dissociate during gating. These dimers assemble with weak interactions, with the ATD dimers sitting in an extended N-shape. Strikingly, intersubunit interactions in the ATD layer promoted by synaptic partners like neurexin are essential for delta receptor ion channel activity (Carrillo et al., 2021). The necessity of a physical restraint in a compact arrangement was demonstrated by cross-linking.

16.7.5 Subunit Gating

For AMPA receptors, binding of two glutamate molecules is enough to generate a sub-level opening, and subsequent binding events increase conductance and open probability (Rosenmund et al., 1998). This behavior is quite distinct from other iGluRs and, for example, glycine and muscle nicotinic receptors, where partially bound openings are to the same amplitude as fully bound ones. Although sublevels are present in NMDA receptors, they have no relation to agonist concentration. Further, open time distributions for NMDA receptors are not concentration-dependent (Schorge et al., 2005), a strong suggestion that only receptors that are saturated by two glycine and two glutamate molecules are able to open (Figure 16.3). The basis of this differential coupling is unknown.

In AMPA and NMDA receptors, particular subunits adopt preferred positions in the heterotetramer. In native AMPARs GluA2 subunit fills the B and D positions, distal to the pore axis in the LBD layer, and thus has a greater leverage to open the pore (Yu et al., 2021). Likewise, the GluN2 subunits adopt the B and D positions, consistent with a principal role in gating for this glutamate-binding subunit (Lü et al., 2017; Chou et al., 2020). Compelling evidence shows that glutamate binding to only the GluK5 subunit can activate heteromeric kainate receptors, but desensitization only occurs at high concentrations that saturate GluK2 (Fisher and Mott, 2011). Structural studies of kainate receptor heteromers, where GluK2 is found in the B and D positions, show that the GluK2 subunit (but not GluK5) undergoes a major conformational change during desensitization (Khanra et al., 2021). This asymmetry between kainate receptor subunits inverts the response to agonist concentration, allowing receptors to be preferentially activated by low concentrations of spillover glutamate.

16.8 Regulation (Second Messengers, Mechanisms, Related Physiology)

Glutamate receptors are subject to a myriad of posttranslational modifications and regulation, and in turn regulate the function of other neurotransmitters and enzymes. Most reports of iGluR modulation by mGluRs and mAChRs relate to kinase signaling (Rojas and Dingledine, 2013), but direct effects are also seen (Rossi et al., 1996) with the most compelling being mGluR-dependent short-term plasticity that relies on slower glutamate diffusion and the postsynaptic protein Homer (Sylantyev et al., 2013).

16.8.1 Phosphorylation

As in many branches of biology, kinases and phosphatases oppose each other in continuous balance, regulating the activity of glutamate receptors. In homosynaptic LTD, persistent phosphatase activation is required to maintain depression. Dephosphorylation of the NMDA receptor happens independently of ion flux and leads to inhibition. Serial induction of LTP and LTD, and vice versa, indicates that the biochemical changes underlying plasticity are reversible, but the changes in the degree of receptor phosphorylation in these experiments are usually small.

GluA1 is phosphorylated at positions 831 and 845 by kinases including PKC, PKA and CaMKII. Kinase action or mutations that approximate phosphorylation (for example, serine to aspartate) alter gating of homomeric receptors in expression systems (Kristensen et al., 2011). But acute reversal is not reported and unsurprisingly any effects on native heteromers are less pronounced. Whether these modifications are the causal events in the potentiation of the synaptic response remains an open question. GluA1 homomers are transiently involved in the generation of some forms of synaptic plasticity, which confounds any simple interpretation. Knock-in mice with ablated phosphorylation sites in the GluA1 C-terminus lack hippocampal NMDA-receptor dependent LTD, but still exhibit LTP. Further, the idea that specific phosphorylation of AMPA receptor C-tails is essential for plasticity has faded, because the identity of the C-tail (or even the glutamate receptor subtype) seems unimportant for LTP (Granger et al., 2012). In general, kinases that modify the AMPA receptor C-tails also affect receptor trafficking and localization through strong effects on auxiliary proteins, which have PDZ binding motifs.

16.8.2 Other Posttranslational Modifications

Conjugation of SUMO to GluK2, coincident with activation (for example, by chronic application of kainate) drives endocytosis of kainate receptors. SUMOylation of synaptic proteins (but not necessarily AMPARs themselves) is related to chemically induced LTP (Jaafari et al., 2013). GluN2 subunits have two palmitoylation clusters in their C-termini indicating that multiple discontinuous parts of the large C-terminus of NMDA receptors are probably associated with the plasma membrane (Hayashi et al., 2009). Orthogonal approaches using fluorescence spectroscopy indicate that the short AMPA receptor C-tails are also close to the membrane (Zachariassen et al., 2016). Palmitoylation is also reported to alter plasma membrane accumulation of homomeric GluK2 receptors (Pickering et al., 1995).

How much do these many modifications impact brain function? Determining the phosphorylation status of individual subunits in a complex, or even of a population at a single synapse, is inconceivable with present methods. Such restrictions limit experimental investigation to population changes across millions of neurons, which are then connected to plasticity events and/or behavior that might be dependent on an ensemble of just a few hundred synapses (Hayashi-Takagi et al., 2015). Many studies have deployed peptide inhibitors based on kinase substrates, either in brain slices or directly in the brain, but these have since proven to be unreliable. For example, the "zeta inhibitory peptide" can robustly reverse LTP in wild-type mice, but also in mice lacking the kinase, which is its presumed target (PKM-zeta) (Volk et al., 2013; Lee et al., 2013). In fact, the effect of this peptide (and perhaps others) is to act as a massive arginine donor for nitric oxide synthase (Bingor et al., 2020), promoting nitric oxide-dependent depression of AMPA receptor activity.

16.8.3 Glutamate Receptor Action without Ion Flux

NMDA receptors and kainate receptors have "metabotropic" function, activating downstream targets without current flow. Convincing experiments demonstrate that KARs can activate other channels like G protein-coupled receptors do (Negrete-Díaz et al., 2006). To date no molecular mechanism, either from within the kainate receptor or its known partner subunits, is proposed for the necessary G protein activation. The GluK2 subunit seems essential for this action, whereas GluK4 and GluK5 subunits are dispensable. Exogenously application of kainate promotes endocannabinoid signaling and alteration of inhibitory transmission. But depression of inhibitory transmission persists in the presence of CB1 blockers, and also if the potential of the presynaptic cell is held in paired recordings. These results illustrate that depolarization of axons through glutamatergic agonist application, or parallel metabotropic effects, are not excluded, complicating the interpretation.

16.9 Cell Biology (Assembly, Trafficking, Associated Proteins)

16.9.1 Assembly

For AMPA receptors, the ATDs might assemble into dimers first, while the rest of the receptor remains to be synthesized by the ribosome. Certain ER-resident auxiliary proteins (such as ABHD6, FRRS1l and CPT1c) selectively associate with monomeric and dimeric forms (Schwenk et al., 2019). The LBD cannot form as a monomer until the majority of the TMD is already synthesized, because of the interdigitated pore region. Mutations that greatly favor LBD dimer association produce mostly dead-end dimeric intermediates. Notably, although homomeric GluA1 and GluA4 tetramers form readily, homomeric GluA2 can reach the surface of cells only when shepherded by auxiliary proteins (see later). Complexes of GluA2 (R) and Stargazin have slow kinetics and a paradoxical, conductive desensitized state (Coombs et al., 2019).

GluN1 subunits assemble into stable intracellular dimers, but only reach the cell surface as part of a heteromeric complex with an alternating N1–N2–N1–N2 arrangement. In mammalian cell lines, paired expression of GluN1 with GluN3A or GluN3B produces only very small currents. Coapplication of CGP78608 blocks the profound desensitization of diheteromeric GluN1-N3A receptors, revealing strong surface expression of this subunit pair (Grand et al., 2018). Coexpression of GluN1–GluN3A–GluN3B receptors in HEK cells also produces substantial currents (Smothers and Woodward, 2007). GluK5 does not reach the cell surface alone, due to poorly matched interfaces in the ATD layer and intracellular ER retention motifs. Instead, a stable association between GluK2 and GluK5 allows this pair of subunits to form the predominant kainate receptor in neurons. As for native AMPA and NMDA receptors, assembly of triheteromeric kainate receptors is reported, and even tetraheterotetramers of GluK1, GluK2, GluK3 and GluK5 subunits were detected using cell-free single molecule techniques (Selvakumar et al., 2021).

Pore editing of the GluA2 subunit is implicated in assembly and trafficking with R-containing subunits being ER-retained (Greger et al., 2002). Mutant glutamate receptors can get trapped in the ER, apparently connecting functional properties to trafficking (Penn et al., 2008). Forward trafficking of kainate, AMPA and NMDA receptors can all be inhibited by mutants that disrupt agonist binding. These general observations have evolved into the idea that a competent binding domain might be needed for proper surface expression of iGluRs. Experimental manipulations designed to link subunits (for example, artificial cysteine bridges) also restrict expression, unsurprising given the complex architecture of iGluRs. Whether receptors actually gate (or more likely desensitize) at intracellular sites, or if the binding of glutamate has a chaperone effect, remains an open question.

A bewildering complement of intracellular proteins act as adaptors between glutamate receptors and kinases and cytoskeletal elements (Henley et al., 2011). Some adaptors are suggested to determine the sign and the magnitude of synaptic plasticity, linking AMPA receptor abundance to stabilization and biochemical transformation at the hands of kinases and other modifiers. In some cases, it is difficult to ascertain precisely when and where these proteins associate with receptors. All trafficking that is related to plasticity is not created equal; intracellular pooling of AMPA receptors due to NMDA and mGluR activation may involve the binding of distinct proteins. Elimination of PICK1 might destabilize intracellular stocks of AMPA receptors, but the time scales of trafficking (~15 minutes), combined with the lack of tools to attack intracellular proteins acutely, make unequivocal statements difficult. Overall, while adaptors such as NSF, GRIP1 and GRIP2 (also called ABP) participate in synaptic plasticity, the complexity of the cascades leaves some details unresolved.

Mass spectrometry of isolated, intact iGluR complexes from native tissue has expanded the zoo of AMPA receptor auxiliary proteins, including highlighting brain-region-specific complexes (Schwenk et al., 2014). Cryo-EM of native complexes has revealed a congruent picture with up to three types of auxiliary proteins (gamma-8, CNIH2 and SynDIG4) collected within a single hippocampal complex (Yu et al., 2021).

16.9.2 Transmembrane AMPA Receptor-Associated Proteins (TARPs)

On account of having been found first, TARPs are by far the best investigated of all auxiliary subunits. What do they do? Two central effects are an increase in single-channel conductance and an apparent increase in the channel opening rate. TARPs also change channel block by spermine and related ligands, by lowering the affinity of the receptor for intracellular polyamines. Several studies propose that AMPA–TARP complexes dissociate rapidly, as a consequence of desensitization (Morimoto-Tomita et al., 2009). But other measurements suggest dissociation likely takes minutes (Baranovic and Plested, 2018). Further, different conformational states (including high activity of partially liganded receptors), and not dissociation, underlie "autoinactivation" (Coombs et al., 2017). An intriguing property related to gating of TARP–AMPA receptor properties is the rebound of steady-state current during long (~1 s) applications of saturating glutamate, termed resensitization or superactivation (Kato et al., 2010; Carbone and Plested, 2016). It proved challenging to connect this slow rebound current to neuronal AMPA receptor gating, but slow AMPA currents identified in hippocampal neurons, the cerebellum and beyond (Pampaloni and Plested, 2021) share the same pharmacological profile (resistance to NBQX inhibition) with AMPA–TARP complexes in heterologous expression (Devi et al., 2016).

Interactions between TARPs and AMPA receptor subunits within the membrane domain govern association and polyamine block (Soto et al., 2014; Ben-Yaacov et al., 2017; Hawken et al., 2017). Further work was stimulated by structures of claudin proteins (related to TARPs) demonstrating that the external domains were compact and structured (Figure 16.4B), with two short loops and one variable length loop that can reach up to the LBD layer. Altering these loops, or their targets, the linkers between the AMPAR TMD and the LBD abolishes TARP modulation without changing association (Riva et al., 2017). Mutagenesis also revealed the base of the LBD as a crucial modulatory site (Dawe et al., 2016).

16.9.3 Neto

Both Neto1 and Neto2 profoundly change the biophysical properties of kainate receptors, lengthening activations and reducing polyamide block. The presence of Neto is enough to drive kainate receptor accumulation at synapses (Copits et al., 2011) suggesting Netos may be expression chaperones. Neto1 has a PDZ ligand and therefore seems equally likely to act as a synaptic anchor for kainate receptors. The hallmark summation of serial responses by kainate receptors is due at least in part to Neto1–kainate receptor complexes (Straub et al., 2011). Cryo-EM structures of kainate receptor–Neto complexes show that their extracellular domains also interact with the LBD layer of kainate receptors (He et al., 2021), much like TARPs do with AMPA receptors.

16.9.4 Synaptic Trapping and Auxiliary Proteins

Transmembrane auxiliary subunits may hold the key to understanding selective deployment of glutamate receptor complexes. The variant functional properties of TARPs and, presumably, different avidity for synaptic sites, likely permits diversity in synaptic

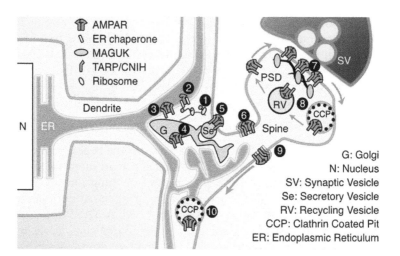

FIGURE 16.6

The glutamate receptor trafficking cycle. Glutamate receptors (AMPAR, green) are synthesized from mRNA at or near to dendritic branch points. (2) Individual subunits and dimers associate with ER chaperones. Assembled tetramers pass through highly complex reticular structures near to spines, where they likely assemble with auxiliary subunits (pink, 3). Their glycans are pruned within Golgi outposts (4) and complexes are secreted to the plasma membrane (5). Diffusion through the spine neck (6) leads into a perisynaptic recycling loop. About half of these AMPA receptor complexes are trapped stably in nanodomains within the postsynaptic density (PSD; 7) but the rest diffuse rapidly within the spine and may be recycled by immediately adjacent clathrin pits (8), which are tethered to Homer and possibly other PSD components. Receptors may also escape the spine (9) to be reclaimed by clathrin-coated pits at dendritic or somatic sites (10), presumably entering the endolysosomal pathway for degradation.

signaling by members of the AMPAR family. All membrane proteins that are not attached to cytomatrix diffuse laterally within plasma membranes (Figure 16.6). AMPA receptors reach synapses by diffusing laterally (Ashby et al., 2006; Penn et al., 2017), and may be trapped by the "necks" of dendritic spines. Auxiliary proteins are expected to move with them, but subcellular exchange between different auxiliary partners is not excluded. It is conceivable that some glutamate receptors in the brain lack TARPs, because they are not an absolute requirement for synaptic clustering (Bats et al., 2012).

The postsynaptic density (PSD) itself is a web of proteins that connects to glutamate receptors through sparse interactions. The macromolecular architectures that underlie trapping of glutamate receptors at synapses remain opaque, due to both their complexity and the dynamic nature of the synapse. Membrane-associated guanylate kinases (MAGUKs), such as PSD-95, are only associated with the membrane through their palmitoylation, and directly bind NMDA receptors through the ramified C-tails of the GluN2 subunits. PSD-95 also directly binds neuroligins, but not AMPA receptors. Instead, the interaction between the PDZ domain of Stargazin (and other TARPs) and PSD-95 stabilizes receptors at the PSD. Other auxiliary proteins (such as Neto1) also interact with PSD components.

16.10 Channelopathies and Disease Mechanisms

Advanced genomic techniques have revealed numerous single-nucleotide polymorphisms and inherited mutations in AMPA and NMDA receptors associated with neurodevelopmental and cognitive disorders (Salpietro et al., 2019; Endele et al., 2010; Trubetskoy et al., 2022; Singh et al., 2022). Biophysical alterations to channel function, such as reduced activation or changes to current rectification arise from mutations in the upper bundle crossing gate (resembling the Lurcher mutation in mice) or to the region around the selectivity filter. The aggressive behavior of the GluA3 knockout mouse has stimulated investigation of GluA3 deficits and their associated to aggression and violence in mouse and in human (Adamczyk et al., 2012). In a broader sense, excitatory synapses are a major focus of disease research, because inherited mutations in synaptic proteins (e.g., neurexin, MUNC13 and Shank) cause cognitive defects and are risk factors for schizophrenia, ALS and autism.

Due to their ubiquity, glutamate receptors have been targets of intensive (but until recently, not particularly successful) research as targets for therapeutic agents. However, some agents approved by the US Food and Drug Administration are becoming available. Perampanel, developed from the prototype noncompetitive AMPA antagonist, GYKI-52466, is approved for epilepsy treatment as an adjunct to existing therapies (Krauss et al., 2012). Drugs that target the gamma-8 auxiliary subunit (Yu et al., 2021) have the great advantage of preferentially inhibiting forebrain AMPA receptors, thus sparing respiratory rhythm. These agents show promise in epilepsy treatment. Memantine is used as a therapeutic drug in Alzheimer's disease (Lipton, 2005). It blocks the pore of NMDARs, perhaps preferentially attacking extrasynaptic receptors, perhaps going some way to explain the tolerance. Ketamine may have application in depression, but at subanesthetic doses that spare NMDA receptors from block. Subtype-specific drugs targeting the NMDA receptor subunit GluN2B have been developed that have increased availability at acidic pH, in order to specifically target receptors in ischemic tissue, in order to have a selective neuroprotective action.

During brain injuries such as stroke, glutamate is released by "reversed uptake" because ion gradients driving glutamate transporters collapse (Rossi et al., 2000). The lack of desensitization and higher glutamate affinity of NMDA receptors renders them especially susceptible in this situation. NMDA and AMPA receptors are subject to autoantibodies that produce schizophrenia and encephalitis (Rogers et al., 1994; Kreye et al., 2016), which can also be associated with tumors (Dalmau et al., 2007).

Nitric oxide produced by neuronal nitric oxide synthase mediates most neurotoxicity due to NMDA receptor overactivation (Dawson et al., 1991). Synaptic NMDA receptor activation may in fact be neuroprotective, on the basis that it stimulates separate antioxidant pathways. This notion relies on the idea that replacing extracellular magnesium by MK801 preferentially removes synaptic receptors from action while sparing extrasynaptic receptors.

The regulation of NMDA receptors by zinc appears important in pain, because genetically targeted mice with a point mutation abolishing zinc inhibition of NMDA receptors are subject to increased sensitivity to painful stimuli (Nozaki et al., 2011). Src kinase attaches to the NMDA receptor complex via ND2, a mitochondrial protein, and disrupting this interaction with a targeted peptide is effective in reducing neuropathic pain (Gingrich et al., 2004). Extending these ideas, it is conceivable that crosstalk between NMDA receptors and the mu-opioid receptor leads to opioid tolerance (Marek et al., 1991), through pathways including stimulation of NO production and PKC by the latter (Chen and Huang, 1991).

Suggested Reading

This chapter includes additional bibliographical references hosted only online as indicated by citations in blue color font in the text. Please visit https://www.routledge.com/9780367538163 to access the additional references for this chapter, found under "Support Material" at the bottom of the page.

Amin, J. B., Gochman, A., He, M., Certain, N., and Wollmuth, L. P. (2021). NMDA receptors require multiple pre-opening gating steps for efficient synaptic activity. *Neuron 109*(3), 488–501.e4.

Armstrong, N., and Gouaux, E. (2000). Mechanisms for activation and antagonism of an AMPA-sensitive glutamate receptor: Crystal structures of the GluR2 ligand binding core. *Neuron 28*(1), 165–181.

Armstrong, N., Sun, Y., Chen, G. Q., and Gouaux, E. (1998). Structure of a glutamate-receptor ligand-binding core in complex with kainate. *Nature 395*(6705), 913–917.

Ashby, M. C., Maier, S. R., Nishimune, A., and Henley, J. M. (2006). Lateral diffusion drives constitutive exchange of AMPA receptors at dendritic spines and is regulated by spine morphology. *J Neurosci 26*(26), 7046–7055.

Bannerman, D. M., Bus, T., Taylor, A., Sanderson, D. J., Schwarz, I., Jensen, V., Hvalby, Ø., Rawlins, J. N. P., Seeburg, P. H., and Sprengel, R. (2012). Dissecting spatial knowledge from spatial choice by hippocampal NMDA receptor deletion. *Nat Neurosci 15*(8), 1153–1159.

Bowie, D., and Mayer, M. L. (1995). Inward rectification of both AMPA and kainate subtype glutamate receptors generated by polyamine-mediated ion channel block. *Neuron 15*(2), 453–462.

Branco, T., Clark, B. A., and Häusser, M. (2010). Dendritic discrimination of temporal input sequences in cortical neurons. *Science 329*(5999), 1671–1675.

Burada, A. P., Vinnakota, R., and Kumar, J. (2020a). Cryo-EM structures of the ionotropic glutamate receptor GluD1 reveal a non-swapped architecture. *Nat Struct Mol Biol 27*(1), 84–91.

Burnashev, N., Zhou, Z., Neher, E., and Sakmann, B. (1995). Fractional calcium currents through recombinant GluR channels of the NMDA, AMPA and kainate receptor subtypes. *J Physiol 485*(2), 403–418.

Carrillo, E., Gonzalez, C. U., Berka, V., and Jayaraman, V. (2021). Delta glutamate receptors are functional glycine- and D-serine-gated cation channels in situ. *Sci Adv 7*(52), eabk2200.

Cheng, Q., Du, M., Ramanoudjame, G., and Jayaraman, V. (2005). Evolution of glutamate interactions during binding to a glutamate receptor. *Nat Chem Biol 1*(6), 329–332.

Granger, A. J., Shi, Y., Lu, W., Cerpas, M., and Nicoll, R. A. (2012). LTP requires a reserve pool of glutamate receptors independent of subunit type. *Nature 493*(7433), 495–500.

Herbrechter, R., Hube, N., Buchholz, R., and Reiner, A. (2021). Splicing and editing of ionotropic glutamate receptors: A comprehensive analysis based on human RNA-Seq data. *Cell Mol Life Sci 78*(14), 5605–5630.

Hollmann, M., and Heinemann, S. (1994). Cloned glutamate receptors. *Annu Rev Neurosci 17*, 31–108.

Isaac, J. T., Nicoll, R. A., and Malenka, R. C. (1995). Evidence for silent synapses: Implications for the expression of LTP. *Neuron 15*(2), 427–434.

Karakas, E., and Furukawa, H. (2014). Crystal structure of a heterotetrameric NMDA receptor ion channel. *Science 344*(6187), 992–997.

Kohda, K., Wang, Y., and Yuzaki, M. (2000). Mutation of a glutamate receptor motif reveals its role in gating and delta2 receptor channel properties. *Nat Neurosci 3*(4), 315–322.

Mayer, M. L., Westbrook, G. L., and Guthrie, P. B. (1984). Voltage-dependent block by Mg2+ of NMDA responses in spinal cord neurones. *Nature 309*, 261–263.

Pampaloni, N. P., Riva, I., Carbone, A. L., and Plested, A. J. R. (2021). Slow AMPA receptors in hippocampal principal cells. *Cell Rep 36*(5), 109496.

Paoletti, P., Perin-Dureau, F., Fayyazuddin, A., Le Goff, A., Callebaut, I., and Neyton, J. (2000). Molecular organization of a zinc binding N-terminal modulatory domain in a NMDA receptor subunit. *Neuron 28*(3), 911–925.

Penn, A. C., Zhang, C. L., Georges, F., Royer, L., Breillat, C., Hosy, E., Petersen, J. D., Humeau, Y., and Choquet, D. (2017). Hippocampal LTP and contextual learning require surface diffusion of AMPA receptors. *Nature 549*, 384–388.

Rosenmund, C., Stern-Bach, Y., and Stevens, C. F. (1998). The tetrameric structure of a glutamate receptor channel. *Science 280*(5369), 1596–1599.

Salpietro, V., Dixon, C. L., Guo, H., Bello, O. D., Vandrovcova, J., Efthymiou, S., Maroofian, R., Heimer, G., Burglen, L., Valence, S., Torti, E., Hacke, M., Rankin, J., Tariq, H., Colin, E., Procaccio, V., Striano, P., Mankad, K., Lieb, A., Chen, S., Pisani, L., Bettencourt, C., Männikkö, R., Manole, A.,

Brusco, A., Grosso, E., Ferrero, G. B., Armstrong-Moron, J., Gueden, S., Bar-Yosef, O., Tzadok, M., Monaghan, K. G., Santiago-Sim, T., Person, R. E., Cho, M. T., Willaert, R., Yoo, Y., Chae, J. H., Quan, Y., Wu, H., Wang, T., Bernier, R. A., Xia, K., Blesson, A., Jain, M., Motazacker, M. M., Jaeger, B., Schneider, A. L., Boysen, K., Muir, A. M., Myers, C. T., Gavrilova, R. H., Gunderson, L., Schultz-Rogers, L., Klee, E. W., Dyment, D., Osmond, M., Parellada, M., Llorente, C., Gonzalez-Peñas, J., Carracedo, A., Van Haeringen, A., Ruivenkamp, C., Nava, C., Heron, D., Nardello, R., Iacomino, M., Minetti, C., Skabar, A., Fabretto, A., SYNAPS, S. G., Raspall-Chaure, M., Chez, M., Tsai, A., Fassi, E., Shinawi, M., Constantino, J. N., De Zorzi, R., Fortuna, S., Kok, F., Keren, B., Bonneau, D., Choi, M., Benzeev, B., Zara, F., Mefford, H. C., Scheffer, I. E., Clayton-Smith, J., Macaya, A., Rothman, J. E., Eichler, E. E., Kullmann, D. M., and Houlden, H. (2019). AMPA receptor GluA2 subunit defects are a cause of neurodevelopmental disorders. *Nat Commun* 10(1), 3094.

Schwenk, J., Baehrens, D., Haupt, A., Bildl, W., Boudkkazi, S., Roeper, J., Fakler, B., and Schulte, U. (2014). Regional diversity and developmental dynamics of the AMPA-receptor proteome in the mammalian brain. *Neuron 84*(1), 41–54.

Silver, R. A., Traynelis, S. F., and Cull-Candy, S. G. (1992). Rapid-time-course miniature and evoked excitatory currents at cerebellar synapses in situ. *Nature 355*(6356), 163–166.

Sobolevsky, A. I., Rosconi, M. P., and Gouaux, E. (2009). X-ray structure, symmetry and mechanism of an AMPA-subtype glutamate receptor. *Nature 462*(7274), 745–756.

Twomey, E. C., Yelshanskaya, M. V., Grassucci, R. A., Frank, J., and Sobolevsky, A. I. (2017a). Channel opening and gating mechanism in AMPA-subtype glutamate receptors. *Nature*.

Yu, A., Salazar, H., Plested, A. J. R., and Lau, A. Y. (2018). Neurotransmitter funneling optimizes glutamate receptor kinetics. *Neuron 97*(1), 139–149.e4.

Yu, J., Rao, P., Clark, S., Mitra, J., Ha, T., and Gouaux, E. (2021). Hippocampal AMPA receptor assemblies and mechanism of allosteric inhibition. *Nature 594*, 448–453.

Zhang, D., Watson, J. F., Matthews, P. M., Cais, O., and Greger, I. H. (2021). Gating and modulation of a hetero-octameric AMPA glutamate receptor. *Nature 594*, 454–458.

17

5-HT₃ Receptors

Susanne M. Mesoy and Sarah C.R. Lummis

CONTENTS

17.1 Introduction

The 5-HT₃ receptor is a cation-selective ligand-gated ion channel, and is structurally and functionally distinct from the other six classes of 5-HT receptors, which are G protein-coupled receptors. 5-HT₃ receptors are members of the Cys-loop and pentameric ligand-gated ion channel (pLGIC) families, and, like other pLGICs, are predominantly located in the nervous system but are also in other tissues. 5-HT₃ receptors play a role in the brain, especially in areas involved in the vomiting reflex, and are important for information transfer in the gastrointestinal tract where they regulate gut motility, secretion and peristalsis. 5-HT₃ receptor function can be modulated by a wide range of compounds including anesthetics, opioids and alcohols. Disturbances within the 5-HT₃ receptor system may contribute to a range of neurological, gastrointestinal and immunological disorders.

DOI: 10.1201/9781003096276-17

17.2 Subunit Diversity

The first cDNA clone encoding a 5-HT$_3$ receptor subunit, the mouse 5-HT$_3$A receptor subunit, was isolated by functional screening of a mouse neuroblastoma (NCB20) cDNA library (Maricq et al., 1991). Subsequently, full-length cDNAs for orthologous 5-HT$_3$A receptor subunits have been cloned from a range of species including, among others, humans, guinea pigs, ferrets and dogs. Somewhat unusually for pentameric ligand-gated ion channels, 5-HT$_3$A subunits can readily form functional homomeric receptors. This suggests they are perhaps evolutionarily ancient, but the lack of 5-HT$_3$ receptor homologues in invertebrates indicates they are more recently evolved than at least some other Cys-loop receptors, such as those for acetylcholine and GABA. An early observation indicating that 5-HT$_3$ receptor populations are not all the same was the different conductances between 5-HT$_3$A receptors and native 5-HT$_3$ receptors in rabbit nodose ganglion. In 1999 a second subunit, the 5-HT$_3$B subunit, was identified (Davies et al., 1999; Dubin et al., 1999). Coexpression of this subunit with the 5-HT$_3$A subunit resulted in functional receptors with properties that more closely represented those found in some native receptors. Since then three other subunits (5-HT$_3$C, 5-HT$_3$D and 5-HT$_3$E; see Figure 17.1) have been discovered (Niesler et al., 2003), and genes for these proteins have now been shown to exist in a range of species, although not in rodents.

The stoichiometry of heteromeric receptors is still not clear, although the presence of at least one 5-HT$_3$A subunit appears to be obligatory for heteromeric receptor assembly (Holbrook et al., 2009; Niesler, 2011). Atomic force microscopy studies suggested 5-HT$_3$AB receptors had a BABBA arrangement, but other work demonstrated the presence of an AA interface; thus two (or possibly more) subunit arrangements may occur. The arrangement and number of 5-HT$_3$C, 5-HT$_3$D and 5-HT$_3$E subunits in functional receptors has not yet been determined, although there are many possibilities (Figure 17.1). The repertoire of 5-HT$_3$ receptor subunits is further increased by a number of different isoforms and promoters, alternative splicing, single-nucleotide polymorphisms (SNPs), and posttranslational modifications.

17.3 Structure

Each 5-HT$_3$ receptor subunit has a large extracellular domain (ECD) with six loops (A–F) contributing to the binding site, a transmembrane domain (TMD) consisting of four membrane-spanning α-helices (M1–M4), and an intracellular domain (ICD) between M3 and M4 that contains two peripheral helices MX and MA (Figure 17.2A, B). The first high-resolution structure of a 5-HT$_3$ receptor was the X-ray structure of the mouse 5-HT$_3$A homopentamer (Hassaine et al., 2014). Since then, structures have been obtained for the 5-HT$_3$A receptor with and without orthosteric ligands (Figure 17.2C) and various allosteric modulators (Basak et al., 2018a, 2018b; Polovinkin et al., 2018; Zarkadas et al., 2020; Zhang et al., 2021). The ion channel pores in these structures show a variety of dimensions (Figure 17.2D), possibly reflecting different states (e.g., open, closed, preopen), and/or the use of different preparation procedures (e.g., the receptor in lipid bilayer discs revealed a more condensed and tightly packed structure than structures solved using detergent-solubilized receptors).

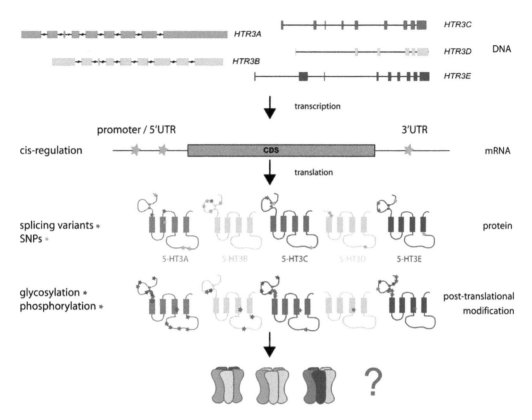

FIGURE 17.1

Molecular basis of the functional and pharmacological diversity in the human 5-HT₃ receptor gene family. There are five distinct genes, with different exons (solid bars), and gene products can vary due to tissue specific promoters and alternative splice sites; there are also a range of naturally occurring variants (mostly SNPs). Post translational modifications can further increase the potential variability. The subunit composition of functional heteromeric receptors is not yet clear. Reprinted from *Pharmacology and Therapeutics*, 128/1, Walstab, Rappold, Niesler, 5-HT₃ receptors: Role in disease and target of drugs, 146–169, Copyright 2010, with permission from Elsevier.

17.3.1 The Extracellular Domain

The ECD contains the agonist-binding site, which is located at the interface of two adjacent subunits and is formed by three loops (A–C) from one (the principal) subunit and three β-strands (referred to as loops D–F) from the adjacent or complementary subunit. Key residues that contribute to the binding pocket in these loops have been identified from a range of studies. In loop A substitutions in the sequence [128]AsnGluPhe[130] modify receptor function, with Glu129 forming a hydrogen bond with the hydroxyl of the agonist 5-HT. Loop B is very sensitive to modification, with substitution of many residues ablating function, suggesting loop B is an obligate rigid structure with an extensive hydrogen bond network. The loop B Trp residue plays an especially critical role as it forms a cation–π interaction with the primary amine of 5-HT. Loop C is important in determining the species specificity of various ligands, with multiple regions of the loop being important. Loops C and D both contribute a Trp to the "aromatic box" (the 4/5 aromatic residues located in the binding pocket of all Cys-loop receptors), and loop D also contributes an Arg to the binding

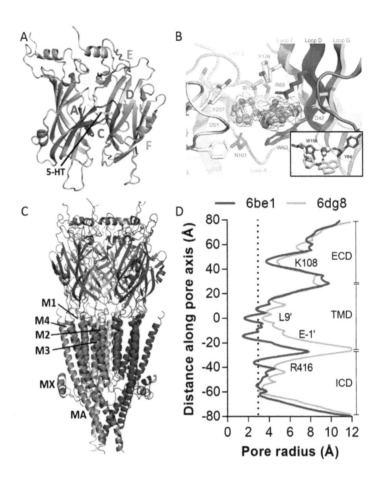

FIGURE 17.2
A) Two adjacent 5-HT$_3$ receptor subunits (PDB 6DG8), showing 5-HT (cyan) bound, and binding loops A-F. B) Palonosetron fits snugly into the 5-HT$_3$ receptor binding pocket consistent with its high affinity. Side chains within 4 Å of the drug are shown. The cryo-EM density for palonosetron is depicted as a gray mesh. Inset: Representative snapshot illustrating the water mediated H-bond network linking palonosetron to W156 and Y64. Reprinted from *Structure*, 28/10, Zarkadas, Zhang, Cai, Effantin, Perot, Neyton, Chipot, Schoehn, Dehez, Nury, The Binding of palonosetron and other antiemetic drugs to the serotonin 5-HT$_3$ receptor, 1131–1140.e4, Copyright 2020, with permission from Elsevier. C) Structure of the 5-HT$_3$ receptor with each subunit in a different colour (PDB 6BE1). The ECD is predominately β-sheet, and the transmembrane domain predominantly α-helix. Much of the ICD is unstructured and is not shown, apart from 2 α-helical regions, MA and MX. D) The pore radius in the 5-HT$_3$ receptor transmembrane domain and flanking regions. The plot shows the different pore radii in the closed (6BE1, red) and open (6DG8, blue) 5-HT$_3$A receptor structures, with the radius of a hydrated Na$^+$ indicated by the dotted line. The smallest radii are at L9′ and E-1′ in the closed structure.

site. Many loop E residues, and at least three loop F residues have been shown to contribute to receptor binding/or function, and residues in an additional loop (loop G) have also been shown to contribute to the binding of some ligands in some species.

17.3.2 The Transmembrane Domain

The four α-helices of the TMD of each 5-HT$_3$ receptor subunit are joined by short loops between M1 and M2 (intracellular), and M2 and M3 (extracellular), and a longer loop between M3 and M4. The M2 α-helices line the ion pore, with the central leucine (L9′)

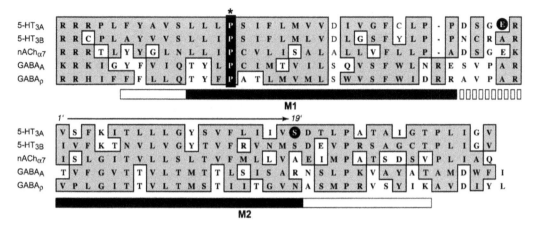

FIGURE 17.3

Sequence alignment of M1, M2 (indicated by bars) and flanking regions of representative pLGIC subunits. The white bars indicate regions of the helix that protrudes above the membrane. The dashed line is the M1-M2 loop; the conserved Pro in M1 is highlighted, as are the two resides, E and S, that result in a reversal in ion selectivity when substituted with A and R respectively. In M2 the prime (') numbering system is shown.

and the M1–M2 loop glutamate (E-1) forming the narrowest points of the channel pore in closed structures (Figures 17.2D and 17.3). M1' and M3 form a ring around the M2 helices, and studies show that these helices contribute to channel function. The conserved proline in M1, for example, is essential for activation, and the receptor is expressed but cannot function when this proline is replaced by alanine, glycine or leucine (Dang et al., 2000). However, substitution with *trans*-3-methyl-proline, pipecolic acid or leucic acid yields active channels similar to wild-type receptors. The commonality between these residues and proline is the lack of hydrogen bond donor activity; thus the data suggest this is a key element in channel gating, possibly because of the resulting flexibility in secondary structure in this region of M1. Finally, the M4 helices form an outer ring beyond M1/M3, separating the rest of the TMD from the lipid bilayer. Studies show alterations to this helix can modify channel function, demonstrating that M4 is not purely a structural component, but is involved in receptor function.

17.3.3 The Intracellular Domain

The ICD is formed by the large M3–M4 intracellular loop, and includes the MX and MA α-helices. This region plays a role in trafficking and also modulates channel function, although deletion studies reveal the ICD is not essential, as the 5-HT₃A receptor subunit ICD can be replaced by the heptapeptide M3–M4 linker of GLIC and retain function (Jansen et al., 2008). The 5-HT₃A receptor ICD can function as a separate domain as shown by studies where it was added to GLIC, resulting in modification of function by the intracellular chaperone RIC-3.

ICD structural details are sparse, which is probably because – like the nACh receptor (Bondarenko et al. 2022) - it is mostly unstructured, but the MA helix contributes to openings (so-called portals) just below the level of the membrane, providing a lateral route in or out of the receptor pore for ions (Kelley et al., 2003). These are not just passive apertures, as blocking relative movement of the MA helices affects channel conductance (Stuebler and Jansen, 2020). The residues that line these portals also contribute to conductance: when 5-HT₃A subunit MA residues are replaced with those found in the 5-HT₃B subunit, the single-channel conductance is increased to that of the heteromeric 5-HT₃AB receptor (Kelley et al., 2003).

17.4 Physiological Roles and Expression

5-HT$_3$ receptors are located in many brain areas including the hippocampus, entorhinal cortex, frontal cortex, cingulate cortex, dorsal horn ganglia, amygdala, nucleus accumbens, substantia nigra and ventral tegmental area (Barnes et al., 2009). The dorsal vagal complex in the brain stem, which is key to the vomiting reflex and contains the area postrema and nucleus tractus solitarius, has the highest receptor levels, consistent with the potent antiemetic properties of 5-HT$_3$ receptor antagonists.

5-HT$_3$ receptor activation in the central nervous system can modulate the release of a variety of neurotransmitters, including dopamine, cholecystokinin, GABA, substance P and acetylcholine. In rat brain, 5-HT$_3$ receptors are expressed by subsets of inhibitory interneurons in the CA1 area and dentate gyrus of the hippocampus, and layer I of the cerebral cortex. Presynaptic 5-HT$_3$ receptor activation at these interneurons likely results in sufficient Ca^{2+} entry to influence GABA release and cause an increase in the frequency of GABA$_A$ receptor-mediated spontaneous inhibitory postsynaptic currents (sIPSCs). Presynaptic 5-HT$_3$ receptors also facilitate the release of glutamate onto dorsal vagal preganglionic, nucleus tractus solitarius and area postrema neurons.

Postsynaptic 5-HT$_3$ receptor activation contributes to fast excitatory synaptic transmission in a range of locations including the rat lateral (but not basolateral) amygdala, nucleus tractus solitarius, ferret visual cortex, and rat neocortical GABAergic interneurons that contain cholecystokinin and vasoactive intestinal peptide. Indeed 5-HT$_3$ receptors are often colocalized with cholecystokinin and also with CB1 cannabinoid receptors: approximately half of the neurons of the rat telencephalon, anterior olfactory nucleus, cortex, hippocampus, dentate gyrus and amygdale expressing the 5-HT$_3$A receptor subunit also expressed CB1 transcripts, and 5-HT$_3$A/CB1-expressing neurons also contained GABA.

5-HT$_3$ receptors have long been known to play a role in the emetic response and thus it is not surprising that they are involved in information transfer in the gastrointestinal tract. In the enteric nervous system they regulate gut motility and peristalsis (Galligan et al., 2002). They also play an important role in the urinary tract, and expression of constitutively active hypersensitive 5-HT$_3$ receptors in mice leads to excitotoxic neuronal cell death, resulting in fatal uropathy.

Functional and expression studies show 5-HT$_3$ receptor mRNA and/or protein is present in many nonneuronal tissues, including peripheral and sensory ganglia, the spleen, colon, small intestine, and kidney. In addition, expression of the receptor has been reported in immune cells such as monocytes, chondrocytes, T-cells, synovial tissue and platelets.

17.5 Biophysical Properties

17.5.1 Receptor Activation

5-HT$_3$ receptors mediate rapidly activating and desensitizing inward currents. 5-HT$_3$A receptors expressed in oocytes have an EC$_{50}$ for 5-HT of 1–4 µM, as do 5-HT$_3$AC, 5-HT$_3$AD and 5-HT$_3$AE receptors, while 5-HT$_3$AB receptors have an EC$_{50}$ approximately threefold higher (Price et al., 2017). Structural data suggest that closure of loop C over the binding site (Figure 17.4) is one of the first conformational changes following agonist binding (as in other pLGICs), and this change is transduced through the protein via the ECD/TMD interface (involving the β1–β2 and β8–β9 loops, the β10 strand, and the M2–M3 linker) to M2. The different pore dimensions can be seen when comparing the open and closed structures, with the smallest pore radii at L9' and E-1' (Figures 17.2 and 17.3). Other regions of the TMD also undergo conformational changes: The M1 helix moves outward from the closed to the open conformation, while M3 remains more stationary, though the top of M3 bends toward the pore; M4 undergoes an upward movement by almost the height of a helical turn, accompanied by a loosening of the α-helix at the M4–MA boundary (compare the identical heights of R416 in MA to movement of Y441 in M4; Figure 17.4).

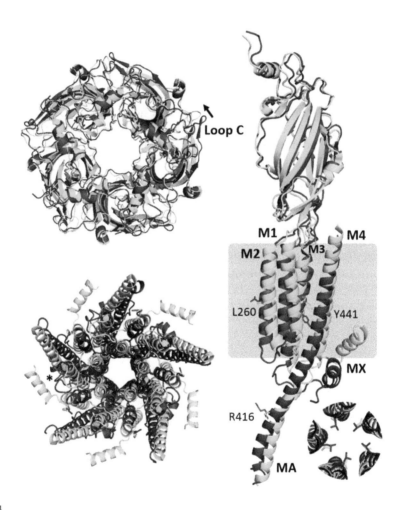

FIGURE 17.4
Movement of the 5-HT₃ receptor upon receptors activation is shown by comparison of open (blue, 6DG8) and closed (purple, 6BE1) structures. Top left: the extracellular domain seen from the top, showing movement of loop C (arrow). Bottom left: A top down view of the transmembrane domain shows movement of the MX helix and the M2-M3 loop (denoted by *), as well as change in M4 angle. Right panel: a single subunit, showing movement of MX and also M4 (note the different relative positions of residue Y441). The grey box indicates the lipid bilayer. Inset: the pore-lining M2 helices seen from the extracellular end, showing the movement of L9' (L260).

17.5.2 Ions and Ionic Selectivity

The 5-HT$_3$ receptor pore is a relatively nonselective cation channel. Currents are primarily carried by Na$^+$ and K$^+$ ions, although divalent and small organic cations are also permeable. 5-HT$_3$A receptors are almost equally permeable to monovalent and divalent cations (P_{Ca}/P_{Cs} = 1.0 – 1.4; although 5-HT$_3$AB receptors have lower Ca^{2+} permeability (P_{Ca}/P_{Cs} = 0.6).

Ionic selectively is predominantly mediated by the –1′ glutamate residue, while a triple mutant receptor with a proline insertion at –1′, and the substitutions E-1′A and V13′T resulted in an anion-permeable 5-HT$_3$A receptor. Subsequent studies showed that the replacement of only two residues (E-1′A, S19′R) was needed to invert ion selectivity (Figure 17.3), and E-1′A alone resulted in non-selective channels.

17.5.3 Single-Channel Conductance

The single-channel conductance of the 5-HT$_3$A receptor is low: values of 0.4–0.76 pS have been reported (Gunthorpe et al., 2000; Kelley et al., 2003). The presence of Lys in the M2 regions was originally considered a possible explanation, but substitutions here revealed this was not the case (Gunthorpe et al., 2000). Subsequent work revealed that the low conductance was due to Arg residues located in the MA helix of the M3–M4 loop (Kelley et al., 2003). These data explain the higher conductance of 5-HT$_3$AB receptor, as these Arg residues are replaced by neutral or negative charged residues in the 5-HT$_3$B subunit.

17.5.4 The Channel Gate

The activation gate of the 5-HT$_3$ receptor channel is located centrally within M2, consistent with the "hydrophobic girdle" model of channel gating; the hydrophobic girdle being a region in the center of M2 that is less than 3.5 Å diameter over a distance of ~8 Å. The residues that face the pore here are hydrophobic (V13′ and L9′) making it effectively impermeable to ions in the closed conformation. Data consistent with this hypothesis include substitution of V13′ residues in the 5-HT$_3$A receptor by threonine or serine, which causes an increase in agonist potency and/or spontaneous channel openings, and substitutions of L9′ by a range of amino acids, which affects agonist potency and desensitization rates.

17.6 Pharmacology

17.6.1 5-HT$_3$ Receptor Agonists

5-HT$_3$ receptor agonists bind to the orthosteric ligand-binding site (Figure 17.5). They have in common a basic amine, an aromatic ring, a hydrophobic group and two hydrogen bond acceptors. Active compounds include 2-methyl-5-HT, phenylbiguanide and m-chlorophenylbiguanide. These agonists are relatively small compounds which cause the C loop to close over the binding site, initiating the gating process.

17.6.2 5-HT$_3$ Receptor Antagonists

5-HT$_3$ receptor competitive antagonists, which bind at the orthosteric (agonist) binding site, are usually larger than agonists; they require an aromatic part, a basic moiety and an intervening hydrogen bond acceptor. Most antagonists have a rigid aromatic or

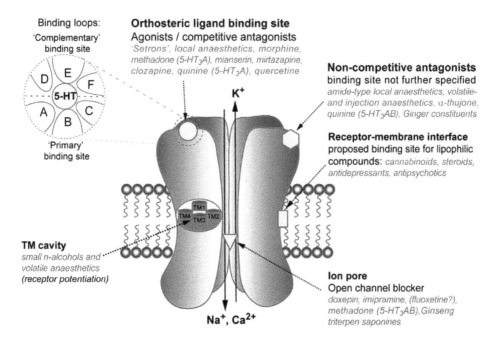

FIGURE 17.5

5-HT₃ receptor drug binding sites. The binding sites of agonist, antagonist, and modulators are shown. The *setrons* include ondansetron, granisetron, tropisetron, bemesetron, dolasetron, and palonosetron. Reprinted from *Pharmacology and Therapeutics*, 128/1, Walstab, Rappold, Niesler, 5-HT₃ receptors: Role in disease and target of drugs, 146–169, Copyright 2010, with permission from Elsevier.

heteroaromatic ring system, a basic amine, and a carbonyl group (or isosteric equivalent) that is coplanar to the aromatic system. There are slightly longer distances between the aromatic and amine group when compared to the agonist pharmacophore. Only small substituents, such as a methyl group, can be accommodated on the charged amine. Many potent antagonists of 5-HT₃ receptors have heterocyclic rings, and the most potent compounds contain an aromatic six-membered ring. Morphine and cocaine were the first antagonists used to characterize the 5-HT₃ receptor, with more selective and potent 5-HT₃ antagonists being developed later.

Antagonists for other channels may also act by binding in the 5-HT₃ receptor pore. Picrotoxin, which was originally considered to be relatively specific as a GABA_A receptor noncompetitive inhibitor, blocks the 5-HT₃ receptor channel; the binding of picrotoxinin (the active component of picrotoxin) has been localized to the 6' position of M2. Compounds structurally similar to picrotoxin, such as the ginkgolides and bilobalide, act similarly. Diltiazem, which blocks voltage-gated calcium channels, also blocks the 5-HT₃ receptor pore, and acts close to the 7' and/or 12' residues in homomeric receptors. Morphine and its analogue methadone, and the antimalarial compounds quinine and mefloquine may also exert their inhibitory effects by binding in the pore, highlighting the common mechanisms that many of these drugs share, and also the promiscuity that many of these compounds display.

17.6.3 5-HT₃ Receptor Modulators

There are a number of allosteric modulators that affect 5-HT₃ receptor function including n-alcohols, anesthetics, antidepressants, cannabinoids, opioids, steroids and natural compounds; these can inhibit or enhance receptor activity and many also modulate other

Cys-loop receptors, although not always in the same direction (see reviews by Davies, 2011; Walstab et al., 2010). Specific binding sites for these compounds have mostly not yet been confirmed, although some may bind in an intersubunit binding cavity at the top of the TMD. The effects of these compounds have mostly been studied on 5-HT$_3$A receptors, although alcohols and inhalational anesthetics have reduced potency at 5-HT$_3$AB receptors, while the effects of etomidate, propofol and pentobarbital are similar at 5-HT$_3$A and 5-HT$_3$AB receptors. Lipids can also modulate the receptor and these are discussed in more detail in the next section.

17.7 Regulation

5-HT$_3$ receptors can undergo posttranslational modification through phosphorylation at kinase consensus sites that have been identified in the ICD. While phosphorylation has been observed at a putative protein kinase A (PKA) site at S409 in the guinea pig 5-HT$_3$ receptor, most studies have been indirect. Activation of PKA accelerates desensitization of 5-HT$_3$ receptors in mouse neuroblastoma and HEK-293 cells, and activators of protein kinase C (PKC) increase the amplitude of 5-HT-activated currents. Unexpectedly, point mutations of putative PKC sites did not affect the sensitivity of the mutant receptors to PKC potentiation, and it may be that enhancement of 5-HT-elicited responses results from increased cell surface expression of the receptor due to structural rearrangements of F-actin with which the receptor clusters, a conclusion that is consistent with observations that 5-HT$_3$A receptors colocalize with F-actin-rich membrane domains. As neurotransmitter release can be regulated through an actin-dependent mechanism, and 5-HT$_3$A receptors can modulate neurotransmitter release, it seems possible that PKC enhancement of 5-HT$_3$ receptor function may play a role in modulating the efficacy of 5-HT$_3$ receptor transmission.

5-HT$_3$ receptor activity, as for other pLGICs (Thompson and Baenziger, 2020), is also regulated by the membrane lipid environment: 5-HT-elicited currents are decreased in cholesterol-depleted cells, and lipid moieties, in particular a cholesterol molecule adjacent to M4, are likely to stabilize the tightly packed conformation observed in receptor structures determined using lipid nanodiscs (Zhang et al, 2021). These structures have a compressed TMD compared to detergent-based structures and, due to more tilted TMD α-helices, a wider open channel. This modulation by lipids has not yet been extensively explored but could explain some unexpected data, e.g., the fact that an 5-HT$_3$ receptor M4 mutation (Y441A) resulted in nonfunctional receptors when expressed in oocytes but not when expressed in HEK cells, which have different membrane compositions (Crnjar et al., 2021).

17.8 Cell Biology

5-HT$_3$ receptors have been followed from "birth" to "death" in elegant studies from Vogel's group (Ilegems et al., 2004). 5-HT$_3$A receptors formed in the endoplasmic reticulum (ER) and Golgi are trafficked in vesicle-like structures along microtubules to the plasma membrane. They aggregate in specific subcellular sites which can be disrupted by F-actin depolymerization. These aggregations have been identified in a range of cells, including neurons, consistent with the precise anatomical location that is required to mediate fast synaptic neurotransmission. Agonist interaction with cell surface 5-HT₃ receptors can result in their internalization and destruction or recycling.

The other subunits have been less well studied, but 5-HT₃B receptor subunits when expressed alone fail to exit the ER, providing a possible explanation as to why these subunits cannot form functional homomeric receptors. ER retention is due, at least in part, to a CRAR retention motif that forms part of the M1–M2 intracellular loop (Boyd et al., 2003). Coexpression with the 5-HT₃A subunit may shield this ER retention motif allowing heteromeric 5-HT₃AB receptors to reach the cell surface. There is some evidence that the 5-HT₃B subunit forces a preference for expression of the heteromeric 5-HT₃ receptor, as coexpression of the 5-HT₃A and 5-HT₃B subunits in tsA-201 cells did not indicate the presence of homomeric 5-HT₃A receptors.

The human 5-HT₃A subunit has four consensus sequence *N*-glycosylation sites in the N-terminal ECD domain and all can be *N*-glycosylated. *N*-glycosylation is essential for export from the ER, cell surface expression and radioligand binding, although it is not necessary to preserve a ligand-binding site once the receptor has matured (Boyd et al., 2003). Three of the four *N*-glycosylation sites are conserved between a range of species (N104, N170 and N186) and appear to be critical, whereas the *N*-glycosylation site at residue 28 is less important and indeed absent in rodents.

BiP, calnexin, and RIC-3 have been identified as ER chaperone proteins that associate with the 5-HT₃ receptor, and are likely to promote correct folding, oligomerization, post-translational modification and/or export from the ER (Boyd et al., 2003). RIC-3 has been the most widely studied but has different effects depending on the subunits, the species they originated from and the expression system. Expression of human 5-HT₃A receptors in transfected mammalian cells, for example, is enhanced by RIC-3, but it causes inhibition of 5-HT₃AB and mouse 5-HT₃A receptor expression. This apparent discrepancy may be due to other proteins that could influence 5-HT₃ receptor expression. Cyclophilin A, for example, promotes 5-HT₃A receptor expression in the cell membrane via an integral peptidyl prolyl isomerase activity, and there may be a range of other proteins yet to be identified that can modify 5-HT₃ receptor expression.

17.9 Channelopathies and Therapeutic Potential

Studies implicate the malfunction of 5-HT₃ receptors in a range of neurological and gastrointestinal disorders (Niesler, 2011; Walstab et al., 2010). Some SNPs have been identified in patients with bipolar affective disorder (BPAD) and schizophrenia, disorders that segregate with cytogenetic abnormalities involving a region on chromosome 11 that harbors the HTR3A gene. Two SNPs have been found in schizophrenic patients (R344H and P391R), and a significant association was found with BPAD with a P16S mutation in the 5-HT₃A subunit, with reporter constructs indicating this mutant could modulate expression levels. Additional SNPs in the HTR3A gene result in the A33T and M257I subunit variants, both of which are associated with reduced levels of cell surface expression.

In the 5-HT₃B subunit there has been an extensive investigation into a very common SNP, Y129S, which is linked both to BPAD and major depression in women (Krzywkowski et al., 2008). The Y129S variant is a gain-of-function mutation, as 5-HT₃AB(Y129S) receptors have an increased maximal response to 5-HT, decreased desensitization and deactivation kinetics, and a sevenfold increase in mean channel open time in comparison to hetero-meric receptors containing the wild-type 5-HT$_3$B subunit. An intermediate effect is apparent for receptors assembled from a mixture of wild-type 5-HT$_3$A, wild-type 5-HT$_3$B and 5-HT$_3$B(Y129S) subunits, suggesting that signaling via the 5-HT$_3$AB receptor in heterozygous, as well as homozygous, individuals is altered by this SNP.

Studies in the more recently discovered HTR3C, HTR3D and HTR3E genes also indicate possible involvement in disease as these subunits have been associated with a number of clinical conditions including chemotherapy-induced nausea and vomiting (CINV), irritable bowel syndrome (IBS), depression, and psychiatric disorders (Fakhfouri et al., 2019; Walstab et al. 2010). A SNP in the 5-HT$_3$C subunit (N163K), for example, has been correlated with IBS, and expression studies suggest it causes an increase in receptor density. Increased expression has also been associated with a SNP in the 3UTR of the HTR3E gene, which inhibits the binding of a micro RNA, and is also associated with IBS,

5-HT$_3$ receptor antagonists are potent antiemetics, originally used for radiotherapy and CINV, but now also widely prescribed as a more general treatment for emesis (Sanger and Andrews 2018). Often-used drugs include ondansetron, granisetron and palonosetron, and are collectively known as the setrons. Palonosetron is the most potent, and structural data reveal that both its rigidity and its interaction with a trapped water molecule in the binding site contribute to its high affinity binding to the orthosteric site (Figure 17.2; Zarkadas et al., 2020). Studies suggest that a wide range of other diseases have the potential to be treated with 5-HT$_3$ receptor selective drugs, including depression, schizophrenia, drug withdrawal, addiction, pruritis, emesis, migraine, chronic heart pain, bulimia, and neurological phenomena (such as anxiety, psychosis, nociception and cognitive function) (e.g., Fakhfouri et al., 2019; Sanger and Andrews, 2018). The identification of 5-HT$_3$ receptors in immune cells also suggests a possible role of 5-HT$_3$ receptors in immunological processes and inflammation, including atherosclerosis, tendomyopathies and fibromyalgia.

Suggested Reading

Barnes, N. M., T. G. Hales, S. C. Lummis, and J. A. Peters. 2009. 'The 5-HT$_3$ receptor--The relationship between structure and function', *Neuropharmacology*, 56(1): 273–84.

Basak, S., Y. Gicheru, S. Rao, M. S. P. Sansom, and S. Chakrapani. 2018. 'Cryo-EM reveals two distinct serotonin-bound conformations of full-length 5-HT$_3$A receptor', *Nature*, 563(7730): 270–74.

Basak, S., Y. Gicheru, A. Samanta, S. K. Molugu, W. Huang, M. Fuente, T. Hughes, D. J. Taylor, M. T. Nieman, V. Moiseenkova-Bell, and S. Chakrapani. 2018. 'Cryo-EM structure of 5-HT$_3$A receptor in its resting conformation', *Nat Commun*, 9(1): 514.

Bondarenko, V., M. M. Wells, Q. Chen, T. S. Tillman, K. Singewald, M. J. Lawless, J. Caporoso, N. Brandon, J. A. Coleman, S. Saxena, E. Lindahl, Y. Xi, P. Tang. 2022. Structures of highly flexible intracellular domain of human alpha7 nicotinic acetylcholine receptor. *Nat Commun* 13(1): 793.

Boyd, G. W., A. I. Doward, E. F. Kirkness, N. S. Millar, and C. N. Connolly. 2003. 'Cell surface expression of 5-hydroxytryptamine type 3 receptors is controlled by an endoplasmic reticulum retention signal', *J Biol Chem*, 278(30): 27681–7.

Crnjar, A., S. M. Mesoy, S. C. R. Lummis, and C. Molteni. 2021. 'A single mutation in the outer lipid-facing helix of a pentameric ligand-gated ion channel affects channel function through a radially-propagating mechanism', *Front Mol Biosci*, 8: 644–720.

Dang, H., P. M. England, S. S. Farivar, D. A. Dougherty, and H. A. Lester. 2000. 'Probing the role of a conserved M1 proline residue in 5-hydroxytryptamine₃ receptor gating', *Mol Pharmacol*, 57(6): 1114–22.

Davies, P. A. 2011. 'Allosteric modulation of the 5-HT₃ receptor', *Curr Opin Pharmacol*, 11(1): 75–80.

Davies, P. A., M. Pistis, M. C. Hanna, J. A. Peters, J. J. Lambert, T. G. Hales, and E. F. Kirkness. 1999. 'The 5-HT₃B subunit is a major determinant of serotonin-receptor function', *Nature*, 397(6717): 359–63.

Dubin, A. E., R. Huvar, M. R. D'Andrea, J. Pyati, J. Y. Zhu, K. C. Joy, S. J. Wilson, J. E. Galindo, C. A. Glass, L. Luo, M. R. Jackson, T. W. Lovenberg, and M. G. Erlander. 1999. 'The pharmacological and functional characteristics of the serotonin 5-HT₃A receptor are specifically modified by a 5-HT₃B receptor subunit', *J Biol Chem*, 274(43): 30799–810.

Fakhfouri, G., R. Rahimian, J. Dyhrfjeld-Johnsen, M. R. Zirak, and J. M. Beaulieu. 2019. '5-HT₃ receptor antagonists in neurologic and neuropsychiatric disorders: The iceberg still lies beneath the surface', *Pharmacol Rev*, 71(3): 383–412.

Galligan, J. J. 2002. 'Ligand-gated ion channels in the enteric nervous system', *Neurogastroenterol Motil*, 14(6): 611–23.

Gunthorpe, M. J., J. A. Peters, C. H. Gill, J. J. Lambert, and S. C. Lummis. 2000. 'The 4'lysine in the putative channel lining domain affects desensitization but not the single-channel conductance of recombinant homomeric 5-HT₃A receptors', *J Physiol*, 522(2): 187–98.

Hassaine, G., C. Deluz, L. Grasso, R. Wyss, M. B. Tol, R. Hovius, A. Graff, H. Stahlberg, T. Tomizaki, A. Desmyter, C. Moreau, X. D. Li, F. Poitevin, H. Vogel, and H. Nury. 2014. 'X-ray structure of the mouse serotonin 5-HT₃ receptor', *Nature*, 512(7514): 276–81.

Holbrook, J. D., C. H. Gill, N. Zebda, J. P. Spencer, R. Leyland, K. H. Rance, H. Trinh, G. Balmer, F. M. Kelly, S. P. Yusaf, N. Courtenay, J. Luck, A. Rhodes, S. Modha, S. E. Moore, G. J. Sanger, and M. J. Gunthorpe. 2009. 'Characterisation of 5-HT₃C, 5-HT₃D and 5-HT₃E receptor subunits: Evolution, distribution and function', *J Neurochem*, 108(2): 384–96.

Ilegems, E., H. M. Pick, C. Deluz, S. Kellenberger, and H. Vogel. 2004. 'Noninvasive imaging of 5-HT₃ receptor trafficking in live cells: from biosynthesis to endocytosis', *J Biol Chem*, 279(51): 53346–52.

Jansen, M., M. Bali, and M. H. Akabas. 2008. 'Modular design of Cys-loop ligand-gated ion channels: Functional 5-HT₃ and GABA rho1 receptors lacking the large cytoplasmic M3M4 loop', *J Gen Physiol*, 131(2): 137–46.

Kelley, S. P., J. I. Dunlop, E. F. Kirkness, J. J. Lambert, and J. A. Peters. 2003. 'A cytoplasmic region determines single-channel conductance in 5-HT₃ receptors', *Nature*, 424(6946): 321–4.

Krzywkowski, K., P. A. Davies, P. L. Feinberg-Zadek, H. Brauner-Osborne, and A. A. Jensen. 2008. 'High-frequency HTR3B variant associated with major depression dramatically augments the signaling of the human 5-HT₃AB receptor', *Proc Natl Acad Sci U S A*, 105(2): 722–7.

Maricq, A. V., A. S. Peterson, A. J. Brake, R. M. Myers, and D. Julius. 1991. 'Primary structure and functional expression of the 5HT₃ receptor, a serotonin-gated ion channel', *Science*, 254(5030): 432–7.

Niesler, B. 2011. '5-HT₃ receptors: Potential of individual isoforms for personalised therapy', *Curr Opin Pharmacol*, 11(1): 81–6.

Niesler, B., B. Frank, J. Kapeller, and G. A. Rappold. 2003. 'Cloning, physical mapping and expression analysis of the human 5-HT₃ serotonin receptor-like genes HTR3C, HTR3D and HTR3E', *Gene*, 310: 101–11.

Polovinkin, L., G. Hassaine, J. Perot, E. Neumann, A. A. Jensen, S. N. Lefebvre, P. J. Corringer, J. Neyton, C. Chipot, F. Dehez, G. Schoehn, and H. Nury. 2018. 'Conformational transitions of the serotonin 5-HT₃ receptor', *Nature*, 563(7730): 275–79.

Price, K. L., Y. Hirayama, and S. C. R. Lummis. 2017. 'Subtle differences among 5-HT₃ AC, 5-HT₃ AD, and 5-HT₃ AE receptors are revealed by partial agonists', *ACS Chem Neurosci*, 8(5): 1085–91.

Sanger, G. J., and P. L. R. Andrews. 2018. 'A history of drug discovery for treatment of nausea and vomiting and the implications for future research', *Front Pharmacol*, 9: 913.

Stuebler, A. G., and M. Jansen. 2020. 'Mobility of lower MA-helices for ion conduction through lateral portals in 5-HT $_{3A}$ receptors', *Biophys J*, 119(12): 2593–603.

Thompson, M. J., and J. E. Baenziger. 2020. 'Structural basis for the modulation of pentameric ligand-gated ion channel function by lipids', *Biochim Biophys Acta Rev Biomembr*, 1862(9): 183304.

Walstab, J., G. Rappold, and B. Niesler. 2010. '5-HT$_3$ receptors: Role in disease and target of drugs', *Pharmacol Ther*, 128(1): 146–69.

Zarkadas, E., H. Zhang, W. Cai, G. Effantin, J. Perot, J. Neyton, C. Chipot, G. Schoehn, F. Dehez, and H. Nury. 2020. 'The binding of palonosetron and other antiemetic drugs to the serotonin 5-HT$_3$ receptor', *Structure*, 28(10): 1131–40.e4.

Zhang, Y., P. M. Dijkman, R. Zou, M. Zandl-Lang, R. M. Sanchez, L. Eckhardt-Strelau, H. Kofeler, H. Vogel, S. Yuan, and M. Kudryashev. 2021. 'Asymmetric opening of the homopentameric 5-HT$_3$A serotonin receptor in lipid bilayers', *Nat Commun*, 12(1): 1074.

18

GABA_A Receptors

Trevor G. Smart and Martin Mortensen

CONTENTS

18.1 Introduction

To enable coherent and controlled excitatory brain activity, and ultimately behavior, the central nervous system relies upon neuronal inhibition to limit and sculpt the extent of neuronal excitability. In the central nervous system (CNS), inhibition is mediated largely by the ubiquitous inhibitory neurotransmitter GABA that increases the cell membrane conductance and electrically shunts excitatory activity. This has two consequences: hyperpolarizing the cell membrane to reduce the likelihood of action potential firing; and increasing membrane conductance to reduce (shunt) excitatory synaptic potentials preventing the threshold for spike firing being breached. To initiate inhibition, GABA rapidly activates ionotropic GABA_A receptors (GABA_ARs), either transiently or persistently, enabling transmembrane Cl⁻ flux. It also activates metabotropic GABA_B receptors that diffusely signal via G protein activation to numerous downstream effectors. Here, we focus on the GABA_AR family (Olsen and Sieghart, 2009; Sieghart and Sperk, 2002).

GABA_ARs belong to the pentameric ligand-gated ion channel family (formerly known as the Cys-loop receptor family) that includes nicotinic acetylcholine receptors, type-3 5-hydroxytryptamine receptors, glycine receptors, the Zn^{2+}-activated cation channel, bacterial homologues from *Gloeobacter violaceus* (GLIC) and *Erwinia chrysanthemi* (ELIC), and the glutamate-activated Cl⁻ channel from *Caenorhabditis elegans* (Corringer et al., 2012).

DOI: 10.1201/9781003096276-18

18.2 Receptor Subunit Diversity and Structure

For GABA$_A$Rs, eight discrete subunit families have been identified: α1–6, β1–3, γ1–3, δ, ε, π, θ and ρ1–3, totaling 19 subunits (Sigel and Steinmann, 2012). This provides built-in diversity for receptor composition, which could, theoretically, be very extensive, but in reality there are far fewer naturally occurring receptor subunit assemblies expressed in neurons (Olsen and Sieghart, 2009). Preeminent among these are receptors containing αβγ and αβδ subunits, which populate inhibitory synapses and also extrasynaptic domains, together with less frequent αβ isoforms (Olsen and Sieghart, 2009) and those receptors containing ρ subunits. The properties of GABA$_A$Rs are strongly influenced by the inclusion of specific subunits. Consequently, in regard to their functional properties, they are categorized into distinct clusters depending on whether they contain α1–3 and α5 subunits, or α4 and α6 subunits. Furthermore, the presence of β subunits is obligatory for expressing GABA$_A$Rs at the cell surface.

The structure of GABA$_A$Rs at near atomic-level resolution has been achieved using X-ray crystallography and cryo-electron microscopy (cryo-EM) of prototypic inhibitory synaptic receptors, composed of α1β1-3γ2 subunits (Laverty et al., 2019; Phulera et al., 2018; Kim et al., 2020; Kim and Hibbs, 2021). The structures confirm a largely consistent stoichiometry of two α, two β and a single γ subunit in the pentamer with mostly defined subunit positions. When αβγ receptors are viewed from above the plane of the postsynaptic membrane, and read counter-clockwise, this arrangement is β–α–β–α–γ (Figure 18.1). The extracellular domain (ECD) and ion channel vestibule comprise most of the receptor's residues (~60%) arranged in two antiparallel arrays of inner and outer sheets of β-strands. The transmembrane domain (TMD; ~20% of residues) houses the ion channel, the receptor activation and desensitization gates, and the ion selectivity filter, formed from clustered bundles of α-helices drawn from each of the five subunits (Figure 18.1). The intracellular domain (ICD; ~20% residues) remains structurally poorly resolved and is the site of receptor phosphorylation and interactions with receptor-associated molecules. Other notable structural signatures of GABA$_A$Rs include a central, contiguous ion conducting pathway through the ECD and TMD, traversing channel activation and desensitization gates before reaching the ICD; and the Cys-loop located at the base of the ECDs, which, although not present in all pentameric ligand-gated ion channels, plays a vital role in communicating neurotransmitter binding to ion channel activation (Miller and Smart, 2010).

18.3 Receptor Trafficking and Clustering

Synaptic GABA$_A$Rs are clustered at inhibitory synapses and this requires receptor-associated molecules (Figure 18.2) to provide scaffold/anchorage support near the cell surface membrane. Receptor clusters are dynamic, and by posttranslational modification of GABA$_A$Rs and/or their scaffolds, receptor expression levels and trafficking can be regulated to affect both function and plasticity (Luscher et al., 2011; Araud et al., 2010).

Gephyrin is a self-assembling scaffold protein, which by anchoring and clustering proteins, provides an ordered signaling postsynaptic density framework for glycinergic and GABAergic synapses. It secures GABA$_A$Rs at inhibitory synapses by forming a lattice structure connecting receptor subunits to the cytoskeleton (Tyagarajan and Fritschy,

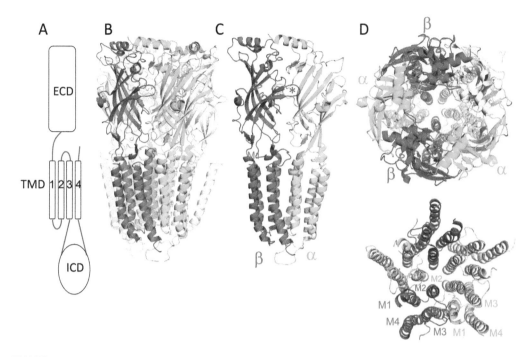

FIGURE 18.1

GABA$_A$R structure. (A) Pictogram showing the modular structure for a typical GABA$_A$R subunit with its extracellular domain (ECD), transmembrane domain (TMD; membrane spanning α-helices, 1–4) and intracellular domain (ICD). Linker regions of amino acids are shown as blue lines. (B) Human α1β3γ2 synaptic-type GABA$_A$R pentamer structure viewed from the side. The α subunit is green, β is red and γ is yellow. The structure of the ICD is unknown and not shown apart from the M1–M2 linker. (C) Juxtaposed β–α subunits shown containing the GABA binding site. Loop C of the β subunit is centrally marked by an asterisk (*). (D) Top panel: GABA$_A$R viewed from the perspective of a presynaptically released GABA molecule. The ion channel is the central pore lined by M2 from each subunit. Bottom panel: Same view as in panel A but with the ECD removed revealing the TMD. Structures are based on Protein Data Bank structure 6I53.

2010). As such, it is a reliable marker for locating inhibitory synapses. GABA$_A$R α1–3 subunits bind directly to gephyrin (Tretter et al., 2008) providing a plausible explanation as to why some GABA$_A$Rs are clustered at synapses (Figure 18.3), but not others, e.g., α4β8. Moreover, α5 subunits can link to another cytoskeletal scaffold molecule, radixin, and this enables α5βγ receptors to be localized outside synapses only diffusing into the synapse once the α5-receptor–radixin link is severed. Several other proteins can also associate with GABA$_A$Rs including, GABA$_A$R-associated protein (GABARAP), Plic-1, and Huntingtin-associated protein 1 (HAP-1). Such proteins regulate the surface stability and trafficking of GABA$_A$Rs (Luscher et al., 2011) by interacting with the ICD of GABA$_A$R subunits (Figure 18.2). Another molecule for anchoring GABA$_A$Rs at inhibitory synapses is provided by the transmembrane GABA$_A$R regulatory lipoma fusion partner-like protein 3/4 (GARHLs) that acts like an accessory protein for GABA$_A$Rs and binds directly to the γ2 subunit's TMD in the cell membrane and to the transsynaptic structural molecule that populates inhibitory synapses, neuroligin-2 (Figure 18.3).

For extrasynaptic GABA$_A$Rs, knowledge of their binding partners (if any) is sparse. The one exception concerns α5-GABA$_A$Rs, which link to radixin to slow their diffusion in the cell membrane. A cell adhesion molecule, neuroplastin-65, also colocalizes with α1 and α2, but not α3, subunits at inhibitory synapses. It also interacts with α5 subunits in the

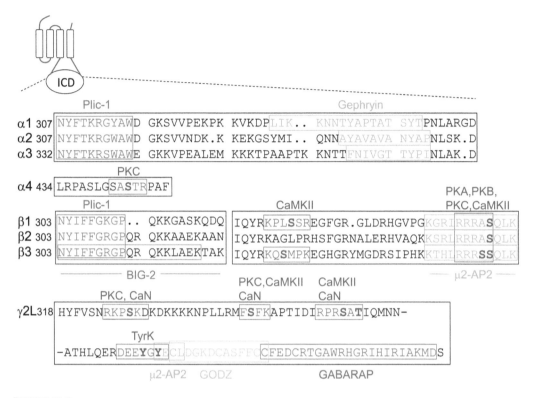

FIGURE 18.2
GABA$_A$R ICD binding sites for interacting proteins. Selected segments of the ICD primary sequences for GABA$_A$R α, β and γ subunits are shown. These illustrate the consensus sequences identified for protein kinase phosphorylation of receptor subunits and the binding motifs that are critical for GABA$_A$R interacting proteins. PK(A-C), protein kinases; CaMKII, Ca^{2+}/calmodulin protein kinase II; TyrK, tyrosine kinase; CaN, calcineurin; GABARAP, GABA$_A$R-associated protein; GODZ, Golgi-specific DHHC (Asp-His-His-Cys) zinc finger protein; BIG 2, brefeldin-A-inhibited GDP/GTP exchange factor 2; AP2, clathrin-adaptor protein 2; Plic-1, protein linking integrin-associated protein with the cytoskeleton-1. (Adapted from Luscher et al., 2011.)

extrasynaptic domain, suggesting it may anchor receptors in a subtype-specific manner. Although GABA$_A$Rs show a preference for locating to synaptic and extrasynaptic domains in a subunit-dependent manner (Figure 18.3), they are also laterally mobile in the plane of the membrane, which serves as an important pathway for replenishing receptors at inhibitory synapses (Thomas et al., 2005).

Protein interactions with GABA$_A$Rs are also regulated by phosphorylation with protein kinases targeting receptor subunits and their associated molecules to regulate receptor trafficking, assembly and cell surface stability. Consensus sites for serine/threonine and tyrosine kinase phosphorylation of GABA$_A$Rs have been identified in the ICDs of α4, β1–3 and γ2 subunits (Figures 18.2 and 18.3) (Brandon et al., 2002). Regulating GABA$_A$R trafficking by phosphorylation is an important pathway for determining cell surface receptor numbers and consequently the extent of synaptic and tonic inhibition.

Considering the cell surface distribution of receptors, the consensus view is that GABA$_A$Rs composed of α1-3βγ subunits mainly populate inhibitory synapses, while the extrasynaptic zone will house α4βδ or α6βδ, α5βγ, and αβ isoforms (Figure 18.3). Their lateral mobility in the surface membrane means that αβγ receptors will also be found in the extrasynaptic zone. Conversely, the less abundant α4βγ and α6βγ receptors can also be

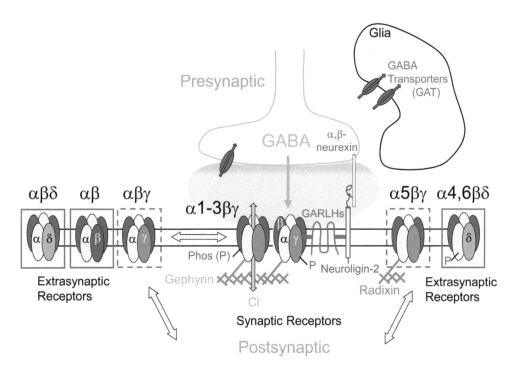

FIGURE 18.3

GABAergic inhibitory synapse. Schematic showing the major elements of a GABA synapse and extrasynaptic membrane. GABA is released (blue plume) onto GABA$_A$Rs at synapses that are predominantly composed of α1-3βγ subunits. Extrasynaptic receptors will comprise α4 or α6βδ and αβ subunit combinations (boxes). For α5βγ (and other αβγ) receptors (dashed boxes), both extrasynaptic and synaptic locations are evident. Sites for phosphorylation (Phos) are indicated on α4, β and γ subunits. Green arrows indicate trafficking routes for the receptor into and out of, and laterally within, the cell membrane. GARLHs, GABA$_A$R-regulatory lipoma HMGIC fusion partner-like 3 and 4 auxiliary subunits. Molecules anchoring pre- and postsynaptic structures are neuroligins and neurexins.

found within selected inhibitory synapses. Overall, although it is difficult to be precise about the relative weightings of GABA$_A$R isoforms at synaptic and extrasynaptic locations, generally, αβγ receptors predominate at synaptic sites and αβδ are located at extrasynaptic sites. In addition to varying expression of GABA$_A$R isoforms at the cellular level, there are also significant differences in subunit expression patterns across the CNS and during neurodevelopment, especially for α subunits (Pirker et al., 2000; Hortnagl et al., 2013).

18.4 Receptor Activation and GABA Binding Sites

The main purpose of GABA-gated ion channels is to rapidly convert the process of GABA binding to ion channel opening. GABA binds at two sites on the receptor that are defined by the interfaces between neighboring β and α subunits. These binding sites involve ECD structures known as binding loops, of which six have been identified (denoted A–F) with three forming the principal "face" (P or +) of the site (loops A–C, on the β subunit) and the remaining three (D–F) comprising the complementary face (C or –) provided by the

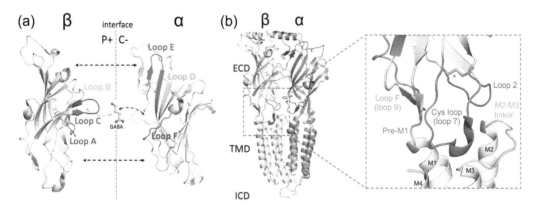

FIGURE 18.4

GABA$_A$R orthosteric site binding loops and structural domains for activation. (A) Part of the extracellular domains of juxtaposed β and α GABA$_A$R subunits. The interface with its principal (P, +) and complementary (C, –) sides is displayed in a separated format showing antiparallel β-sheet arrays and the six neurotransmitter binding loops (A–C and D–F), with a bound GABA molecule. (B) The interfacial region between the ECD and TMD of a GABA$_A$R is critical for activation following ligand binding. This includes loops 2, 7 and 9 from the ECD; and the pre-M1 linker, and the M2–M3 linker region from the TMD. An asterisk (*) denotes the Cys-Cys bridge that characterizes these pentameric receptors.

adjacent α subunit (Figure 18.4A) (Miller and Smart, 2010). Mutating residues within these loops significantly affect the potency of GABA and the activation state of the receptor with some mutations to loops A, D and E enabling GABA$_A$Rs to activate spontaneously in the absence of GABA.

Cryo-EM structures reveal that the orthosteric GABA binding sites are formed from a "box structure" of aromatic residues coordinating the position of GABA molecules that engage in cation-pi, H-bonding, electrostatic, and salt bridge formation with juxtaposed residues. Following GABA/agonist binding to the orthosteric site, loop C closes toward the receptor interface securing the bound agonist molecule. This precedes the anticlockwise rotation of subunit ECDs and compaction of the GABA binding β–α subunit interfaces, leading to opening of the activation gate in the ion channel. By comparison, for a bound competitive antagonist, particularly of large volume, such as bicuculline, which occupies the neurotransmitter binding site, loop C becomes displaced away from the receptor and no ECD rotation or compaction of the β–α interfaces occurs, resulting in a lack of receptor activation.

Although we lack precise high-resolution structural details for "open, activated" GABA$_A$Rs, comparison with other ligand-gated ion channel structures suggest the coupling of the GABA binding loops exert conformational changes to enable receptor activation, and thus the opening of the channel. This involves a key interface between the receptor's ECD and TMD. Here loops 2, 7 and 9 of the ECD are likely to interact with the pre-M1 region and the M2–M3 linker of the TMD to enable channel opening (Figure 18.4B). This functional linkage underpinning receptor operation, formed by the ECD and the TMD interface, relies on both hydrophobic and electrostatic interactions. For αβγ subunit GABA$_A$Rs at inhibitory synapses, these will be activated by presynaptically released GABA attaining millimolar concentrations to cause transient synaptic inhibition (Farrant and Nusser, 2005). However, in the extrasynaptic zone, GABA$_A$Rs (e.g., αβδ subunits) will be exposed to much lower (nanomolar) GABA concentrations, which cause a persistent tonic inhibition (Brickley and Mody, 2012).

18.5 Ion Channel Domain: Conductance and Ion Selection

The GABA ion channel is formed by a ring of five M2 α-helices, one helix contributed by each receptor subunit Figure 18.1D). Residues that comprise the ion channel are denoted by a numerical prime notation. Starting with a conserved arginine residue at the intracellular end of the channel designated as 0', the channel extends toward the external portal at around 20'–22'. This notation facilitates comparison among the pentameric ligand-gated ion channel family (Miller and Smart, 2010). Located near the middle of the GABA ion channel there is a constriction at 9' formed by concentric rings of hydrophobic residues (leucines, valines) forming an activation gate. GABA-induced channel opening will involve an outward retraction of the M2 α-helices toward M1 and M3 widening the 9' leucine (ring) activation gate to facilitate anion flux. The conductance of the GABA ion channels is constant, a feature also noted with other members of the ligand-gated ion channel family. For αβγ and αβδ receptors, single-channel conductance is approximately 28–30 pS. By contrast, for αβ heteromers, the single-channel conductance is reduced by approximately 50% to 12–16 pS. Subconductance states (~50% of the main state) are also observed, but overall for most GABA$_A$Rs, these are relatively rare events compared to the main channel conductance states.

Located at each end of the channel there are rings of charged residues (glutamates, arginines) that together will have a critical role in determining ion selectivity in GABA channels (Keramidas et al., 2004; Nemecz et al., 2016). The GABA$_A$R ion channel is selective for anions enabling permeability (P) to Cl$^-$ and HCO$_3^-$ under physiological conditions with P$_{Cl}$ > P$_{HCO3}$ by approximately fivefold. Ion selection was originally thought to occur deep into the ion channel at 2' to –2' with key roles for a proline residue (–2') and alanine residue (–1'), an area displaying overall electropositivity. However, anion selection may also occur much earlier in the ion permeation pathway, e.g., at 13' (threonine residues) in TM2, and even earlier in the channel's external vestibule that forms part of the ECD.

The near intracellular selectivity filter for ions also overlays a constriction in the ion channel that is ascribed to a "desensitization gate," composed of hydrophobic residues (Gielen et al., 2015), which closes when GABA$_A$Rs are persistently activated. This closed state of the receptor is frequently represented in atomic-level cryo-EM structures (Laverty et al., 2019), suggesting it is likely to be a kinetically stable state.

18.6 Biophysical Properties of GABA$_A$Rs: Influence of Subunit Composition

The biophysical properties of GABA$_A$Rs depend upon their subunit composition. Physiologically, the sensitivity to GABA, and the extent of receptor deactivation (channel closure after unbinding GABA) and desensitization (channel closure with GABA persistently bound) all vary with the α subunit isoform assembled into αβγ GABA$_A$Rs (Figure 18.5). Assessing the potency of GABA in causing receptor activation reveals three groups: those containing α6 subunits show the highest sensitivity to GABA followed by intermediate sensitivity for α1, α4 and α5 subunits; and receptors containing α2 and α3 subunits being least sensitive (Mortensen et al., 2011) (Figure 18.5A). Receptor β subunits display a high degree of homology, even so, αβγ receptors containing β3 subunits are approximately

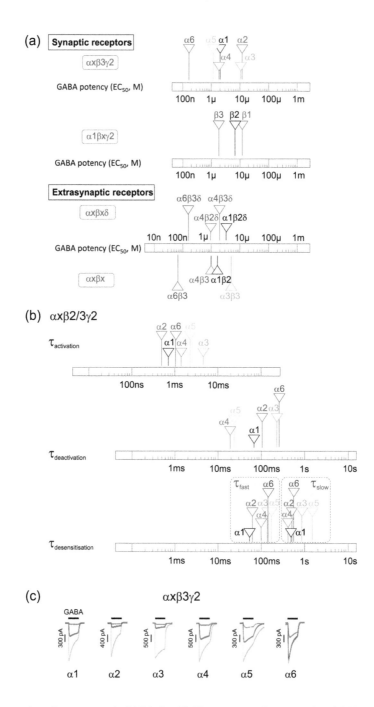

FIGURE 18.5

Properties of synaptic and extrasynaptic GABA$_A$Rs. (A) "Tape measure" pictographs of GABA potency (defined as the EC$_{50}$; n, nano; μ, micro; m, millimolar) in activating a selection of synaptic- and extrasynaptic-type GABA$_A$Rs. (B) Pictographs of biophysical properties of GABA$_A$Rs. Measuring tapes show the time constants (τ) for GABA current activation (from 10%–90% max current, top), deactivation (GABA unbinding, middle) and desensitization (GABA bound closed state, lower) for synaptic-type receptors comprising αβγ subunits. (C) Whole-cell GABA current profiles activated by 0.1 (orange), 1 (red) and 100 μM (blue) GABA rapidly applied to HEK293 cells expressing αxβ3γ2 GABA$_A$ receptors. Note the different desensitization and deactivation profiles. (Currents adapted from Mortensen et al., 2011.)

fivefold more sensitive to GABA than either β1 or β2 subunit-containing receptors (Figure 18.5A). For extrasynaptic δ-subunit-containing GABA$_A$Rs, or just αβ diheteromeric combinations, GABA potency is also dependent upon the identity of the α subunit, with α6 receptors (α6βxδ, α6βx) exhibiting the highest sensitivity to GABA followed by α4 and α1 receptors, with α3 receptors being least sensitive where X = 1–3.

The α subunit also influences the kinetic parameters for GABA$_A$R activation, deactivation and desensitization. For activation, α1 and α2 subunit receptors are generally fast at submillisecond speeds, whereas α4 and α6 receptors require 1–2 ms, and α3 and α5 receptors activate between 2 and 10 ms (Figure 18.5B). The rate of deactivation reflecting GABA unbinding is invariably an order of magnitude slower than activation, with α4 and α5 subunit receptors deactivating fastest followed by an intermediate group containing α1 and α2 subunits, and α3 and α6 being slowest to deactivate (Figure 18.55B). Entry into desensitized states is a familiar characteristic of GABA$_A$Rs. Usually, rates of desensitization are best described by two exponential time constants with receptors containing α1, α2 or α4 subunits desensitizing fastest, followed by α3 and α6 subunit receptors, with α5 subunit receptors being the slowest to desensitize (Figure 18.5B).

18.7 Modulation of GABA$_A$Rs

Ligands, many with current or potential therapeutic value, can modulate GABA$_A$Rs from an array of discrete binding sites including acting directly at the neurotransmitter binding site or proceeding via binding to one or more allosteric sites. Three major geographic binding site categories on the receptor can be distinguished: the ECD, TMD and the "ion channel." These categories are not mutually exclusive, and distinguishing binding site residues from signal transduction pathways is often not trivial.

18.7.1 ECD Interfacial Binding Sites: Agonists

The neurotransmitter binding site is a classic example of an interfacial binding site being located between β and α subunits (Figure 18.6A). This is the site of action for numerous GABA agonists. For synaptic-type αβγ receptors, GABA, muscimol and isoguvacine are full agonists (with similar relative macroscopic efficacies of 1) with potencies in the order of muscimol > GABA > isoguvacine. Other agonists can be classified into several categories of partial agonist with different potencies and macroscopic efficacies. For example, THIP and isonipecotic acid are less potent than GABA with relative macroscopic efficacies of ~0.8 compared to GABA. Piperidine-4-sulphonic acid and imidazole acetic acid attain ~0.4 relative efficacy with similar or reduced potency to GABA, respectively. Very weak partial agonists are represented by Thio-4-PIOL and 4-PIOL with low potencies, and low relative efficacies of 0.01 – 0.05.

18.7.2 Interfacial Binding Sites: Antagonists

Competitive antagonists such as bicuculline and gabazine both bind to the GABA (orthosteric) site on the GABA$_A$R. However, both antagonists will inhibit GABA$_A$R activity initiated by either a neurosteroid (e.g., alphaxolone) or a barbiturate (e.g., pentobarbitone) that

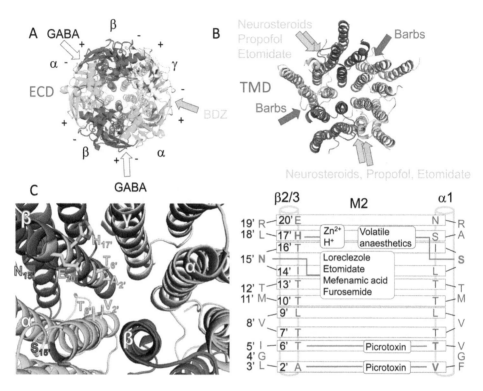

FIGURE 18.6

GABA$_A$R ligand binding sites. (A) Plan view of an αβγ synaptic GABA$_A$R showing the ECD interfacial binding sites for GABA and benzodiazepines (BDZ; arrows). (B) Plan view of TMD interfacial binding sites for neurosteroids, barbiturates (barbs), propofol and etomidate. The ECD has been removed. (C) TMD and ion channel binding and transduction residues. Left panel depicts a plan view of the TMD M2 α-helices. Subunits are color-coded in red (β), green (α) and yellow (γ). Selected ion channel residues are shown at a single β–α interface. These residues map to those shown in the right panel schematic for β and α subunit M2 segments (based on Protein Data Bank structures 6I53 and 6DW0, and using prime number notation). Elements are also color coded: O, red; N, deep blue. Right panel: Residues affecting drug and ion modulation are indicated here in bold red and are linked to ion/ligand panels by red lines. These residues variously form part of binding or coordination sites, and/or are involved in signal transduction.

bind to sites distinct from the GABA site. Since these antagonists do not compete for the steroid and barbiturate sites, they are capable of allosteric inhibition (termed negative allosteric modulators), suggesting they are not simply blocking, by competition with GABA, the orthosteric binding site.

18.7.3 Interfacial Binding Sites: Positive Allosteric Modulators

Pentameric αβγ GABA$_A$Rs contain five subunit–subunit interfaces; of these four are distinct by virtue of the juxtaposed subunit pairings (two β–α; and one each for α–β, β–γ and α–γ; Figure 18.6A). The α–γ interface is significant because benzodiazepines (BDZs) bind here, potentiating receptor function but unable to activate the receptor. Given the interfacial position of the BDZ site, it is unsurprising that the γ subunit is a key requirement for BDZ sensitivity of GABA$_A$Rs. However, there is also a dependence on the type of α subunit present with α1–3 and α5 displaying BDZ sensitivity, while α4 and α6 subunit receptors do not. This occurs because BDZ sensitivity depends on a histidine residue in the ECD of α subunits 1–3 and 5 (H101 in α1) that is replaced by arginine in α4 and α6. Cryo-EM GABA$_A$R structures with bound diazepam reveal not just binding at the α+/γ– interface but also a presence in the TMD at each β+/α– interface, and at the γ+/β– interface overlapping with a barbiturate binding site in the TMD (Masiulis et al., 2019; Kim et al., 2020). The functional consequences of these other sites remain to be seen but could involve stabilizing the receptor protein's structural organization.

18.7.4 Interfacial Binding Sites: The Non-GABA Binding Interface

Another interface, where GABA does not bind and for which we know very little, lies between α and β subunits (α+β–). A screen of numerous compounds has identified an anxiolytic pyrazoloquinoline-based compound, CGS9895, which acts as an antagonist at the α+/γ– BDZ site, and also as an allosteric modulator potentiating GABA currents by binding to the α+β– interface.

Structural analogues of CGS9895 also act at this site and, given the unusual nature of the interface, it is predicted that modulation initiated by occupancy of this site may depend on the identity of the β subunit as well as the α subunit. Indeed, specific modulators have been reported for α6β2/3γ2 over α1–5β2/3γ2 receptors, by binding to the α6+β– interface. This discovery indicates that drugs acting at the α+β– interface will, in principle, be able to modulate αβ, αβγ and αβδ receptors (i.e., major synaptic- and extrasynaptic-type GABA$_A$Rs) without causing direct receptor activation. Thus, they will achieve a broader therapeutic span, in excess of that attributed to the BDZs, which require the inclusion of the γ subunit (along with α and β subunits) for functional modulation.

18.7.5 Transmembrane Domain Binding Sites: Intrasubunit and Interfacial

Another major region on GABA$_A$Rs that is targeted by drugs is the TMD. It is here that several classes of inhalational and intravenous anesthetics, barbiturates and endogenous modulatory neurosteroid molecules have their binding sites (Figure 18.6B) (Hosie et al., 2006; Yamakura et al., 2001; Kim et al., 2020).

Cryo-EM structures of α1β2/3γ2 GABA$_A$Rs reveal that for phenobarbital (barbiturate), clear binding is evident at two TMD interfaces: α+/β– and γ+/β–, but not at β+/α– interfaces as previously thought from structure–function studies (Kim et al., 2020). The barbituric acid group that characterizes barbiturates is interposed between the M3 and M1 α-helices of adjacent subunits causing barbiturate molecules to bind at the level of 15' in M2, within a pocket formed by M1, lying just below the M2–M3 linker, a region that is important for receptor activation. Molecules binding in this vicinity of the TMD differ from the BDZs by not only potentiating receptor function but can also cause direct activation in the absence of GABA. Traditionally, these two effects were thought to proceed via two distinct binding sites, but it is now apparent that a single site could mediate both effects.

To locate anesthetic binding sites, a bacterial homologue of the Cys-loop receptors, GLIC, was first used (Corringer et al., 2012). Crystal structures of GLIC with the anesthetic propofol or desflurane bound in situ, indicate a binding site for anesthetics located within a single subunit (intrasubunit). For GABA$_A$Rs, a diazirine-based photolabel attached to propofol identified residues in the ion channel of the β subunit, quite distant from the propofol binding site in GLIC, and from the photolabeled binding site for another anesthetic, azietomidate, previously identified in GABA$_A$Rs. Using cryo-EM structures with bound propofol or etomidate revealed a shared binding site in the TMD (Kim and Hibbs, 2021), but only at the β+/α– interfaces. Binding of these anesthetics was identical at either interface, and the phenyl ring of etomidate was juxtaposed to N265 in the β subunit at the level of 15′ – a residue known to affect anesthetic activity.

Deeper into the TMD, near the inner leaflet of the cell membrane, lie binding sites for the endogenous neurosteroid molecules in the brain, which can potentiate, directly activate and also inhibit GABA$_A$R function. Using X-ray crystallography and photolabeling, tetrahydro-deoxycorticosterone and allopregnanolone (both positive allosteric modulators of GABA$_A$Rs), were found to bind at the β+/α– interface. Neurosteroids link M3 and M1 α-helices from adjacent subunits involving H-bonding with the β subunit F301 and α subunit Q241 (Laverty et al., 2017; Miller et al., 2017). For inhibitory neurosteroids, such as pregnenolone sulfate, which inhibit GABA$_A$R function, a discrete site was tentatively identified in the receptor TMD involving M3 and M4 α-helices (Laverty et al., 2017).

18.7.6 Transmembrane Binding Sites: The Ion Channel Pore

The third major area for drug action on GABA$_A$Rs involves the ion channel pore (Figure 18.6C). Here, residues lining the ion channel lumen, which can bind ligands and ions, can be readily identified.

In the β subunit, a histidine residue (17′) is a major coordinating site for Zn^{2+}. This divalent cation acts as a receptor subtype-selective inhibitor for GABA$_A$Rs, potently and reversibly inhibiting αβ, αβδ receptor function (nanomolar-low micromolar concentrations) over αβγ isoforms (>300 μM). Mutation of this histidine significantly reduces Zn^{2+} potency, but complete abolition of Zn^{2+} sensitivity requires the mutation of additional residues in the ECD at the α+β– subunit interface (E147 and H141 in α1 subunits, and E182 in β subunits) (Hosie et al., 2003). The reduced sensitivity of αβγ receptors to Zn^{2+} inhibition is a consequence of replacing a β subunit in αβ diheteromers (β–α–β–α–β) with a γ subunit (β–α–β–α–γ) that lacks the Zn^{2+} binding residues.

Protons will modulate GABA$_A$Rs causing both potentiation and inhibition of currents depending on receptor subunit composition (Krishek et al., 1996). The same histidine residue that is important for Zn^{2+} inhibition also plays a key role in the proton sensitivity of GABA$_A$Rs, though other residues, notably in the M2–M3 linker of β subunits and the ECD, are also important.

Traveling farther into the channel, another key residue in β2/3 subunits is an asparagine (N) at 15′ (Figure 18.6C). This is replaced by serine in β1, which is sufficient to remove modulation of GABA$_A$Rs by several therapeutic and experimental drugs: loreclezole (anticonvulsant), methyl 6,7-dimethoxy-4-ethly-β-carboline (DMCM, a benzodiazepine negative allosteric modulator), etomidate (anesthetic), furosemide (diuretic) and mefenamic acid (anti-inflammatory). The importance of this 15′ residue, which is not readily accessible from the channel lumen, is emphasized by loreclezole potentiating αβγ GABA$_A$R function when the receptor contains β2 or β3, but not β1, subunits (Wingrove et al., 1994). Inserting a 15′ asparagine into β1 confers a sensitivity to loreclezole on αβ1γ receptors, a feature noted also with DMCM, which is thought to bind at the benzodiazepine site α+γ– subunit interface, but clearly its modulation is affected by N290 in β2/3.

The diuretic furosemide also shows β subunit selectivity dependent on 15′N, though there are residues elsewhere in the GABA$_A$R (e.g., in M1) that are important for its modulatory action. Another member of this quintet of drugs, mefenamic acid, is a modulator of GABA$_A$Rs that is also dependent on 15′N causing potentiation at α1β2/3γ and α1β2/3 receptors, but inhibition at α1β1 receptors. Finally, the anesthetic etomidate, a potentiator at GABA$_A$Rs, similarly displays a preference for receptors with β2/3. Substitution of 15′N in β3 reduced both direct etomidate-induced currents and its GABA-modulatory actions at αβ3γ2 receptors.

In the absence of structural context, the natural question arising from these observations is whether the 15' asparagine in β2/3 subunits constitutes part of a generic binding site or represents a crucial region for signal transduction. Comparing the chemical structures of mefenamic acid, etomidate and loreclezole suggests their hydrophobic and electronegative moieties can be overlaid in minimum energy conformations, which would be in accord with sharing a similar binding site. However, other evidence argues against a single binding site. Photolabeling with azietomidate, and also cryo-EM data, indicate that the anesthetic binds to β+/α− interfacial sites on GABA$_A$Rs. Furthermore, furosemide is critically dependent on other residues in M1 and shows a preference toward α6-subunit-containing receptors. Overall, the evidence suggests that 15' asparagine is unlikely to form part of a binding site and probably participates in a transduction pathway.

Deeper into the GABA ion channel pore, toward the internal portal, reveals other critical residues that are conserved across all GABA$_A$R α subunits, namely, threonine at 6' and valine at 2' (Figure 18.6C). These residues play an important role in the non-competitive-mixed antagonism of GABA$_A$Rs by the sesquiterpenoid convulsant picrotoxinin and quite likely by some other ligands such as t-butyl-bicyclophosphorothionate (TBPS). Cryo-EM structures of α1β2/3γ2 receptors reveal picrotoxinin located in the ion channel bound beyond the 9' activation gate in a region 6' to 2', deep within the ion channel. This reflects a typical ion channel blocking mechanism, it is also located close to the desensitization gate (Gielen et al., 2015) which, when closed, may compromise picrotoxinin binding. Close inspection of the channel location for picrotoxinin reveals that its isopropenyl group approaches the 9' leucine ring with the exocyclic oxygen atoms engaged in H-bonding with the 6' ring residues (Masiulis et al., 2019). Picrotoxinin can also become "trapped" in the ion channel when the channel closes. However, the mode of inhibition for picrotoxinin is more complex than simple open channel block, with several studies proposing an allosteric inhibitory role. Future structural studies will no doubt cast light on this and other binding site debates.

18.8 Conclusions

A variety of subunits and GABA$_A$R isoforms, with their innate differences in physiological properties, provides a diverse signaling platform from which to influence the activity patterns of individual neurons and neural networks. Exquisite control can be exercised over excitability by expressing receptors at key points on neuronal cell surface membranes, and by controlling their subunit composition to alter receptor physiology. How these various receptor isoforms function within single cells, and their relative importance, remains a difficult task to address. Having a greater armamentarium of specific pharmacological agents will be helpful. The GABA$_A$R is sensitive to a myriad of drugs and some of these do show receptor subtype selectivity. An alternative approach to studying GABA$_A$Rs is to develop photochemical ligands. Such analytical tools can be used to explore the impact of receptor activation in membrane domains that are exposed to photons. Principally, photosensitive ligands, whose conformation can be reciprocally changed by light, can rapidly affect receptor function. In addition, inactive GABA molecules can be "uncaged" by light to map GABA$_A$R density and clustering along neuronal processes. Overall, it is evident, given the therapeutic portfolio of the GABA$_A$R, that the development of selective ligands to target-specific receptor isoforms is a laudable goal.

Suggested Reading

Araud T, Wonnacott S, & Bertrand D (2010). Associated proteins: The universal toolbox controlling ligand gated ion channel function. *Biochem Pharmacol* **80**(2), 160–169.

Brandon N, Jovanovic J, & Moss S (2002). Multiple roles of protein kinases in the modulation of gamma-aminobutyric acid(A) receptor function and cell surface expression. *Pharmacol Ther* **94**(1–2), 113–122.

Brickley SG & Mody I (2012). Extrasynaptic GABA$_A$ receptors: Their function in the CNS and implications for disease. *Neuron* **73**(1), 23–34.

Corringer PJ, Poitevin F, Prevost MS, Sauguet L, Delarue M, & Changeux JP (2012). Structure and pharmacology of pentameric receptor channels: From bacteria to brain. *Structure* **20**(6), 941–956.

Farrant M & Nusser Z (2005). Variations on an inhibitory theme: Phasic and tonic activation of GABA$_A$ receptors. *Nat Rev Neurosci* **6**(3), 215–229.

Gielen M, Thomas P, & Smart TG (2015). The desensitization gate of inhibitory Cys-loop receptors. *Nat Commun* **6**. doi: 10.1038/ncomms7829.

Hortnagl H, Tasan RO, Wieselthaler A, Kirchmair E, Sieghart W, & Sperk G (2013). Patterns of mRNA and protein expression for 12 GABAA receptor subunits in the mouse brain. *Neuroscience* **236**, 345–372.

Hosie AM, Dunne EL, Harvey RJ, & Smart TG (2003). Zinc-mediated inhibition of GABA$_A$ receptors: Discrete binding sites underlie subtype specificity. *Nat Neurosci* **6**(4), 362–369.

Hosie AM, Wilkins ME, da Silva HMA, & Smart TG (2006). Endogenous neurosteroids regulate GABA$_A$ receptors through two discrete transmembrane sites. *Nature* **444**(7118), 486–489.

Keramidas A, Moorhouse AJ, Schofield PR, & Barry PH (2004). Ligand-gated ion channels: Mechanisms underlying ion selectivity. *Prog Biophys Mol Biol* **86**(2), 161–204.

Kim JJ, Gharpure A, Teng J, Zhuang Y, Howard RJ, Zhu S, Noviello CM, Walsh RM, Lindahl E, & Hibbs RE (2020). Shared structural mechanisms of general anaesthetics and benzodiazepines. *Nature* **585**(7824), 303–308.

Kim JJ & Hibbs RE (2021). Direct structural insights into GABA$_A$ receptor pharmacology. *Trends Biochem Sci* **46**(6), 502–517.

Krishek BJ, Amato A, Connolly CN, Moss SJ, & Smart TG (1996). Proton sensitivity of the GABA$_A$ receptor is associated with the receptor subunit composition. *J Physiol* **492**(2), 431–443.

Laverty D, Desai R, Uchariski T, Masiulis S, Stec WJ, Malinauskas T, Zivanov J, Pardon E, Steyaert J, Miller KW, & Aricescu AR (2019). Cryo-EM structure of the human 132 GABA$_A$ receptor in a lipid bilayer. *Nature* **565**(7740), 516–522.

Laverty D, Thomas P, Field M, Andersen OJ, Gold MG, Biggin PC, Gielen M, & Smart TG (2017). Crystal structures of a GABA$_A$-receptor chimera reveal new endogenous neurosteroid-binding sites. *Nat Struct Mol Biol* **24**(11), 977–985.

Luscher B, Fuchs T, & Kilpatrick C (2011). GABA$_A$ receptor trafficking-mediated plasticity of inhibitory synapses. *Neuron* **70**(3), 385–409.

Masiulis S, Desai R, Uchariski T, Serna Martin I, Laverty D, Karia D, Malinauskas T, Zivanov J, Pardon E, Kotecha A, Steyaert J, Miller KW, & Aricescu AR (2019). GABA$_A$ receptor signalling mechanisms revealed by structural pharmacology. *Nature* **565**(7740), 454–461.

Miller PS, Scott S, Masiulis S, De Colibus L, Pardon E, Steyaert J, & Aricescu AR (2017). Structural basis for GABA$_A$ receptor potentiation by neurosteroids. *Nat Struct Mol Biol* **24**(11), 986–992.

Miller PS & Smart TG (2010). Binding, activation and modulation of Cys-loop receptors. *Trends Pharmacol Sci* **31**(4), 161–174.

Mortensen M, Patel B, & Smart TG (2011). GABA potency at $GABA_A$ receptors found in synaptic and extrasynaptic zones. *Front Cell Neurosci* **6**, 1–10.

Nemecz A, Prevost MS, Menny A, & Corringer PJ (2016). Emerging molecular mechanisms of signal transduction in pentameric ligand-gated ion channels. *Neuron* **90**(3), 452–470.

Olsen RW & Sieghart W (2009). $GABA_A$ receptors: Subtypes provide diversity of function and pharmacology. *Neuropharmacology* **56**(1), 141–148.

Phulera S, Zhu H, Yu J, Claxton DP, Yoder N, Yoshioka C, & Gouaux E (2018). Cryo-EM structure of the benzodiazepine-sensitive 112S tri-heteromeric $GABA_A$ receptor in complex with GABA. *eLife* **7**, e39383. doi: 10.7554/eLife.39383.

Pirker S, Schwarzer C, Wieselthaler A, Sieghart W, & Sperk G (2000). $GABA_A$ receptors: Immunocytochemical distribution of 13 subunits in the adult rat brain. *Neuroscience* **101**(4), 815–850.

Sieghart W & Sperk G (2002). Subunit composition, distribution and function of $GABA_A$ receptor subtypes. *Curr Top Med Chem* **2**(8), 795–816.

Sigel E & Steinmann ME (2012). Structure, function, and modulation of $GABA_A$ receptors. *J Biol Chem* **287**(48), 40224–40231.

Thomas P, Mortensen M, Hosie AM, & Smart TG (2005). Dynamic mobility of functional $GABA_A$ receptors at inhibitory synapses. *Nat Neurosci* **8**(7), 889–897.

Tretter V, Jacob TC, Mukherjee J, Fritschy JM, Pangalos MN, & Moss SJ (2008). The clustering of $GABA_A$ receptor subtypes at inhibitory synapses is facilitated via the direct binding of receptor α2 subunits to gephyrin. *J Neurosci* **28**(6), 1356–1365.

Tyagarajan SK & Fritschy JM (2010). $GABA_A$ receptors, gephyrin and homeostatic synaptic plasticity. *J Physiol* **588**(1), 101–106.

Wingrove PB, Wafford KA, Bain C, & Whiting PJ (1994). The modulatory action of loreclezole at the gamma-aminobutyric acid type A receptor is determined by a single amino acid in the beta 2 and beta 3 subunit. *Proc Natl Acad Sci U S A* **91**(10), 4569–4573.

Yamakura T, Bertaccini E, Trudell JR, & Harris RA (2001). Anesthetics and ion channels: Molecular models and sites of action. *Annu Rev Pharmacol Toxicol* **41**, 23–51.

19

Glycine Receptors

Josip Ivica and Lucia Sivilotti

CONTENTS

19.1 Introduction

Glycine was first identified as an inhibitory transmitter of the mammalian central nervous system in the 1960s. Glycine was later found to be also important for glutamatergic excitatory signaling via NMDA-type glutamate ion channels, which require both glutamate and glycine for their activation (see Chapter 16). The focus of this chapter is on the *inhibitory* glycine receptors (GlyRs). These are strychnine-sensitive, anion-permeable receptors belonging to the superfamily of pentameric ligand-gated ion channels (pLGIC), a group that includes GABA receptors (GABARs), muscle nicotinic acetylcholine receptors (nAChRs) and serotonin type 3 receptors (5-HT$_3$Rs) (see Chapters 18, 15 and 17, respectively). GlyRs are particularly important in the caudal areas of the nervous system, especially the spinal cord and brain stem. In man, heritable mutations of proteins found at glycinergic synapses, particularly GlyR subunits, cause startle disease/hyperekplexia, a very rare neurological disorder.

19.2 Subunit Diversity and Basic Structural Organization

The GlyR was the first pLGIC to be isolated from mammalian brain and this led directly to the cloning of the genes of the subunits that form the adult synaptic receptor: α1 and β. Five GlyR subunit genes are known in mice, α1–α4 and β (GLRA1-4 and GLRB; Figure 19.1a). In man and rat, one subunit, α4, is probably a pseudogene, because of a stop codon

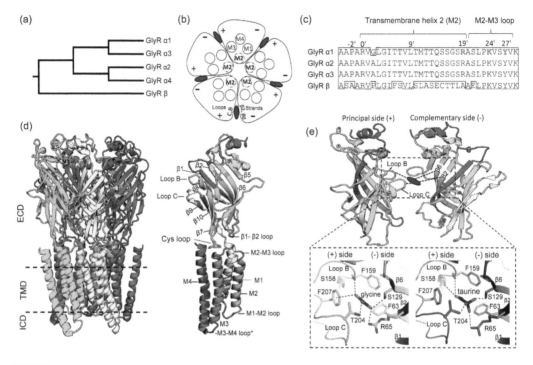

FIGURE 19.1
(a) Phylogram of the α and β GlyR subunit genes from mouse. (b) Organization of GlyR pentamers, showing the agonist binding sites at subunit interfaces (schematically shown as dark blue ovals; + and – denote the principal and complementary sides, respectively) and the transmembrane helices, with M2 lining the pore. (c) Sequence alignment of the M2 helix and the M2–M3 loop (conserved residues are enclosed in a box). M2 numbering follows the pLGIC convention, with the highly conserved L in the middle of M2 at 9′ (261 in the α1 mature subunit). The –2′ to 2′ residues are at the narrowest part of the open channel, and 2′ G (blue) is a determinant of conductance of homomeric α1 GlyR. The residues in red show the position of the earliest characterized startle disease α1 mutations. (d) Cryo-EM structure of zebrafish α1 GlyR in the apo resting state (Protein Data Bank: 6PXD). Right panel shows a single subunit with annotations of strands, loops and transmembrane helices. This structure is from constructs where the long M3–M4 loop was replaced with a short tripeptide linker. (e) ECD structures showing the position of the agonist binding site (dark blue oval). The two subunits are spaced farther apart than they are in the receptor for clarity of display. Note that strands β2 and β6 are also known as loops D and E. The expanded panel shows glycine and taurine bound to zebrafish α1 GlyR (Protein Data Bank: 6PM6 and 6PM2; residue numbering as in mature subunit, no extra spacing between the two α subunits; Yu et al., 2021).

in transmembrane helix 4 (M4; Figure 19.1b). Sequence similarity is high (~85%) across the different α subunits, especially between α1 and α3, between α2 and α4, and between species for the same subunit, but the similarity between α subunits and β is lower (~60%; for M2 see Figure 19.1c).

Additional diversity comes from alternative splicing. One notable example is in the β1–β2 loop (at the bottom of the extracellular domain; Figure 19.1d), where α2A has VT and α2B has IA in positions 58–59. These residues align to IA 51–52 in α1, where they are important for signal transduction: in α1, A52S causes a loss of function channelopathy (*spasmodic* mouse; Ryan et al., 1994). The second notable splice site is in the intracellular loop between M3 and M4, where long and short variants are known for α1 and α3, differing by 8 and 15 amino acids, respectively. The relative abundance of long and short variant transcripts and subunits depends on the species, but their functional significance is unclear.

In heterologous expression, α subunits form functional homomeric receptors very efficiently, whereas β alone cannot form functional GlyRs (Bormann et al., 1993). If α β heteromeric GlyRs are required, it is advisable to transfect an excess of β DNA constructs to reduce contamination by α homomers.

Structural evidence from heteromeric GlyRs extracted from native mammalian tissue shows a stoichiometry of 4 α to 1 β subunits (Zhu and Gouaux, 2021).

19.2.1 Structural Organization

Several homomeric GlyRs structures have been resolved, including those of zebrafish α1 GlyR (Du et al., 2015; Kumar et al., 2020; Yu et al., 2021) and of human α3 GlyR (Huang et al., 2015). These structures show the usual pLGIC features (Figure 19.1d). The N-terminal extracellular domain (ECD) is made of ten β strands connected by loops, with the highly conserved Cys-loop between β strands 6 and 7. The agonist binding sites are at the ECD subunit interfaces (Figure 19.1b, e), with the principal (+) side in the subunit anticlockwise of the pocket when the channel is viewed from above the ECD. The most important segments that contribute to the binding site are (+)-side loops B and C and (–)-side strands β2 and β6 (also called loops D and E; Figure19.1d). Other residues near the binding pocket probably stabilize its shape. The receptor was resolved in resting state when either unbound (apo) or bound to the competitive antagonist strychnine (which binds to the agonist/orthosteric site, but keeps it in an expanded state). Structures of GlyRs bound to agonists (with or without modulators) can be closed (either preopen or desensitized) or open (Du et al., 2015; Huang et al., 2017; Kumar et al., 2020; Yu et al., 2021).

The ECD is followed by the four helices of the transmembrane domain (M1–M4; Figure19.1b, d): M2 is highly conserved across α subunits and has the greatest effect on permeability and conductance because it lines the pore (Figure 19.1b, c; see Section 19.5). In the middle of M2 we find the canonical pLGIC Leu residue in both α and β subunits (for ease of comparison across subunits, M2 residues are numbered from the intracellular to the extracellular side – in this prime numbering system, the middle Leu is 9′ in all pLGICs). In the resting closed state, the M2 helices of different subunits are roughly parallel and the channel is held closed by an activation gate formed by hydrophobic residues 9′–13′ as in other pLGICs. The M2 helix tilts with its top outward to open the pore and remains tilted in the desensitized channel, where the pore is held closed by deeper M2 residues.

The loops between the transmembrane helices are short, except for the M3–M4 loop (about 80 residues), which makes up the intracellular domain (ICD). This domain is mostly intrinsically disordered and its structure is largely unresolved. The ICD is important for synaptic localization, as it mediates the interaction of GlyRs with the scaffolding protein gephyrin, which binds to the M3–M4 loop of the β subunit (Meyer et al., 1995), and it is also the substrate for GlyR regulation by phosphorylation. Drastic shortening of the ICD (common in constructs engineered for structural work) enhances channel gating (Ivica et al., 2021).

19.3 Physiological Roles

GlyRs are most abundant in the retina and in the caudal areas of the adult central nervous system (spinal cord, brain stem, medulla oblongata), where they contribute to spinal motor reflexes, respiratory rhythm regulation and sensory processing, especially in the spinal cord and in auditory pathways.

Embryonic neurons express mostly α2 GlyR subunits, probably as extrasynaptic homomeric receptors. Their activation by ambient glycine or taurine causes depolarization (because of the high intracellular chloride in the embryonic central nervous system) and therefore Ca^{+2} entry, which may be important for synaptic maturation (Kneussel and Betz, 2000).

In adults, postsynaptic GlyR are largely α1β heteromers, mediating fast inhibitory currents that are typically faster than GABAergic ones.

In the rat spinal cord, the switch occurs at two weeks after birth (Malosio et al., 1991; Singer et al., 1998), when the distinctive higher conductance and slower kinetics of α2 homomers are replaced by the smaller amplitude and shorter openings of α1β heteromeric GlyR (Figure 19.2a; Takahashi et al., 1992).

Levels of α3 also increase after birth, but this subunit remains less abundant than α1 and is distinctive in its discrete localization in the superficial layers of the spinal cord gray matter, especially lamina II, where glycinergic transmission gates pain processing and where α3 may play a role in changes in nociception brought on by inflammation (Harvey et al., 2004). In the adult central nervous system some α homomeric GlyR are found on presynaptic terminals, such as the calyx of Held (Hruskova et al., 2012; Turecek and Trussell, 2001).

β is the most widely expressed of the GlyR subunits, even where there is little or no α present. The reason for this is not clear, as β homomers are not functional (Bormann et al., 1993).

During development it is common for presynaptic terminals to release both glycine and GABA, each to act on its own receptor. Within the resulting mixed population of synaptic events, glycinergic ones can be identified by their faster kinetics and their sensitivity to low concentrations of the antagonist strychnine (Jonas et al., 1998). Another form of glycine and GABA co-release is found in auditory pathways, where the decay of glycinergic inhibitory currents is made faster than GlyR channel deactivation by the co-release of glycine and GABA, both to act on GlyRs (Lu et al., 2008). The addition of GABA, a weak partial agonist of GlyR, makes macroscopic current responses to glycine faster (and somewhat smaller) than responses to glycine alone (Figure 19.2b).

When comparing the kinetics of single-channel and synaptic events, it is important to check on the recording conditions, because GlyR or GABAR deactivation is slowed by high intracellular chloride. High chloride is commonly used in intracellular solutions to increase signal amplitude in whole-cell recordings, but it is not physiological, as in intact adult neurons intracellular chloride is in the low millimolar range. The decay of glycinergic currents is slowed by about twofold by increasing intracellular chloride from 10 to 130 mM, a concentration similar to the extracellular one (Pitt et al., 2008).

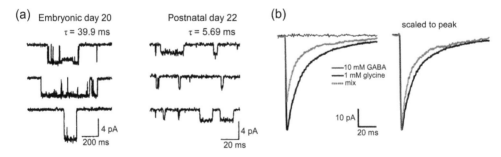

FIGURE 19.2

(a) Changes in the properties of GlyR with development. Single-channel currents elicited by 3–10 µM glycine in outside-out patches from rat spinal neurons at embryonic day 20 (left panel) and postnatal day 22 (right panel). Note the larger amplitude and the longer openings in the recordings from embryonic spinal neurons (From Takahashi et al., 1992.) (b) Co-application of the partial agonist GABA speeds the decay of glycinergic currents evoked in an outside-out patch from the rat medial nucleus of the trapezoid body by 1 ms agonist applications. Recordings are in the presence of the GABA antagonist SR-95531, 20 µM. Currents evoked by the combination of glycine and GABA (red trace) had a smaller peak and faster deactivation (clearer when responses were scaled to peak, right panel) (From Lu et al., 2008.)

19.4 Gating

GlyR isoforms containing the α1 subunit have proven ideal for single-channel kinetics work aimed at measuring agonist efficacy and understanding quantitatively channel activation by fitting mechanisms directly to the data. GlyRs have relatively high conductance that gives good signal-to-noise ratios for channel openings, high glycine concentrations do not block the open GlyR pore (as agonists do in nAChR), and it is therefore straightforward to estimate efficacy directly from the channel maximum open probability, without correcting for block.

Figure 19.3a shows another property of the single-channel activity of α1β heteromeric GlyRs that is favorable to kinetic analysis, namely, its strong dependence on agonist concentration. At low glycine concentrations, openings are short and occur either in isolation or in groups of few openings. These groups, termed "bursts," occur while agonist is bound (open probability is negligible in unbound GlyRs). The shut intervals between bursts of

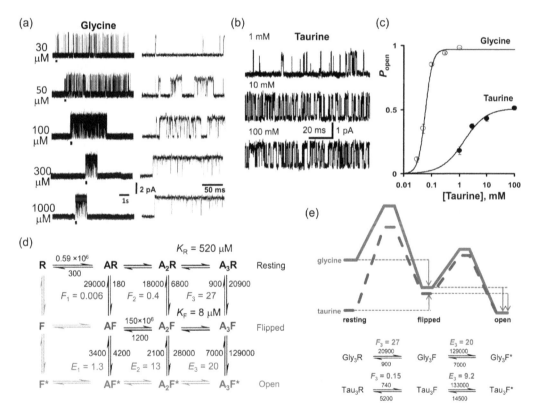

FIGURE 19.3
GlyR single-channel kinetics. (a and b) Concentration dependence of cell-attached α1β GlyR single-channel activity elicited by glycine or taurine. Recordings are from HEK293 transfected with α1 and β GlyR subunits, nominal holding potential +100 mV, opening upward. (c) Open probability–concentration curves from the records in (a) and (b) fitted with the Hill equation. (d) Detailed "flip" activation mechanism fitted to single-channel glycine data for α1β GlyR with the rate constant and equilibrium constant values for each step. Units for rate constants units are s⁻¹ or M⁻¹s⁻¹. Gray arrows denote steps where rate constants could not be estimated. (e) Energy diagrams drawn from the rate and equilibrium constants values (bottom of panel) for GlyRs fully bound to three molecules of either glycine (red) or taurine (blue), aligned at their open state. (Data from Burzomato et al., 2004, and Lape et al., 2008.)

openings reflect the time that it takes for the channel to rebind agonist and they become shorter as agonist concentration increases. Bursts then merge into clusters of openings, which are isolated by long, desensitized shut times and represent the activity of a single-channel molecule. In Figure 19.3a, clusters can be defined from 30 μM, and their open probability reaches a maximum at about 300 μM (see the concentration–open probability curve in Figure 19.3c). This maximum is high (96%) for glycine, which is a full agonist that, when bound, keeps the channel either open or desensitized. A similar behavior is produced by taurine, but taurine is a partial agonist and the maximum cluster open probability is much lower, at 54% (Figure 19.3b, c).

GlyRs open more effectively as more agonist molecules bind. α1 homomeric and α1β heteromeric channels reach their maximum open probability when three of the five possible binding sites are occupied. A detailed description of GlyR activation is given by the "flip" mechanism (Figure 19.3d; Burzomato et al., 2004) where an extra set of shut preopening intermediate states, the "flipped" F states, link the resting and the open states and bind to the agonist with higher affinity than the resting states. Entry into the intermediate states is an important determinant of agonist efficacy: the low efficacy of taurine when compared with glycine is due to its poor ability to drive the channel into these intermediate states (Lape et al., 2008). In the energy diagram in Figure 19.3e, this is shown by the height of the barriers (activation energy) and by the equilibrium positions (levels of the horizontal bars) for the two agonists compared to the different functional states. Once the intermediate is reached, the energy barrier to opening is fairly similar for the two agonists, as shown by the values of the rate and equilibrium constants for GlyRs fully bound to glycine or taurine: the main difference between the two agonists is in the value of the forward equilibrium constants for flipping.

Possible structures of intermediate states have been described in zebrafish GlyRs (Yu et al., 2021; Ivica et al. 2022): in the presence of high concentrations of the full agonists glycine and amino methanesulfonate, single-particle cryo-EM detects only two channel states (open or desensitized, as in the single-channel records), but in the presence of partial agonists, such as taurine or GABA, the agonist-bound GlyR populates an additional structural class, where the binding site has tightened onto the agonist, but the channel is still closed. This may correspond to the shut times within the clusters of openings evoked by partial agonists (see Figures 19.3b and 19.4b). In GlyRs, fitting single-channel recordings with the detailed scheme in Figure 19.3d suggested that the flip state is short (less than 10 μs), but it is possible that this very brief state is just the last of several intermediates before opening, and that these have longer lifetimes (as suggested by their detectability in structural work), but are grouped together with the resting shut state by the kinetic analysis.

Evidence for preopening intermediates have also been identified in other GlyR subtypes, and in other pLGICs such as ELIC, nACh, 5-HT3 and GABA$_A$ receptors.

19.5 Ion Permeability and Conductance

GlyR are permeable to chloride and bicarbonate anions, and have high selectivity for anions over cations. In GlyR from spinal neurons in culture (probably heteromeric), the potassium-to-chloride permeability ratio is lower than 0.05 (Bormann, 1987).

The *permeability* sequence is SCN$^-$ > NO$_3^-$ > I$^-$ > Br$^-$ > Cl$^-$ > F$^-$, an order that follows the decreasing size of the hydrated anions. In measurements of GlyR *conductance* (with one species of permeant anion), the sequence is the inverse of the permeability sequence, Cl⁻ > Br⁻ > I⁻ >NO_3^- > SCN⁻, suggesting that permeant anions bind in the channel, probably at two binding sites (Bormann et al., 1987). Note that when the concentration of permeant ions is high, their binding in the pore can slow channel kinetics (see Section 19.3). The largest organic anion permeant through GlyRs is acetate, suggesting a pore diameter of ~5.2 Å (Bormann et al., 1987). In symmetrical chloride, GlyR conductance is a hyperbolic function of chloride concentration, with a half-maximum conductance at 108 mM chloride and a maximum reached only at concentrations above the physiological chloride extracellular level (Bormann et al., 1987).

As in other pLGICs, the main conductance determinants are in the M2 helix, with some contribution by the extracellular vestibule. The narrowest part of the open channel pore is the stretch from –2' to 2' of the M2 helix (Figure 19.1c). Note that the structure of the open GlyR pore has been controversial. Early structures were too "wide open" and it may be that a physiologically relevant open pore structure (rather than a "wide open" structure) can be obtained only when GlyRs are imaged in the presence of lipids (Yu et al., 2021).

A key feature of single GlyR channels is that homomeric and heteromeric receptors differ in their conductance, which is much larger for homomers, e.g., 86–111 pS in 145 mM symmetrical chloride, compared to ~50 pS for heteromers (Bormann et al., 1993). This stems from differences in the M2 of α and β (Figure 19.1c). Among homomeric GlyR, the main conductance is somewhat higher for α2 and α3 (111 and 105 pS, respectively, versus 86 pS for α1), because of a residue difference in 2' M2 (Ala and Gly, respectively; Figure 19.1c; Bormann et al., 1993).

GlyR single-channel openings show multiple subconductances. For reasons that are not clear, subconductances are most common in homomeric channels and in outside-out recordings, and are scarce in cell-attached recordings of heteromeric GlyRs (Bormann et al., 1987; Bormann et al., 1993; Beato et al. 2004; Burzomato et al., 2004)

19.6 Pharmacology

The pharmacology of GlyRs is relatively simple, and no drug that acts primarily on GlyRs is in therapeutic use.

The best GlyR agonist is the neurotransmitter itself, glycine, which is a full agonist on all isoforms and elicits a maximum open probability close to 100% when the channel is not desensitized. Several small amino acids are also GlyR agonists. Efficacy, high for amino methanesulfonate, L-alanine, L-serine and β-alanine, decreases sharply in larger molecules, and taurine and GABA are partial agonists, as shown by their lower maximum open probability and the pronounced shut time within clusters (Figure 19.4b, c).

The interface (orthosteric) binding pocket of GlyR has typical pLGIC features, including the aromatic box. Figure 19.1e shows the binding poses of glycine and taurine in a zebrafish homomeric GlyR structure viewed from above, highlighting the contributions of loops B and C from the principal (+) subunit, and β strands 2 (loop D) and 6 (loop E) in the complementary (–) subunit (Yu et al., 2021). This picture is very similar to that from α3 GlyR structures (Huang et al., 2017), with a network of hydrogen bonds and cation-π interactions between agonist and receptor. A feature that is consistent across agonists is the ionic bond between the side chain of a conserved β2/loop D Arg (R65 in Figure 19.1e) and the negatively charged moiety of the agonist molecule.

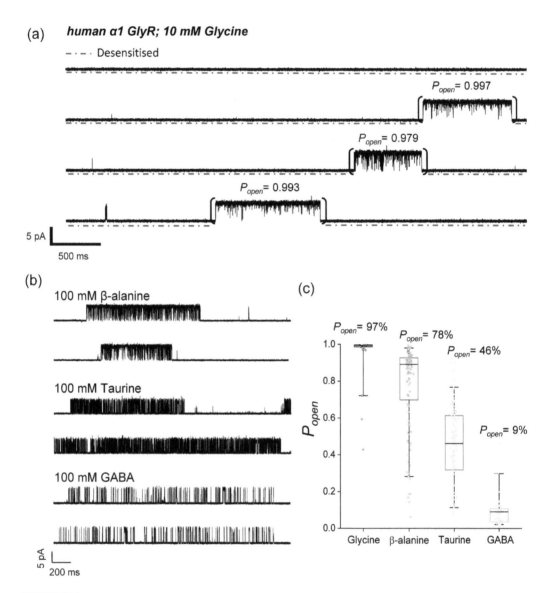

FIGURE 19.4

Agonists of GlyR and their efficacy. (a) This cell-attached single-channel recording from human α1 GlyR in the presence of 10 mM glycine shows how clusters are selected. (b) Representative single-channel currents from human α1 GlyR activated by 100 mM β-alanine, taurine and GABA. Currents were recorded at +100 mV from transfected HEK293 cells, openings are upward. (c) Boxplot showing maximum open probability elicited by the different agonists on the human α1 GlyR (one point per cluster). Boxes show 25–75 percentiles; black line represents median and whiskers 5–95 percentiles. (Data from Ivica et al., 2021.)

Distinguishing receptors containing different forms of α subunits because of their pharmacology is not straightforward. Across GlyR subtypes, the rank order of *potency* of the different agonists is not sufficiently different to help in receptor classification. There are some differences in the *efficacy* of agonists on different GlyR (cf. human and zebrafish α1 or the 2A/2B splice variants of α2), but these seem to be similar across agonists and are likely due to differences in the general propensity of the channel isoform to open (i.e., its allosteric constant; Ivica et al., 2021; Miller et al., 2004).

An important note of caution: Many commercially supplied amino acid agonists are contaminated by glycine, even at the highest purity available, and if high agonist concentrations must be used, they should be purified further.

The classical competitive antagonist, strychnine, is potent (affinity below 20 nM) and selective for GlyRs versus other channels, but again it not useful in classifying GlyRs.

Picrotoxin (an equimolar mixture of picrotin and picrotoxinin) binds in the pore of GlyRs and is much more potent in blocking homomeric GlyRs than heteromeric GlyRs (about twentyfold against EC_{50} concentrations of glycine) and is therefore useful to check whether heteromeric assembly is complete in heterologous expression work.

α2 is relatively resistant to the pore blocker cyanophenyltriborate, but the slow kinetics of GlyRs containing this subunit is probably more distinctive.

GlyR activity can be modulated by a range of agents, some of which are endogenous (e.g., zinc, extracellular pH and cannabinoids). Modulation by zinc has been studied in depth and may be of physiological importance. Nanomolar zinc concentrations enhance submaximal GlyR responses, but micromolar zinc concentrations inhibit GlyRs, especially those that contain α1. The inhibition and potentiation sites of zinc in the ECD are different, and additional residues that transduce the effects of zinc have been identified (Miller et al., 2008). The potential of positive allosteric modulators of GlyR as analgesics is as yet unrealized, despite promising results with a lead compound in rodent models (Huang et al., 2017).

As other pLGICs, GlyR are modulated by ethanol and by general anesthetics, inhalational or intravenous, which bind in the transmembrane domain, but GlyRs are unlikely to be the primary target of the action of these drugs. The antiparasitic drug ivermectin also binds to the transmembrane domain of GlyRs: it both enhances the effect of glycine and activates the GlyR by itself, more or less irreversibly.

19.7 Trafficking, Clustering and Regulation

Effective synaptic transmission requires channels to be clustered in a precise postsynaptic location. For the GlyR, this process requires the presence of gephyrin, a microtubule-associated protein (Dumoulin et al., 2009), which connects the intracellular loop of the β subunit with the postsynaptic cytoskeleton. Gephyrin molecules interact with each other to form a hexagonal lattice of subsynaptic binding sites for GlyRs (Meyer et al., 1995). Homomeric α GlyRs diffuse more quickly in the membrane because they cannot associate with gephyrin (Meier et al., 2001). Gephyrin may also enhance GlyR trafficking to the membrane, by associating with channels as they are assembled in the endoplasmic reticulum (Maas et al., 2006).

Phosphorylation consensus sequences are found in the M3–M4 intracellular loop of GlyR. For instance, the α1 subunit has a protein kinase C (PKC) consensus site and – in its long isoform – also a protein kinase A (PKA) site. Both enhancement and depression of function have been attributed to phosphorylation, depending on the origin of the receptor (recombinant or native), and its subunit composition and phosphorylation may play a role in changes in nociception brought on by inflammation (Harvey et al., 2004). β subunit phosphorylation may depress the GlyR–gephyrin interaction and thus increase receptor diffusion and decrease postsynaptic density (Specht et al., 2011).

19.8 Channelopathies and Disease

GlyRs are notable for the number of inherited mutations that cause the human channelopathy hyperekplexia (startle disease). This rare disease affects only a few hundred patients worldwide, but the multiple locations of the startle mutations, their molecular phenotype and the existence of startle disease in mouse strains gave early insights into the function of the different pLGIC domains.

Inherited major hyperekplexia (OMIM #149400) typically starts at birth, with generalized stiffness and an exaggerated motor response to non-noxious sensory stimuli (tapping the patient's nose). Stiffness wanes early in life, but the excessive startle response persists and is followed by episodic stiffness, commonly causing falls. This classical form of the disease is autosomal dominant, is not associated with neurodevelopmental problems and responds to treatment with the benzodiazepine clonazepam.

Linkage of hyperekplexia to the GLRA1 gene was quickly followed by the identification of the first startle mutation, a missense mutation at 19′ in M2 in the α1 subunit (Shiang et al., 1993). In recombinant GlyRs, this mutation, R271L or Q (numbering as in the mature protein, without the signal peptide, red residues in the M2 alignment in Figure 19.1c) causes a complex loss of function, with both a reduction in single-channel conductance and a typical molecular "startle phenotype". The latter consists of a decrease in the potency of glycine and a profound loss in the efficacy of partial agonists, and strongly suggests impairment of channel gating. Other startle mutations in the M2–M3 loop that were characterized soon after that (24′ K276E and 27′ Y279C; Figure 19.1c) do not change conductance, but otherwise resemble R271L/Q in producing the "startle phenotype" and identified the M2-M3 loop as a key area for signal transduction in the superfamily (Lewis et al., 1998; Lynch et al., 1997).

More than two decades of work have not exhausted the topic and confirmed that most startle disease mutations are in α1, but a few are found in β (GLRB) and in other proteins important to the glycinergic synapse, such as the neuronal glycine transporter GlyT2 (SCLC6A5), or scaffolding proteins such as gephyrin (GPHN) and collybistin (ARHGEF9). Of the 45 GLRA1 missense mutations known to date, about half are autosomal dominant (mostly those in the transmembrane domain). All but one of the seven missense mutations in β are inherited in a recessive pattern. Recessive inheritance (or compound heterozygosity) is overall the most common pattern and is the rule for nonsense mutations or for deletions in GlyR subunits, both α (13) or β (11), and for mutations in other glycinergic proteins. Recessively inherited mutations often have an "extended phenotype", with developmental delays and episodic apnea accompanying the startle reaction. It is unclear whether this difference in the patient phenotype is due to a difference in the molecular phenotype or to the association between recessive inheritance and parents' consanguinity.

When startle mutations are expressed in recombinant GlyRs, most missense mutations cause a loss of function, which in whole-cell recordings leads to a decrease in the potency of glycine and in the maximum response to partial agonists relative to glycine, although a few mutations cause gain of function and spontaneous GlyR channel openings.

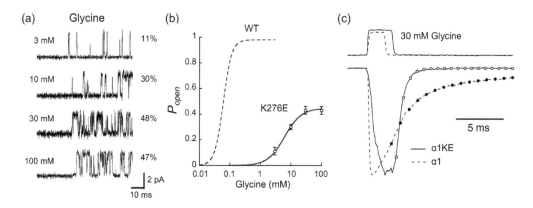

FIGURE 19.5

Effects of a startle mutation in the M2–M3 loop on GlyR function, (a) Cell-attached single-channel recordings from rat K276Eα1β GlyR activated by glycine. (b) Open probability–concentration curve for K276Eα1β GlyR from recordings in (a); note the marked decrease in maximum open probability and in glycine potency. (c) Example of inward currents elicited by 30 mM glycine pulses to outside-out patches from HEK 293 cells expressing wild-type α1 (dashed) or α1 K276E mutant (solid); averages of at least five responses (From Lape et al., 2012.)

Relatively few mutants have been investigated in conditions relevant to those at the synapse, where glycine reaches 2.2–3.5 mM concentrations (Beato, 2008). The examples in Figure 19.5a, b shows that the startle mutation α1K276E (position 24′ in the M2–M3 loop; see Figure 19.1c) reduces the glycine maximum open probability of heteromeric GlyRs. Clusters evoked by glycine in the mutant look like those of a partial agonist, with a maximum open probability of about 45%, which is less than half of wild type, and is reached at much higher glycine concentrations. When glycine is applied in quasi-synaptic conditions with a piezo stepper to outside-out patches (Figure 19.5c), the current response in the mutant is slower in rising and faster in decaying than in the wild-type receptor (note that the activation of wild-type GlyR is so fast that the rate of rise of currents reflects solution exchange, rather than receptor activation, even with a fast exchange of the order of 0.15–0.2 ms). The traces in Figure 19.5c are normalized to their peak and do not show what is an even more important effect of this startle mutation, namely, that it reduces by about 99% the peak of the synaptic current (this is calculated from the activation mechanism obtained from single-channel data and data on the time course of glycine concentration at the synapse; Lape et al., 2012).

19.9 Conclusions

GlyRs are anion permeable and their activation causes inhibitory effects at the synapse, where they are aggregated by the scaffolding protein gephyrin. GlyRs remain largely unexploited in terms of human therapeutics, but their high conductance and the availability of agonists with different efficacy have made them a useful subject for investigating channel activation and the basis of agonist efficacy with single-molecule electrophysiology, and for coupling this work to structural investigations. Such work could help our understanding of all pLGICs.

Suggested Reading

Beato, M. 2008. "The time course of transmitter at glycinergic synapses onto motoneurons." *J. Neurosci.* 28(29):7412–25. doi: 10.1523/JNEUROSCI.0581-08.2008.

Beato, M., P. J. Groot-Kormelink, D. Colquhoun, and L. G. Sivilotti. 2004. "The activation mechanism of α1 homomeric glycine receptors." *J. Neurosci.* 24(4):895–906. doi: 10.1523/JNEUROSCI.4420-03.2004.

Bormann, J., O. P. Hamill, and B. Sakmann. 1987. "Mechanism of anion permeation through channels gated by glycine and gamma-aminobutyric acid in mouse cultured spinal neurones." *J. Physiol.* 385:243–86. doi: 10.1113/jphysiol.1987.sp016493.

Bormann, J., N. Rundstrom, H. Betz, and D. Langosch. 1993. "Residues within transmembrane segment M2 determine chloride conductance of glycine receptor homo- and hetero-oligomers." *EMBO J.* 12(10):3729–37.

Burzomato, V., M. Beato, P. J. Groot-Kormelink, D. Colquhoun, and L. G. Sivilotti. 2004. "Single-channel behavior of heteromeric α1β glycine receptors: An attempt to detect a conformational change before the channel opens." *J. Neurosci.* 24(48):10924–40. doi: 10.1523/JNEUROSCI.3424-04.2004.

Du, J., W. Lu, S. Wu, Y. Cheng, and E. Gouaux. 2015. "Glycine receptor mechanism elucidated by electron cryo-microscopy." *Nature* 526(7572):224–9. doi: 10.1038/nature14853.

Dumoulin, A., A. Triller, and M. Kneussel. 2009. "Cellular transport and membrane dynamics of the glycine receptor." *Front. Mol. Neurosci.* 2:28. doi: 10.3389/neuro.02.028.2009.

Harvey, R. J., U. B. Depner, H. Wassle, S. Ahmadi, C. Heindl, H. Reinold, T. G. Smart, K. Harvey, B. Schutz, O. M. Abo-Salem, A. Zimmer, P. Poisbeau, D. Welzl, D. P. Wolfer, H. Betz, H. U. Zeilhofer, and U. Muller. 2004. "GlyR α3: An essential target for spinal PGE2-mediated inflammatory pain sensitization." *Science* 304(5672):884–7. doi: 10.1126/science.1094925.

Hruskova, B., J. Trojanova, A. Kulik, M. Kralikova, K. Pysanenko, Z. Bures, J. Syka, L. O. Trussell, and R. Turecek. 2012. "Differential distribution of glycine receptor subtypes at the rat calyx of Held synapse." *J. Neurosci.* 32(47):17012–24. doi: 10.1523/JNEUROSCI.1547-12.2012.

Huang, X., H. Chen, K. Michelsen, S. Schneider, and P. L. Shaffer. 2015. "Crystal structure of human glycine receptor-α3 bound to antagonist strychnine." *Nature* 526(7572):277–80. doi: 10.1038/nature14972.

Huang, X., P. L. Shaffer, S. Ayube, H. Bregman, H. Chen, S. G. Lehto, J. A. Luther, D. J. Matson, S. I. McDonough, K. Michelsen, M. H. Plant, S. Schneider, J. R. Simard, Y. Teffera, S. Yi, M. Zhang, E. F. DiMauro, and J. Gingras. 2017. "Crystal structures of human glycine receptor α3 bound to a novel class of analgesic potentiators." *Nat. Struct. Mol. Biol.* 24(2):108–13. doi: 10.1038/nsmb.3329.

Ivica, J., R. Lape, V. Jazbec, J. Yu, H. Zhu, E. Gouaux, M. G. Gold, and L. G. Sivilotti. 2021. "The intracellular domain of homomeric glycine receptors modulates agonist efficacy." *J. Biol. Chem.* 296:100387. doi: 10.1074/jbc.RA119.012358.

Ivica, J., H. Zhu, R. Lape, E. Gouaux, and L. G. Sivilotti. 2022. "Aminomethanesulfonic acid illuminates the boundary between full and partial agonists of the pentameric glycine receptor." *elife* 11:e79148:1–19. DOI: https://doi.org/10.7554/eLife.79148

Jonas, P., J. Bischofberger, and J. Sandkuhler. 1998. "Corelease of two fast neurotransmitters at a central synapse." *Science* 281(5375):419–24. doi: 10.1126/science.281.5375.419.

Kneussel, M., and H. Betz. 2000. "Clustering of inhibitory neurotransmitter receptors at developing postsynaptic sites: The membrane activation model." *Trends Neurosci.* 23(9):429–35. doi: 10.1016/s0166-2236(00)01627-1.

Kumar, A., S. Basak, S. Rao, Y. Gicheru, M. L. Mayer, M. S. P. Sansom, and S. Chakrapani. 2020. "Mechanisms of activation and desensitization of full-length glycine receptor in lipid nanodiscs." *Nat. Commun.* 11(1):3752. doi: 10.1038/s41467-020-17364-5.

Lape, R., D. Colquhoun, and L. G. Sivilotti. 2008. "On the nature of partial agonism in the nicotinic receptor superfamily." *Nature* 454(7205):722–7. doi: 10.1038/nature07139.

Lape, R., A. J. Plested, M. Moroni, D. Colquhoun, and L. G. Sivilotti. 2012. "The 1K276E startle disease mutation reveals multiple intermediate states in the gating of glycine receptors." *J. Neurosci.* 32(4):1336–52. doi: 10.1523/JNEUROSCI.4346-11.2012.

Lewis, T. M., L. G. Sivilotti, D. Colquhoun, R. M. Gardiner, R. Schoepfer, and M. Rees. 1998. "Properties of human glycine receptors containing the hyperekplexia mutation α1(K276E), expressed in *Xenopus* oocytes." *J. Physiol.* 507(1):25–40. doi: 10.1111/j.1469-7793.1998.025bu.x.

Lu, T., M. E. Rubio, and L. O. Trussell. 2008. "Glycinergic transmission shaped by the corelease of GABA in a mammalian auditory synapse." *Neuron* 57(4):524–35. doi: 10.1016/j.neuron.2007.12.010.

Lynch, J. W., S. Rajendra, K. D. Pierce, C. A. Handford, P. H. Barry, and P. R. Schofield. 1997. "Identification of intracellular and extracellular domains mediating signal transduction in the inhibitory glycine receptor chloride channel." *EMBO J.* 16(1):110–20. doi: 10.1093/emboj/16.1.110.

Maas, C., N. Tagnaouti, S. Loebrich, B. Behrend, C. Lappe-Siefke, and M. Kneussel. 2006. "Neuronal cotransport of glycine receptor and the scaffold protein gephyrin." *J. Cell Biol.* 172(3):441–51. doi: 10.1083/jcb.200506066.

Malosio, M. L., B. Marqueze-Pouey, J. Kuhse, and H. Betz. 1991. "Widespread expression of glycine receptor subunit mRNAs in the adult and developing rat brain." *EMBO J.* 10(9):2401–9.

Meier, J., C. Vannier, A. Serge, A. Triller, and D. Choquet. 2001. "Fast and reversible trapping of surface glycine receptors by gephyrin." *Nat. Neurosci.* 4(3):253–60. doi: 10.1038/85099.

Meyer, G., J. Kirsch, H. Betz, and D. Langosch. 1995. "Identification of a gephyrin binding motif on the glycine receptor beta subunit." *Neuron* 15(3):563–72. doi: 10.1016/0896-6273(95)90145-0.

Miller, P. S., R. J. Harvey, and T. G. Smart. 2004. "Differential agonist sensitivity of glycine receptor α2 subunit splice variants." *Br. J. Pharmacol.* 143(1):19–26. doi: 10.1038/sj.bjp.0705875.

Miller, P. S., M. Topf, and T. G. Smart. 2008. "Mapping a molecular link between allosteric inhibition and activation of the glycine receptor." *Nat. Struct. Mol. Biol.* 15(10):1084–93. doi: 10.1038/nsmb.1492.

Pitt, S. J., L. G. Sivilotti, and M. Beato. 2008. "High intracellular chloride slows the decay of glycinergic currents." *J. Neurosci.* 28(45):11454–67. doi: 10.1523/JNEUROSCI.3890-08.2008.

Ryan, S. G., M. S. Buckwalter, J. W. Lynch, C. A. Handford, L. Segura, R. Shiang, J. J. Wasmuth, S. A. Camper, P. Schofield, and P. O'Connell. 1994. "A missense mutation in the gene encoding the α1 subunit of the inhibitory glycine receptor in the spasmodic mouse." *Nat. Genet.* 7(2):131–5. doi: 10.1038/ng0694-131.

Shiang, R., S. G. Ryan, Y. Z. Zhu, A. F. Hahn, P. Oconnell, and J. J. Wasmuth. 1993. "Mutations in the α-1 subunit of the inhibitory glycine receptor cause the dominant neurologic disorder, hyperekplexia." *Nat. Genet.* 5(4):351–8. doi: 10.1038/ng1293-351.

Singer, J. H., E. M. Talley, D. A. Bayliss, and A. J. Berger. 1998. "Development of glycinergic synaptic transmission to rat brain stem motoneurons." *J. Neurophysiol.* 80(5):2608–20. doi: 10.1152/jn.1998.80.5.2608.

Specht, C. G., N. Grunewald, O. Pascual, N. Rostgaard, G. Schwarz, and A. Triller. 2011. "Regulation of glycine receptor diffusion properties and gephyrin interactions by protein kinase C." *EMBO J.* 30(18):3842–53. doi: 10.1038/emboj.2011.276.

Takahashi, T., A. Momiyama, K. Hirai, F. Hishinuma, and H. Akagi. 1992. "Functional correlation of fetal and adult forms of glycine receptors with developmental changes in inhibitory synaptic receptor channels." *Neuron* 9(6):1155–61. doi: 10.1016/0896-6273(92)90073-m.

Turecek, R., and L. O. Trussell. 2001. "Presynaptic glycine receptors enhance transmitter release at a mammalian central synapse." *Nature* 411(6837):587–90. doi: 10.1038/35079084.

Yu, J., H. Zhu, R. Lape, T. Greiner, J. Du, W. Lu, L. Sivilotti, and E. Gouaux. 2021. "Mechanism of gating and partial agonist action in the glycine receptor." *Cell* 184(4):957–968.e21. doi: 10.1016/j.cell.2021.01.026.

Zhu, H., and E. Gouaux. 2021. "Architecture and assembly mechanism of native glycine receptors." *Nature* 599(7885):513–7. doi: 10.1038/s41586-021-04022-z.

20

Acid-Sensing Ion Channels

Yangyu Wu and Cecilia M. Canessa

CONTENTS

20.1 Introduction

Acid-sensing ion channels (ASICs) are proton-activated ionotropic receptors that belong to the epithelial Na channel/degenerins (ENaC/DEG) family of ion channels. All members of this class share a common protein structure consisting of three identical or homologous pore-forming subunits. In spite of a highly similar structure, DEG/ENaC channels differ markedly in their functional roles and means of activation; some are constitutively active (ENaC), while others are gated by protons (ASIC), peptides (FaNaCh), hormones (PPK23 and PPK29 respond to pheromones in *Drosophila*), osmolality (water taste in *Drosophila*), mechanical forces (touch receptor in *Caenorhabditis elegans*), and in many more instances the agonists are not yet known.

There are four ASIC genes in the mammalian genome, ACCN1–ACCN4. By differential splicing, the four genes *ACCN1–ACCN4* give rise to seven distinct proteins. ASIC1a and ASIC1b are spliced variants of the *ACCN2* gene. Similarly, ASIC2a and ASIC2b are spliced forms of the *ACCN1* (Figure 20.1). Homotrimers of ASIC2a form proton-activated channels but those of ASIC2b do so only when ASIC2b associates with other subunits to form heteromeric channels. Splicing of the *ACCN3* gene produces two isoforms that vary in the distal intracellular carboxy terminus; all three ASIC3 proteins produce channels with identical functional properties. The single product of ACCN4 is the subunit ASIC4, which does not produce proton-activated currents when expressed as a homotrimer.

DOI: 10.1201/9781003096276-20

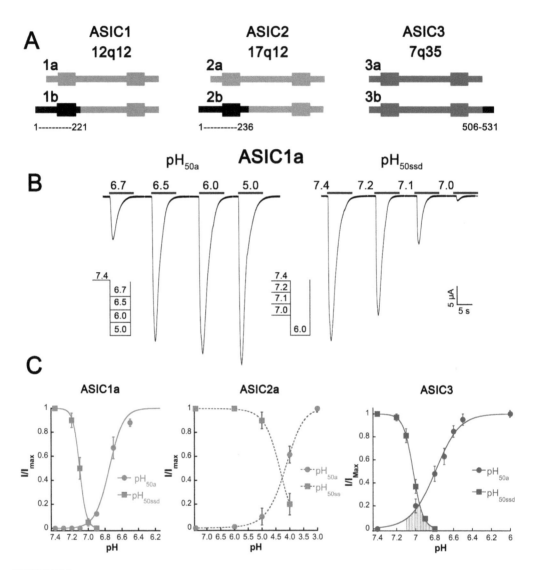

FIGURE 20.1
(A) ACCN genes and their localization in human chromosomes. ASIC proteins and their spliced isoforms are represented with transmembrane domains shown as squares: TM1 and TM2. Numbers are the amino acid positions of the spliced exons. (B) Representative examples of current traces of homomeric human ASIC1a evoked by concentrations of external pH from 6.7 to 5.0 administered from a baseline pH of 7.4 to calculate the apparent proton affinity for channel activation (pH_{50a}). Currents evoked by pH 6.0 while the concentration of protons of the conditioning solution was increased by small steps (0.1 pH units) from 7.4 to 7.0 to calculate the apparent proton affinity of the channel's steady-state desensitization (pH_{50ssd}). (C) Plots of apparent pH_{50a} and pH_{50ssd} of hASIC1a, hASIC2a and mASIC3 homomeric channels.

20.2 Tissue Distribution

Functional expression of proton-activated currents mediated by ASIC channels is for the most part limited to neurons and glial cells. The central nervous system expresses all ASIC transcripts except for ASIC1b, which is present only in peripheral neurons. The most abundant subunit in the brain is ASIC1a that forms homomeric channels in most areas of the brain but also heteromeric channels in association with ASIC2a. ASIC1a seems to be essential for the expression of H^+-mediated currents in the brain, as they disappear when ASIC1a is inactivated in mice (Wemmie et al., 2002). Low abundance of ASIC transcripts or proteins have been found in a few peripheral tissues, but robust functional activity has not been detected outside the nervous system. Examples of expression outside the nervous system are ASIC3 in the lung, and ASIC1, ASIC2, and ASIC3 in bone.

Peripheral neurons express all the ASIC isoforms in high abundance; any given neuron contains several types of ASIC subunits that associate in various combinations giving rise to a diversity of channels with different properties and pharmacological responses. This pattern of expression suggest that ASICs are involved in detection of various sensory modalities, for example, activation of ASIC1 and ASIC3 elicit pain.

Within neurons, ASICs reside primarily on the cell surface and not in intracellular organelles. On the plasma membrane, the channels distribute over somata, axon, and dendrites, including postsynaptic densities. ASIC2 also localizes in specialized nerve terminals that form mechanosensitive structures in the skin such as Meissner and Pacinian corpuscles and in hair follicles. The broad expression pattern of ASICs across structures of the brain, and within neurons – it covers most areas of the plasma membrane – is consistent with a functional role of detecting increases in proton concentration around neurons, whereas its presence in specialized structures of the skin suggests additional roles processing various sensory modalities such as nociception and mechanosensation.

20.3 Function

Almost every type of neuron examined to date, independent of its location (brain, spinal cord, dorsal root ganglia) or specialization (excitatory, inhibitory, sympathetic or sensory receptor), expresses at least one type of ASIC protein. The magnitude of the proton-evoked current, however, varies across structures of the nervous system being largest in some sensory neurons. The almost ubiquitous expression of ASIC in central and peripheral nervous systems and the high degree of sequence conservation across vertebrates and invertebrates – spanning millions of years of evolutionary distance – underscore the importance of these channels in the nervous system. However, despite significant progress made in recent years, the biological functions of many ASIC channels remain mostly unknown.

To date no genetic disorders associated with mutations of ASICs have been reported. The present understanding of the physiological and pathological roles of ASICs has been derived chiefly from studies of phenotypes displayed by mice with gene inactivation of ASIC1a, ASIC2a/b or ASIC3. In contrast, there is a dearth of information on the consequences of direct activation of ASIC in the central nervous system (CNS), in large part owing to technical difficulties of changing the external pH in selected areas of the brain. The other difficulty lies on obtaining accurate spatial and temporal measurements of external pH in microdomains of the CNS. This remains a challenge despite improvements in the sensitivity and time resolution of H^+-selective microelectrodes. The large size, invasiveness and relative low temporal response of H^+-selective microelectrodes in comparison to the high-frequency range of neuronal activity are all drawbacks. pH-sensitive fluorescent indicators might offer some advantages, though at present, they require large and complex detection apparatus not easily adapted for in vivo measurements in the mammalian brain.

All ASIC knockout mice exhibit normal gross morphology, fertility and no noticeable aberrant behaviors unless examined under specific settings that uncover subtle defects. The first genetic inactivation of ASIC1a in mice showed absent or markedly diminished proton-evoked currents in all brain structures of the knockout mouse (ASIC1$^{-/-}$), confirming that ASIC1a is the main subunit forming both homomeric and heteromeric ASIC channels in the brain. There is evidence that ASICs modulate synaptic plasticity, including long-term potentiation (LTP) and long-term depression (LTD). Excitatory potentials and NMDA receptor activation during high-frequency stimulation are both reduced in ASIC1$^{-/-}$ mice. Furthermore, ASIC1$^{-/-}$ animals display deficits in hippocampus-dependent spatial learning, cerebellum-dependent eye blinking and amygdala-dependent fear conditioning. However, some of those findings were challenged by results from groups that generated mice with brain-specific inactivation of ASIC1a. The complexity of ASIC function may be the result of differential distribution and function in excitatory and inhibitory neurons, variations in ASIC subunit compositions and expression level of regulatory partners in different brain structures. Further studies are needed to elucidate the still contentious and elusive role of ASICs in modulating synaptic plasticity.

The close structural similarity of ASICs with the degenerins, some of which are ion channels that participate in mechanotransduction in the nematode *C. elegans*, suggested that at least a few members of the ASICs might also transduce mechanical stimuli in vertebrates. Early studies reported abnormal touch and baroreceptor reflexes in ASIC2$^{-/-}$ mice; however, the original studies have not been reproduced. ASIC2$^{-/-}$ mice, however, are not normal; in early age, they exhibit enhanced electroretinograms and with aging the retina degenerates suggesting that ASIC2 and also ASIC3 may be negative modulators of rod phototransduction, and may have a protective role in maintaining integrity of the retina.

Studies of ASICs in peripheral neurons pose fewer challenges than in the CNS, as sensory neuron terminals are accessible to the application of acid solutions, inflammatory agents or mechanical stimuli. Exposure to nociceptors to acid solutions and the snake venom of a coral snake produces severe and long-lasting pain that originates from sustained activation of ASIC1a.

20.4 Agonists, Modulators and Inhibitors

20.4.1 Agonists and Modulators

ASICs are activated by external protons, but the question remains of whether additional endogenous stimuli can also gate these channels. Several lines of evidence support such a possibility. First came the realization that some members of the ENaC/DEG superfamily are ionotropic receptors activated by neuropeptides. FMRFamide Na channel (FaNaCh) is an ion channel activated by the neuropeptide FMRFamide in neurons from several snails. Another peptide-gated channel is hydra Na channel (HyNac) from the freshwater polyp hydra that is gated by the neuropeptides hydra-RFamides I and II.

Evidence relevant to mammalian ASICs has come from the isolation of a toxin of the Texas coral snake (*Micrurus tener tener*) whose bite produces excruciating pain (Bohlen et al., 2011). The toxin, named MitTx-α/β, consists of two separate polypeptides: a Kunitz-like protein (MitTx-α) and a phospholipase-A2-like protein (MitTx-β). When applied together they are potent activators of ASIC channels. Subsequently, screening of small-molecule chemical libraries have been conducted with the aim of finding alternative molecules to open ASIC channels. Screening of libraries enriched in compounds with basic groups that are able to bind to the negative charges of the proton sensor led to the isolation of molecules with a guanidinium group and a heterocyclic ring, exemplified by 2-guanidine-4-methylquinazoline (GMQ), which activate mouse ASIC3 at neutral pH (Yu et al., 2010).

Enhancers of ASIC currents have also been found in the peptide category. The neuropeptide FMRFamide (Ph-Met-Arg-Phe-amide), although specific for the molluscan channel FANaCh, also binds ASIC1, ASIC2 and ASIC3 (EC_{50} ~30 μM) increasing the magnitude of the peak current, slowing the rate of channel desensitization, and inducing a sustained current in the continuous presence of both protons and the peptide (Askwith et al., 2000). FMRFamide is abundant in the nervous systems of invertebrates, where it serves as a neurotransmitter and neuromodulator, but it is not expressed in vertebrates. Three mammalian FMRFamide-related peptides modulate ASICs with low potency: neuropeptide FF, neuropeptide AF and neuropeptide SF. In all cases, these neuropeptides bind to the closed state and slow the kinetics of desensitization. The big dynorphin opioid peptide is the most potent endogenous modifier of ASIC1a (EC_{50} ~26 nM). It displaces the pH_{50ssd} toward a more acidic range enabling proton-mediated activation of ASIC1a at pH values that normally induce desensitization. The binding site has been mapped to the acidic pocket, which also corresponds to PcTx1 binding site, consistent with the observed competition of these two peptides.

Externally applied divalent cations also modulate ASIC currents. Zn^{2+} increases the apparent affinity for H^+ of homomeric ASIC2a, heteromeric ASIC2a/ASIC1a and ASIC2a/ASIC3 channels. The enhancement of ASIC2a currents by Zn^{2+} is likely to be relevant in glutamatergic synapses where neurotransmitter-containing vesicles have high concentrations of both Zn^{2+} and H^+. Co-release of Zn^{2+} and H^+ into the synaptic cleft raises Zn^{2+} concentration from 0.5 μM to approximately ~200 μM, which is sufficient to increase ASIC2a currents: the EC_{50} for Zn^{2+} has been estimated at ~120 μM on ASIC2. The enhancement of proton affinity conferred by Zn^{2+} is potentially relevant since ASIC2a channels are abundant in the CNS but have intrinsic low sensitivity to protons.

ASICs are also modulated by endogenous proinflammatory factors such as arachidonic acid that sensitize channels to protons by shifting the relation between pH and inward currents. Ischemic tissues release arachidonic acid and lactic acid that also potentiates ASIC currents, thus the effect of protons in ischemia is amplified by the presence of additional

TABLE 20.1

Molecules That Open or Enhance Activity of ASICs

Activator/ Modulator	Target	IC_{50}	Type	Reference
MitTx-α/β	ASIC1a	9.4 nM	Snake toxin	Bohlen et al., 2011
	ASIC1b	230 nM		
	ASIC3	830 nM		
GMQ	ASIC3	1 mM	2-Guanidine-4-methylquinazoline	Yu et al., 2010
FMRFamide	ASIC3	50 μM	Neuropeptide	Askwith et al., 2000
Big dynorphin	ASIC1a	33 μM	Endogenous opioid	Sherwood et al., 2009
RFRP-1/2	ASIC1 ASIC3			
Zn^{2+}	ASIC2a	120 μM	Divalent cation	Baron et al., 2001
	ASIC2a+1a	111 μM		
Arachidonic acid	ASIC1	ND	Endogenous lipid	Smith et al., 2007
Spermine	ASIC1		Endogenous polyamine	Babini et al., 2002

endogenous factors that lead to activation of ASICs at pH values higher than that expected from the pH_{50a}.

20.4.2 Inhibitors

The most successful source of potent inhibitors of ASICs has been found in venoms. Psalmotoxin 1 (PcTtx1), isolated from the venom of the tarantula *Psalmopoeus cambridgei*, was originally identified as a highly selective and potent inhibitor of ASIC1a homomeric channels (Escoubas et al., 2000). PcTtx1 is a peptide of 40 amino acids with the cystine knot and triple-strand antiparallel β-sheet structure that defines a family of gating modifiers of ion channels. PcTx1 increases the apparent affinity for H+ and shifts the desensitization curve of mammalian ASIC1a from pH 7.19 to 7.40 leading to desensitization at neutral pH and consequently unresponsiveness to a further decrease in pH. Intrathecal administration of PcTx1 produces analgesia in rodent models of acute pain suggesting that inhibition of ASIC1a in the CNS may serve as a new pathway to reduce chronic pain (Mazzuca et al., 2007). High concentration of PcTx1, in the micromolar range, can open and slow the desensitization rate of ASIC1a. Co-crystallization of PcTx1 bound to the chicken ASIC1a revealed two different open conformations of the pore: one obtained at pH 7.25 and the other at pH 5.5 (Baconguis et al., 2012; Dawson et al., 2012).

Hi1a is disulfide-rich spider venom peptide that comprises two homologous inhibitor cystine knot domains (equivalent to two PcTx1 peptides joined by a few amino acids). Hi1a partially inhibits ASIC1a in a pH-independent and slowly reversible manner. Intracerebroventricular administration of a single low dose of Hi1a in rats significantly

TABLE 20.2

Inhibitors of ASICs; Subunit-Specific Polypeptides and Small Molecules Inhibit ASIC Currents with Various Potency

Inhibitor	Target	IC_{50}	Type	Reference
PcTx1	ASIC1a	2 nM	Peptide toxin	Escoubas et al., 2000
APETx2	ASIC3	63 nM	Peptide toxin	Diochot et al., 2004
Mambalgin 1 and 2	ASIC1a	55 nM	Peptide toxin	Diochot et al., 2012
	ASIC1a+2a	246 nM		
	ASIC1a+2b	61 nM		
	ASIC1b	192 nM		
	ASIC1a+1b	72 nM		
Hi1a	ASIC1a	0.52 nM	Spider peptide	Chassagnon et al., 2017
A-317567	ASIC1	2 μM		Dubé et al., 2005
	ASIC2	29 μM		
	ASIC3	9.5 μM		
Diminazene	ASIC1a, ASIC1b	2.4/0.3 μM	Antiprotozoan	Chen et al., 2010
Nafamostat mesilate	ASIC1a	13.5 μM		Ugawa et al., 2007
	ASIC2a	70 μM		
	ASIC3	2.5 μM		
Flurbiprofen	ASIC1a	350 μM	NSAID	Voilley et al., 2001
Salicylic acid	ASIC3	260 μM	NSAID	
Diclofenac	ASIC3	92 μM	NSAID	

NSAID, nonsteroidal anti-inflammatory drug.

reduces infarct size induced by occlusion of the middle cerebral artery even if it is administered 8 h after the occlusion (Chassagnon et al., 2017).

APETx2 is a toxin from the sea anemone *Anthopleura elegantissima* that inhibits homomeric and heteromeric ASIC3 channels (Diochot et al., 2004). APETx2 is a peptide of 42 amino acids with three disulfide bridges and a structure similar to that of other inhibitory anemone toxins targeting voltage-gated Na^+ and K^+ channels. APETx2 has been used to induce analgesia in a rodent model of postoperative pain where protons derived from local cellular inflammation presumably activate ASIC3.

Mambalgins (mambalgin-1 and mambalgin-2) are 57-residue three-finger toxins isolated from the venom of the black mamba snake that produces analgesia as strong as that of morphine (IC_{50} ~123 nM). They are gate modifiers that reduce proton sensitivity in peripheral and central neurons inhibiting homomeric ASIC1a and heteromeric channels formed by ASIC1 and ASIC2 (Diochot et al., 2012). Mambalgins induce potent analgesia against acute and chronic inflammatory pain whether the toxins are administered centrally or peripherally. A cryo-electron microscopy (cryo-EM) structure of human ASIC1a bound to mambalgin-1 shows one toxin bound to each subunit of the channel in the closed conformation (Figure 20.3) (Sun et al., 2020).

20.5 Biophysical Properties

20.5.1 Structure and Gating

Elucidation of the proton-mediated gating mechanism of ASIC has been advanced by solving structures of ASIC1a in various functional states. The first structure of ASIC1a in the closed state revealed the architecture of the subunits and the trimeric composition of the channel (Jasti et al., 2007). The shape of the subunits has been compared to that of a fist; each subdomain has been assigned to an element of a hand, as illustrated in Figure 20.2. Two transmembrane domains, TM1 and TM2, form the ion pathway that includes the closing gate and the selectivity filter. TM2, with highly conserved sequence among all members of this family of channels, exhibits an unusual structure: an unwound α turn in the middle of the helix splits TM2 into two segments – the upper TM2a makes contact with TM1 of the same subunit, whereas the lower TM2b shifts contact to the adjacent TM1. Most recently the structure of the amino terminal segment proximal to TM1 was resolved by single-particle cryo-EM of cASIC1 in resting and desensitized states (Yoder et al., 2020). The segment forms a loop that enters and fills the wide space between TM1 and TM2. Together they form the lower pore, explaining why the highly conserved $H_{28}G_{29}$ residues play such important roles in ion permeation and kinetics of ASIC1 and other channels of the ENaC/degenerin family. Mutations of the corresponding residue G_{29} in ENaC reduce open probability and cause pseudohypoaldosteronism type 1 (Gründer et al., 1997), whereas various residues in the loop of ASIC1a and ASIC2 change ion selectivity and kinetics (Chen et al., 2021). In contrast, the carboxy terminus is not essential for function or expression, as most of it can be deleted with no apparent consequences; its structure remains unknown due to intrinsic flexibility.

The extracellular domain (ECD) has highly conserved negatively charged residues distributed across various subdomains. The thumb (Asp_{346}, Asp_{350}, Glu_{354}), finger (Glu_{236}, Glu_{238}, Glu_{239}, Glu_{243}, Asp_{260}) and palm (Glu_{178}, Glu_{220}, Asp_{408}) domains form together an

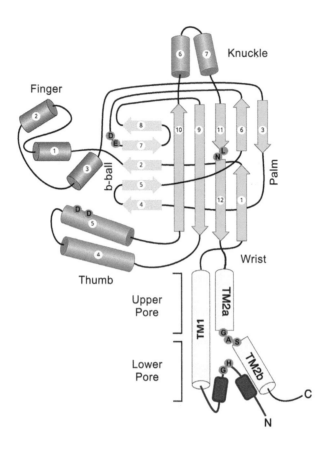

FIGURE 20.2
Cartoon representation of structural domains of a single ASIC subunit. α-Helices are shown as cylinders, β-strands as arrows and coil structures as lines. Four negatively charged residues in the acidic pocket are shown in circles: D346 and D350 in the α5-helix of the thumb and D238 and E239 in a finger domain loop. The cartoon depicts the acidic pocket in the expanded conformation. Residues important for locking channels in the desensitized state (L415 and N416) are shown in the β11–β12 linker. An extended loop, referred as the GAS belt, divides TM2 into two segments (TM2a and TM2b), and the pore into upper and lower sectors. From V16 to L39, the N-terminus folds into the lower pore as a reentrant loop with residues from S23 to H28 lining the entire lower permeation pathway. The HG motif is highly conserved and functionally important for determining ion selectivity and kinetics. The side chain of H28 faces the lumen of the pore and forms a constriction below the GAS belt. The carboxy terminus does not appear in any of the currently available structures. Numbering corresponds to human ASIC1a.

acidic pocket. Hydrogen bonds between pairs of carboxylates in the thumb and finger: Asp_{238}/Asp_{350}, Glu_{239}/Asp_{346} and Glu_{220}/Asp_{408} suggest that those residues may constitute the "proton sensor," the site where binding of protons triggers the initial conformational changes of the ligand-mediated gating process. Notwithstanding the importance of those interactions, the mentioned residues can be substituted by neutral amino acids without abolishing proton gating. Thus, the identity and location of the putative proton sensor remain elusive.

The apparent pH_{50a} of ASICs varies widely according to the subunit composition. This is attributed not to wide differences in affinity of the proton sensor, but to elements that transmit conformational changes from the sensor to the pore in the transmembrane domain (Chen et al., 2020). Homomeric ASIC1a channels exhibit the highest apparent affinity to protons (pH_{50A} from 7.0 and 6.8), whereas the least sensitive are homomeric ASIC2a (pH_{50A} from 4.0 to 5.0), and ASIC2b and ASIC4 do not respond to protons.

Structures of ASIC1 in closed, high pH desensitized (steady-state desensitization), low pH desensitized and open conformations are shown in Figure 20.3. Comparisons of the structures in the four main functional conformations have made it possible to reconstruct

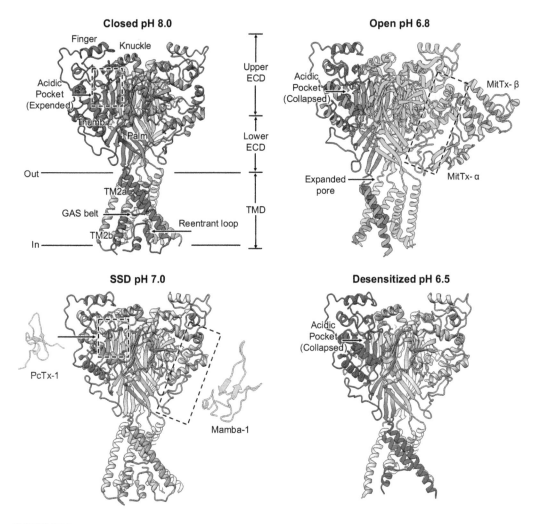

FIGURE 20.3

Structure of cASIC in closed, open, and desensitized conformations and toxin binding sites. Ribbon representation of closed cASICI (Protein Data Bank: 6VTL) (Yoder et al., 2020), open (Protein Data Bank: 4NTW) (Baconguis et al., 2014), steady-state desensitized (SSD) pH 7.0 (Protein Data Bank: 6VTK) (Yoder et al., 2020), and desensitized low pH (pH 6.5) (Protein Data Bank: 4NYK). Closed and desensitized pH 7.0 are cryo-EM structures from single-particle reconstruction. The pore architecture is almost identical in closed (pH 8.0) and SSD pH 7.0 conformations. Pctx-1 binds to the acidic pocket inducing the SSD conformation while MamBa-1 toxin binds over the thumb locking the channel in a closed inactive state. The open structure pH 6.8 was obtained in complex with MitTxα/β, which binds from the top of the thumb to the lower tip of the thumb from each subunit. The open conformation exhibits expansion of the lower palm and widening of the transmembrane domain that removes the gate constriction above the GAS belt and opens the pore. The reentrant loop was not solved in the crystal structure of the open channel. All toxins are shown in orange ribbon representation and their binding sites on the channel are demarcated by dashed boxes.

some of the events of the proton-induced gating mechanism. The most relevant features in the resting state are that the thumb domain is away from the palm, the acidic pocket is expanded and the pore is shut by the closing gate – a narrowing of the transmembrane domain formed by residues $D_{434}I_{435}G_{436}G_{437}$ of the TM2a helix. Increases in proton concentration above baseline moves the thumb near the palm and finger collapsing the acidic pocket. When movement of the thumb is restricted by the formation of a cysteine bond, protons fail to induce currents, suggesting that displacement of the thumb and collapse of the acidic pocket are conformational changes required for opening of the pore. The identity and location of residues that become protonated and drive this conformational change have not been identified yet. Displacement of the thumb initiates conformational changes in the ECD that propagate downward inducing expansion of the lower palm and an iris-type rotation of TM1 and TM2 that opens the gate increasing the pore diameter and enabling ion permeation. The single available structure in the open conformation was captured in association with MitTx, which not only opens channels but also slows their desensitization. The features that distinguish the open conformation are an open pore, collapsed acidic pocket and expansion of the lower palm domain.

20.5.2 Desensitization

All ligand-gated cation channels exhibit some degree of desensitization to protect neurons from prolonged depolarization. Desensitization influences the magnitude and time course of ASIC currents by (1) steady-state desensitization (SSD), (2) rate of decay of the peak current and (3) kinetics of recovery from desensitization.

SSD is an electrically silent process by which concentrations of protons that are too low to effectively open the pore (pH range 7.4 to 7.0) induce an ECD conformational change that makes channels unable to respond to an activating stimulus. To return fully to the resting state, channels have to be exposed back to pH 7.4 for at least 30 seconds, indicating that the change in conformation is much slower than the time required for unbinding of protons. The resting state is highly sensitive to SSD and since the pH_{50SSD} value is near the baseline pH of the CNS, small or slow acidification tends to silence rather than to activate ASIC1. However, various factors modulate the pH_{50SSD} of ASICs; for instance, increased concentration of Ca^{2+} or other divalent cations shifts the pH_{50SSD} to a more acid range, whereas decreased Ca^{2+} has the opposite effect. Organic molecules with positive net charge such as polyamines (Babini et al., 2002) and the neuropeptide big dynorphin (8 positively charged residues out of 32) (Sherwood et al., 2009) shift the pH_{50SSD} by ~0.2 pH units to a more acidic range. Mechanistically, binding of these molecules to the surface of ASIC shields H^+ binding site(s) hampering H^+-induced conformational changes that lead to SSD. Polyamines are primarily cytosolic, but they are released by ischemic neurons into the interstitial space, and the concentration of big dynorphin (~1–7 nM) rises after injury; thus, concentrations of these molecules in the CNS are sufficient to modulate ASIC responses in vivo.

Desensitization from the open state takes place at high proton concentrations; it drives the decay of the peak current back to the zero level in the presence of protons. It is attributed to low stability of the open conformation that rapidly returns the pore to the closed state, while reopening is prevented by a rearrangement of the β11– β12 linker. The linker works as a valve or clutch disengaging the ECD from the pore. Closure of the pore compresses the linker forcing a 180° rotation of the linker residues of L_{415} and N_{416}, locking the pore in the closed conformation (Yoder et al., 2018).

Human ASIC1 evades desensitization from the open state if the train of stimuli is made of brief acid pulses, ~10 ms, shorter than the time required for τD, which is the time constant of desensitization (~770 ms). If the pulse is brief, channels close by inactivation (unbinding of H^+ from the open state) rather than by desensitization (closed state with H^+ bound). These observations indicate that prolonged stimuli silences ASIC1, but channels remain responsive if the stimuli are applied at high frequency, which is characteristic of neuronal activity in the CNS.

The main features of the desensitized conformation, whether it is SSD or desensitization from the open state, are pore closed, collapsed acidic pocket, and flipping of L_{415} and N_{416} in the β11–β12 linker. Channels recover from desensitization when protons are washed out enabling a slow return back to the resting state.

20.5.3 Ion Selectivity

Activation of ASIC1 leads to membrane depolarization, as influx of predominantly sodium ions enters neurons exposed to a transient drop in extracellular pH. The ASIC1a ion selectivity sequence is Li^+~$Na^+ > K^+ > Cs^+$. It also permeates nitrogen-containing cations including NH_4^+ and guanidinium. Initially, high Ca^{2+} permeability was reported, but it has since been shown by several studies that the P_{Na}/P_{Ca} ratio is ≥20 in human and rodent ASIC1a. The increase in intracellular Ca^{2+} observed after activation of ASIC1a is likely indirect owing to membrane depolarization and secondary activation of Ca^{2+}-permeable pathways.

Extensive electrophysiological measurements and mutagenesis indicate that ENaC – the channel most closely related to ASICs – allows passage of only Li^+, Na^+ and H^+, but not of K^+ or larger cations. The segment corresponding to the GAS belt of ENaC has been proposed as the most likely selectivity filter candidate, and the size of the filter as a main determinant of ENaC selectivity. However, validation at the molecular level has not been possible so far because the resolution of the TM domain of the single ENaC structure currently available is low, and the only ASIC structure in open conformation (cASIC1-MitTx) misses one of the three elements that form the pore: the reentrant loop. The reentrant loop is comprised of amino acids from the amino terminus (V_{17} to L_{40}) that fold as two short helical segments separated by a turn containing $H_{28}G_{29}$; the loop enters the pore from the cytosolic side and lines the entire lower section of the ion pathway (Figures 20.2 and 20.3).

In the open conformation of cASIC1-MitTx the GAS belt makes the narrowest constriction of the open pore defined by a ring of carbonyl oxygens of G_{443} with radius of 3.6 Å. The size could accommodate hydrated Na^+ but not hydrated K^+. In closed conformations, both resting and desensitized, the closing gate shuts the pore; below it the GAS belt retains the same radius as in the open conformation. The resting and desensitized structures containing the reentrant loop indicate that the smallest constriction in the closed pore is made by H_{28} right beneath the GAS belt. The His side chain angles toward the lumen of the ion pathway reducing the radius to 2.1–2.6 Å. Whether the ring of three His defines the smallest constriction in the open state rather than the GAS belt has not been determined yet. Computational modeling of hASIC1a in the open conformation predicts similar radius for the GAS belt and H_{28} ring constrictions suggesting that both may have roles in determining ion permeation (Chen et al., 2021). A definitive answer awaits solving the structure of the complete pore in the open conformation.

In addition to size of the selectivity filter, other factors such as electrostatic negative potential and binding sites for ions are likely contributors to ion permeation and selectivity. Electrophysiological and structural evidence support the notion that ion selectivity of ENaC and ASIC follows a barrier mechanism, wherein selectivity is achieved mainly by ion discrimination based on the size of the hydrated ion.

20.6 Perspectives

Since the first description of proton-evoked currents in mammalian neurons by Krishtal and Pidoplichko (1980), almost two decades elapsed until the identity of the molecules that mediate such currents was determined (Waldmann et al., 1997). Thereafter, a rapid pace of discovery has led to great progress in understanding the biophysical properties and architecture of ASIC channels; in particular, a clear picture of the channel structure has emerged thanks to X-ray crystallography and cryo-EM single-particle reconstruction. A remaining challenge in the field of molecular structure is to solve the architecture of the complete pore in the open conformation to gain understanding of ion permeation and selectivity. Given the structural similarities among members of ENaC/Degenerin, the basis of ion conduction could be extended to other channels of this family.

The numerous biophysical and structural developments stand in contrast to the scant progress made in elucidating the physiological roles of ASICs. Thus far, only very small postsynaptic currents mediated by ASIC have been detected, suggesting that the contribution to synaptic transmission is minor. It is possible that in vivo those currents could be amplified by small modulatory molecules expressed in the brain but not present in the experimental assays. Advancements in this area will require developing novel technologies for delivering protons with high spatial and temporal precision in selected structures of the CNS while simultaneously monitoring the functional effects, ideally in conscious animals. Equally important is the search for additional endogenous agonists of ASICs. Finding such molecules has remained elusive despite the prediction that they should be ubiquitously expressed in the brain to match the broad distribution of ASICs across most structures of the CNS.

Intriguing is the marked differences in biophysical properties exhibited by ASICs from closely related vertebrate species. For instance, hASIC3 does not reproduce the response of mouse and rat ASIC3 to protons. At present, it is not known whether these differences reflect particular physiological adaptations to species-specific internal conditions (temperature, baseline pH, endogenous modulators) or are mere curiosities of laboratory experiments without physiological importance.

Amidst the efforts to unravel ASIC functions, it is worth contemplating the possibility that some of the ASICs might also work in ways distinct from ion channels. Without a doubt, protons activate ASICs in vitro and pathological conditions that generate significant acidosis in vivo are likely to gate ASICs in the brain, but gating of these channels by means not yet envisioned remains a viable possibility. Albeit there remains a great deal to be learned about the physiology of ASICs in the nervous system. The rapid pace of progress in recent years illustrated by the discovery of molecules that modify ASIC activity and insightful structure–functional studies indicate that the field is moving toward the kind of detailed understanding that is necessary to enable the design of therapeutic interventions targeting the ASIC family of ion channels.

Suggested Reading

Askwith CC, Cheng C, Ikuma M, Benson C, Price MP, Welsh MJ. 2000. Neuropeptide FF and FMRFamide potentiate acid-evoked currents from sensory neurons and proton-gated DEG/ENaC channels. *Neuron.* 26(1):133–141.

Babini E, Paukert M, Geisler HS, Grunder S. 2002. Alternative splicing and interaction with di- and polyvalent cations control the dynamic range of acid-sensing ion channel 1 (ASIC1). *J Biol Chem.* 277(44):41597–41603.

Baconguis I, Bohlen CJ, Goehring A, Julius D, Gouaux E. 2014. X-ray structure of acid-sensing ion channel-1-snake toxin complex reveals open state of a Na⁺-selective channel. *Cell.* 156(4):717–729.

Baconguis I, Gouaux E. 2012. Structural plasticity and dynamic selectivity of acid-sensing ion channel-spider toxin complexes. *Nature.* 489(7416):400–405.

Baron A, Schaefer L, Lingueglia E, Champigny G, Lazdunski M. 2001. Zn²⁺ and H⁺ are coactivators of acid-sensing ion channels. *J Biol Chem.* 276(38):35361–35367.

Bohlen CJ, Chesler AT, Sharif-Naeini R, Medzihradszky KF, Zhou S, King D, Sánchez EE, Burlingame AL, Basbaum AI, Julius D. 2011. A heteromeric Texas coral snake toxin targets acid-sensing ion channels to produce pain. *Nature.* 479(7373):410–414.

Chassagnon IR, McCarthy CA, Chin Y, Pineda SS, Keramidas A, Mobli M, Pham V, De Silva TM, Lynch JW, Widdop RE, et al. 2017. Potent neuroprotection after stroke afforded by a double-knot spider-venom peptide that inhibits acid-sensing ion channel 1a. *Proc Natl Acad Sci U S A.* 114(14):3750–3755.

Chen X, Qiu L, Li M, Dürrnagel S, Orser BA, Xiong Z-G, MacDonald JF. 2010. Diarylamides: High potency inhibitors of acid-sensing ion channels. *Neuropharmacology.* 58(7):1045–1053.

Chen Z, Kuenze G, Meiler J, Canessa CM. 2020. An arginine residue in the outer segment of hASIC1a TM1 affects both proton affinity and channel desensitization. *J Gen Physiol.* 153(5):e202012802.

Chen Z, Lin S, Xie T, Lin J-M, Canessa CM. 2021. A flexible GAS belt responds to pore mutations changing the ion selectivity of proton-gated channels. *J Gen Physiol.* 154(1):1–19.

Dawson RJ, Benz J, Stohler P, Tetaz T, Joseph C, Huber S, Schmid G et al. 2012. Structure of the acid-sensing ion channel 1 in complex with the gating modifier Psalmotoxin 1. *Nat Commun.* 3:936. doi: 10.1038/ncomms1917.

Diochot S, Baron A, Rash LD, Deval E, Escoubas P, Scarzello S, Salinas M, Lazduncki M. 2004. A new sea anemone peptide, APETx2, inhibits ASIC3, a major acid-sensitive channel in sensory neurons. *EMBO J.* 23(7):1516–1525.

Diochot S, Baron A, Salinas M, Douguet D, Scarzello S, Dabert-Gay AS, Debayle D, Friend V, Alloui A, Lazdunski M, Lingueglia E. 2012. Black mamba venom peptides target acid-sensing ion channels to abolish pain. *Nature.* 490(7421):552–555.

Dubé GR, Lehto SG, Breese NM, Baker SJ, Wang X, Matulenko MA, Honoré P, Stewart AO, Moreland RB, Brioni JD. 2005. Electrophysiological and in vivo characterization of A-317567, a novel blocker of acid sensing ion channels. *Pain.* 117(1–2):88–96.

Escoubas P, De Weille JR, Lecoq A, Diochot S, Waldmann R, Champigny G, Moinier D, Ménez A, Lazdunski M. 2000. Isolation of a tarantula toxin specific for a class of proton-gated Na⁺ channels. *J Biol Chem.* 275(33):25116–25121.

Gründer S, Firsov D, Chang SS, Jaeger NF, Gautschi I, Schild L, Lifton RP, Rossier BC. 1997. A mutation causing pseudohypoaldosteronism type 1 identifies a conserved glycine that is involved in the gating of the epithelial sodium channel. *EMBO J.* 16(5):899–907.

Jasti J, Furukawa H, Gonzales EB, Gouaux E. 2007. Structure of acid- sensing ion channel 1 at 1.9 Å resolution and low pH. *Nature.* 449(7160):316–323.

Krishtal OA, Pidoplichko VI. 1980. A receptor for protons in the nerve cell membrane. *Neuroscience.* 5(12):2325–2327.

Mazzuca M, Heurteaux C, Alloui A, Diochot S, Baron A, Voilley N, Blondeau N, et al. 2007. A tarantula peptide against pain via ASIC1a channels and opioid mechanisms. *Nat Neurosci.* 10(8):943–945.

Sherwood TW, Askwith CC. 2009. Dynorphin opioid peptides enhance acid-sensing ion channel 1a activity and acidosis-induced neuronal death ion channel 1a activity and acidosis-induced neuronal death. *J Neurosci.* 29(45):14371–14380.

Smith ES, Cadiou H, McNaughton PA. 2007. Arachidonic acid potentiates acid-sensing ion channels in rat sensory neurons by a direct action. *J Neurosci.* 16. 145(2):686–698.

Sun D, Liu S, Li S, Zhang M, Yang F, Wen M, Shi P, Wang T, et al. 2020. Structural insights into human acid-sensing ion channel 1a inhibition by snake toxin mambalgin1. *eLife.* 9:e57096.

Ugawa S, Ishida Y, Ueda T, Inoue K, Nagao M, Shimada S. 2007. Nafamostat mesilate reversibly blocks acid-sensing ion channel currents. *Biochem Biophys Res Commun*. 363(1):203–208.

Voilley N, de Weille J, Mamet J, Lazdunski M. 2001. Nonsteroid anti-inflammatory drugs inhibit both the activity and the inflammation-induced expression of acid-sensing ion channels in nociceptors. *J Neurosci*. 21(20):8026–8033.

Waldmann R, Champigny G, Bassilana F, Heurteaux C, Lazdunski M. 1997. A proton-gated cation channel involved in acid-sensing. *Nature*. 386(6621):173–177.

Wemmie JA, Chen J, Askwith CC, Hruska-Hageman AM, Price MP, Nolan BC, Yoder PG, Lamani E, Hoshi T, Freeman JH, Welsh MJ. 2002. The acid-activated ion channel ASIC contributes to synaptic plasticity, learning, and memory. *Neuron*. 34(3):463–477.

Yoder N, Yoshioka C, Gouaux E. 2018. Gating mechanisms of acid-sensing ion channels. *Nature*. 555(7696):397–401.

Yoder N, Gouaux E. 2020. The His-Gly motif of acid-sensing ion channels resides in a reentrant 'loop' implicated in gating and ion selectivity. *eLife*. 9:e56527.

Yu Y, Chen Z, Li WG, Cao H, Feng EG, Yu F, Liu H, Jiang H, Xu TL. 2010. A nonproton ligand sensor in the acid-sensing ion channel. *Neuron*. 68(1):61–72.

21

ENaC Channels

Mike Althaus, Diego Alvarez de la Rosa, and Martin Fronius

CONTENTS

21.1 Introduction

The epithelial sodium channel (ENaC) is a constitutively active, Na^+-selective ion channel and part of the degenerin (DEG)/ENaC superfamily of proteins. ENaCs are only found in vertebrates where their major function is the control of Na^+ and, consequently, electrolyte and water homeostasis. In addition, ENaC proteins are involved in mechanosensory processes. Due to its constitutive activity, ENaC function is meticulously regulated by a multitude of cellular mechanisms and extracellular stimuli. Imbalanced ENaC activity is associated with severe human diseases including hypertension and renal and respiratory diseases. This chapter highlights the role of ENaC in health and disease, elaborates on regulatory mechanisms controlling ENaC function, and summarizes recent insights into channel structure.

21.2 Physiological Roles

ENaC's best understood physiological role relates to its presence and function in epithelial cells where it mediates transepithelial Na^+ absorption in conjunction with the Na^+/K^+-ATPase. Transepithelial Na^+ absorption generates osmotic gradients driving transepithelial water flux (Figure 21.1). ENaC is therefore essential for maintaining Na^+ and water

DOI: 10.1201/9781003096276-21

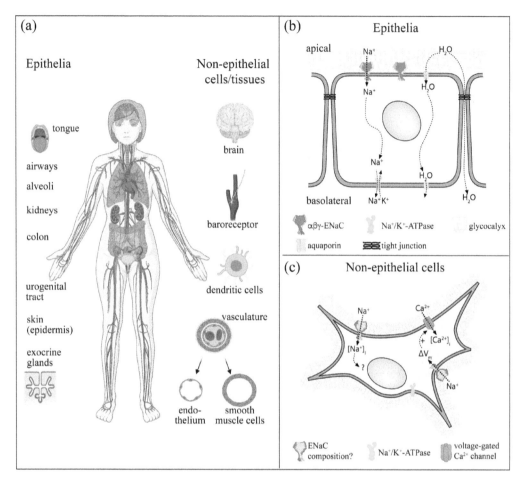

FIGURE 21.1
Physiological roles of ENaC in epithelial and non-epithelial tissues and cells. (a) ENaC function was detected in epithelial and non-epithelial cells and tissues. (b) The function of ENaC in Na^+-absorbing epithelia. ENaC-mediated transepithelial Na^+ reabsorption in collaboration with the Na^+/K^+-ATPase generates osmotic gradients to drive water flux across the epithelia. (c) In non-epithelial cells ENaC subunit combinations other than $\alpha\beta\gamma$- or $\delta\beta\gamma$-ENaC may exist. ENaC-mediated Na^+ influx is suggested to cause depolarization of the membrane potential (ΔV_m) that influences voltage-sensitive proteins such as voltage-gated Ca^{2+} channels. Another possibility is that a change in the intracellular Na^+ concentration ($[Na^+]_i$) may interfere with "Na^+ sensitive" molecules to influence their activity.

homeostasis at the whole-body level and in local environments, as demonstrated in pioneer studies using frog skin epithelia (Koefoed-Johnsen and Ussing, 1958). ENaC's molecular identity was revealed by Canessa and co-workers who cloned three subunits (α, β and γ) that form the trimeric ion channel (Canessa et al., 1994). In addition to ENaC localization in ion-transporting epithelia (Figure 21.1), ENaC subunits are expressed in non-epithelial cells where their function is less established.

21.2.1 ENaC in Epithelial Cells

In the aldosterone-sensitive distal kidney epithelium (distal convoluted tubule and cortical collecting duct), ENaC-mediated Na^+ and water reabsorption matches dietary Na^+ intake with its excretion rate, thereby regulating whole-body Na^+ homeostasis and fluid volume. Its function here is also important for driving H$^+$ and K$^+$ secretion (Alvarez de la Rosa et al., 2013). The body fluid volume regulated via these processes determines cardiac output to maintain and regulate blood pressure (Rossier, 2014). Similarly, ENaC in the distal colon epithelium contributes to Na$^+$ uptake and body fluid volume regulation (Nakamura et al., 2018). Transgenic mice lacking the α-ENaC subunit in lungs die shortly after birth due to defective neonatal lung liquid clearance (Hummler et al., 1996). This demonstrates the importance of ENaC in balancing liquid volumes covering the alveolar epithelium to facilitate diffusion of breathing gases. In the airways, ENaC regulates the volume and viscosity of the airway surface liquid layer (Hollenhorst et al., 2011) to facilitate mucociliary clearance and innate immune defense.

In epithelia of exocrine glands, such as salivary or sweat glands, ENaC prevents excess Na$^+$ and water loss as a consequence of the secretion process. In the female reproductive tract, ENaC plays a role in fertility by maintaining a liquid environment enabling fertilization (Hanukoglu and Hanukoglu, 2016), and in the bladder epithelium ENaC contributes to the sensation of bladder volume, which regulates afferent pathways initiating micturition (Araki et al., 2004).

ENaC is also located in taste receptor cells in taste buds, and involved in the detection of food Na$^+$ and appetitive salt taste in rodents (Chandrashekar et al., 2010). Here, the channel's subunit composition may be different compared to canonical ENaCs and their contribution to human salt taste is unclear.

21.2.2 ENaC in Non-Epithelial Tissues and Cells

ENaC has been detected in the vasculature, specifically in endothelial cells (ECs) and in vascular smooth muscle cells (VSMCs), where it regulates vascular resistance and blood pressure. In ECs, ENaC determines the mechanical properties and the production of the vasodilator nitric oxide (Warnock et al., 2014). The mechanical properties affect the compliance of small vessels and capillaries to influence the peripheral vascular resistance. In VSMCs, ENaC proteins sense membrane stretch and initiate myogenic constriction (Warnock et al., 2014).

In the hypothalamus, ENaC appears to be involved in the release of vasopressin (Tasker et al., 2020). ENaC expression was also detected in neurons of the cortex and hippocampus (Giraldez et al., 2013), suggesting a potential role in modulating neuronal excitability and synaptic transmission.

ENaC proteins were detected in sensory nerve endings. In baroreceptors ENaC mediates afferent inputs that initiate the baroreceptor reflex (Drummond et al., 1998). In sensory nerve endings of the skin, ENaC participates in touch and pain sensation (Geffeney and Goodman, 2012).

ENaC is expressed in dendritic cells of the immune system and mediates immune responsiveness by influencing the production of pro-inflammatory cytokines in a salt-sensitive manner (Barbaro et al., 2017).

21.3 Subunit Diversity and Basic Structural Organization

There are four individual ENaC subunits (α, β, γ and δ), which are encoded by the genes *SCNN1A, SCNN1B, SCNN1G* and *SCNN1D,* respectively. Whereas *SCNN1D* is exclusively found in lobe-finned fish and tetrapod vertebrates (with the exception of Muridae), *SCNN1A, SCNN1B* and *SCNN1G* are found in most vertebrates (with the exception of ray-finned fish) (Wichmann and Althaus, 2020). The cryo-EM-derived structure of heterotrimeric human

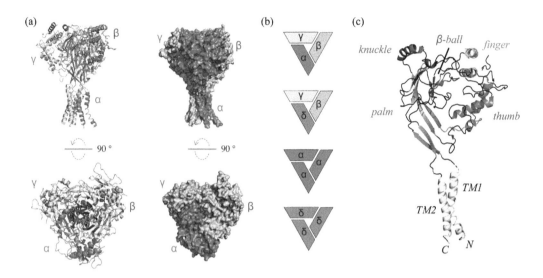

FIGURE 21.2
Basic structural organization and ENaC subunit diversity. (a) Illustration and surface representations of the cryo-EM-derived structure of human αβγ-ENaC (Protein Data Bank: 6BQN). (b) Four ENaC subunits (α, β, γ and δ) can assemble in different combinations to functional ENaC. (c) Structural organization of a single ENaC subunit (here human α-ENaC). The structure of the N- and C-termini is not resolved. TM1/TM2, transmembrane domains (yellow). The extracellular part resembles a clenched hand, holding a "ball of β-sheets" (magenta). The "palm" domain is shown in green, the "knuckle" in red, the "finger" in light blue and the "thumb" in dark blue.

αβγ-ENaC reveals a counterclockwise α–γ–β orientation (Noreng et al., 2018) (Figure 21.2a). Besides αβγ-ENaCs, δβγ-ENaCs form functional ion channels (Figure 21.2b) that consistently generate larger currents than αβγ-ENaCs. Currents generated by homomeric α-ENaCs or δ-ENaCs (Figure 21.2b) are a hundred-fold smaller compared with heterotrimeric ENaCs (Canessa et al., 1994; Waldmann et al., 1995) and the physiological function of such homomeric ENaCs is unknown. In humans, two δ-ENaC subunit splice variants containing a short (1) or long N-terminus (2) were identified (Giraldez et al., 2007). These isoforms form functional δ1βγ-ENaCs and δ2βγ-ENaCs, with similar biophysical properties but different membrane insertion rates (Wesch et al., 2012).

All four ENaC subunits share a common structural organization (Figure 21.2c), with short intracellular N- and C-termini, two transmembrane regions and a large extracellular loop. Structural information on the intracellular N- and C-termini is lacking. The extracellular loop of each ENaC subunit resembles a clenched hand formed by "wrist," "palm," "knuckle," "finger" and "thumb" domains, holding a "ball of β-sheets" (Noreng et al., 2018) (Figure 21.2c). The palm domain is mainly formed by β-sheets and allows interactions with neighboring subunits. The knuckle, finger and thumb domains contain α-helices and play an important role in ENaC gating. The knuckle domains interact with the finger domains of adjacent subunits, thereby forming a "collar" that might aid stabilizing the heterotrimeric protein (Noreng et al., 2018).

A key regulatory site is the gating relief of inhibition by proteolysis (GRIP) domain located at the interface of the finger and thumb domains within each subunit (Noreng et al., 2018) (Figure 21.3b). The α-ENaC and γ-ENaC GRIP domains contain inhibitory peptides that are flanked by protease cleavage sites and keep ENaC in a low open probability (P_O) state (Noreng et al., 2018, 2020). Removal of these peptides by intracellular and extracellular proteases increases ENaC P_O. Interestingly, the β-ENaC GRIP domain does not contain protease cleavage sites (Noreng et al., 2018).

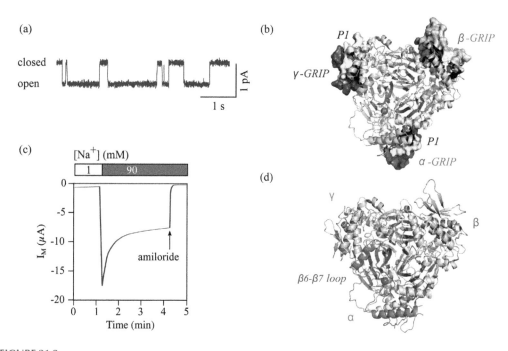

FIGURE 21.3

ENaC gating. (a) Representative cell-attached patch-clamp trace of *Xenopus laevis* αβγ-ENaC recorded at –100 mV. (The single-channel trace is part of data published in L. Wichmann, K. S. Vowinkel, A. Perniss, I. Manzini, and M. Althaus, 2018, "Incorporation of the δ-subunit into the epithelial sodium channel (ENaC) generates protease-resistant ENaCs in *Xenopus laevis*," *J Biol Chem* 293(18):6647–58.) (b) Structure of the extracellular domain of human αβγ-ENaC (Protein Data Bank: 6WTH). The GRIP-domains of each subunit are shown as surface models. The inhibitory P1 peptides within the α- and γ-GRIP domains are shown in yellow. (c) Representative current trace from a two-electrode voltage-clamp experiment showing sodium self-inhibition of human αβγ-ENaC expressed in *Xenopus* oocytes. A rapid switch from low (1 mM) to high (90 mM) Na⁺ concentrations trigger a rapid inward current that declines to a lower plateau. (d) Structure of the extracellular domain of human αβγ-ENaC (Protein Data Bank: 6WTH), with the β6–β7 loop of the acidic cleft of the α-ENaC subunit shown in red. The α2-helix of the same subunit is shown in yellow. A cation (likely Na⁺) is also represented in the structure in purple.

ENaC is a glycosylated protein. In human ENaC, three N-linked glycosylation sites are present in the α-subunit and in the γ-subunit, whereas the β-subunit contains six glycosylation sites (Noreng et al., 2018). The number of glycosylation sites within the δ-ENaC subunit is unknown. Glycosylation is important for proper channel maturation and trafficking (see Section 21.8). Furthermore, extracellular peptidoglycans can form tethers with the extracellular matrix, thereby facilitating activation of human αβγ-ENaC by mechanical stimuli such as laminar shear stress (Knoepp et al., 2020).

21.4 Gating

ENaC is a constitutively active ion channel (Figure 21.3a). There is currently no structural information on open-state ENaC available, thereby hindering definite conclusions on ENaC gating mechanisms. Channel opening/closing is likely the result of movements of the second transmembrane domains that presumably line the pore. This is supported by mutations of a serine within the second transmembrane domain (αS549, βS520 and γS529 in human ENaC) that is conserved across DEG/ENaC family members and has been named the "degenerin site." Replacing the serine in the β-ENaC subunit with a cysteine and exposing the channel to alkylating agents (e.g., methanethiosulfonates) locks human ENaC in an open state ($P_O = 1$) (Haerteis et al., 2009).

ENaC gating is affected by both the extracellular and the intracellular domains. The GRIP domains of the ENaC α- and γ-subunits contain four strands of β-sheets (P1–P4) (Noreng et al., 2020). P1 contains the inhibitory peptides that lock the channels in a low P_O state and such ENaCs have been described as "near-silent channels" (Caldwell et al., 2004) (Figure 21.3b). The inhibitory peptides are flanked by protease cleavage sites. During ENaC maturation in the trans-Golgi network, the protease furin cleaves the α-subunit twice, thereby removing its inhibitory P1 peptide, whereas the γ-subunit is only cleaved once. Such furin-processed ENaCs have an intermediate P_O. Upon additional cleavage of the γ-ENaC subunit of membrane-located ENaCs by extracellular proteases (e.g., prostasin), the inhibitory peptide of the γ-ENaC subunit is removed, thereby further increasing ENaC P_O (Kleyman and Eaton, 2020). Overall, the proteolytic maturation state determines ENaC P_O, likely contributing to the variation in ENaC P_O observed in patch-clamp experiments in native tissues (Palmer and Frindt, 1988).

Sodium self-inhibition (SSI) describes the rapid inhibition of ENaC P_O in response to high extracellular Na^+ concentrations. It is suggested to match ENaC-mediated Na^+ absorption in response to fluctuations in extracellular Na^+ concentrations in the distal nephrons of the kidney (Kleyman and Eaton, 2020). SSI was first observed in frog skin preparations, where an increase in apical extracellular Na^+ concentration induced a transient current that rapidly declined to a lower level (Fuchs et al., 1977). Similar observations are made in *Xenopus* oocytes expressing $\alpha\beta\gamma$-ENaC. High extracellular Na^+ concentrations trigger an inward current that rapidly declines to a lower level due to SSI (Figure 21.3c). The putative Na^+-binding site triggering this mechanism is likely located in an "acidic cleft" that is located between the knuckle and finger domains of the α-ENaC subunit (Noreng et al., 2020). Acidic residues (an EQND motif) within the $\beta6$–$\beta7$ loop of this cleft (between the knuckle and finger domains) provide a cation-binding site that likely mediates the reduction of ENaC P_O by extracellular Na^+ ions (Kashlan et al., 2015; Noreng et al., 2018; Wichmann et al., 2019) (Figure 21.3b). A similar acidic cleft has been identified in *Xenopus laevis* δ-ENaC (Wichmann et al., 2019), but it is not conserved in mammalian δ-ENaCs, potentially explaining the reduced SSI of mammalian δ-ENaC orthologues (Wichmann et al., 2019). SSI is antagonized by protons, thereby leading to activation of $\alpha\beta\gamma$-ENaCs (Collier and Snyder, 2009) and $\delta\beta\gamma$-ENaCs (Wichmann et al., 2019) under acidic extracellular pH conditions.

SSI and proteolytic processing of the GRIP domains influence each other. The structure of human $\alpha\beta\gamma$-ENaC suggests that the P3 strand of the GRIP domain in the α-ENaC subunit stabilizes the interaction between the $\alpha2$ helix and $\beta6$–$\beta7$ loop that contains the putative Na^+-binding site of this subunit (Noreng et al., 2020) (Figure 21.3b). After removal of the inhibitory P1 peptide following furin cleavage, this stabilization might be lost, potentially explaining the reduced SSI after proteolytic processing of the α-ENaC subunit (Noreng et al., 2020). How SSI and proteolytic processing of the GRIP domains are linked to opening/closing of the channel is unknown. However, intersubunit interactions between the palm and thumb domains of neighboring subunits appear to be important for ENaC gating and might transmit structural changes in the acidic cleft and GRIP domain toward the channel pore. Introduction of cross-linkers of varying lengths between ENaC subunits demonstrated that intersubunit distances affect ENaC P_O (Collier et al., 2014). Overall, it appears that a series of conformation changes involving the flexibility of the GRIP domains, cation interactions within the acidic cleft and intersubunit interactions at the palm–thumb interface are involved in ENaC gating.

In addition to the extracellular domains, the intracellular domains affect ENaC gating as well. A conserved histidine–glycine (HG) motif is located within the intracellular N-terminus, and mutations of these residues affect ENaC P_O, including mutations causing pseudohypoaldosteronism type 1 (Grunder et al., 1997). Structural information of the ENaC N-terminus is lacking, but the cryo-EM-derived structure of chicken ASIC1 revealed that the N-terminus of this closely related ion channel forms a "reentrant loop" that extends into the pore and might thereby affect ASIC gating (Yoder and Gouaux, 2020). Whether a similar mechanism accounts for the contribution of the intracellular N-termini to ENaC gating remains unknown.

21.5 Ion Permeability

ENaC is highly selective for Na^+ ions, with Na^+/K^+ permeability ratios >100:1 for $\alpha\beta\gamma$-ENaC (Kellenberger et al., 1999) and >50:1 for $\delta\beta\gamma$-ENaC (Waldmann et al., 1995). The low potassium permeability likely evolved to prevent substantial potassium leak out of the cell due to constitutive ENaC activity (Yang and Palmer, 2018). The single-channel Na^+ conductance (G_{Na+}) of $\alpha\beta\gamma$-ENaCs across vertebrates is consistently 5 pS, whereas a larger G_{Na+} variation has been reported for $\delta\beta\gamma$-ENaCs (5–12 pS). $\alpha\beta\gamma$-ENaCs are more permeable for Li^+ ions than for Na^+ ions (Li^+/Na^+ permeability ratio 2:1; Canessa et al., 1994), whereas human $\delta\beta\gamma$-ENaCs are less permeable for lithium ions (Li^+/Na^+ permeability ratio 1:1.6; Waldmann et al., 1995). Nevertheless, ENaC permeability to small cations (Li^+/Na^+) suggests that ion selectivity is mainly achieved by ion size discrimination.

The selectivity filter has not been resolved in the ENaC structure (Noreng et al., 2018), but mutagenesis studies suggest that it is likely formed by a conserved G/SxS motif within the second transmembrane domain of each subunit (Kellenberger and Schild, 2002). Replacement of the second serine within this G/SxS motif (S589A/C/D) in the human α-ENaC subunit increases ENaC permeability to larger monovalent (K^+, Rb^+, Cs^+) and divalent (Ca^{2+}) cations, suggesting that this residue might contribute to a size-exclusion filter (Kellenberger et al., 1999). Whether additional mechanisms contribute to the high ENaC Na^+/K^+ permeability ratio remains unknown.

21.6 Pharmacology of ENaC

The important role of ENaC in kidney Na^+ reabsorption and K^+ and H^+ excretion makes channel modulators important tools to correct defects in extracellular volume and mineral homeostasis (Alvarez de la Rosa et al., 2013). In addition, its role in the lung, both in the control of airway mucus fluidity and alveolar water transport (Schoenberger and Althaus, 2013), widens its potential as a therapeutic drug target. The prototypical inhibitor of ENaC is amiloride, a pyrazinoylguanidine derivative, originally developed as a K^+-sparing diuretic antagonizing aldosterone actions in the kidney (Kleyman and Cragoe, 1988). In addition to its clinical use, amiloride has been instrumental in defining the physiological roles of ENaC and allowed molecular cloning of the channel subunits (Canessa et al., 1994).

Amiloride blocks canonical ENaC with an $IC_{50} \approx 100$ nM (Canessa et al., 1994), which is much lower than its affinity for other Na^+ transporters such as the Na^+/H^+ exchanger or the Na^+/Ca^{2+} antiporter (IC_{50} in the µM range; Kleyman and Cragoe, 1988). Because of its guanidinium moiety, amiloride is a weak base. At physiological pH, amiloride is a monovalent cation. ENaC block by amiloride depends on pH and membrane potential, with hyperpolarization increasing channel block. In addition, amiloride block is modulated by the extracellular concentration of permeating cations (Na^+ and Li^+), which is all consistent with the action of a pore blocker. Amiloride likely binds to the external mouth of the pore, formed by the three subunits, close to the selectivity filter (G/SxS motif). In this model, the pyrazine ring is positioned at residues αSer583/βGly525/γGly537, while the guanidinium moiety occupies the outer mouth of the selectivity filter tract (Kashlan and Kleyman, 2011). The lack of an open-state ENaC structure precludes a more detailed analysis of amiloride binding. Systematic analysis of amiloride structure–function relationships led to the development of second-generation inhibitors with increased ENaC affinity, including benzamil and phenamil (Kleyman and Cragoe, 1988), which have mainly found use in research applications as more efficient substitutes of amiloride. A third generation of high-affinity ENaC blockers have been developed, mainly prompted by the possible role of ENaC as a drug target to treat cystic fibrosis and other lung conditions, and the poor performance of amiloride in this setting due to rapid clearance of the drug (Moore and Tarran, 2018). Amiloride derivatives with higher affinity and reduced absorption in the lung epithelium have been reported (Hirsh et al., 2008). Development of compounds based on pyrazinoyl quaternary amines chemically unrelated to amiloride also produced novel high-affinity ENaC blockers that potentially have improved pharmacokinetic properties (Alvarez de la Rosa et al., 2013). Selective inhibitors of δβγ-ENaC channels have been proposed, but they generally lack sufficient specificity and selectivity to be useful (Giraldez et al., 2012).

21.7 Regulation of ENaC

Tight control of ENaC activity is essential for whole-body and tissue homeostasis. Regulatory mechanisms can be initiated by hormones, most importantly aldosterone, which plays a key role in extracellular volume and blood pressure regulation and electrolyte homeostasis (Rossier et al., 2015). Other hormones regulate ENaC activity, including vasopressin, atrial natriuretic peptide (ANP), insulin and endothelin (Rotin and Staub, 2021). Paracrine signals such as ATP have been shown to regulate ENaC via P2Y receptors (Deetjen et al., 2000). From a molecular perspective, ENaC regulation functions at multiple levels (Figure 21.4), with mechanisms affecting P_O, the number of active channels in the membrane (N) or both. These mechanisms operate in a tissue-specific fashion (Rotin and Staub, 2021).

ENaC regulation includes fast control of gating by extracellular factors such as Na^+ or mechanical forces (Section 21.4; recently reviewed by Kleyman and Eaton, 2020; Kleyman et al., 2018). In addition, channel P_O increases within a longer time frame by proteolytic processing (Section 21.4). Regulation of ENaC by multiple proteases has been well established

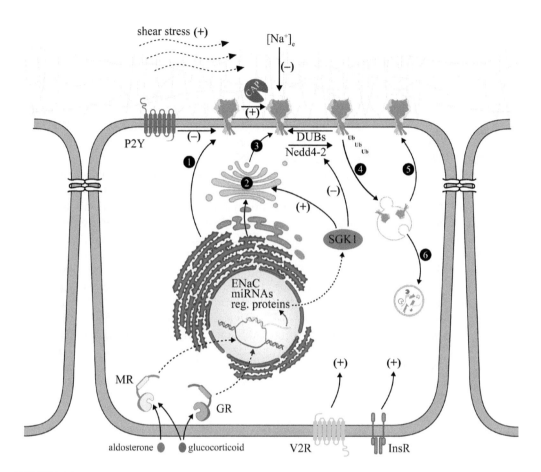

FIGURE 21.4

ENaC regulation. Schematic representation of the main regulatory mechanisms regulating ENaC P_O, gating and membrane abundance. Newly synthesized ENaC can be delivered directly to the plasma membrane from the ER (1) or through the Golgi apparatus (2), where it acquires mature glycosylation and is proteolytically processed. Plasma membrane delivery can be constitutive or regulated from stored vesicle pools (3) in a process induced by vasopressin and insulin signaling. Plasma membrane ENaC is regulated by shear stress, extracellular Na^+ concentration and further proteolytic processing. ENaC plasma membrane stability is regulated by reversible ubiquitylation. Ubiquitin tags the channel for endocytosis (4). Mono-, multi- or polyubiquitylation regulates channel recycling to the membrane (5) or degradation in the lysosome (6). Aldosterone and glucocorticoids regulate transcription of ENaC subunits and regulatory factors, including proteins such as SGK1 and miRNAs. SGK1 negatively regulates ENaC ubiquitylation and likely also affects forward trafficking and P_O. Abbreviations: CAP, channel-activating protease; DUBs, deubiquitinating enzymes; GR, glucocorticoid receptor; InsR, insulin receptor; MR, mineralocorticoid receptor; Nedd4-2, neural precursor cell expressed developmentally downregulated 4-2; P2Y, metabotropic purinergic receptor; SGK1, serum- and glucocorticoid-induced kinase 1; Ub, ubiquitin; V2R, vasopressin receptor 2.

in cultured cells, but its role in vivo has been more difficult to isolate. There is emerging evidence for prostasin controlling ENaC in the lung and colon, while evidence for the in vivo role of other proteases such as kallikrein is lacking (Kleyman and Eaton, 2020). Importantly, increased ENaC proteolytic processing is associated with hormonal stimulation (Frindt and Palmer, 2015; Zheng et al., 2011). Furthermore, multiple studies have associated abnormal protease activity in the human urine to increased ENaC activation during nephrotic syndrome (Kleyman and Eaton, 2020).

ENaC P_O is also modulated by phosphatidylinositides, such as phosphatidylinositol 4,5-bisphosphate (PIP2) and phosphatidylinositol 3,4,5-triphosphate (PIP3) (Kleyman and Eaton, 2020). PIP2 appears necessary to maintain basal ENaC activity. Stimulation of purinergic P2Y receptors activate phospholipase C, which hydrolyzes PIP2, leading to decreased ENaC activity (Deetjen et al., 2000). Stimulation with aldosterone acutely increase membrane PIP3 levels via activation of phosphotidylinositol-3-kinase, increasing ENaC activity (Blazer-Yost et al., 1999). Whether phosphatidylinositides regulate ENaC by direct association or indirect effects is uncertain, but evidence supports direct binding of PIP2 to basic residues in cytosolic domains of ENaC (Archer et al., 2020).

Posttranslational modifications are important in controlling ENaC activity, affecting both P_O and N. These include ubiquitination, which is essential to control ENaC stability at the plasma membrane (Section 21.8), phosphorylation and palmitoylation (Rotin and Staub, 2021). Hormones that activate ENaC, such as aldosterone or insulin, increase subunit phosphorylation in S/T and Y residues in the COOH-termini of the β- and γ-ENaC subunits (Shimkets, Lifton, and Canessa, 1998). Several kinases have been shown to directly phosphorylate ENaC, including PKC, casein kinase II, and serum- and glucocorticoid-induced kinase 1 (SGK1). However, most regulatory phosphorylation processes appear to affect channel activity and trafficking indirectly (Rotin and Staub, 2021). These include, in addition to SGK1, the metabolic sensor AMP-activated protein kinase (AMPK), with no lysine/K kinases (WNKs) or mTORC2 (Rotin and Staub, 2021). Palmitoylation of the β- and γ-ENaC subunits has been demonstrated (Mukherjee et al., 2017), but the role of this modification in vivo remains unclear.

21.8 Cell Biology of ENaC: Biogenesis, Trafficking and Turnover

The main physiological ENaC regulator, the hormone aldosterone, exerts its functions through the mineralocorticoid receptor (MR), a member of the nuclear receptor family of transcription factors. However, ENaC subunit expression is not coordinated and aldosterone effects on subunit mRNA and protein abundance are highly tissue-specific (Masilamani et al., 1999). The main target cells of aldosterone in terms of electrolyte balance, principal cells of the distal nephron/collecting duct, show increased α-subunit, but not β-subunit or γ-ENaC subunit expression upon aldosterone stimulation, together with a dramatic increase in channel surface abundance (Masilamani et al., 1999). Aldosterone-induced proteins involved in the regulation of ENaC trafficking have been identified, most importantly SGK1 (Chen et al., 1999). This, together with the discovery of ENaC cell-surface abundance dysregulation in human congenital hypertension (Firsov et al., 1996), prompted great interest in studying ENaC biogenesis and trafficking, which has been extensively studied (for a recent review, see Ware et al., 2020), particularly in renal epithelial cells. Channel subunits are synthesized at the rough endoplasmic reticulum (ER) to allow membrane insertion (Canessa, Merillat, and Rossier, 1994) and subunit assembly. All ENaC subunits are N-glycosylated. While it appears that ENaC glycosylation is not essential for function (Canessa et al., 1994) and that core-glycosylated, immature subunits can reach the plasma membrane, it has been demonstrated that mature glycosylation probably represents the main conducting pool of channels in the membrane (Frindt et al., 2016).

Once synthesized, ENaC exits the ER and is delivered to the apical membrane of epithelial cells through the Golgi apparatus from stored vesicle pools, a process that can be regulated by cAMP/PKA signaling (Ware et al., 2020), or from the recycling pathway, which may account for the different kinetics of ENaC delivery reported in the literature (Ware et al., 2020) (Figure 21.4). There is evidence indicating that aldosterone promotes apical delivery in the rat kidney (Frindt et al., 2016). This study also suggested that channels with different combinations follow different exit pathways from the ER, with the ENaC heterotrimer following the canonical Golgi pathway, which would include proteolytic processing. This model could account for earlier discrepancies reporting both coordinated and noncoordinated trafficking of the three ENaC subunits.

ENaC plasma membrane residency time is controlled by clathrin-mediated endocytosis (Shimkets et al., 1997), which is promoted by subunit ubiquitylation of K residues in the NH_2-termini by Nedd4-2, a ubiquitin ligase that interacts with PY motifs present in the COOH-termini (Staub et al., 1997) (Figure 21.4). The importance of PY motifs in ENaC regulation was recognized because they are altered in Liddle's syndrome (Shimkets et al., 1994) and produce increased channel activity due to accumulation in the plasma membrane (Firsov et al., 1996). Polyubiquitylation directs ENaC to lysosomal degradation, while mono- and multiubiquitylation controls its endocytosis (Ware et al., 2020). Acetylation of ENaC reduces its ubiquitylation, stabilizing the channel (Butler, Staruschenko, and Snyder, 2015). SGK1 is an aldosterone-induced kinase (Chen et al., 1999) that controls ENaC activity by phosphorylating Nedd4-2, inactivating it and preventing ENaC ubiquitylation, thus stabilizing the channel in the membrane (Debonneville et al., 2001). However, recent evidence obtained from the rat kidney suggests that stimulation of forward trafficking may be a key regulatory step in aldosterone action (Frindt et al., 2020). SGK1 represents a convergence point for hormone-regulated ENaC trafficking, including its role in the effects of insulin and vasopressin (Rotin and Staub, 2021). After channel endocytosis, ENaC can be directed to lysosomal degradation or recycled back to the plasma membrane, depending on the removal of ubiquitin tags by deubiquitinating enzymes (DUBs) (Figure 21.4) and interaction with multimeric protein recycling complexes (Ware et al., 2020). DUBs themselves have been shown to be targets of aldosterone (Fakitsas et al., 2007) and thus ENaC sorting is likely regulated by hormonal control. ENaC endocytosis and trafficking may be modified by other proteins, which in turn have been proposed to be controlled, at least in part, by aldosterone-regulated microRNA expression (Edinger et al., 2014). There is evidence suggesting segment-specific hormonal control of ENaC trafficking in the renal tubule (Nesterov et al., 2016).

Little is known about ENaC trafficking and regulation in non-epithelial tissues, notably in vascular endothelium and smooth muscle cells. ENaC appears to be an important factor regulating vascular tone, but some of its properties in endothelial cells and variable subunit composition suggest distinct regulatory pathways (Warnock et al., 2014).

21.9 Channelopathies and Disease

ENaC channelopathies include both gain of function and loss of function, and are associated with either mutations of ENaC subunit genes or with altered cell signaling pathways involved in ENaC regulation. The best characterized ENaC channelopathies are related to impaired renal function and blood pressure regulation.

Elevated blood pressure (hypertension) is associated with increased ENaC activity (gain of function) (Pavlov and Staruschenko, 2017). An autosomal dominant heritable form of hypertension, known as Liddle syndrome, is caused by gain of function mutations in the β-/γ-ENaC coding genes (Shimkets et al., 1994). These mutations disrupt the PY motifs and cause prolonged presence of channels in the membrane (see Section 21.8), resulting in increased body fluid content and cardiac output. Besides mutations in the PY motifs, several single-nucleotide polymorphisms in α-, β- or γ-ENaC were identified to associate with hypertension, but the underlying mechanisms affecting ENaC function are unknown (Hanukoglu and Hanukoglu, 2016).

Mutations of ENaC subunit genes leading to loss of function are associated with renal salt wasting and low blood pressure (hypotension). The resulting disorder is known as pseudohypoaldosteronism (PHA1) and comprises autosomal recessive mutations within the genes coding for the α-, β- and γ-ENaC subunits (Hanukoglu and Hanukoglu, 2016). Besides low blood pressure due to renal salt and water loss, PHA 1 patients also present with respiratory symptoms, altered sweat and salivary gland function, and impaired fertility (Hanukoglu and Hanukoglu, 2016).

Cystic fibrosis (CF), commonly caused by loss of function mutations of the cystic fibrosis transmembrane conductance regulator (CFTR) gene, is defined by impaired ion and water homeostasis in the airways, causing viscous mucus covering the epithelium. This affects mucociliary clearance and results in chronic infections and progressive decline of lung function. In a cohort of patients with CF-like symptoms but normal CFTR function, gain of function mutations in ENaC encoding genes were identified (Azad et al., 2009) and cause a CF-like phenotype due to elevated Na^+ and water reabsorption across the airway epithelium.

Pulmonary edema is characterized by excess fluid accumulation in the extravascular compartments of the lung, impairing blood oxygenation. Reduced ENaC function associates with the formation of pulmonary edema due to impaired Na^+ and water reabsorption from the alveolar airspace (Hummler et al., 1996). This can be caused by hypoxia (e.g., at high altitudes) or elevated heme levels (e.g., during acute respiratory distress syndrome). While hypoxia seems to reduce ENaC function at the transcriptional level (Hollenhorst, Richter, and Fronius, 2011), heme directly interacts with the channel and reduces P_O (Wang, Publicover, and Gu, 2009). Pulmonary edema due to reduced ENaC activity in patients with COVID-19 could be a consequence of infections with Sars-CoV-2 viruses. This is implied by a sequence in the viral spike protein that is identical with the furin cleavage site in the GRIP domain of α ENaC (Anand et al., 2020).

Recent evidence identified ENaC as a contributing factor for the progression of cancer (Liu et al., 2016). In breast cancer a reduction of ENaC expression associates with an increased cell proliferation rate, a characteristic of a more malignant phenotype (Ware et al., 2021).

21.10 Conclusion

Molecular cloning of ENaC defined a new ion channel family (DEG/ENaC) and quickly led to the identification of inherited human mutations profoundly affecting blood pressure and electrolyte homeostasis. Since then, a great deal has been learned about the structure, expression, regulation and pathophysiological roles of this channel. Current and future challenges include better defining the role of ENaC in non-epithelial tissues and the possible contribution of different subunit combinations to the pathophysiology of the channel. To that end, deeper characterization of ENaC complex regulatory network, along with improved tissue-specific pharmacological tools to control the channel are needed. Finally, important questions remain about ENaC structure, particularly regarding transmembrane and intracellular domains.

Suggested Reading

This chapter includes additional bibliographical references hosted only online as indicated by citations in blue color font in the text. Please visit https://www.routledge.com/9780367538163 to access the additional references for this chapter, found under "Support Material" at the bottom of the page.

Alvarez de la Rosa, D., J. F. Navarro-Gonzalez, and T. Giraldez. 2013. "ENaC modulators and renal disease." *Curr Mol Pharmacol* 6(1):35–43. doi: 10.2174/1874467211306010005.

Anand, P., A. Puranik, M. Aravamudan, A. J. Venkatakrishnan, and V. Soundararajan. 2020. "SARS-CoV-2 strategically mimics proteolytic activation of human ENaC." *eLife* 9:e58603. doi: 10.7554/eLife.58603.

Azad, A. K., R. Rauh, F. Vermeulen, M. Jaspers, J. Korbmacher, B. Boissier, L. Bassinet, Y. Fichou, M. des Georges, F. Stanke, K. De Boeck, L. Dupont, M. Balascakova, L. Hjelte, P. Lebecque, D. Radojkovic, C. Castellani, M. Schwartz, M. Stuhrmann, M. Schwarz, V. Skalicka, I. de Monestrol, E. Girodon, C. Ferec, M. Claustres, B. Tummler, J. J. Cassiman, C. Korbmacher, and H. Cuppens. 2009. "Mutations in the amiloride-sensitive epithelial sodium channel in patients with cystic fibrosis-like disease." *Hum Mutat* 30(7):1093–103. doi: 10.1002/humu.21011.

Canessa, C. M., A. M. Merillat, and B. C. Rossier. 1994. "Membrane topology of the epithelial sodium channel in intact cells." *Am J Physiol* 267(6 Pt 1):C1682–C90. doi: 10.1152/ajpcell.1994.267.6.C1682.

Canessa, C. M., L. Schild, G. Buell, B. Thorens, I. Gautschi, J. D. Horisberger, and B. C. Rossier. 1994. "Amiloride-sensitive epithelial Na+ channel is made of three homologous subunits." *Nature* 367(6462):463–7. doi: 10.1038/367463a0.

Chen, S. Y., A. Bhargava, L. Mastroberardino, O. C. Meijer, J. Wang, P. Buse, G. L. Firestone, F. Verrey, and D. Pearce. 1999. "Epithelial sodium channel regulated by aldosterone-induced protein sgk." *Proc Natl Acad Sci U S A* 96(5):2514–9. doi: 10.1073/pnas.96.5.2514.

Debonneville, C., S. Y. Flores, E. Kamynina, P. J. Plant, C. Tauxe, M. A. Thomas, C. Munster, A. Chraibi, J. H. Pratt, J. D. Horisberger, D. Pearce, J. Loffing, and O. Staub. 2001. "Phosphorylation of Nedd4-2 by Sgk1 regulates epithelial Na(+) channel cell surface expression." *EMBO J* 20(24):7052–9. doi: 10.1093/emboj/20.24.7052.

Firsov, D., L. Schild, I. Gautschi, A. M. Merillat, E. Schneeberger, and B. C. Rossier. 1996. "Cell surface expression of the epithelial Na channel and a mutant causing Liddle syndrome: A quantitative approach." *Proc Natl Acad Sci U S A* 93(26):15370–5. doi: 10.1073/pnas.93.26.15370.

Frindt, G., D. Gravotta, and L. G. Palmer. 2016. "Regulation of ENaC trafficking in rat kidney." *J Gen Physiol* 147(3):217–27. doi: 10.1085/jgp.201511533.

Giraldez, T., J. Dominguez, and D. Alvarez de la Rosa. 2013. "ENaC in the brain--Future perspectives and pharmacological implications." *Curr Mol Pharmacol* 6(1):44–9. doi: 10.2174/1874467211306010006.

Haerteis, S., B. Krueger, C. Korbmacher, and R. Rauh. 2009. "The delta-subunit of the epithelial sodium channel (ENaC) enhances channel activity and alters proteolytic ENaC activation." *J Biol Chem* 284(42):29024–40. doi: 10.1074/jbc.M109.018945.

Hummler, E., P. Barker, J. Gatzy, F. Beermann, C. Verdumo, A. Schmidt, R. Boucher, and B. C. Rossier. 1996. "Early death due to defective neonatal lung liquid clearance in alpha-ENaC-deficient mice." *Nat Genet* 12(3):325–8. doi: 10.1038/ng0396-325.

Kashlan, O. B., and T. R. Kleyman. 2011. "ENaC structure and function in the wake of a resolved structure of a family member." *Am J Physiol Ren Physiol* 301(4):F684–F96. doi: 10.1152/ajprenal.00259.2011.

Kellenberger, S., I. Gautschi, and L. Schild. 1999. "A single point mutation in the pore region of the epithelial Na+ channel changes ion selectivity by modifying molecular sieving." *Proc Natl Acad Sci U S A* 96(7):4170–5. doi: 10.1073/pnas.96.7.4170.

Kleyman, T. R., and D. C. Eaton. 2020. "Regulating ENaC's gate." *Am J Physiol Cell Physiol* 318(1):C150–C62. doi: 10.1152/ajpcell.00418.2019.

Knoepp, F., Z. Ashley, D. Barth, J. P. Baldin, M. Jennings, M. Kazantseva, E. L. Saw, R. Katare, D. Alvarez de la Rosa, N. Weissmann, and M. Fronius. 2020. "Shear force sensing of epithelial Na(+) channel (ENaC) relies on N-glycosylated asparagines in the palm and knuckle domains of αENaC." *Proc Natl Acad Sci U S A* 117(1):717–26. doi: 10.1073/pnas.1911243117.

Masilamani, S., G. H. Kim, C. Mitchell, J. B. Wade, and M. A. Knepper. 1999. "Aldosterone-mediated regulation of ENaC alpha, beta, and gamma subunit proteins in rat kidney." *J Clin Invest* 104(7):R19–R23. doi: 10.1172/JCI7840.

Noreng, S., A. Bharadwaj, R. Posert, C. Yoshioka, and I. Baconguis. 2018. "Structure of the human epithelial sodium channel by cryo-electron microscopy." *eLife* 7:e39340. doi: 10.7554/eLife.39340.

Noreng, S., R. Posert, A. Bharadwaj, A. Houser, and I. Baconguis. 2020. "Molecular principles of assembly, activation, and inhibition in epithelial sodium channel." *eLife* 9:e59038. doi: 10.7554/eLife.59038.

Rossier, B. C. 2014. "Epithelial sodium channel (ENaC) and the control of blood pressure." *Curr Opin Pharmacol* 15:33–46. doi: 10.1016/j.coph.2013.11.010.

Rossier, B. C., M. E. Baker, and R. A. Studer. 2015. "Epithelial sodium transport and its control by aldosterone: The story of our internal environment revisited." *Physiol Rev* 95(1):297–340. doi: 10.1152/physrev.00011.2014.

Rotin, D., and O. Staub. 2021. "Function and regulation of the epithelial Na(+) channel ENaC." *Compr Physiol* 11(3):1–29. doi: 10.1002/cphy.c200012.

Schoenberger, M., and M. Althaus. 2013. "Novel small molecule epithelial sodium channel inhibitors as potential therapeutics in cystic fibrosis – A patent evaluation." *Expert Opin Ther Pat* 23(10):1383–9. doi: 10.1517/13543776.2013.829454.

Shimkets, R. A., R. Lifton, and C. M. Canessa. 1998. "In vivo phosphorylation of the epithelial sodium channel." *Proc Natl Acad Sci U S A* 95(6):3301–5. doi: 10.1073/pnas.95.6.3301.

Shimkets, R. A., D. G. Warnock, C. M. Bositis, C. Nelson-Williams, J. H. Hansson, M. Schambelan, J. R. Gill Jr., S. Ulick, R. V. Milora, and J. W. Findling. 1994. "Liddle's syndrome: Heritable human hypertension caused by mutations in the beta subunit of the epithelial sodium channel." *Cell* 79(3):407–14. doi: 10.1016/0092-8674(94)90250-x.

Staub, O., I. Gautschi, T. Ishikawa, K. Breitschopf, A. Ciechanover, L. Schild, and D. Rotin. 1997. "Regulation of stability and function of the epithelial Na+ channel (ENaC) by ubiquitination." *EMBO J* 16(21):6325–36. doi: 10.1093/emboj/16.21.6325.

Waldmann, R., G. Champigny, F. Bassilana, N. Voilley, and M. Lazdunski. 1995. "Molecular cloning and functional expression of a novel amiloride-sensitive Na+ channel." *J Biol Chem* 270(46):27411–4. doi: 10.1074/jbc.270.46.27411.

Ware, A. W., S. R. Rasulov, T. T. Cheung, J. S. Lott, and F. J. McDonald. 2020. "Membrane trafficking pathways regulating the epithelial Na(+) channel." *Am J Physiol Ren Physiol* 318(1):F1–FF13. doi: 10.1152/ajprenal.00277.2019.

Warnock, D. G., K. Kusche-Vihrog, A. Tarjus, S. Sheng, H. Oberleithner, T. R. Kleyman, and F. Jaisser. 2014. "Blood pressure and amiloride-sensitive sodium channels in vascular and renal cells." *Nat Rev Nephrol* 10(3):146–57. doi: 10.1038/nrneph.2013.275.

Wichmann, L., and M. Althaus. 2020. "Evolution of epithelial sodium channels: Current concepts and hypotheses." *Am J Physiol Regul Integr Comp Physiol* 319(4):R387–R400. doi: 10.1152/ajpregu.00144.2020.

22

TRPC Channels

Jin Bin Tian and Michael X. Zhu

CONTENTS

22.1 Introduction

TRP channels got their names because of their homology to the *Drosophila* TRP (Montell et al., 2002), a protein so named because its mutations caused a transient receptor potential (*trp*) phenotype that impairs an insect's phototransduction. The term "receptor potential" describes the light response of an animal to light stimulus measured by using electro-retinography. The *trp* mutants lack the sustained phase in the electroretinogram (Figure 22.1A). Among the TRP homologs, TRP canonical (or TRPC) refers to those TRP proteins displaying a high (~35% identity in the overall amino acid sequences) sequence homology to the prototypical *Drosophila* TRP, which in insects has three isoforms (TRP, TRP-like and TRPγ) and in mammals has seven isoforms (TRPC1–C7) (Figure 22.1B). In humans, TRPC2 is a pseudogene. The genes coding for these proteins were all identified during a period of ten years, from 1989 to 1999. The *Drosophila trp* gene was first isolated by positional cloning using genetic approaches in 1989 (Montell and Rubin, 1989), and TRP-like, a calmodulin-binding protein that displayed a high homology to TRP, was subsequently discovered in 1992 by screening a cDNA expression library. Functional studies during this early time period demonstrated that both TRP and TRP-like mediate light-induced currents in the photoreceptor cells of insects, establishing their fundamental roles in the vision of insects. However, the mammalian TRPCs are typically not involved in phototransduction, despite the fact that they were all cloned based on homology screening using the sequences of *Drosophila* TRP and TRP-like (Zhu et al., 1996).

DOI: 10.1201/9781003096276-22

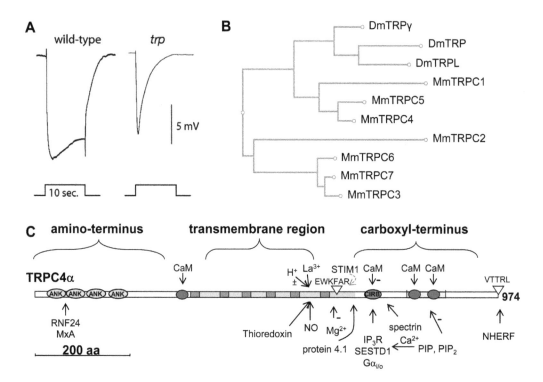

FIGURE 22.1
General background about TRPCs. (A) Diagram depicting receptor potentials measured by electroretinogram from wild-type and *trp* mutant fruit flies. The brackets below the traces indicate light exposure. (B) Phylogenetic tree showing the relationships among the three *Drosophila* (Dm) and seven murine (Mm) TRPC proteins. (C) Structural organization and some of the key regulatory sites of TRPC4α. Ankyrin-like repeats (ANKs), transmembrane segments (blue boxes) and calmodulin (CaM)-binding sites (red circles) are labeled. EWKFAR indicates the TRP motif. VTTRL indicates the NHERF-binding domain. Experimental evidence for some of the features indicated has only been reported for TRPC5. −, negative regulation; ±, both positive and negative regulation. See Tian et al. (2014) and Wang et al. (2020) for more detail.

22.2 Physiological Roles

The *Drosophila* TRP and TRP-like channels are not directly activated by light. Rather, they are activated downstream from a G_q protein that responds to the light activation of rhodopsin in the insect's photoreceptor cells. The G_q protein is coupled to β isoforms of phospholipase C (PLCβ), which in turn break down phosphatidylinositol 4,5-bisphosphate (PIP_2) on the plasma membrane to produce two second messengers – inositol 1,4,5-trisphosphate (IP_3) and diacylglycerol (DAG) – with protons (H^+) as a by-product (Huang et al., 2010). It has been shown that the decrease in PIP_2, and also its precursors, PIP and PI, and local acidification at the cytoplasmic side of the channel are important for the activation of native TRP and TRP-like in *Drosophila* photoreceptor cells (Huang et al., 2010).

Similarly, the mammalian TRPCs are also activated downstream of the PLC pathways, including the $G_{q/11}$–PLCβ and receptor tyrosine kinase–PLCγ pathways. These pathways are widely present in all mammalian cells and responsible for IP_3-evoked Ca^{2+} mobilization from the endoplasmic reticulum (ER) stores. Intriguingly, the release of Ca^{2+} from the ER always induces Ca^{2+} influx into the cell. This process had been coined capacitative Ca^{2+} entry or store-operated Ca^{2+} entry (Putney, 1986). When the *Drosophila* TRP was first discovered, its activation downstream from PLC and its Ca^{2+} permeability drew attention, leading to the hypothesis that its mammalian homolog might underlie the molecular basis of store-operated Ca^{2+} entry, which fueled the main drive on searching for TRP homologs in mammalian species and the functional characterization of these homologs as store-operated channels during the early era of TRPC research (Birnbaumer et al., 1996). However, these channels did not exhibit the high Ca^{2+} selectivity and very small unitary conductance as the best characterized native store-operated channel, i.e., the Ca^{2+} release-activated Ca^{2+} (CRAC) channel. Instead, the TRPC channels are cation nonselective with excellent permeability to not only Ca^{2+} but also Na^+ and K^+, and they displayed single-channel conductance ranging from 14 pS to ~70 pS. This conundrum was not resolved until 2006 when ORAI proteins were identified to form CRAC channels, which are opened by the binding of STIMs, trans-ER membrane proteins that transmit the Ca^{2+} store-depletion signal to the plasma membrane-localized ORAI channels (Prakriya et al., 2006; see also Chapter 25). Interestingly, evidence exists that STIM also directly interacts with at least some TRPC isoforms. This may underpin the store-operated activation of TRPC channels, although the currents are distinct from that of the ORAI channels in terms of ion selectivity and conductance. It is worth noting that native store-operated nonselective currents with unitary conductance similar to that of TRPC channels have been reported before, meaning there are multiple store-operated Ca^{2+} entry pathways; some are encoded by ORAIs and some by TRPCs (Wang et al, 2020).

All TRPC channels can also be viewed as receptor-operated channels because of their activation downstream from receptors that stimulate PLC function. In essence, ER Ca^{2+} store depletion is an inevitable consequence of PLC activation due to IP_3-mediated Ca^{2+} mobilization; hence, it is not always distinguishable whether the activation of TRPC resulted from store depletion or other constituents of the PLC pathway when the initial trigger is a receptor agonist, which represents the most reliable approach to induce TRPC currents. In more strict conditions when store depletion is induced in the absence of receptor activation, the possibility exists that ORAI channels might cooperate with the TRPC through either functional or physical interactions to make the TRPC "appear" to be store-operated, although in cases of direct TRPC activation by STIM, ORAI is not assumed to play a role. At present, there is still no consensus as to whether and how TRPC channels are activated by Ca^{2+} store depletion.

Whether or not store-operated, the receptor-operated TRPC channel activation brings Na^+ and Ca^{2+} into the cell, leading to membrane depolarization and cytosolic Ca^{2+} concentration ($[Ca^{2+}]_c$) elevation, respectively. These also represent the main physiological roles of these channels. Numerous studies have demonstrated the involvement of TRPC channels in various physiological and pathophysiological settings, different cell types and organs. These span from neurotransmission, neurodegeneration, cardiac contractility, arrhythmia, hypertrophy, endothelial cell permeability, vasodilator release, cell migration, angiogenesis, and atherogenesis to smooth muscle contraction and proliferation, as well as cancer growth and metastasis. The diverse functions may arise from differences in the expression of TRPC isoforms in different cell types, the spaciotemporal patterns of Na^+ and Ca^{2+} signals produced by the individual TRPC channels, and unique protein–protein interactions that may exist between specific TRPC isoforms and their protein partners (Jeon et al., 2021).

22.3 Subunit Diversity and Basic Structural Organization

Based on sequence homology, mammalian TRPCs are subdivided into four groups: TRPC1, TRPC2, TRPC3/C6/C7 and TRPC4/C5. The latter two groups share about 75% and 65% amino acid identities among members within the groups, respectively. All TRPC members contain an absolutely conserved sequence, EWKFAR, termed TRP domain, at their carboxyl (C)-termini (Figure 22.1C). They also have ankyrin-like repeats at their amino (N)-termini and another conserved motif, namely, calmodulin (CaM) and IP_3 receptor-binding (CIRB) site at their C-termini (Zhang et al., 2001). In addition, individual TRPC isoforms also interact with a diverse range of other proteins, e.g., Homer, Junctate, NHERF and STIM1 (Wang et al., 2020) (Figure 22.1C).

Like other TRP channels, TRPCs form tetramers composed of either identical or related subunits (Figure 22.2A). Each TRPC subunit is composed of six transmembrane segments (S1–S6) with both the N and C termini located at the cytoplasmic side (Figure 22.2B). The S1–S4 segments are referred to as voltage sensing-like domains. Their domain-swapped arrangement with the S5–S6 pore domain resembles that of other TRP channels and voltage-gated K^+ (Kv) channels. The ion conducting pore is formed by S5 and S6 and an intervening pore (P) loop in a symmetric arrangement of the four protomers. A four-helical bundle formed at the C-terminal portion of S6 represents the narrowest part or channel gate in nearly all the cryo-EM TRPC structures. The amino acids involved are isoleucine (I), asparagine (N) and glutamine (Q) from the highly conserved sequence VLLNMLIAMXNXSXQ. However, in some structures, the most restricted path is formed by L654 and I658 (in TRPC3), or L723, I727, and F731 (in TRPC6) (Fan et al., 2018; Bai et al., 2020), which are one residue before the commonly found I, N and Q, respectively. For selectivity filter, the narrowest part has a conserved glycine, which is three residues below the glutamate (E618 in TRPC3 and E687 in TRPC6) or asparagine (N580 in TRPC4 and N584 in TRPC5) at the upper entrance. Both TRPC3-E618 and TRPC5-N584 have been shown to be critical for the channel's Ca^{2+} permeability (see review in Wang et al., 2020).

At the cytoplasmic side, the four ankyrin-like repeats make up the outskirt, which surrounds the helical bundle formed by the distal C-terminal α-helix called CH2. CH2 is

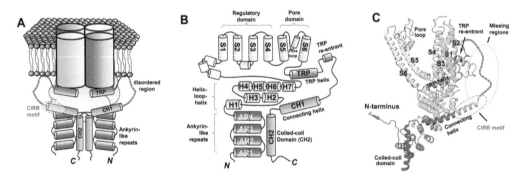

FIGURE 22.2

Structure features of TRPC4/C5 channels. (A) Diagram of assembled tetrameric channel. For the cytoplasmic area, only two subunits are shown. Note the crossover between CH1 and CH2 helices. The disordered region between TRP reentrant (labeled in panel B) and CH1 is highlighted by the red chain. The CIRB motif that encompasses parts of CH1 and the disordered region is indicated by the oval shade. (B) Main structural features of a single subunit. Cylinders highlight α-helices in the transmembrane domain (S1–S6), amino (N) and carboxyl (C) terminal domains. AR1–AR4 indicate ankyrin-like repeats and H1–H7 indicate helix–loop–helix domains in the N-terminus. A TRP helix or TRP domain is present in most TRP channels. (C) Ribbon diagram side view of a single subunit of closed mouse TRPC4 channel.

connected to CH1 via a short loop that crosses over the adjacent protomer. In this way, the CH2 faces the ankyrin repeats of the same protomer (Figure 22.2A). The four crossovers, together with the long helical bundle composed of individual CH2 domains from the four protomers, help stabilize the tetrameric structure. CH1 is a long C-terminal α-helix more proximal than CH2, which runs in an antiparallel direction with the helix that contains the conserved TRP domain. The TRP helix sits right underneath the membrane and is connected with CH1 via a TRP reentrant loop followed by a disordered region that contains the CIRB site (Figure 22.2A, C). This region is completely exposed to the membrane–cytosolic interface, where multiple TRPC-interaction proteins may gain access to the channel for functional modulation. CH1 is separated from TRP helix by N-terminal linker helices of a neighboring protomer. The TRP helix is connected to the C-terminal side of the S6 transmembrane segment, allowing a direct effect on gating (Figure 22.2C). This overall architecture at the cytoplasmic side is similar to that of TRPMs and NOMPC.

Also in common with TRPMs and TRPA1 is the cation-binding site coordinated by two residues (E417, Q420 of TRPC4; E418, E421 of TRPC5) in the last two helical turns of S2 and another two residues (N435, D438 of TRPC4; N436, D439 of TRPC5) at the beginning two turns of S3. In TRPC4 and C5, either a sodium or a calcium ion is bound to this position (Figure 22.3), while in TRPMs and TRPA1, the site is occupied by Ca^{2+} and involved in both facilitation and inhibition (at least for TRPA1) of the channel function by Ca^{2+} (Zhao et al.,

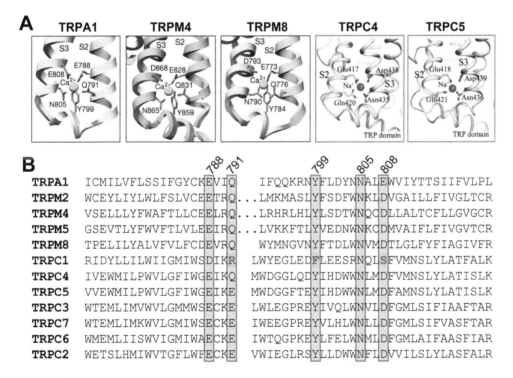

FIGURE 22.3

A common Ca^{2+}-binding site in TRPA, TRPCs and TRPMs. (A) Structural details on the conserved Ca^{2+}-binding site found in TRPA1, TRPM4, TRPM8 and TRPC4/C5. In TRPC4/C5, the cation can be either a Na^+ or Ca^{2+}. S2 and S3 depict the second and third transmembrane helices, respectively. Although not depicted here, the closely apposed TRP helix shown in TRPC4/C5 is also found in other TRP channels. (B) Amino acid sequence alignment of S2–S3 regions of murine TRPA1, TRPM2/M4/M5/M8 and TRPC1-C7, showing the conservation of key residues involved in Ca^{2+} binding (red boxes), except for TRPC1. Numbers indicate the positions in TRPA1. The substitutions (blue letters on gray background) suggest that this binding site is crippled in TRPC1.

2020). However, it remains to be determined in which way this site is involved in the Na^+-dependent or Ca^{2+}-dependent regulation of TRPC channels. The functional significance of this site has been highlighted by recent cryo-EM structures showing its involvement in binding to certain TRPC inhibitors (Bai et al., 2020).

22.4 Gating

Although the recent cryo-EM TRPC structures have provided clues as to where the gates are and how gating might occur based on differences between structures, locations of ligand-binding sites, and comparison with other related channels, exactly how TRPC channels are gated remains unclear. TRPC channels are generally believed to be activated downstream from stimulation of PLC. At least three constituents of the PLC pathway have been implicated in TRPC channel gating: PIP_2, IP_3 and DAG. Among them, IP_3 may exert its effect indirectly by activating IP_3 receptors, which then physically interact with TRPC at the conserved CIRB site. However, the CIRB motif is largely disordered in cryo-EM structures and is also quite remote from the S6 gate described earlier, making it difficult to know how IP_3 receptor binding affects gating. For PIP_2, a putative site for phospholipids has been reported between S4 and the S4–S5 linker in cryo-EM structures of TRPC6 (Bai et al., 2020), which partially overlaps with where a TRPC6 antagonist, BTDM, binds (Song et al., 2021). Whether this represents a PIP_2-binding site remains to be determined, although in TRPV1, phosphoinositide and vanilloid also bind to the similar region. For DAG, a putative binding site has recently been reported for TRPC5 in the outer layer of the transmembrane region wedged in between the pore helix and the S6 segment of the neighboring protomer (Song et al., 2021). Contained in this region is a conserved glycine behind the selectivity filter, which is accessible through a subunit-joining fenestration and has also been shown to be critical for DAG binding in TRPC3 (Lichtenegger et al., 2018). Interestingly, a synthetic TRPC6 agonist, AM-0883 is also bound in the equivalent pocket (Bai et al., 2020). The compound interacts with F675 and W680 on the pore helix and Tyr705, Val706 and Val710 on S6 of the neighboring subunit. In all these cases, DAG, or its synthetic analog OAG, or AM-0883, seems to gain access from the extracellular side, but the endogenously generated DAG via PIP_2 hydrolysis should first appear in the inner leaflet of the plasma membrane. Therefore, whether this represents an authentic site for the action of endogenous DAG produced from PLC signaling and whether it is conserved among all TRPC channels warrant further investigation.

Ca^{2+}, as a part of PLC signaling pathway resulting from IP_3 receptor-mediated ER Ca^{2+} release and Ca^{2+} entry through plasma membrane Ca^{2+}-permeable channels including the TRPC channels themselves, has also been implicated in TRPC channel gating (Wang et al., 2020). These include both the stimulatory and inhibitory effects, and can occur either by the calcium ions acting directly on the channel or through other Ca^{2+}-binding proteins, for example, calmodulin and PLCδ1 (Thakur et al., 2016). Based on the conservation of the cation-binding site between S2 and S3 among TRPA1, TRPCs and TRPMs, and the findings that a number of TRPC antagonists occupy the same region, it is plausible that binding to this site represents a common mechanism of TRP channel gating by Ca^{2+} (Zhao et al., 2020; Song et al., 2021). Possibly, the conformation changes caused by substrate binding to this site act similarly to how voltage affects the conformations in the voltage-gated channels, leading to changes in the S4–S5 linker and the rotation and shift of the

S6 four- helix bundle to open the lower gate. In this sense, the voltage-sensing domains of voltage-gated channels and the voltage-sensing-like domains of TRP channels may indeed share a common conserved mechanism in channel gating. Yet, the detailed mechanism by which the voltage-sensing-like domain, which contains the S2–S3 cation-binding site, gates each TRPC subtype remains to be elucidated.

In addition to the second messengers, a number of proteins have also been implicated in the activation of TRPC channels. These include STIM1, an ER transmembrane protein that senses store depletion; the $G\alpha_{i/o}$ subunits of the heterotrimeric G proteins; and SESTD1, a protein that contains a SEC14-like lipid-binding domain and two spectrin domains, and binds PIP and PIP_2 in the presence of Ca^{2+}. Interestingly, both $G\alpha_{i/o}$ and SESTD1 bind to the previously described CIRB motif, which also binds to calmodulin and the IP_3 receptors, and is mostly disordered in the cryo-EM structures. STIM1, on the other hand, can bind to TRPCs via an electrostatic interaction between two conserved aspartates/glutamates (D639, D640 of TRPC1; E648, E649 of TRPC4) at the TRPC C-terminus and two lysine residues at the C-terminal end of STIM1. These residues in the TRPC are located right at the border of the TRP helix and TRP reentrant, very close to the disordered portion of the CIRB motif (Wang et al., 2020) (Figure 22.1C). Of note, using the recombinant STIM–Orai1-activating region (SOAR) of STIM1, which does not contain the C-terminal lysines, Asanov et al. (2015) successfully activated single-channel currents of TRPC1, C4, and C5, but not C3 and C6, in excised inside-out patches. Combined with single-molecule fluorescence imaging, these researchers also demonstrated that two SOAR molecules are needed for one TRP tetramer and the activity was antagonized by four calmodulin molecules. Precisely where the SOAR fragment binds and how it gates the TRPC channels are unclear at this point.

Other gating mechanisms of TRPCs include lysophospholipids, redox state changes, arachidonic acid and its metabolite 20-HETE, nitric oxide, mechanical stimulation, and cooling temperature. Details on these regulations can be found in previous review articles (Tian et al., 2014; Wang et al., 2020). For the most part, except for the s-nitrosylation sites of TRPC5 by nitric oxide, the structural basis of the regulation is still lacking, and controversy exists on whether the regulation is direct or indirect.

22.5 Ion Permeability

All homomeric TRPCs are nonselective cation channels that permeate Ca^{2+} and related divalent cations. In fact, the permeation to Sr^{2+}, Ba^{2+} and Mn^{2+} had been used to assess TRPC channel activity in expression systems. The reported ion selectivity values, based on reversal potential measurement, vary considerably. The PCa^{2+}/PNa^+ (or Cs^+) ratios for TRPCs have been summarized to range from ~1 to 10, with a rank order of TRPC5 > C6 > C7 > C2 > C3 > C4 > C1 (Gees et al., 2010). So far there has been no report on fractional Ca^{2+} current measurement for TRPCs. Under physiological conditions, the main ion conducted by TRPCs is Na^+. The total cationic influx ($INa^+ + ICa^{2+}$) mediated by TRPC leads to depolarization, which in excitable cells can activate voltage-gated Na^+ and Ca^{2+} channels.

The Na^+ influx mediated by TRPC3 has also been shown to trigger Na^+/Ca^{2+} exchangers (NCX) in "reverse mode," extruding Na^+ in exchange for Ca^{2+} entry and thereby indirectly enhancing the Ca^{2+} signal (Eder et al., 2005). It is possible that TRPC channels may facilitate or alter other Na^+-dependent transport activities via the increase in local $[Na^+]_c$.

The Ca^{2+} influx through TRPC1, however, has been shown to activate large conductance Ca^{2+}-activated K^+ (BK) channels in salivary glands and vascular smooth muscle cells, leading to hyperpolarization instead of depolarization. Physical association between TRPC1 and BK was demonstrated, suggesting that the coupling is local and efficient (Kwan et al. 2009). Therefore, though the immediate local effect of TRPC activation is a depolarizing Na^+ and Ca^{2+} influx, depending on association with other proteins, global TRPC activation can lead to either depolarization or hyperpolarization. Clearly, TRPC channels function not only as mediators for $[Ca^{2+}]_c$ changes but also as key modulators of membrane potential.

In excitable cells, Ca^{2+} influx through TRPC channels may have special functional significance despite the notion that the main Ca^{2+} entry pathway appears to rely on voltage-gated Ca^{2+} channels (VGCCs), activated in response to TRPC-mediated membrane depolarization. For instance, a mutation at the TRPC3 selectivity filter, E633Q, disrupted its Ca^{2+} permeability without affecting the monovalent conductance or its functional coupling to L-type VGCCs in atrial myocytes, but it completely abolished TRPC3-dependent and calcineurin-mediated nuclear translocation of nuclear factor of activated T cells (NFAT) (Poteser et al., 2011).

22.6 Pharmacology

Past study of TRPC channels had been hampered by the lack of specific agonists and antagonists. Until recently, only nonspecific blockers, such as SKF96365, 2-aminoethoxydiphenyl borate (2-APB) and flufenamic acid, were available to assess the involvement of TRPC channels in native systems. These drugs do not allow distinction among different TRPCs, and at the concentrations used they also have many non-TRPC targets. The situation has improved dramatically in the last few years, with the recent development of new small molecular selective agonists and antagonists of TRPC channels from both academic labs and the pharmaceutical industry. For a comprehensive discussion of the current TRPC small molecular probes, readers are referred to the review article by Wang et al. (2020). In general, all the new TRPC probes show relative selectivity between the TRPC1/4/5 and TRPC3/6/7 subgroups. Some of them even specifically target TRPC5 but not the closely related TRPC4 or show preference for a specific TRPC isoform among the TRPC3/6/7 subgroup members. However, the information in selectivity should be taken with caution, as most of the drugs have not been thoroughly tested for their off-target effects. Also, currently, there is no specific drug for TRPC1 owning to the lack of a specific assay for this TRPC isoform in heterologous systems. In most screening assays, TRPC1 is tested together with TRPC4 in the form of the TRPC1–C4 heteromeric complex.

22.7 Regulation

Some aspects of TRPC channel regulation are covered in the section on gating. Here we discuss the modulatory actions that affect the activity of these channels. The first one is phosphorylation. TRPC3/6/7 are inhibited by PKA, PKC and PKG to varying degrees, and some of the identified sites are conserved among the three channels and can overlap between PKA and PKG. It was also shown that PKC could inhibit these channels through stimulating PKG. For other TRPCs, the inhibition by PKC is also common. However, the native heteromeric TRPC1/C5 channels are activated by PKC. PKA and members of Src family of nonreceptor tyrosine kinases, e.g., Fyn, have also been shown to regulate trafficking of some TRPC channels. Phosphorylation may also affect TRPC interaction with its partner proteins. More details on TRPC phosphorylation and other posttranslational modifications can be found in Tian et al. (2014) and Liu et al. (2020).

An effect that distinguishes TRPC4/C5 from other TRPC isoforms is the dependence on $G_{i/o}$ proteins. For TRPC4, $G_{i/o}$ is required for receptor-operated channel gating, in which the $G_{q/11}$–PLC pathway can only stimulate this channel very weakly. However, if $G_{i/o}$ is activated, TRPC4 undergoes robust activation, likely mediated by $G\alpha_{i/o}$–GTP through a direct binding to the CIRB site (Jeon et al., 2012). In this scenario, even $G_{q/11}$–PLCβ is no longer necessary, although it can still facilitate the process by accelerating the functional coupling between TRPC4 and PLCδ1, a PLC isoform that is activated solely by Ca^{2+} and is essential for $G_{i/o}$-dependent TRPC4 activation (Thakur et al., 2016, 2020). By contrast, TRPC5 can be activated by either $G_{q/11}$ or $G_{i/o}$ and is not absolutely dependent on PLCδ1 (Thakur et al., 2016).

Cations affect the activity of various TRPC channels in different ways. In addition to Ca^{2+}, H^+ (pH), lanthanides (La^{3+} and Gd^{3+}), Na^+, Mg^{2+} and Hg^{2+} also exert effects on TRPC4/C5. While strong extracellular acidosis generally inhibits TRPC channels, mild acidosis potentiates TRPC4/C5 activities. These dual effects of extracellular protons are also evident in lanthanides, which potentiate TRPC4/C5 activation at concentrations ranging from 10 to 300 μM. Notably, both protons and lanthanides only potentiate TRPC4/C5 when the channel activity is relatively weak. When the channels are strongly activated by other stimuli, they become inhibitory. The inhibition may be explained by the effect of H^+ and lanthanides on reducing the unitary conductance of the channel. The potentiation effect is due to an increase in the channel's open probability and is dependent on glutamate residues in the pore loop. Interestingly, extracellular protons and lanthanides share the same glutamate residues for their potentiation effect. Furthermore, H^+ also facilitates the activation of TRPC4 from the intracellular side by promoting Ca^{2+} activation of PLCδ1 (Thakur et al., 2020). Since H^+ is a by-product of PLC-mediated hydrolysis of PIP_2 (Huang et al., 2010) and Ca^{2+} arises from IP_3-mediated ER Ca^{2+} mobilization, by acting as both the products and activators of PLCδ1, Ca^{2+} and H^+ help sustain a positive feedback loop that supports a self-propagating activation of PLCδ1. Since Ca^{2+} is quickly exhausted from internal stores, the coupling to TRPC4, which mediates Ca^{2+} influx from extracellular space and increases $[Ca^{2+}]_c$ near plasma membrane where PLCδ1 functions to break down PIP_2, the additional recruitment of TRPC4 further sustains the feedback loops for a robust and long-lasting activation of both PLCδ1 and TRPC4 (Thakur et al., 2020) (Figure 22.4).

Although TRPC4/C5 currents are detected in both Na^+-based and Cs^+-based bath solutions, there are conditions when the currents are very small in the Na^+ bath but become markedly larger upon switching to the Cs^+ bath (Jeon et al., 2012). This suggests an inhibitory effect of Na^+ on TRPC4/C5, unrelated to the selectivity difference between the two monovalent cations, as the difference is very small, if any. The inhibitory effect of Na^+ is not obvious when the channels are robustly activated through receptor stimulation (Thakur et al., 2016), indicating that the Na^+-sensitive state represents weakly activated TRPC4/C5 channels. The mechanism for this inhibition, however, remains to be elucidated.

Mercury compounds potently activate TRPC4/C5, but not other TRPCs. The two cysteine residues in the TRPC5 pore loop, which were identified for nitric oxide and thioredoxin

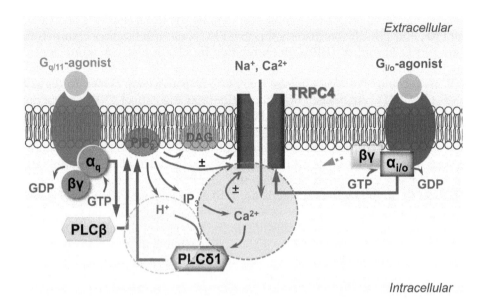

FIGURE 22.4

Mechanism of TRPC4 channel activation by G protein signaling. Schematic drawing of factors critical for receptor-operated TRPC4 channel activation in response to agonists that simultaneously activate both $G_{q/11}$-coupled
and $G_{i/o}$-coupled receptors. The light yellow and light pink circles highlight the propagating H^+ and Ca^{2+} signals, respectively, that support continued PLCδ1 activation via positive feedback mechanisms, which can be initiated by stimulating $G_{q/11}$–PLCβ. The propagating PLCδ1 activation needs simultaneous $G_{i/o}$ activation, likely
through $G\alpha_{i/o}$–GTP, in order to support robust TRPC4 activation. Red arrows, protein-based regulation; blue
arrows, regulation by ions and small diffusible messengers. ±, indicates that the effects (of PIP_2 and Ca^{2+}) on
TRPC4 activation can be both positive and negative.

sensing, were shown to be critical for mercury stimulation. The mercury effect was not
mimicked by Cd^{2+}, Ni^{2+} and Zn^{2+}. It was proposed that TRPC4/C5 may be responsible for
the cytotoxic effect of mercurial compounds on neurodevelopment (Xu et al., 2012).

The current–voltage (I–V) relationships of TRPCs are not linear in the range of –120 to
+100 mV, indicating voltage-dependent changes in conductance (Figure 22.5). There are at
least three reasons for these changes. One is the block by divalent ions, Ca^{2+} and Mg^{2+}. A
Mg^{2+} block was found to account for the unusual "flat" segment or negative slope between
+10 and +40 mV on the I–V curves of TRPC5 currents. Asp633 situated at the end of the S6
transmembrane helix was shown to mediate the cytosolic Mg^{2+} block of outward current
through TRPC5. This residue also affects unitary conductance of TRPC5 at negative potentials (Obukhov and Nowycky, 2005). Likewise, the unitary current amplitude of the native
TRPC-like current (mI_{CAT}) in ileal myocytes is also reduced by extracellular Mg^{2+} and Ca^{2+}
(Dresviannikov et al., 2006), resembling that of extracellular protons and lanthanides.

Very few studies have examined the intrinsic voltage dependence of TRPC channels. Because the currents are quite unstable over extended time periods, they are typically detected using repetitive voltage ramps or gap-free recordings at a fixed potential.
Since the ramps are often quite fast (<0.5 s) and run from negative to positive potentials,
the obtained I–V curves can be rather distorted from the true steady-state I–V relation.
Nevertheless, the I–V relationships of TRPC4 and C5 often change from U- or V-shaped, or
flat, to linear at negative potentials depending on the degree of activation (Figure 22.5A).
Using slow (6 s) voltage ramps from positive to negative potential, a negative shift of the
conductance–voltage curve can be clearly demonstrated during activation of the native

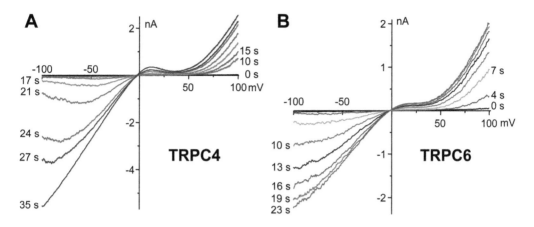

FIGURE 22.5

Current–voltage (I–V) relationships during activation of TRPC4 (A) and TRPC6 (B). (A) Whole-cell currents elicited by voltage ramps from +100 mV to –100 mV in a HEK293 cell that coexpressed TRPC4β and M2 muscarinic receptor during the course of stimulation by carbachol (5 μM). The time points (in seconds) following application of the agonists are indicated. Note the I–V curves at negative potentials, which changed from V-shaped to linear with time. (B) Similar to panel A, but the cell coexpressed TRPC6 and M5 muscarinic receptor, and the stimulation was carbachol (30 μM). The changes in the shapes of the I–V curves are less dramatic as compared to panel A, but the trend is similar.

TRPC-like current in ileal myocytes by receptor agonist or intracellular dialysis of GTPγS (Dresviannikov et al., 2006), indicating that a negative shift of the voltage dependence to physiological potentials constitutes a part of TRPC activation.

Heteromultimerization can also affect voltage dependence. This has been shown for TRPC1–C4 and C1–C5 heteromeric channels, which when activated have stronger outward rectification and about sevenfold smaller unitary conductance at negative potentials than homomeric TRPC4 and C5 channels (Strübing et al., 2001). The measured unitary conductance of homomeric TRPCs vary significantly in the range of 25 to 130 pS. The values can vary at different voltages for the same channel (Obukhov and Nowycky, 2005). The unitary conductance of TRPC1-C5 heteromer, however, was determined to be ~5 pS between potentials –20 to –70 mV (Strübing et al., 2001).

Finally, TRPC trafficking to plasma membrane can also be regulated. Ca^{2+} influx and/ or receptor stimulation have been shown to promote vesicular trafficking to the plasma membrane, and as a result, TRPCs sequestered in vesicular membranes are brought to the plasma membrane, leading to increases in channel density and function. The Ca^{2+} influx that triggers TRPC trafficking may be mediated by a different channel, e.g., ORAI-mediated Ca^{2+} entry was shown to provide the local Ca^{2+} signal to enhance TRPC1 insertion into plasma membrane of cultured human salivary gland cells (Cheng et al., 2011). Additional mechanisms that enhance TRPC channel trafficking include pathways involving PKA, tyrosine kinases, PI3 kinase and cysteine oxidation (Wang et al., 2020).

22.8 Channelopathies and Disease

A large number of missense TRPC6 mutations (~30), mostly gain-of-function mutations but also some loss-of-function ones, have been reported to be linked to familial focal segmental glomerulosclerosis (FSGS). These mutations affect different aspects of TRPC6 function, expression and trafficking that may or may not impair the ion conductance of the channel. The end results are the deregulation of TRPC6 that leads to cell death or cytoskeleton disorganization of podocytes in the kidney, resulting in slit diaphragm disruption and proteinuria.

A gain-of-function TRPC3 mutation (T635A) has been reported in mice with inherited cerebellar ataxia. For TRPC4, diabetic patients with a missense single-nucleotide polymorphism (SNP), I957V, was found to have lower risk of myocardial infarction. SNPs in the promoter and intronic region of TRPCs have also shown association with diseases, including cancer. A more detailed review on TRPC channelopathy can be found in Liu et al. (2020).

22.9 Conclusions

Being the most closely related to the prototypical *Drosophila* TRP channel, mammalian TRPCs have been extensively studied for their involvements in store- and receptor-operated Ca^{2+} entry and membrane potential regulation downstream from activation of PLC pathways following ligand stimulation of G protein-coupled receptors and receptor tyrosine kinases. While all TRPCs can be directly activated by DAG, TRPC4 and TRPC5 also respond to $G_{i/o}$ proteins. The breakdown of PIP_2 and generation of Ca^{2+} and H^+ signals contribute to channel activation. All TRPC channels are Ca^{2+} permeable, but the inward currents at negative potentials are mostly carried by Na^+, eliciting membrane depolarization and/or facilitating Na^+-dependent transport activities. These functions are linked to Ca^{2+} entry and/or membrane potential changes mediated by TRPC channels activated downstream from PLC pathways. The presence of multiple factors that affect the relative response levels of TRPC channels suggests that these channels may serve as coincidence detectors of signaling by divergent inputs, through which they fine-tune the outputs of targeting cells. Recent high-resolution cryo-EM structures of ligand-free and ligand-bound TRPC channels have unveiled great details on TRPC channel assembly and structural organization, as well as the key regulatory sites for gating and ligand interaction. The pharmacology for TRPCs is improving, which together with genetic studies, is expected to greatly improve our understanding of the physiological functions of TRPC channels and their contributions to disease.

Suggested Reading

Asanov, A., A. Sampieri, C. Moreno, J. Pacheco, A. Salgado, R. Sherry, and L. Vaca. 2015. "Combined single channel and single molecule detection identifies subunit composition of STIM1-activated transient receptor potential canonical (TRPC) channels." *Cell Calcium* 57(1): 1–13. doi: 10.1016/j. ceca.2014.10.011.

Bai, Y., X. Yu, H. Chen, D. Horne, R. White, X. Wu, P. Lee, et al. 2020. "Structural basis for pharmacological modulation of the TRPC6 channel." *eLife* 9: e53311. doi: 10.7554/eLife.53311.

Birnbaumer, L., X. Zhu, M. Jiang, G. Boulay, M. Peyton, B. Vannier, D. Brown, et al. 1996. "On the molecular basis and regulation of cellular capacitative calcium entry: Roles for Trp proteins." *Proc. Natl. Acad. Sci. U. S. A.* 93(26): 15195–15202. doi: 10.1073/pnas.93.26.15195.

Cheng, K. T., X. Liu, H. L. Ong, W. Swaim, and I. S. Ambudkar. 2011. "Local Ca^{2+} entry via Orai1 regulates plasma membrane recruitment of TRPC1 and controls cytosolic Ca^{2+} signals required for specific cell functions." *PLOS Biol.* 9(3): e1001025. doi: 10.1371/journal.pbio.1001025.

Dresviannikov, A. V., T. B. Bolton, and A. V. Zholos. 2006. "Muscarinic receptor-activated cationic channels in murine ileal myocytes." *Br. J. Pharmacol.* 149(2): 179–187. doi: 10.1038/sj.bjp.0706852.

Eder, P., M. Poteser, C. Romanin, and K. Groschner. 2005. "Na^+ entry and modulation of Na^+/Ca^{2+} exchange as a key mechanism of TRPC signaling." *Pflugers Arch.* 451(1): 99–104. doi: 10.1007/s00424-005-1434-2.

Fan, C., W. Choi, W. Sun, J. Du, and W. Lü. 2018. "Structure of the human lipid-gated cation channel TRPC3." *eLife* 7: e36852. doi: 10.7554/eLife.36852.

Gees, M., B. Colsoul, and B. Nilius. 2010. "The role of transient receptor potential cation channels in Ca^{2+} signaling." *Cold Spring Harb. Perspect. Biol.* 2(10): a003962–a003962.

Huang, J., C. H. Liu, S. A. Hughes, M. Postma, C. J. Schwiening, and R. C. Hardie. 2010. "Activation of TRP channels by protons and phosphoinositide depletion in *Drosophila* photoreceptors." *Curr. Biol.* 20(3): 189–197. doi: 10.1016/j.cub.2009.12.019.

Jeon, J., F. Bu, G. Sun, J. B. Tian, S. M. Ting, J. Li, J. Aronowski, et al. 2021. "Contribution of TRPC channels in neuronal excitotoxicity associated with neurodegenerative disease and ischemic stroke. *Front. Cell Dev. Biol.* 8: 618663. doi: 10.3389/fcell.2020.618663.

Jeon, J. P., C. Hong, E. J. Park, J. H. Jeon, N. H. Cho, I. G. Kim, H. Choe, et al. 2012. "Selective $G\alpha_i$ subunits as novel direct activators of transient receptor potential canonical (TRPC)4 and TRPC5 channels." *J. Biol. Chem.* 287(21): 17029–17039. doi: 10.1074/jbc.M111.326553.

Kwan, H. Y., B. Shen, X. Ma, Y. C. Kwok, Y. Huang, Y. B. Man, S. Yu, X. Yao. 2009. "TRPC1 associates with BK(Ca) channel to form a signal complex in vascular smooth muscle cells." *Circ. Res.* 104(5): 670–678. doi: 10.1161/CIRCRESAHA.108.188748.

Lichtenegger, M., O. Tiapko, B. Svobodova, T. Stockner, T. N. Glasnov, W. Schreibmayer, D. Platzer, et al. 2018. "An optically controlled probe identifies lipid-gating fenestrations within the TRPC3 channel." *Nat. Chem. Biol.* 14(4): 396–404. doi: 10.1038/s41589-018-0015-6.

Liu, X., X. Yao, and S. Y. Tsang. 2020. "Post-translational modification and natural mutation of TRPC channels." *Cells* 9(1): 135. doi: 10.3390/cells9010135.

Montell, C., L. Birnbaumer, V. Flockerzi, R. J. Bindels, E. A. Bruford, M. J. Caterina, D. E. Clapham, et al. 2002. "A unified nomenclature for the superfamily of TRP cation channels." *Mol. Cell* 9(2): 229–231. doi: 10.1016/s1097-2765(02)00448-3.

Montell, C. and G. M. Rubin. 1989. "Molecular characterization of the *Drosophila trp* locus: A putative integral membrane protein required for phototransduction." *Neuron* 2(4): 1313–1323. doi: 10.1016/0896-6273(89)90069-x.

Obukhov, A. G. and M. C. Nowycky. 2005. "A cytosolic residue mediates Mg^{2+} block and regulates inward current amplitude of a transient receptor potential channel." *J. Neurosci.* 25(5): 1234–1239. doi: 10.1523/JNEUROSCI.4451-04.2005.

Poteser, M., H. Schleifer, M. Lichtenegger, M. Schernthaner, T. Stockner, C. O. Kappe, T. N. Glasnov, et al. 2011. "PKC-dependent coupling of calcium permeation through transient receptor potential canonical 3 (TRPC3) to calcineurin signaling in HL-1 myocytes." *Proc. Natl. Acad. Sci. U. S. A.* 108(26): 10556–10561. doi: 10.1073/pnas.1106183108.

Prakriya, M., S. Feske, Y. Gwack, S. Srikanth, A. Rao, and P. G. Hogan. 2006. "Orai1 is an essential pore subunit of the CRAC channel." *Nature* 443(7108): 230–233. doi: 10.1038/nature05122.

Putney Jr., J. W. 1986. "A model for receptor-regulated calcium entry." *Cell Calcium* 7(1): 1–12. doi: 10.1016/0143-4160(86)90026-6.

Song, K., M. Wei, W. Guo, L. Quan, Y. Kang, J. X. Wu, and L. Chen. 2021. "Structural basis for human TRPC5 channel inhibition by two distinct inhibitors." *eLife* 10: e63429. doi: 10.7554/eLife.63429.

Strübing, C., G. Krapivinsky, L. Krapivinsky, and D. E. Clapham. 2001. "TRPC1 and TRPC5 form a novel cation channel in mammalian brain." *Neuron* 29(3): 645–655. doi: 10.1016/s0896-6273(01)00240-9.

Thakur, D. P., J. B. Tian, J. Jeon, J. Xiong, Y. Huang, V. Flockerzi, and M. X. Zhu. 2016. "Critical roles of $G_i/_o$ proteins and phospholipase C-δ1 in the activation of receptor-operated TRPC4 channels." *Proc. Natl. Acad. Sci. U. S. A.* 113(4): 1092–1097. doi: 10.1073/pnas.1522294113.

Thakur, D. P., Q. Wang, J. Jeon, J. B. Tian, and M. X. Zhu. 2020. "Intracellular acidification facilitates receptor-operated TRPC4 activation through PLCδ1 in a Ca²⁺-dependent manner." *J. Physiol.* 598(13): 2651–2667. doi: 10.1113/JP279658.

Tian, J., D. P. Thakur, and M. X. Zhu. 2014. "TRPC channels." In: *Handbook of Ion Channels*, edited by Zheng, J. and M. C. Trudeau, 411–426, CRC Press.

Wang, H., X. Cheng, J. Tian, Y. Xiao, T. Tian, F. Xu, X. Hong, M. X. Zhu. 2020. "TRPC channels: Structure, function, regulation and recent advances in small molecular probes." *Pharmacol. Ther.* 209: 107497. doi: 10.1016/j.pharmthera.2020.107497.

Xu, S. Z., B. Zeng, N. Daskoulidou, G. L. Chen, S. L. Atkin, and B. Lukhele. 2012. "Activation of TRPC cationic channels by mercurial compounds confers the cytotoxicity of mercury exposure." *Toxicol. Sci.* 125(1): 56–68. doi: 10.1093/toxsci/kfr268.

Zhang, Z., J. Tang, S. Tikunova, J. D. Johnson, Z. Chen, N. Qin, A. Dietrich, et al. 2001. "Activation of Trp3 by inositol 1,4,5-trisphosphate receptors through displacement of inhibitory calmodulin from a common binding domain." *Proc. Natl. Acad. Sci. U. S. A.* 98(6): 3168–3173. doi: 10.1073/pnas.051632698.

Zhao, J., J. V. Lin King, C. E. Paulsen, Y. Cheng, and D. Julius. 2020. "Irritant-evoked activation and calcium modulation of the TRPA1 receptor." *Nature* 585(7823): 141–145. doi: 10.1038/s41586-020-2480-9.

Zhu, X., M. Jiang, M. Peyton, G. Boulay, R. Hurst, E. Stefani, and L. Birnbaumer. 1996. "*trp*, a novel mammalian gene family essential for agonist-activated capacitative Ca²⁺ entry." *Cell* 85(5): 661–671. doi: 10.1016/s0092-8674(00)81233-7.

23

TRPM Channels

David D. McKemy

CONTENTS

23.1 Introduction

TRPM channels are a genetically and functionally diverse subfamily of the transient receptor potential family of ion channels and are comprised of eight members (Figure 23.1). Few characteristics are commonplace between these channels, but some share significant functional similarity (Huang et al., 2020). For example, TRPM2, TRPM6 and TRPM7 are

DOI: 10.1201/9781003096276-23

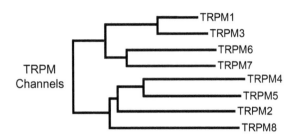

FIGURE 23.1
The phylogenetic tree of TRPM channels organized based on homology at the amino acid level.

unique among known ion channels in that they also contain enzymatically active protein domains in their C-termini. Moreover, TRPM6 and TRPM7 are functionally similar to each other, as are TRPM4 and TRPM5, yet a divergent expression profile between each member of these pairings leads to significantly different physiological roles.

TRPM1 (otherwise known as melastatin) was the first TRPM identified and is considered a marker for metastasized melanomas. TRPM2 has a unique nudix hydrolase domain that is homologous to the ADP pyrophosphatase NUDT9, which serves as a binding site for channel activation by adenine nucleotides. TRPM3 is an enigmatic channel in that it has several alternatively spliced variants. TRPM4 and TRPM5 are the only members of the TRP channel family that are selective for only monovalent cations and do not permeate Ca^{2+}. However, both are activated by intracellular Ca^{2+}, serving as transduction channels downstream of increased cellular Ca^{2+} levels. Like TRPM2, TRPM6 and TRPM7 possess both ion channel and enzymatic domains, in this case atypical protein kinases that are structurally similar to PKA. These channels appear to serve as sensors of the levels of intracellular Mg^{2+} and are important for general Mg^{2+} homeostasis. TRPM8 was first identified in prostate, in which it is proposed to be an androgen-responsive channel, but the channel chiefly serves as a sensor of cold temperatures in the peripheral nervous system. Thus, TRPM channels are a diverse array of ion channels key in a number of cellular processes. The following addresses salient points of each channel, putting them in the context of molecular and structural differences between each protein, as well as describes what is known of their basic cellular functions at this time.

23.2 TRPM1

TRPM1 is the founding member of the TRPM subfamily of TRP ion channels and was originally termed "melastatin" when first identified in a screen for genes involved in melanoma metastasis (Duncan et al., 1998). In this screen, TRPM1 was found to be robustly downregulated in highly metastatic cell variants, consistent with subsequent analyses of expression patterns in melanocytic tumors where there is an inverse correlation with TRPM1 expression and melanoma metastasis, melanocytic tumor progression, and melanoma tumor thickness. Thus, the expression of TRPM1 serves as a marker for the progression of melanoma metastasis. In addition to melanocytes, TRPM1 transcripts were also found in the mouse eye (Duncan et al., 1998), specifically in retinal bipolar ON cells with recent evidence suggesting that it mediates a cationic current that modulates activity of retinal bipolar ON cells (Morgans et al., 2009).

23.2.1 Molecular Structure

It is notable that as the founding member of the TRPM channel subfamily, the initial transcript identified in mouse B16 cells only encodes a protein of 542 amino acids that lacks transmembrane domains and a putative pore region of the channel (Duncan et al., 1998). However, the human clone subsequently identified in retina encoded a protein of 1533 amino acids with six predicted transmembrane domains and homology to other channels of the TRP family (Figure 23.2). The full-length cDNAs encode 1516 to 1643 residue proteins suggesting that the shorter melastatin cDNA represents an amino (N)-terminal splice variant of a full-length isoform. Indeed, several splice variants of TRPM1 have been identified. For example, a short N-terminal variant directly interacts with the full-length channel and inhibits its translocation to the plasma membrane, thereby suppressing channel function. In addition to melanocyte and retina expression, TRPM1 has also been detected in brain and heart tissues, albeit at very low levels. The TRPM1 gene also encodes a microRNA, termed miR-211, that is also expressed in both the skin and retina, and proposed to regulate function in these tissues.

23.2.2 Cellular Function

Remarkably, despite years of research since the discovery of TRPM1 in melanocytes, the role of the channel in melanoma advancement is poorly understood. Within the TRPM1 promoter region, binding sites for the microphthalmia transcription factor (MITF), a helix–loop–helix leucine zipper transcription factor, were identified. This finding is important because MITF has been shown to regulate expression of several melanocyte-specific genes, including TRPM1 (Miller et al., 2004). Expression of many predicted targets for miR-211 were reduced in melanoma cell lines ectopically expressing this miRNA, and MITF is required for miR-211 expression, suggesting a potential tumor-suppressor mechanism. One significant confound in the study of TRPM1 in melanocyte function is the absence of robust animal models as rodents do not have melanocytes. In humans, melanocytes generate the pigment melanin in organelles, termed "melanosomes," which are then transferred to keratinocytes where they protect cells from ultraviolet damage. Activation of the melanocortin-1 receptor (MC1R), a Gαs protein-coupled receptor, regulates the synthesis of proteins needed to produce melanin, a process mediated by MITF expression, suggesting the involvement of TRPM1. TRPM1 expression correlates with melanin content, and signaling via metabotropic glutamate receptor subtype 6 (mGluR6) and changes in Ca^{2+} signaling in melanocytes regulated melanin content. Furthermore, inhibition of cellular TRPM1 expression reduces melanin content in melanocytes indicating the channel plays a role in normal melanocyte function and regulates pigmentation.

In the retina, photoreceptors make synapses with ON and OFF bipolar cells, which transmit visual contrast to ganglion cells. ON cells detect increases in light intensity and are depolarized subsequent to photoreceptor hyperpolarization, which reduces the rate of glutamate release that occurs in response to light. The ON bipolar cell response is mediated by mGluR6 that, when activated, couples to the Go class of trimeric G proteins that regulate a depolarizing current in these cells. While this pathway was well appreciated, the identity of the channel responsible for the depolarizing current was unknown. TRPM1 expression was found in retinal cells isolated via mGluR6 expression and functional evidence suggested the mGluR6-dependent current was likely to be a TRP channel, supporting a role for TRPM1 channels in light-evoked responses of ON bipolar cells. This hypothesis was confirmed by analyses of TRPM1$^{-/-}$ mice (Morgans et al., 2009). Electroretinogram

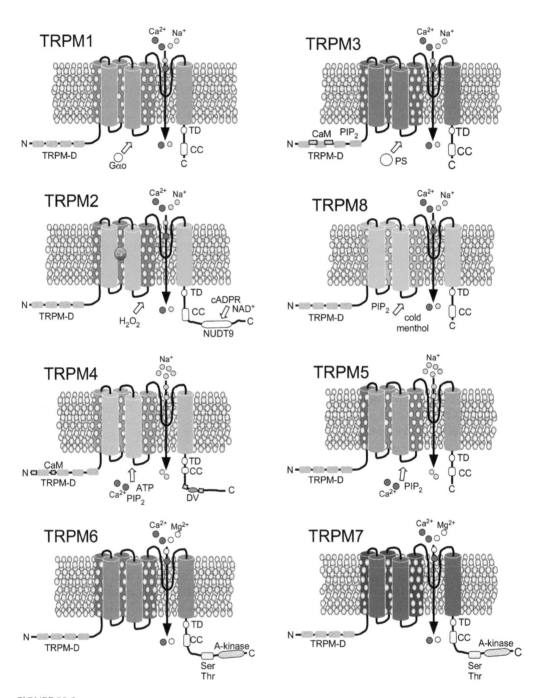

FIGURE 23.2
General structures localized to TRPM channels. Each channel has up to four TRPM homology domains (TRPM-D) as well as the TRP (TD) and coiled-coil domains (CC). TRPM1 channels are sensitive to the G protein Gαo. TRPM2 channels contain an enzymatic nudix homology domain in the C-terminus (NUDT9), similar to the A-kinase domains in TRPM6 and TRPM7, with the former containing several ATP (not shown) and vanadate binding domains (DV). TRPM3 and TRPM4 channels contain Ca^{2+}/calmodulin-binding domains (CaM). Most all channels are sensitive to either PIP2 directly or as a consequence of PIP2 cleavage.

analysis of TRPM1⁻/⁻ mice found that b-waves associated with bipolar cell responses were abolished, yet these animals retained normal a-waves. Moreover, whole-cell patch-clamp recordings from ON bipolar cells in retinal slices from TRPM1⁻/⁻ mice found that light-evoked responses were abolished. These and other results identified TRPM1 as a visual transduction channel in mammalian retina.

23.3 TRPM2

One the distinguishing features of certain members of the TRPM subfamily is that, in addition to having ion-permeating channel domains, they also have intracellular regions that have enzymatic function. TRPM2 is one such multifunctioning protein in that it is a nonselective cation channel, but also has an enzymatic domain in its C-terminus that is important for gating of the channel (Sumoza-Toledo and Penner, 2011). In addition, TRPM2 channels are important functionally in a variety of different cell types, including pancreatic β-cells, and immune, endothelial, neuronal, and cardiac cells, supporting functions ranging from insulin release to the production of cytokines, cell motility to cell death. Unlike other TRPMs, TRPM2 channels show little voltage dependence with a linear current–voltage relationship that reverses at 0 mV (Perraud et al., 2001). TRPM2 is critical for Ca^{2+} influx in these cells, as well as is localized intracellularly to the lysosome where it serves as a Ca^{2+} release channel. Last, TRPM2 is considered an intriguing target for diseases associated with oxidative stress, such as inflammation and neurodegeneration, as it is activated by reactive oxygen species (ROS) such as hydrogen peroxide.

23.3.1 Molecular Structure

TRPM2 is one of several TRPM channels in which the protein structure has been solved (Huang et al., 2020; Yu et al., 2021). TRPM2 is unique among TRPM channels in that, like TRPM6 and TRPM7 (see Sections 23.7 and 23.8, respectively), it is considered a "chanzyme" in that it contains an NUDT9 adenosine diphosphate ribose (ADPR) pyrophosphatase domain (nudix-like homology domain) of >300 amino acids in its C-terminus (Figure 23.2) (Perraud et al., 2001). The human TRPM2 gene encodes a protein of 1503 amino acids consisting of six transmembrane domains and the C-terminus comprising the NUDT9 domain. Additionally, there are four TRPM homology domains (MHD) and a calmodulin-binding IQ-like domain in the N-terminus, and the characteristic TRP box and coiled-coil region between the sixth transmembrane domain and the NUDT9 domain.

Several splice variants have been identified, and similar to TRPM1, a truncated form containing only residues corresponding to the N-terminus is suggested to have a dominant negative role in TRPM2 channel activity in tissues such as brain, marrow and the vasculature. Others include those lacking residues in both the N- and C-terminus, variants that have proven informative regarding channel function. Specifically, channel function is abolished in the N-terminal variant, which is missing 20 amino acids corresponding to an IQ-like calmodulin-binding site as well as two PxxP protein–protein interaction motifs. However, mutation of individual residues within these regions had no effect on channel function, suggesting that individually they are dispensable for TRPM2 activation. The C-terminal splice variant lacks 34 residues that are within the NUDT9 domain and thus has been shown to be insensitive to ADPR, but responds normally to H_2O_2. However, as with the N-terminal variant, no specific residues were shown to be necessary for ADPR-induced gating, leading to speculation that the sequences deleted represent a spacer segment that stabilizes this region. Last, while ADPR is an agonist, its hydrolysis is not required for channel gating.

23.3.2 Cellular Function

TRPM2 channels are activated by ADPR, cyclic ADPR, nicotinamide dinucleotide (NAD^+), and ROS such as H_2O_2 and other substances that generate reactive oxygen and nitrogen species; and TRPM2 channel activation requires Ca^{2+} (McHugh et al., 2003). The channel functions in both Ca^{2+} influx through the plasma membrane and as a Ca^{2+} release channel in lysosomes. In addition, TRPM2 requires the presence of the phospholipid phosphatidylinositol-4,5-bisphosphate (PIP2). Based on the profile of agonists activating TRPM2, the channel is thought to be important in cell death associated with oxidative stress. For example, expression of TRPM2 confers susceptibility of heterologous cells to H_2O_2, leading to cell death that is associated with an increase in intracellular Ca^{2+}. Moreover, downregulation of TRPM2 expression significantly reduced cell death induced by oxidative stress in several cell lines. In pancreatic β-cells, H_2O_2 and alloxan induce TRPM2-mediated Ca^{2+} release from lysosomes inducing apoptotic cell death. However, TRPM2$^{-/-}$ mice have higher levels of blood glucose that was found to associate with impaired glucose-stimulated insulin secretion in both the whole animal and at the cellular level, also suggesting a potential role for TRPM2 in diabetes.

In addition, oxidative stress is critical in neural degeneration, and TRPM2 channels are most robustly expressed in the brain where they have been linked to Alzheimer's and Parkinson's disease, as well as amyotrophic lateral sclerosis. Ca^{2+} excitotoxicity plays a key role in the neuropathology associated with Alzheimer's, and it has been shown that H_2O_2-induced Ca^{2+} entry in cortical and basal ganglia neurons involves TRPM2 channels. In line with the linkage of TRPM2 and oxidative stress, ADPR is also generated by the breakdown of ADPR polymers by poly-ADPR glycohydrolase (PARG), an enzyme whose activity increases as a result of DNA damage. Thus, the channel plays a key role in these processes and will be a target of much needed research in the future.

23.4 TRPM3

TRPM3 is one of the least understood members of the TRPM subfamily, as the gene encodes at least 23 splice variants, thereby creating ambiguity in the study of channel function due to the uncertainty into the physiologically relevant isoforms. Moreover, the endogenous activators of TRPM3 have yet to be elucidated, but the channel is activated by the neurosteroid pregnenolone sulfate (PS), which has enabled characterization of channel expression and function in various cell types. TRPM3 expression has been detected in an array of different tissues including the brain (dentate gyrus; choroid plexus), as well as in the pituitary, pancreas, ovaries, testis, sensory ganglia and spinal cord. However, the functional relevance of the channel is most well understood in pancreatic β-cells, primary sensory neurons, the retina and in various cancers.

23.4.1 Molecular Structure

As with other TRPM channels, TRPM3 has six transmembrane domains, a C-terminal coiled-coil region, as well as the TRPM homology domain located in the N-terminus of the channel

(Figure 23.2). Moreover, two calmodulin-binding sites are also present in the N-terminus, and it has been shown that calmodulin can bind to heterologous proteins containing these sequences, suggesting regulation of the channel by Ca^{2+} (see later). As with most TRPM channels, there is a PIP2 binding site that interacts with N-terminal calmodulin domains.

The TRPM3 gene is remarkably large (850 kb in mouse) with introns separating the first few exons spanning over 250 kb of DNA sequence. Thus, it is not surprising that TRPM3 encodes for the largest number of splice variants of TRP channels, presenting a divergent population of channels with different functional domains and gating properties. In most cases, these variants are derived from alternative splicing mainly in the 5 region of the transcript, as well as in the putative pore domain. For example, coding sequences corresponding to exon 2 are only found in one putative transcript in humans. Several splice variants alter sequences near the putative pore of the channel, and it has been found that cation permeability is affected in two mouse variants in exon 24 of the channel that differ only in the pore region. In most studies, TRPM3α2 is the variant of choice and lacks 12 amino acid residues within the pore region but is highly permeable to divalent cations. Another defining feature of the TRPM3 gene is that, like TRPM1 (see earlier), a microRNA, miR-204, is located in intron 8 of the mouse sequence. miR-204 is one of the most extensively studied miRNA in the eye and is known to be physiological in human retinal pigmented epithelium, although its exact role is not understood.

23.4.2 Cellular Function

To date, analyses of TRPM3 channel function, regardless of the variant tested, find that the channel is constitutively active when heterologously expressed, and is permeable to divalent cations such as Ca^{2+} and Mg^{2+}. TRPM3 channels are also blocked by gadolinium and lanthanum ions and by 2-axminoethoxydiphenyl borate (2-APB), an inhibitor of IP3 receptors, store-operated channels (SOC), and several TRP channels. The pore is also permeable to manganese, and channel activity is blocked by hypertonic solutions and increased by hypotonicity, suggesting that TRPM3 channels in the kidney are involved in osmoregulation.

A number of studies have shown that TRPM3 channels are activated by PS in pancreatic β-cells and the INS-1 cell line, two cells types shown to express a PS-sensitive channel. Indeed, the majority of functional data related to TRPM3 is based on cellular currents activated by PS. TRPM3 is also activated by nifedipine, a dihydropyridine inhibitor of voltage-gated Ca^{2+} channels. Conversely, cholesterol and mefenamic acid block TRPM3 channels, as do rosiglitazone and troglitazone, two agonists for peroxisome proliferator-activated receptors (PPARg), and a number of citrus fruit flavanones. These agonists and antagonists have been instrumental in determining the potential roles of TRPM3 in physiology.

In pancreatic cells, PS-potentiated glucose induced insulin secretion from islets, effects that were blocked by mefenamic acid. Consistent with these data, PS increases intracellular Ca^{2+} in these cells, a cellular effect that is also blocked by mefenamic acid, which was ineffective in altering classical K_{ATP}-dependent pathways, thereby suggesting that TRPM3 channels work independent of glucose-mediated insulin secretion. It has been proposed that activation of TRPM3 in the pancreas leads to sufficient depolarization such that L-type Ca^{2+} channels open and change the levels of intracellular Ca^{2+} such that changes in gene expression are induced, likely through the Raf and ERK protein kinase pathways and increased production of the zinc finger transcription factor Egr-1. However, TRPM3-deficient mice exhibit normal resting levels of blood glucose, and insulin release is also normal, suggesting a minor role for the channel in glucose homeostasis (Vriens et al., 2011).

In neuronal tissues, TRPM3 is expressed in oligodendrocytes, and activation of the channel leads to increased intracellular Ca^{2+}, suggesting involvement in myelination. Similarly,

PS has been found to regulate the formation of glutamatergic synapses in the developing nervous system by enhancing glutamate release in synapses on Purkinje cells. However, the most well-characterized role for neuronally expressed TRPM3 is in small-diameter neurons of the dorsal root and trigeminal ganglia where it was shown to mediate aversive responses to PS in mice (Vriens et al., 2011). TRPM3 channels are activated by heat and TRPM3-deficient mice show deficits in avoidance of noxious heat, as well as the development of heat hypersensitivity associated with inflammation. TRPM3 expression in the eye is also robust and detected in the retinal pigment epithelia, iris and retina. Again, PS sensitivity was used to show functional TRPM3 channels in retinal ganglion cells, although to date there are no reported deficiencies in visual processing in mice lacking TRPM3 channels. However, a point mutation in TRPM3 discovered in humans is linked to high-tension glaucoma and autosomal dominant cataracts.

23.5 TRPM4

Of the TRPM subfamily, TRPM4 (as well as TRPM5) is distinct in that the channel is not permeable to Ca^{2+}. However, Ca^{2+} is the primary regulator of channel function in that it can be gated by an increase in intracellular Ca^{2+}. Indeed, even though the channel does not permit Ca^{2+} entry directly, its activity is closely linked to store-operated Ca^{2+} entry (SOCE), as Na^+ entry via TRPM4 depolarizes the membrane thereby decreasing the driving force for Ca^{2+} entry via the ORAI1/STIM1 complex. SOCE is linked to a variety of cellular functions including those associated with many cancers. Another distinction of TRPM4 is that the channel appears to be expressed ubiquitously, and due to its activation by increased intracellular Ca^{2+}, the channel plays an important regulatory role in Ca^{2+}-dependent processes, such as cardiac function and constriction of cerebral arteries, insulin secretion, activity of inspiratory neurons, the immune response, and keratinization disorders.

23.5.1 Molecular Structure

TRPM4 channels are most closely related to TRPM5 in structure (~40% homologous at the amino acid level), but as with other TRPM channels described, the TRPM4 gene does encode for several splice variants, with TRPM4b considered the predominant channel variant and has been the primary transcript studied (Figure 23.2). Human and mouse TRPM4b genes encode proteins of 1214 and 1213 amino acids, respectively, and as with other members of the TRPM subfamily TRPM4 contains four TRPM homology domains in the N-terminus, a coiled-coil domain on the C-terminus, as well as putative calmodulin-binding sites in both the N- and C-terminus. TRPM4 also contains motifs known to interact with one another, including two ATP-binding cassettes (ABC) transporter-like motifs, in addition to four Walker nucleotide binding domains (NBD) that are localized in close proximity (Ullrich et al., 2005). Last, TRPM4 channels have a PIP2 binding site, homologous to pleckstrin homology (PH) domains that confer the channel's sensitivity to phosphoinositides (Autzen et al., 2018).

As described, TRPM4 is fairly ubiquitously expressed in a number of tissues, including expression of mRNA detected by Northern blot analysis in pancreas, skeletal muscle, heart, liver, kidney, placenta, colon, lung, thymus, spleen, prostate and small intestine from human tissues (Launay et al., 2004). Intracellular Ca^{2+} plays a significant role in most all physiological responses and thus is strongly regulated by many cellular elements. Thus, within these tissues, TRPM4 channels serve as a sensor of intracellular Ca^{2+}. Although the exact location of the Ca^{2+} binding site has yet to be determined, it is thought to reside in the first four transmembrane domains and thereby comprises a Ca^{2+}-activated nonselective cation channel (NSC_{Ca}). TRPM4 channel activation enables Na^+ entry into the cell, altering the driving forces for Ca^{2+} and other ions by changing the cellular membrane potential. Specifically, it has been shown that TRPM4 affects the activity of a variety of cellular ionic conductances such as voltage-gated and non-voltage-gated Ca^{2+} channels.

Channel function of both TRPM4 and TRPM5 is robustly sensitive to the phosphoinositide PIP2. Depletion of PIP2 in both excised membrane patches and by intracellular manipulation reduces channel activity, effects that are reversed by application of exogenous PIP2. Both channels are activated by Ca^{2+} in excised patches, but their Ca^{2+} sensitivity decreases with time. This desensitization can be reversed by the application of PIP2. TRPM4 is more sensitive to PIP2 in comparison to TRPM5 channels, and the levels of this phosphoinositide alters Ca^{2+} sensitivity and shifts the voltage sensitivity of each channel.

23.5.2 Cellular Function

TRPM4 is expressed in several cell types in the heart, including sinoatrial node pacemaking cells, and ventricular and atrial myocytes. In a rat model of hypertrophy, TRPM4 expression is increased in ventricular myocytes, consistent with the correlation of channel expression and cell capacitance, the latter a marker of hypertrophy. Arrhythmias related to hypertrophy involve a calcium-activated transient inward current (Iti) responsible for both delayed afterdepolarizations and early afterdepolarizations, and it has been suggested that TRPM4 channels may participate.

In the immune system, TRPM4 appears to function to lessen the immune response by preventing Ca^{2+} overload in both T-lymphocytes and dendritic cells (Barbet et al., 2008). Dendritic cells fail to migrate to lymphoid organs after being stimulated by a bacterial infection in TRPM4-deficient mice. TRPM4 is also implicated in inflammation and the allergic response due to increased activation of mast cells in TRPM4-deficient mice, as well as a decrease in mast cell migration during inflammation. Thus, TRPM4 channels play a key role in immune responses associated with inflammation and other disorders.

Like many TRP ion channels, TRPM4 is expressed in keratinocytes, and missense mutations in the channel are reported to lead to hyperactivity and enhanced intracellular calcium that are associated with a form of progressive symmetric erythrokeratodermia (Wang et al., 2019). Similarly, the release of insulin from pancreatic β-cells in response to a glucose challenge is dependent on a rise in intracellular Ca^{2+}. TRPM4 channels are expressed in β-cells and it has been shown that glucose-stimulated insulin release from an immortalized β-cell line is reduced if TRPM4 channel activity is inhibited. However, in TRPM4-deficient mice, glucose and insulin homeostasis were normal in vivo and in vitro, calling into question the role of the channel in the pancreas. These, and other results, show that TRPM4 is involved in many cellular processes, a diversity in expression that has hampered interpretation of the role of this channel.

23.6 TRPM5

TRPM5, like TRPM4, is a Ca^{2+}-activated cation channel selective for monovalent ions and impermeable to divalent cations. However, unlike TRPM4, TRPM5 expression is restricted to only a few tissues, and therefore its role in physiology is similarly constrained. TRPM5 is found at high levels in the tongue, small intestine and stomach, as well as in pancreatic β-cells where it appears to localize with insulin. TRPM5 channel activity is strongly regulated by voltage, plasma membrane phosphoinositides, acidity and temperature, and is required for transduction of bitter, sweet and umami tastes. Moreover, TRPM5-deficient mice display impaired glucose homeostasis. Thus, like TRPM4, TRPM5 channels mediate depolarizing currents that regulate excitable cells, but in a more defined set of physiological responses.

23.6.1 Molecular Structure

TRPM5 was originally identified in a screen for genes with homology to TRPM1, and encodes for a protein of 1165 amino acids in humans, and a slightly smaller peptide of 1158 amino acids in the mouse (Figure 23.2). To date, only a single functional transcript of TRPM5 has been identified and studied, and structurally TRPM5 does not contain many of the different motifs found in TRPM4. For example, TRPM5 lacks an ABC-transporter-like motif and possesses only one Walker B site that is located in a region of the channel that is inaccessible to ATP (Ullrich et al., 2005). These sequence differences equate to function as TRPM4 is sensitive to block by decavanadate, presumably through the ABC motifs, whereas TRPM5 activity was unaffected.

For both TRPM4 and TRPM5, a direct comparison of Ca^{2+}-sensitivity of each channel found that TRPM5 is activated at much lower Ca^{2+} concentrations than TRPM4, but that both channels strongly desensitize when studied in excised membrane patches. TRPM5 displays multimodal $Ca2+$ sensitivity with activation requiring membrane depolarization at low Ca^{2+} concentrations, whereas the voltage dependence of the channel is essentially linear at high Ca^{2+} concentrations. TRPM5 channels are not sensitive to regulation by the Ca^{2+}-binding protein calmodulin, but application of phosphoinositides such as PIP2 diminish desensitization, suggesting coordination in the intracellular events regulating channel function (Liu and Liman, 2003). TRPM5 has been most extensively studied in taste cells where it is activated downstream of G protein-coupled receptors and PLC-mediated production of inositol trisphosphate (IP3), which is generated by the cleavage of PIP2, thereby implicating negative-feedback inhibitory pathway regulating channel function.

23.6.2 Cellular Functions

TRPM5 is robustly expressed in taste receptor cells and is coexpressed with G protein-coupled receptors for bitter, sweet and umami tastants, as well as genes essential for taste transduction including the G protein gustducin and PLCβ2 (Perez et al., 2002). The role of TRPM5 in taste was established when it was shown that TRPM5$^{-/-}$ mice were insensitive to these taste modalities, but still detect sour and salty tastants (Zhang et al., 2003). Mechanistically, TRPM5 serves as the downstream target of tastant receptor activation, which leads to the activation of PLC and the hydrolysis of PIP2 into diacylglycerol (DAG) and IP3, the latter inducing the liberation of Ca^{2+} from intracellular stores via IP3 receptors. As earlier, Ca^{2+} directly gates TRPM5 channels, and tastant-evoked responses are blocked when intracellular Ca^{2+} is buffered or IP3 receptors are inhibited (Liu and Liman, 2003).

As stated earlier, in comparison to TRPM4, TRPM5 is less broadly expressed, but like TRPM4, it is found in pancreatic islets and predominantly in cells containing insulin, suggesting expression in insulin-secreting β-cells (Colsoul et al., 2010). β-cells express a Ca^{2+}-sensitive nonselective cation channel, a current that is significantly reduced in TRPM5$^{-/-}$ mice, suggesting that TRPM5 channels contribute to the electrical activity of β-cells under conditions of glucose-induced stimulation. Consistent with these cellular data, pancreatic islets obtained from TRPM5$^{-/-}$ mice exhibit deficits in glucose-induced insulin release, and these animals perform poorly in glucose tolerance tests and have lower plasma insulin levels.

Last, TRPM5 is also expressed in the gut in a chemosensory cell type that releases hormones such as glucagon-like peptide 1 (GLP-1). The channel, as well as other molecules normally associated with taste cells, is found in a cohort of intestinal solitary brush cells and endocrine cells of the duodenal glands, suggesting that the channel may serve to detect chemicals in the gut lumen. In a model of type 2 diabetes, TRPM5 expression inversely correlated with blood glucose concentration, and enteroendocrine cells in mouse duodenum that express TRPM5 also express β-endorphin and other peptides that may have local roles in regulating intestinal function when under high glucose or hyperosmotic conditions.

23.7 TRPM6

Like TRPM4/TRPM5, the channels TRPM6 and TRPM7 are closely related, with TRPM7 ubiquitously expressed, whereas TRPM6 expression is restricted to the intestine and kidney. These channels are also unique in that they possess kinase domains residing in their C-termini, making them, like TRPM2, multifunctional channels or chanzymes. Physiologically, TRPM6 and TRPM7 are critical for magnesium homeostasis, with a genetic defect in the former found to underlie hypomagnesemia with secondary hypocalcemia (HSH), an autosomal recessive disorder that is characterized by low serum magnesium (Walder et al., 2002). Thus, TRPM6 channels have a clear physiological role, but the mechanisms underlying how they regulate magnesium levels and if they form functional channels in the absence of TRPM7 are unclear. Despite these parallels, the two channels do not appear to compensate for each other physiologically.

23.7.1 Molecular Structure

TRPM6 encodes for a protein of 2022 amino acids in humans and 2028 residues in mouse, shares 50% homology with TRPM7 (Figure 23.2), and was originally cloned in a screen for transcripts with homology to elongation factor kinase 2. Structurally, TRPM6 contains many similarities to other TRPM channels, but as described earlier contains an A-kinase domain in its C-terminus, and is reported to form both homotetrameric and heterotetrameric channels, the latter with TRPM7. Indeed, the ubiquitous nature of TRPM7 expression suggests that TRPM6 may form functional channels when oligomerized with TRPM7, the latter suggested to be necessary for trafficking of the channel to the plasma membrane. However, the oligomerization of TRPM6 is still under debate, with evidence suggesting that there are different channel properties (i.e., conductance, ion permeation and pharmacological sensitivity) when each channel is individually expressed, compared to when both TRPM6 and TRPM7 are coexpressed (Li et al., 2006). TRPM6 currents are strongly outwardly rectifying with the small amount of current passed at negative membrane potentials made up primarily of Mg^{2+} or Ca^{2+}, but the channel also permeates a range of cations with little evidence that monovalent cations are conducted through the pore.

23.7.2 Cellular Functions

Due to the channel's relatively exclusive expression in the kidney and intestine, as well as its link to HSH, the majority of what is known of TRPM6 function at the cellular level arises from analysis of Mg^{2+} homeostasis. Mg^{2+} is essential for energy metabolism and any defects in Mg^{2+} balance affects a number of physiological processes. Mg^{2+} concentration is regulated by dietary absorption in the intestine and kidney excretion, as well as from bone. Individuals with HSH have low levels of both Mg^{2+} and Ca^{2+}, a result of poor Mg^{2+} absorption in the intestine and altered renal excretion, but can be easily treated by supplemental Mg^{2+}. The gene locus for HSH was mapped to TRPM6 and shown that loss-of-function mutations of the channel were causative for this disease (Walder et al., 2002). TRPM6 channels are regulated by receptors such as epidermal growth factor receptor (EGFR), whose activation leads to translocation of TRPM6 channels to the membrane of kidney cells.

Consistent with its role in Mg^{2+} homeostasis, TRPM6 was shown to be expressed on the apical membrane of the renal distal convoluted tubule as well as along the brush-border membrane of the small intestine. Channel function is regulated by the levels of intracellular Mg^{2+} and enhanced by extracellular acidification. The A-kinase domain on TRPM6 has been shown to be nonessential for channel activity, but does appear to play a regulatory role in modulating channel function. The channel can be autophosphorylated, as well as increases phosphate incorporation in TRPM7 channels. However, the channel is inhibited by RACK1, a receptor for activated C-kinase 1 that interacts with the A-kinase domain of TRPM6 when autophosphorylated. Similarly, repressor of estrogen receptor activity (REA) also binds to this region and inhibits channel function in a kinase-dependent manner. Autophosphorylation regulates the ability of Mg^{2+} to inhibit TRPM6 channel function, thereby encompassing a feedback inhibitory mechanism to prevent Mg^{2+} overload.

23.8 TRPM7

As described earlier, TRPM7 channels are closely related to TRPM6, but its tissue distribution is much wider and has been implicated in a number of cellular functions, including Mg^{2+} homeostasis, cell death, proliferation, neuronal degeneration and the cell cycle. Moreover, and likely due to this diversity in expression, considerably more research has been performed on TRPM7 channel function. Deletion of TRPM7 in mice results in embryonic lethality, further evidence as to the general importance of this channel. TRPM7 channels are enhanced when intracellular Mg^{2+} concentration plummets and when placed in acidic environments, whereas channel function is diminished when intracellular levels of PIP2 are reduced in a PLC-dependent manner.

23.8.1 Molecular Structure

The TRPM7 gene encodes for a protein of 1865 amino acids in humans, and a slightly smaller protein of 1863 residues in mice (Figure 23.2) (Runnels et al., 2001). As noted earlier, TRPM7 is closely homologous to TRPM6 (49%–52% sequence identity between these two proteins) and contains much the same molecular architecture as TRPM6, including the C-terminal A-kinase domain. TRPM7 is a ubiquitously expressed protein with expression found even in embryonic stem cells and is required for embryonic development. Moreover, native TRPM7 currents have been recorded in every cell type examined to date.

TRPM7 channels are strongly outwardly rectifying and permeable to mainly divalent cations such as Zn^{2+}, Ni^{2+}, Mg^{2+}, Ca^{2+} and Mn^{2+}, in that order, yet is normally inhibited under normal physiological conditions (Ryazanova et al., 2010). However, when intracellular divalent concentrations are reduced, TRPM7 channel activity increases, mainly conducting K^+ and Na^+ currents. Moreover, physiological concentrations of Mg^{2+}-bound nucleotides, such as Mg-ATP, serve as a source of Mg^{2+} and are also reported to inhibit TRPM7 channels. The crystal structure of TRPM7 revealed that partially hydrated Mg^{2+} ions locate to the center of the pore of the channel (Duan et al., 2018). As with TRPM6, TRPM7 channels are also pH sensitive, in this case increased protons lead to an increase in inward current comprised mostly of monovalent cations as the extracellular environment acidifies.

23.8.2 Cellular Function

Due to its ubiquitous expression, there is an overabundance of data examining TRPM7 function in various cell and tissue types, but these studies are hampered by the embryonic lethality of TRPM7-null mice (Jin et al., 2008). Moreover, mice in which only the kinase domain of TRPM7 is lacking are also embryonic lethal, but supplemental Mg^{2+} can restore the viability of embryonic stem cells from these mice, suggesting the importance of Mg^{2+} homeostasis as the underlying cause of death. However, deletion of TRPM7 from thymocytes does not alter Mg^{2+} homeostasis, but does block development of these cells. Moreover, induced deletion of TRPM7 in adult mice resulted in no observable phenotype, suggesting that its role in development is critical for viability (Jin et al., 2012). Indeed, TRPM7 function has been linked to cell survival in a number of tissues, including mast cells, vascular smooth muscle, hepatocytes, and various carcinomas.

It is still controversial as to whether the kinase domain is involved in channel gating, as inactivating mutations in the kinase domain do not affect gating. Moreover, mice expressing a kinase dead mutant do not exhibit a development or Mg^{2+} homeostasis phenotype (Kaitsuka et al., 2014). That being said, TRPM7 has been shown to phosphorylate a number of substrates, and expression of truncated kinase domain in vivo does affect Mg^{2+} homeostasis (Ryazanova et al., 2010), suggesting the kinase domain can modulate channel function. The relevance of TRPM7 in Mg^{2+} is critical, as this ion has a diverse array of functions in development and cell viability. Global knockout of TRPM7 is embryonic lethal in mice, whereas a cell-specific deletion in cardiac cells altered heart development (Jin et al., 2008, 2012).

23.9 TRPM8

TRPM8 channels are most closely related to TRPM2 in sequence, but lack the enzymatic domains located on the latter's C-terminus. TRPM8 was first identified in a screen for genes upregulated in cancerous prostate epithelial cells, but then later found to be the receptor for menthol, the cooling ingredient in mint. Consistent with the psychophysical sensation of cold induced by menthol, TRPM8 channels are also temperature sensitive and directly gated when temperatures drop below 26°C (McKemy et al., 2002). Thus, TRPM8 falls into the class of TRP channels that serve as temperature detectors in the peripheral nervous system. TRPM8 mediates both innocuous cool and noxious cold perception, heightened responses to cold after injury or disease, as well as is critical for pain-relieving aspects of cooling in conditions of neuropathic and inflammatory pain. Moreover, TRPM8 serves an as yet undefined role in several malignant cell types, as well as is critical in controlling thermoregulatory responses.

23.9.1 Molecular Structure

Similar to other TRPM channels, TRPM8 has six transmembrane domains as well as the characteristic TRP and coiled-coil domains in the C-terminal region (Figure 23.2). The coiled-coil domains in the C-terminus are reported to be necessary for tetramer assembly and channel trafficking (Diver et al., 2019). TRPM8 is activated by several cooling compounds including menthol and icilin, and regions in the channel important for each agonist have been identified. Icilin activates TRPM8 in a manner that requires a coincident increase in intracellular Ca^{2+}, a requirement not needed for menthol or cold gating of the channel. Icilin's actions are localized to the linker region between the second and third TM with residues N799, D802 and G805 shown to be critical. However, how Ca^{2+} and icilin interact to gate TRPM8 is unclear.

A key modulator of TRPM8 channel activity is the phosphoinositide PIP2 that has been shown to be obligatory for TRPM8 channel function. TRPM8 channels adapt to prolonged stimulation in a manner that requires the presence of external Ca^{2+}, and channel activity in excised membrane patches rapidly runs down after excision, but can be recovered by the presence of exogenous PIP2 on the cytoplasmic face of the channel (Rohacs et al., 2005). Activation of PLC also leads to channel adaptation, suggesting that Ca^{2+} entry through the TRPM8 pore activates Ca^{2+} sensitive PLCδ isoforms to cleave PIP2 and thereby adapt TRPM8 channels. Moreover, residues in the proximal region of the C-terminus near the TRP domain of TRPM8 have been proposed to interact with PIP2.

The mechanisms underlying gating TRPM8 channels by cold are still unresolved, as is the case for all thermosensitive TRP channels. TRPM8 channels are weakly voltage-sensitive and show characteristic outward rectification. This voltage sensitivity, along with the topological similarity between TRP channels and voltage-gated K^+ channels, suggested that activation by temperature and voltage might be linked. However, the closest TRPM8 homologue TRPM2 has an identical pattern of putative-charged residues involved in cold gating as TRPM8, but is not temperature- or voltage-sensitive. Moreover, there is evidence that temperature-, agonist- and voltage-dependent gating are independent processes since distinct activation domains for each have been identified, suggesting that the effect of one gating mechanism acts on another in an allosteric fashion. Last, purified TRPM8 channels have been reconstituted into a planar lipid bilayer and shown to be activated by cold, the most convincing evidence that this channel is directly gated by temperature.

23.9.2 Cellular Function

In sensory ganglia, TRPM8 is expressed in <15% of small-diameter (~20 μm) sensory neurons, consistent with the proportion of neurons shown to be cold- and menthol-sensitive in neuronal cultures. Biophysically, cold- and menthol-evoked TRPM8 currents have surprisingly similar properties to conductances recorded in native cells, including ion selectivity, menthol potency and voltage dependence. More remarkably, TRPM8 currents are also evoked by temperature decreases with an activation temperature threshold of ~26°C, with activity increasing in magnitude down to 8°C. Interestingly, this broad range spans what are considered both innocuous cool (~30°C–15°C) and noxious cold temperatures (<15°C). Moreover, a number of cold-mimetic compounds activate TRPM8 channels and will induce characteristic shivering or "wet dog" shakes when given intravenously in rodents, a process that requires TRPM8 channels and neurons. Similarly, a number of antagonists have been identified and are shown to induce a dramatic hypothermic response in mice, likely due to inhibition of cold-sensing afferents. Similarly, TRPM8 is functionally present in brown adipose tissue (BAT) and stimulate brown adipocytes with menthol, upregulated UCP1 expression, a regulator of thermogenesis.

Menthol and cooling are commonly used as topical analgesic, thereby suggesting that activation of TRPM8 may lead to pain relief. Topical application of cold or cooling compounds produces a temporary analgesic effect mediated by TRPM8-expressing afferents. In a rodent model of neuropathic pain, paw withdrawal latencies in response to mechanical or heat stimuli were significantly attenuated in animals first treated with cold or cooling compounds such as icilin. Analgesia persists for over 20–30 minutes after which the animals regained their hypersensitivity similar to before the cool stimuli were applied and at levels of those not pretreated with cold or cooling compounds. Only modest cooling and low doses of cooling compounds produced analgesia and is dependent on TRPM8 channels and neurons. Together these data indicate that TRPM8 is mediating the analgesia provided by cool temperatures and cooling compounds, suggesting that modest activation of TRPM8 afferent nerves can serve as an endogenous mechanism to promote pain relief.

TRPM8 also serves other biological roles in addition to neuronal thermal sensing. Along with expression in normal prostate, TRPM8 is expressed in other nonprostatic tumors such as breast, colon, lung, and skin, as well as in the bladder and male genital tract (Tsavaler et al., 2001). TRPM8 transcripts have been found in the gastric fundus and cooling as a result of the consumption of cold foods inducing contraction of gastrointestinal smooth muscles, resulting in a short-lived gastric voiding. These and other data strongly suggest that TRPM8 may have diverse biological functions outside of the peripheral nervous system.

Acknowledgments

This work was supported by US National Institutes of Health grants (NS106888 and NS0118852).

Suggested Reading

Autzen, H. E., A. G. Myasnikov, M. G. Campbell, D. Asarnow, D. Julius, and Y. Cheng. 2018. "Structure of the human TRPM4 ion channel in a lipid nanodisc." *Science* 359(6372):228–32. doi: 10.1126/science.aar4510.

Barbet, G., M. Demion, I. C. Moura, N. Serafini, T. Leger, F. Vrtovsnik, R. C. Monteiro, R. Guinamard, J. P. Kinet, and P. Launay. 2008. "The calcium-activated nonselective cation channel TRPM4 is essential for the migration but not the maturation of dendritic cells." *Nat Immunol* 9(10):1148–56. doi: 10.1038/ni.1648.

Colsoul, B., A. Schraenen, K. Lemaire, R. Quintens, L. Van Lommel, A. Segal, G. Owsianik, K. Talavera, T. Voets, R. F. Margolskee, Z. Kokrashvili, P. Gilon, B. Nilius, F. C. Schuit, and R. Vennekens. 2010. "Loss of high-frequency glucose-induced Ca2+ oscillations in pancreatic islets correlates with impaired glucose tolerance in *Trpm5$^{-/-}$* mice." *Proc Natl Acad Sci U S A* 107(11):5208–13. doi: 10.1073/pnas.0913107107.

Diver, M. M., Y. Cheng, and D. Julius. 2019. "Structural insights into TRPM8 inhibition and desensitization." *Science* 365(6460):1434–40. doi: 10.1126/science.aax6672.

Duan, J., Z. Li, J. Li, R. E. Hulse, A. Santa-Cruz, W. C. Valinsky, S. A. Abiria, G. Krapivinsky, J. Zhang, and D. E. Clapham. 2018. "Structure of the mammalian TRPM7, a magnesium channel required during embryonic development." *Proc Natl Acad Sci U S A* 115(35):E8201–E10. doi: 10.1073/pnas.1810719115.

Duncan, L. M., J. Deeds, J. Hunter, J. Shao, L. M. Holmgren, E. A. Woolf, R. I. Tepper, and A. W. Shyjan. 1998. "Down-regulation of the novel gene melastatin correlates with potential for melanoma metastasis." *Cancer Res* 58(7):1515–20.

Huang, Y., R. Fliegert, A. H. Guse, W. Lu, and J. Du. 2020. "A structural overview of the ion channels of the TRPM family." *Cell Calcium* 85:102111. doi: 10.1016/j.ceca.2019.102111.

Jin, J., B. N. Desai, B. Navarro, A. Donovan, N. C. Andrews, and D. E. Clapham. 2008. "Deletion of Trpm7 disrupts embryonic development and thymopoiesis without altering Mg2+ homeostasis." *Science* 322(5902):756–60. doi: 10.1126/science.1163493.

Jin, J., L. J. Wu, J. Jun, X. Cheng, H. Xu, N. C. Andrews, and D. E. Clapham. 2012. "The channel kinase, TRPM7, is required for early embryonic development." *Proc Natl Acad Sci U S A* 109(5):E225–E33. doi: 10.1073/pnas.1120033109.

Kaitsuka, T., C. Katagiri, P. Beesetty, K. Nakamura, S. Hourani, K. Tomizawa, J. A. Kozak, and M. Matsushita. 2014. "Inactivation of TRPM7 kinase activity does not impair its channel function in mice." *Sci Rep* 4:5718. doi: 10.1038/srep05718.

Launay, P., H. Cheng, S. Srivatsan, R. Penner, A. Fleig, and J. P. Kinet. 2004. "TRPM4 regulates calcium oscillations after T cell activation." *Science* 306(5700):1374–7. doi: 10.1126/science.1098845.

Li, M., J. Jiang, and L. Yue. 2006. "Functional characterization of homo- and heteromeric channel kinases TRPM6 and TRPM7." *J Gen Physiol* 127(5):525–37. doi: 10.1085/jgp.200609502.

Liu, D., and E. R. Liman. 2003. "Intracellular Ca2+ and the phospholipid PIP2 regulate the taste transduction ion channel TRPM5." *Proc Natl Acad Sci U S A* 100(25):15160–5. doi: 10.1073/pnas.2334159100.

McHugh, D., R. Flemming, S. Z. Xu, A. L. Perraud, and D. J. Beech. 2003. "Critical intracellular Ca2+ dependence of transient receptor potential melastatin 2 (TRPM2) cation channel activation." *J Biol Chem* 278(13):11002–6. doi: 10.1074/jbc.M210810200.

McKemy, D. D., W. M. Neuhausser, and D. Julius. 2002. "Identification of a cold receptor reveals a general role for TRP channels in thermosensation." *Nature* 416(6876):52–8. doi: 10.1038/nature719.

Miller, A. J., J. Du, S. Rowan, C. L. Hershey, H. R. Widlund, and D. E. Fisher. 2004. "Transcriptional regulation of the melanoma prognostic marker melastatin (TRPM1) by MITF in melanocytes and melanoma." *Cancer Res* 64(2):509–16. doi: 10.1158/0008-5472.can-03-2440.

Morgans, C. W., J. Zhang, B. G. Jeffrey, S. M. Nelson, N. S. Burke, R. M. Duvoisin, and R. L. Brown. 2009. "TRPM1 is required for the depolarizing light response in retinal ON-bipolar cells." *Proc Natl Acad Sci U S A* 106(45):19174–8. doi: 10.1073/pnas.0908711106.

Perez, C. A., L. Huang, M. Rong, J. A. Kozak, A. K. Preuss, H. Zhang, M. Max, and R. F. Margolskee. 2002. "A transient receptor potential channel expressed in taste receptor cells." *Nat Neurosci* 5(11):1169–76. doi: 10.1038/nn952.

Perraud, A. L., A. Fleig, C. A. Dunn, L. A. Bagley, P. Launay, C. Schmitz, A. J. Stokes, Q. Zhu, M. J. Bessman, R. Penner, J. P. Kinet, and A. M. Scharenberg. 2001. "ADP-ribose gating of the calcium-permeable LTRPC2 channel revealed by Nudix motif homology." *Nature* 411(6837):595–9. doi: 10.1038/35079100.

Rohacs, T., C. M. Lopes, I. Michailidis, and D. E. Logothetis. 2005. "PI(4,5)P2 regulates the activation and desensitization of TRPM8 channels through the TRP domain." *Nat Neurosci* 8(5):626–34. doi: 10.1038/nn1451.

Runnels, L. W., L. Yue, and D. E. Clapham. 2001. "TRP-PLIK, a bifunctional protein with kinase and ion channel activities." *Science* 291(5506):1043–7. doi: 10.1126/science.1058519.

Ryazanova, L. V., L. J. Rondon, S. Zierler, Z. Hu, J. Galli, T. P. Yamaguchi, A. Mazur, A. Fleig, and A. G. Ryazanov. 2010. "TRPM7 is essential for Mg(2+) homeostasis in mammals." *Nat Commun* 1:109. doi: 10.1038/ncomms1108.

Sumoza-Toledo, A., and R. Penner. 2011. "TRPM2: A multifunctional ion channel for calcium signalling." *J Physiol* 589(7):1515–25. doi: 10.1113/jphysiol.2010.201855.

Tsavaler, L., M. H. Shapero, S. Morkowski, and R. Laus. 2001. "Trp-p8, a novel prostate-specific gene, is up-regulated in prostate cancer and other malignancies and shares high homology with transient receptor potential calcium channel proteins." *Cancer Res* 61(9):3760–9.

Ullrich, N. D., T. Voets, J. Prenen, R. Vennekens, K. Talavera, G. Droogmans, and B. Nilius. 2005. "Comparison of functional properties of the Ca2+-activated cation channels TRPM4 and TRPM5 from mice." *Cell Calcium* 37(3):267–78. doi: 10.1016/j.ceca.2004.11.001.

Vriens, J., G. Owsianik, T. Hofmann, S. E. Philipp, J. Stab, X. Chen, M. Benoit, F. Xue, A. Janssens, S. Kerselaers, J. Oberwinkler, R. Vennekens, T. Gudermann, B. Nilius, and T. Voets. 2011. "TRPM3 is a nociceptor channel involved in the detection of noxious heat." *Neuron* 70(3):482–94. doi: 10.1016/j.neuron.2011.02.051.

Walder, R. Y., D. Landau, P. Meyer, H. Shalev, M. Tsolia, Z. Borochowitz, M. B. Boettger, G. E. Beck, R. K. Englehardt, R. Carmi, and V. C. Sheffield. 2002. "Mutation of TRPM6 causes familial hypomagnesemia with secondary hypocalcemia." *Nat Genet* 31(2):171–4. doi: 10.1038/ng901.

Wang, H., Z. Xu, B. H. Lee, S. Vu, L. Hu, M. Lee, D. Bu, X. Cao, S. Hwang, Y. Yang, J. Zheng, and Z. Lin. 2019. "Gain-of-function mutations in TRPM4 activation gate cause progressive symmetric erythrokeratodermia." *J Invest Dermatol* 139(5):1089–97. doi: 10.1016/j.jid.2018.10.044.

Yu, X., Y. Xie, X. Zhang, C. Ma, L. Liu, W. Zhen, L. Xu, J. Zhang, Y. Liang, L. Zhao, X. Gao, P. Yu, J. Luo, L. H. Jiang, Y. Nie, F. Yang, J. Guo, and W. Yang. 2021. "Structural and functional basis of the selectivity filter as a gate in human TRPM2 channel." *Cell Rep* 37(7):110025. doi: 10.1016/j.celrep.2021.110025.

Zhang, Y., M. A. Hoon, J. Chandrashekar, K. L. Mueller, B. Cook, D. Wu, C. S. Zuker, and N. J. Ryba. 2003. "Coding of sweet, bitter, and umami tastes: Different receptor cells sharing similar signaling pathways." *Cell* 112(3):293–301. doi: 10.1016/s0092-8674(03)00071-0.

24

TRPV Channels

Tamara Rosenbaum

CONTENTS

24.1 Introduction

Transient receptor potential vanilloid (TRPV) channels are composed of a protein subfamily with six members (TRPV1–TRPV6) that form tetramers with varying cation selectivity. These channels are polymodal as they are gated by a wide diversity of mechanisms and serve as hygro- and mechanosensors in insects; chemical and mechanical sensors in nematodes; and temperature, chemical and osmotic sensors in several organs of vertebrates. As will be discussed, their importance is exemplified by the multiple roles they play in physiology.

24.2 Physiological Roles

24.2.1 TRPV1

The transient receptor potential vanilloid 1 (TRPV1) is an extensively studied member of the TRP family of ion channels. It is mainly expressed by small-diameter neurons within

DOI: 10.1201/9781003096276-24

sensory ganglia such as the dorsal root (DRG) and trigeminal ganglia (TG). TRPV1 is poly-modal and responds to a variety of stimuli including noxious heat (>43°C), small chemical agonists and animal peptide toxins (Table 24.1) (Caterina and Julius, 2001). Also, extracel-lular protons activate TRPV1 and, since they are released during tissue injury, ischemia and inflammation, they play important roles in producing pain through the activation of TRPV1. Protons also sensitize activation of the channel by capsaicin and heat. Although protons function as endogenous positive regulators of TRPV1 activation by potentiating gating of the channel, it has also been shown that they inhibit permeation through the open channel (Lee and Zheng, 2015).

TRPV1 responds to inflammatory mediators in the organism and plays an important role in signaling pathways associated with pain and inflammation, since its activation is one of the first molecular events that lead to the perception of pain. Hence, TRPV1 has become a promising target for pain-relieving therapies (Lee and Zheng, 2015).

Nonetheless, it must be noted that, although the roles of TRPV1 have been far better explored in neuronal cells, this channel also participates in the physiology of non-neu-ronal tissues. For example, TRPV1 plays important roles in the function of respiratory airways, and chronic cough patients exhibit higher sensitivity to capsaicin and increased channel expression levels in their airways (Emir, 2017). In smooth muscle and endothelial cells, TRPV1 is also expressed and thought to contribute to vasculature control induced by chronic hypoxia (Du et al., 2019).

TRPV1 is also expressed in keratinocytes, which are cells that sustain the integrity of the immune response in the skin. Several studies have shown that TRPV1 is expressed in human epidermal keratinocytes and, while no direct measurements of TRPV1 currents have been reported in these cells, increases in Ca^{2+} levels in cultured human epidermal keratinocytes have been observed in response to acidification and capsaicin (and blocked by capsazepine). Moreover, epidermal proliferation in response to ATP release in skin, which is also blocked by capsazepine, has led to the proposal that TRPV1 and purinergic receptors participate in low pH-promoted skin proliferation (Denda et al., 2001).

24.2.2 TRPV2

TRPV2 in sensory neurons is activated by changes in temperature (>52°C), acts as a lipid and cannabinoid receptor, and responds to mechanical and osmotic stimuli. TRPV2 is expressed in several cell types and in particular in intracellular membranes (i.e., endo-somes) (Shibasaki, 2016).

This channel is found in the brain, indicating that it may be implicated in autonomic regulation (i.e., appetite and cardiovascular regulation), cerebellum, retina, and trigeminal and dorsal root ganglia. In skeletal and cardiac muscles and in stomach neurons, TRPV2 responds to stretch and shear stress and to hypotonicity. In myocyte intercalated discs, TRPV2 possibly functions as a stretch-activated channel, playing important roles in the heart during mechanical stimulation of cardiomyocytes. Cardiomyocytes from TRPV2-deficient mice present Ca^{2+}-handling defects and reduced contractile responses during electrical stimulation, reduced survival, and severe decline in cardiac contractility (Falcón et al., 2019).

The expression of TRPV2 in Aδ and Aβ fibers points to a nociceptive role in detection of harmful chemical stimuli and transduction of high thermal pain, and its expression level is upregulated after nerve injury. On the other hand, in some cells this channel regulates axonal outgrowth when activated in a stretch-dependent manner. It is also expressed in epithelial and innate immune cells, in the stomach, and astrocytes (Emir, 2017).

TABLE 24.1

Agonists and Antagonists of TRPV Channels

Channel	Agonists				Antagonists		
TRPV1	**Endogenously Produced**						
	LPA	5(S)-HPETE	18:1 NOE/OEA	OAG	RA	RvE1	Thapsigargin
	LTB4	12(S)-HEPE	N-acylsalsolinols	PEA	OA	RvD2	Yohimbine
	OLDA	15-(S)-HPETE	Prostaglandins	NADA	EPA	Adenosine	α-Spinasterol
	20-HEPE	9-HODE	Spermine	2-AG			
	CPA 18:1	13-HODE	Oxytocin	NAT			
	AEA		NAE				
	Exogenously Produced						
	Capsaicin	Evodiamine	Gingerol	DkTx	Capsazepine	SB-452533	AG489
	RTX	Cannabidiol	Zingerone	RhTx	JNJ17203212	JYL1421	AG505
	THCV	Camphor	Shogaol	BmP01	SB-705498	ABT-102	APHC1
	Geraniol	Alicin	Citronellol	VaTx1-3	AMG9810	SB-782443	APHC3
	Olvanil	Eugenol	Polygodial	2-APB	SB-366791	AMG517	HCRG21)
	Phar	Piperine	BrP-LPA		A-425619	DD161515	5-iodoRTX
TRPV2	**Exogenously Produced**						
	Cannabidiol	2-APB	THC		Probenicid	Tranilast (SKF96365)	
TRPV3	**Endogenous**						
	FPP	ETYA	DPBA	IPP	PI(4,5)P2	17R-RvD1	
	Exogenously Produced						
	Camphor	Carveol	Menthol (murine)	THCV	2-APB	GRC 15133	
	Thymol	Vanillin	Incensole acetate	Δ(9)-THC DPBA	DPTHF	GRC 17173	
	Carvacrol	Citral		Cannabidiol			
	Eugenol	2-APB					
TRPV4	**Endogenously Produced**						
	DMAPP	14,15-EET	N-arachidonoyl taurine	Gd³⁺	La³⁺		
	5,6-EET	AA	AEA				
	Exogenously Produced						
	4α-PDD	BAA	GSK1016790A Cannabidiol	Citral	RN-1734	GSK2798745	
	4α-LPDD	Apigenin	Cannabidivarin	Oxazinin A	RN-9893	SiNP	
	4α-PD	Eugenol		Onydecalin A	GSK205	Crotamiton	
	PMA	PDDHV		Aintennol	HC-067047	Piperidine-benzimidazole	
	THCV	RN-1747		Butamben	GSK3527497		

(Continued)

TABLE 24.1 (CONTINUED)

Agonists and Antagonists of TRPV Channels

Channel	Agonists		Antagonists		
TRPV5	Endogenously Produced				
	OAG PI(4,5)P2	β-glucuronidase Klotho	Mg^{2+}		$Pb^{2+} = Cu^{2+} > Zn^{2+} > Co^{2+} > Fe^{2+}$
			tTG		80K-H protein
	Exogenously Produced				
			Econazole	THCV	ZINC17988990
			Miconazole	ZINC9155420	TH-1177
TRPV6	Endogenously Produced				
	PI(4,5)P2 1,25[OH]2D3		Mg^{2+}	Gd^{3+}	2-APB
	Exogenously Produced				
			Econazole	Xestospongin C	TH-1177 THCV
			Miconazole	ZINC9155420	Tamoxifen Soricidin

Sources: Reviewed in Holzer and Izzo (2014) and Blair et al. (2019).Abbreviations: Trigeminal ganglio (TG), dorsal root ganglia (DRG), 5-(S)-hydroperoxyeicosatetraenoic acid (5(S)-HPETE), 12-(S)-hydroxyeicosapentaenoic acid (12(S)-HEPE), 15-(S)-hydroperoxyeicosatetraenoic acid [15-(S)-HPETE], leukotriene B4 (LTB4), 20-hydroxyeicosapentaenoic acid (20-HEPE), 9-hydroxyoctadecadienoic acids (9-HODE), 13-hydroxyoctadecadienoic acids (13-HODE), N-arachidonoyl taurine (NAT), lysophosphatidic acid (LPA), N-oleoyl ethanolamine (18:1 NOE or OEA), palmitoylethanolamide (PEA), N-arachidonoyl dopamine (NADA), 2-arachidonoylglycerol (2-AG), N-oleoyldopamine (OLDA), anandamide (AEA), N-acylethanolamines (NAE), hydrogen sulfide (H_2S), cyclic phosphatidic acid (CPA 18:1), 1-oleoyl-2-acetyl-sn-glycerol (OAG), resiniferatoxin (RTX), Δ9-tetrahydrocannabivarin (THCV), Δ(9)-tetrahydrocannabinol (Δ(9)-THC), vanillotoxins from *Psalmopoeus cambridgei* (VaTx1-3), double-knot toxin (DkTx), 1-bromo-3(S)-hydroxy-4-(palmitoyloxy)butyl-phosphonate (BrP-LPA), 2-aminoethoxyldiphenyl borate (2-APB), ricinoleic acid (RA), oleic acid (OA), eicosapentaenoic acid (EPA), resolvin E1 (RvE1), resolvin D2 (RvD2), *Heteractis crispa* RG 21 (HCRG21), (–)-trans-Δ⁹-tetrahydrocannabidol (THC), phenylborate (DPBA), farnesyl pyrophosphate (FPP), 5,8,11,14-eicosatetraynoic acid (ETYA), diphenylboronic anhydride (DPBA), isopentenyl pyrophosphate (IPP), phosphatidylinositol 4,5-bisphosphate (PI(4,5)P2), 17(R)-resolvin D1 (17R-RvD1), 2, 2-diphenyltetrahydrofuran (DPTHF), dimethylallyl pyrophosphates (DMAPP), 5,6-epoxyeicosatrienoic acid (5,6-EET), 14,15-epoxyeicosatrienoic acid (14, 15-EET), arachidonic acid (AA), 4α-phorbol 12,13-didecanoate (4α-PDD), 4α-lumiphorbol didecanoate (4α-LPDD), 4α-phorbol monoester (4α-PD), phorbol 12-myristate 13-acetate (PMA), 2-phorbol 12,13-didecanoate 20-homovanillate (PDDHV), bisandrographolide A (BAA), silica nanoparticles (SiNP), 1-oleoyl-acetyl-sn-glycerol (OAG), tissue transglutaminase (tTG), 1,25-dihydroxyvitamin D3 (1,25[OH]2D3).

24.2.3 TRPV3

The TRPV3 channel participates in warm sensation and pain, and is mainly expressed in skin, but also in other epithelia in the nose, palate, hair follicle, tongue, colon, and sensory ganglia, and in the brain, spinal cord, and testes. This channel exhibits use-dependent and hysteretic temperature-dependent activation since initial activation requires temperatures above 50°C and then it can be opened by lower temperatures (around 30°C) (Nilius et al., 2014).

TRPV3 has been targeted as a key molecule for treatment of itch and other skin conditions, for its involvement in skin barrier formation and function, hair development, and cutaneous sensations that also include pain. TRPV3-coupled signaling mechanisms participate in cutaneous nociception in epidermal keratinocytes, since overexpression of TRPV3 is also accompanied by increased release of prostaglandin E2 (PGE2), activating adjacent sensory afferents. Moreover, activation of TRPV3 in keratinocytes leads to the release of ATP and nitric oxide that can also influence pain states (Nilius et al., 2014).

Supporting evidence for the role of TRPV3 in nociception also comes from the effects of resolvins, molecules linked to anti-inflammatory and antinociceptive effects, such as 17(R)-resolvin D1 that inhibits this channel by affecting its voltage dependence and decreasing its activation.

24.2.4 TRPV4

TRPV4 is a polymodal protein activated by temperatures around 27°C, hypoosmotic conditions, mechanical stress and some compounds (Liedtke, 2005; Nilius et al., 2007), with ubiquitous expression (Rosenbaum et al., 2020).

The influx of Ca^{2+} from the extracellular milieu is extremely relevant for the function of the endothelium and TRPV4 is one of the molecules that participates in this process. In vascular endothelial cells it contributes to relaxing effects of endocannabinoids on the vascular tone and blood pressure regulation and the supply of glucose and oxygen.

TRPV4's mechanotransducing roles are significant within Müller cells, ganglion cell soma-dendrite and microglia in the retina, and the channel has been highlighted as a molecular effector in glaucoma.

In the respiratory system, TRPV4 is found in several types of cells, where it maintains homeostasis of osmotic pressure and integrates diverse stimuli into Ca^{2+} signals, regulating relaxation of the main pulmonary artery and vasoconstriction of pulmonary circulation. Its function can alter the integrity of the lung walls, influence the severity of chronic asthma and play an important role in pulmonary injury induced by the mechanical force of ventilators used to treat respiratory failure.

In the kidney, TRPV4 is present in the distal convoluted tubule in regions that do not exhibit apical water permeability and where transcellular osmotic gradients can develop. In the thick ascending limb, cell swelling can occur, which activates TRPV4, regulating the osmotic balance by modifying water secretion in the kidney. TRPV4 is expressed in regions of the nephrons, which are water-impermeant, allowing detection of osmotic stimuli and regulation of blood pressure in the presence of increased salt intake (Emir, 2017).

In osteoblasts, which are important for skeletal homeostasis, as well as in chondrocytes, TRPV4 expression is required for mechanotransduction. TRPV4 also contributes to skin homeostasis by preserving the skin barrier, regulating hair follicle growth, and by participating in immunological and itch responses in some pathologies (Emir, 2017).

24.2.5 TRPV5 and TRPV6

Among TRP channels, TRPV6 (formerly known as CaT1 and ECaC2) retains the highest homology with TRPV5 (previously known as ECaC1), although the former is found in a wider variety of cells.

Both channels also differ from other members of the vanilloid subfamily in that they are not ligand-activated or thermosensitive, they respond to 1,25-dihydroxyvitamin D_3, and they are localized at the apical membrane of Ca^{2+}-transporting epithelia. Hence, these channels play pivotal roles in the regulation of blood Ca^{2+} levels in higher organisms, as discussed later.

TRPV5 is important for systemic Ca^{2+} homeostasis by functioning as a central protein for fine-tuning of Ca^{2+} reabsorption in the kidney. Mice with genetic deletion of TRPV5 exhibit phenotypes related to decreased active Ca^{2+}-reabsorption such as decreased trabecular and cortical bone thickness with disturbed bone morphology and increased intestinal Ca^{2+} absorption as a result of a compensatory mechanism for impaired renal Ca^{2+} reabsorption (Emir, 2017).

While TRPV5 is mostly expressed in the distal convoluted tubules and connecting tubules of the kidney, TRPV6 is found in the intestine, kidney and exocrine tissues. Hence, TRPV5 participates in determining the level of urinary Ca^{2+} excretion and TRPV6 participates in intestinal Ca^{2+} absorption but also likely contributes to transcellular Ca^{2+} transport in certain parts of the tubular system. TRPV6 also participates in Ca^+-dependent sperm maturation and its malfunction leads to reduced sperm motility and fertility (Yelshanskaya et al., 2021).

24.3 Subunit Diversity and Basic Structural Organization

Similar to other members of the voltage-gated superfamily of ion channels, TRPV channels are tetramers composed of monomers that encompass intracellular N- and C-termini, six transmembrane helices (S1–S6), and a pore region (S5–P–S6). As distant relatives of voltage-gated potassium (Kv) channels, these proteins contain an S1–S4 region that is considered a voltage-sensor-like domain (VSLD). The S5 and S6 stretch out away from this domain and interact with the VSLD of a neighbor monomer to give rise to the tetrameric pore. The selectivity filter is formed by a loop and helix between the S5 and S6, and the TRPV1, TRPV2 and TRPV4 channels exhibit a pore turret formed by an unstructured loop region with 15–25 residues between S5 and the pore helix. It has been demonstrated that the S5–S6 region experiences conformational changes that lead to the opening of the pore and both temperature gating and ligand gating seem to lead to the same conformational change in this activation gate region in the S6 (Cao, 2020; Gao et al., 2016).

TRPV channels monomers contain six N-terminal ankyrin repeats forming a region termed the ankyrin repeat domain (ARD) and usually each repeat is composed of 33 residues (Figure 24.1). These are involved in subunit–subunit interactions and can bind molecules such as allicin that modulate their activity. The N-terminal region contains the N-linker with a helix-turn-helix motif and a helix preceding the S1. Intersubunit interactions are possible through contacts of the ARDs with the β-sheet region (composed of two β-sheets from the N-linker and one β-sheet from the C-terminus) of neighboring monomers.

The C-terminal region contains the TRP domain, a bent α-helix that runs parallel to the membrane and that lodges into the S1–S4 bundle. This helix converts into a loop and a

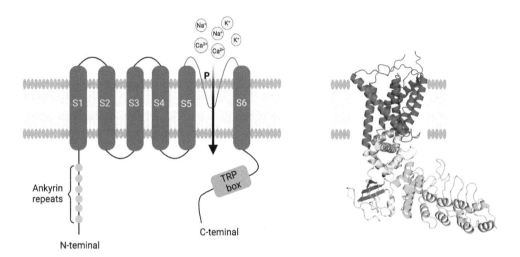

FIGURE 24.1

Structure of TRPV channels. For all TRPV channels, the N-terminal and C-termini contain six ankyrin repeats and a TRP box, respectively (right). Structure of a TRPV1 subunit solved by cryo-EM (Protein Data Bank ID 7LQY, squirrel TRPV1 apo, left). The figure shows the intracellular N- and C-termini, transmembrane domains (S1–S6), where the S1–S4 region is a weak voltage-sensing-like domain (red), and the pore region is formed between S5–P–S6 (blue). (Created with BioRender.com.)

β-strand and interacts with some of the ankyrin repeats of an adjacent subunit. Compared to the C-terminus of Kv channels, this arrangement in TRP channels is unique since in Kv's it is continuous with the S6 segment and does not interact with the N-terminus.

The S4–S5 linker is in a privileged position to regulate channel gating, as it interacts with the S1–S4 bundle, and the TRP and a pore domain from an adjacent monomer. A domain-swapped pore domain entails S5 and S6, which give rise to the central pore and an intracellular gate.

24.4 Gating by Temperature and Ligands

Thermoreceptors are peripheral afferent fibers that respond to temperature changes. The molecular basis of this response are the thermoTRP channels. Among TRP channels, there are nine members that have been highlighted as sensitive to temperature (thermoTRPs) in mammalian cells: TRPV1–4, TRPM2 and TRPM3 respond to heat temperatures (including noxious heat); TRPM8 and TRPC5 to cold and TRPA1 to cold or heat (depending on the species).

ThermoTRPs not only exhibit polymodal activation by being sensitive to the binding of several agonists, but they also share activators, which function as nonspecific agonists of these proteins. They also display activation by covalent modification of intracellular cysteines that leads to the opening of the pore (i.e., TRPV1 and TRPA1) and occurs in the absence of other stimuli, constituting an allosteric mechanism of coupling to pore opening that is yet to be understood.

In canonical voltage-gated ion channels, a voltage-sensing region (with charge z), typically the S4, is displaced through the electrical field produced by the potential difference (V) across the plasma membrane. Although thermoTRP channels show voltage dependence in their activation, the value of z is rather low (~1 e_o) as compared to voltage-gated ion channels such as Kv channels (9–13 e_o). Their S4 lacks the repeating motif (R/KXXR/K; where X is a hydrophobic residue). A possible mechanism for gating by voltage in thermoTRP channels involves the movement of the partial charges of aromatic residues found in the S4 of these proteins in the electrical field, or movement of charged residues near the selectivity filter (Zheng, 2013).

Other differences between gating of thermoTRPs and Kv channels include that the mechanisms leading to heat or cold activation and voltage-dependent opening interact. Although it was first suggested that the same conformational transition (with large enthalpic [ΔH] and entropic changes [ΔS]) was directed by temperature and voltage, it is now more accepted that temperature and voltage influence different conformational changes that might be coupled via allosteric mechanisms.

ThermoTRPs can be directly gated by temperature changes, meaning that they can absorb heat turning it into a conformational change with large enthalpy (ΔH). As temperature increases, the open probability of the channel continuously changes as heat is absorbed (Yao et al., 2010). This increase in open probability as a function of temperature can be quantified by the temperature coefficient or Q_{10}, and a higher value of this parameter signifies a larger enthalpy change linked to the change in open probability. The Q_{10} describes the fold change in current amplitude (I) produced by a 10°C change in temperature (T), as follows:

$$Q_{10} = \left(\frac{I_2}{I_1} \right)^{\frac{10}{(T_2 - T_1)}} \tag{24.1}$$

Gating by other stimuli such as ligand binding or voltage can be described with the same formalism.

Channel opening can occur in the absence of stimulus (C_o–O_2 transition), and heat absorption happens through the C_o–C_1 transition, controlled by the equilibrium constant, K(T). A simple allosteric model for activation by temperature is

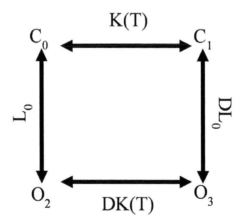

Transition C_1–O_3 leads to temperature-dependent opening of the channel and channel opening is augmented by an allosteric factor D to DL_0. Thus, in the presence of a temperature stimulus, the open probability is increased to

$$\frac{DL_0}{1+DL_0} \tag{24.2}$$

This occurs because, in general, although L_0 may be very small, D is very a large number. When D is a large number, then the open probability can be very close to 1.

For activation by temperature, the equilibrium constant K(T) is given by the quotient of forward and backward rate constants for the C_0–C_1 transition:

$$\frac{k1}{k_{-1}} = \frac{\theta_1 \cdot \exp^{-\left(\Delta H_1^{\ddagger} - T\Delta S_1^{\ddagger}\right)/RT}}{\theta_{-1} \cdot \exp^{-\left(\Delta H_{-1}^{\ddagger} - T\Delta S_{-1}^{\ddagger}\right)/RT}} \tag{24.3}$$

In this equation, θ_i corresponds to the diffusion-limited rate constants at 0 K, $H_i^{\pm}$ and $S_i^{\pm}$ are the activation enthalpies and entropies, R is the gas constant, and T is the temperature in kelvins.

The steady-state open probability as a function of temperature for this allosteric model is given by

$$p(T) = \frac{L_0 + K(T)DL_0}{1 + K + L_0 + K(T)DL_0} \tag{24.4}$$

where K(T) is the temperature-dependent equilibrium constant and L_0 is the temperature-independent equilibrium constant for opening of the channel.

If we assume that only K(T) is a temperature-dependent equilibrium constant and that it represents a particular conformational change propelled by the change in temperature, then Equation 24.1 envisages that the open probability change occurs with a steepness associated to the heat (ΔH) required for the transition and ΔS determines the temperature range over which the transition takes place. Moreover, D, which is the value of the allosteric coupling factor between the temperature-sensitive transition and the pore module, also contributes to determine the steepness of the activation curve.

Whether the channel is activated by increases or decreases in temperature depends on the sign of the entropic and enthalpic changes in the K(T) transition: If ΔH and $\Delta S > 0$, then the channel is a heat-activated channel, but if ΔH and $\Delta S < 0$, then it will be activated by cold temperatures (Islas, 2017). Other modules can be included so that they interact with the temperature sensing modules (i.e., voltage sensitivity) to obtain more complex allosteric models.

For TRPV1, repetitive activation by heat results in a temperature-dependent inactivation process, which constitutes an irreversible transition that was suggested to be due to a partial or complete unfolding of one or several regions of the channel protein during heat absorption. It was specifically shown that reduction of current amplitude in the face of repetitive stimulation is accompanied by a large change in the sensitivity, determined by the apparent activation of ΔH, of channels to temperature. As the channel is challenged with repetitive heat stimuli, a reduction of the ΔH associated with opening is observed. Importantly, this inactivated or desensitized state in response to heat remains if the channel is presented with capsaicin (Figure 24.2a, b). While heat could be absorbed by several regions of the channel and result in its opening, some regions could absorb heat through a mechanism analogous to partial polypeptide denaturation, resulting in channels with a decreased apparent enthalpy of activation. These transitions would render these proteins less efficient at absorbing or converting heat into a conformational change and, after opening, they would enter the inactivated state irreversibly (Sánchez-Moreno et al., 2018).

Ligand-dependent activation in TRP channels has been more widely studied for TRPV1. Capsaicin binds to the vanilloid-binding pocket (VBP) constituted by residues in S3, S4 and part of the S5 transmembrane α-helices as well as the S4–S5 linker in the mammalian channel. Capsaicin binding in the VBP, in a "head-down" orientation, stabilizes the S4–S5 linker in a position that allows the S6 gate to be in the open conformation (Cao, 2020).

In TRPV1 it has been shown that a site on its outer surface binds sodium ions and modifies its gating, allowing it to be activated by potentially damaging temperatures. The TRPV1 channel contains external sodium-binding sites that must be occupied so that the channel can stay closed at physiological temperatures, tuning its response (Jara-Oseguera et al., 2016).

24.5 Ion Permeability and Pharmacology

Most TRPVs are nonselective cation channels that permeate monovalent and divalent cations (i.e., K^+, Na^+ and Ca^{2+}), except TRPV5 and TRPV6 (see later). For most TRP channels, there is no unifying selectivity filter sequence (as the one typically observed for K^+ channels), and they exhibit selectivity filters with sizes stretching from 3.2 to 10.6 Å in diameter. Ion selectivity of TRP channels is determined by interaction energy of the permeable cations with dynamic selectivity filters. A negatively charged ring is formed by four aspartate residues to attract and coordinate cations. Also, backbone carbonyl oxygen atoms of some selectivity filter residues face the pores of TRPV channels to presumably coordinate cations.

Notably, gating of TRPV1 by lysophosphatidic acid (LPA) produces an open state with a larger single-channel conductance (Figure 24.2c, d), with no change in selectivity, as compared to activation by capsaicin. Hence, this phenomenon by which different full agonists open the channel with a high probability (i.e., capsaicin and LPA) can produce different open states. This means that there could be more efficient generation of action potentials by LPA and of pain through neurons that express TRPV1 (Canul-Sánchez et al., 2018; Nieto-Posadas et al., 2011).

TRPV2, TRPV3 and TRPV4 are also nonselective cation channels that allow permeation of monovalent and divalent cations, with high selectivity for Ca^{2+} and strong outward rectification (Owsianik et al., 2006). TRPV4 stands out because it has the widest selectivity filter in diameter (~10.6 Å), through which fully hydrated cations could permeate, also allowing for its large conductance (~280–310 pS) (Cao, 2020) when Ca^{2+} is included as a charge carrier and activated by osmotic changes. TRPV4 behaves as an inward or outward rectifier, depending on whether Ca^{2+} is absent or present, respectively.

The biophysical properties of TRPV5 and TRPV6 channels are different from other TRP channels since they exhibit very high permeability for Ca^{2+} (P_{Ca}/P_{Na} >100:1). Removal of extracellular divalent cations results in the permeation of monovalent ions. In TRPV6, Mg^{2+} functions as a permeant pore blocker, influencing its strong inward rectification. Since these two channels can transport Ca^{2+} into the cell at negative membrane potentials, this allows for substantial calcium permeation at physiological membrane potentials (Owsianik et al., 2006).

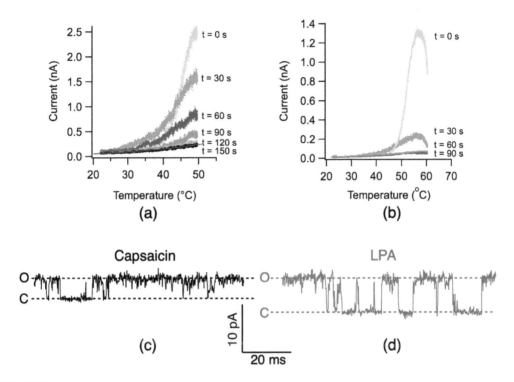

FIGURE 24.2
Activation of TRPV1 by temperature and LPA. (a) Inside-out patch expressing TRPV1 channels where the activation (up) part of the response to a temperature ramp is shown at 60 mV; 49.7°C was the maximum temperature. Peak current and slope of activation are decreased with every subsequent ramp-activated current. At 150 s all TRPV1 current has been lost. (b) At a higher temperature of 59°C, loss of current is also observed during the up ramp before it has reached its maximum temperature, and the following responses are very decreased at this higher temperature. (c) Single-channel recordings of TRPV1 (+60 mV) in the presence of capsaicin and in the presence of (d) LPA. Single-channel current amplitude is increased in the presence of LPA. (Panels (a) and (b) were taken from Sánchez-Moreno et al., 2018, in accordance with a Creative Commons Attribution License (CC BY 4.0).)

As mentioned before, TRPV channels are mostly polymodal and activated and inhibited by several molecules including both exogenous and endogenous substances, which are summarized in Table 24.1 (Holzer and Izzo, 2014; Blair et al., 2019).

24.6 Regulation

The activity of TRPV1 is regulated by phosphatidylinositol-4,5-biphosphate (PIP$_2$) and, in fact, this phospholipid activates all of TRPV channels but one, TRPV3 (Gamper and Rohacs, 2012).

Phosphorylation of several key residues in TRPV1 leads to an increased sensitivity to both chemical and thermal stimuli. Protein kinase C (PKC), PKA, calmodulin (CaM)-dependent kinase and Src kinase increase the activity of TRPV1 activity, but calcineurin-mediated dephosphorylation at the same sites produces TRPV1 desensitization, becoming refractory to activation (Yao et al., 2005).

The function of TRPV2 can also be rapidly regulated by phosphorylation/dephosphorylation by PKA or PI3-kinase or desensitization by Ca^{2+}. A Ca^{2+}-dependent binding site for CaM in the C-terminal fragment has been identified for this channel, and depletion of PIP_2 from membranes causes desensitization of TRPV2 in a Ca^{2+}-dependent manner.

In TRPV3, hydrolysis of PIP_2 leads to potentiation of the currents by a shift in the voltage dependence of opening and also an apparent reduction in the threshold for thermal activation. Hence, a decrease in PIP_2 levels, as well as activation of PKC, leads to cutaneous inflammation or thermal hyperalgesia through an increase in TRPV3 activity and its sensitization to warm temperatures. This channel is sensitized by repetitive ligand stimulations, and changes in extracellular and intracellular Ca^{2+} levels play an important role in this process, in which they cause slow activation of TRPV3 at positive potentials and strong deactivation at negative potentials. Specifically, this sensitization of TRPV3 depends on a reduction of extracellular Ca^{2+} that binds to residues in the pore loop and can be eliminated by CaM.

PIP_2 is an endogenous activity modulator of TRPV4, allowing for activation by temperature, osmotic changes and 5,6-epoxyeicosatrienoic acid (5,6-EET; a product of the metabolism of arachidonic acid) (Nilius et al., 2007). PIP_2 also regulates the constitutive activity of both TRPV5 and TRPV6, as follows. While fast desensitization occurs through the binding of CaM, interaction of PIP_2 with TRPV5 is crucial for channel opening. For TRPV5, activators that bind directly to the channel may allosterically enhance conductance during PIP_2 activation and/or decrease CaM-induced desensitization. In contrast, inhibitors of the channel probably act by (1) directly blocking PIP_2 binding, (2) allosterically limiting the PIP_2-induced conformational changes, (3) blocking ion flow by binding in the pore and/or (4) increasing CaM-dependent desensitization.

PIP_2 has also been shown to be necessary for TRPV6 function and, in planar lipid bilayers, it directly interacts with this channel. Recently, interaction of this phospholipid with cationic residues in the inner S5 helix and C-terminus of TRPV6 have been suggested to be necessary for activation of the channel by PIP_2. TRPV5 and TRPV6 also present inactivation and current decay that serve as Ca^{2+}-dependent feedback regulation mechanisms and that are controlled, in part, by Ca^{2+}-dependent binding of CaM to the base of the pore of the channels.

24.7 Biogenesis, Trafficking and Turnover

TRPV1 can be expressed in its glycosylated and unglycosylated forms. The N604 residue at the extracellular linker between the S5 transmembrane helix and the pore loop is subject to N-linked glycosylation, but glycosylation at this site is not essential for correct folding and targeting of the protein to the plasma membrane. Nonetheless, it is important for capsaicin regulation (i.e., desensitization) and permeability.

In TRPV4, the residue N651 adjacent to the pore-forming loop is an N-linked glycosylation site that influences trafficking. For TRPV5, the Klotho protein cleaves an N-linked oligosaccharide (at residue N358 within an extracellular loop between the first and second transmembrane segments) and traps the channel in the apical plasma membrane in the kidneys.

Translocation of a TRP channel to the plasma membrane was first shown for TRPV2 using stimulation of the insulin growth factor (IGF)-1 receptor that resulted in increases of

TRPV2 levels in the plasma membrane. The activity of the PI 3-kinase was demonstrated to be pivotal for the translocation of TRPV2 and of TRPV1 into the plasma membrane. Moreover, protein kinase C and Src kinase, downstream from PI 3-kinase, promote protein phosphorylation, which is an important step for exocytosis of TRPV1 induced by hormones. Moreover, a protein localized in the perinuclear Golgi apparatus and endoplasmic reticulum, RGA or gene-activating protein, interacts with the N-terminus of TRPV2, and its overexpression increases cell surface levels and may play a role in the early biosynthesis and trafficking of TRPV2 among the endoplasmic reticulum and the compartments of the Golgi apparatus (Ferrandiz-Huertas et al., 2014).

TRPV1 interacts with synaptotagmin, a molecule involved in Ca^{2+}-dependent exocytosis, and with snapin and tubulin dimers as well as with polymerized microtubules, allowing for TRPV1 tetramerization and preservation of the functionality of TRPV1 in the membrane. Disassembly of microtubules results in receptor self-aggregation and partial loss of activity as well as in interference with the vesicular trafficking of TRPV1.

Another protein, Cdk5, positively controls TRPV1 membrane transport and promotes TRPV1 anterograde transport in vivo after inflammation, regulating heat sensitivity and reduction of capsaicin-evoked calcium influx in primary sensory neurons.

The E3-ubiquitin ligase Myc-binding protein-2 (MYCBP2) also regulates TRPV1 internalization through inhibition of P38MAPK signaling, and altered receptor trafficking in MYCBP2-deficient DRGs and capsaicin-induced desensitization are absent in these cells due to a decrease in TRPV1 internalization (Ambudkar, 2007).

TRPV1 has also been shown to interact with the sigma-1 receptor, a chaperone that is antagonized by progesterone. Such an interaction regulates TRPV1 expression levels in the plasma membrane (Figure 24.3) as well as pain responses in animal experimental models (Ortíz-Rentería et al., 2018).

PACSIN proteins also bind to dynamin, synaptojanin 1 and N-WASP, and regulate synaptic vesicular trafficking and endocytosis. For TRPV4 it is known that the C-terminal SH3 domain of PACSIN 3 interacts with the N-terminal proline-rich domain of the ion cannel, increasing its cell surface expression. The interaction of microtubule-associated protein 7 (MAP7) and the C-terminus of TRPV4 also increases cell surface expression of this ion channel, and MAP7 also interacts with actin microfilaments, contributing to epithelial polarity and increasing trafficking of TRPV4 to the cell surface when cell shape changes occur. Protein–protein interactions that control the level and insertion of TRPV4 into the cell surface include those with the reticulum-associated protein OS-9 that interacts with the N-terminal tail of TRPV4 and facilitates proper folding and tetramer formation by binding to monomers and immature variants of TRPV4, and protecting the channel from precocious ubiquitination and associated degradation. The family of lysine-deficient protein kinases (WNK) also influence the physiology of TRPV4: WNK4 and WNK1 downregulate the function of TRPV4 by decreasing its expression in the plasma membrane, highlighting a role for these WNKs in the trafficking of TRPV4 to the cell surface (Ferrandiz-Huertas et al., 2014).

Small G proteins of the Rab family, such as Rab11, are also involved in trafficking TRPV5 and TRPV6 to the plasma membrane. Finally, TRPV5 and TRPV6 interact with the S-100–annexin 2 complex, which also binds membrane lipids.

The expression of splice variants of these channels can affect their oligomerization and their trafficking. For example, it has been shown that some splice variants of TRPV4 exhibit deletions on their N-termini (i.e., ARDs). Study of these splice variants helped determine that biogenesis of TRPV4 depends on core glycosylation, and that the ARDs are necessary for oligomerization of the channel and, when there is no oligomerization, the channels are retained in the endoplasmic reticulum (Ferrandiz-Huertas et al., 2014).

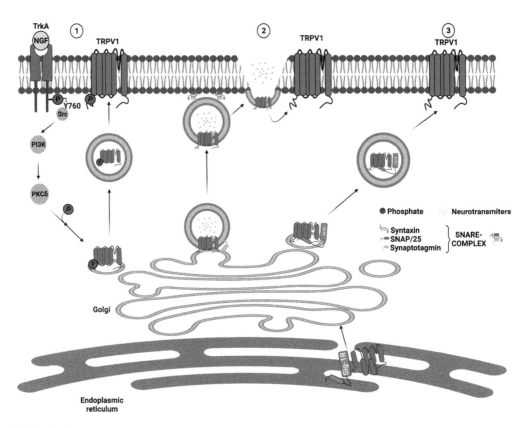

FIGURE 24.3
Trafficking pathways of TRPV1 channels. Three major pathways are shown as follows: (1) The signaling pathways that connect TrkA to sensitization of TRPV1 through the actions of nerve growth factor (NGF). This pathway is initiated by autophosphorylation of the Y760 site of TrkA, followed by activation of Src and PI 3-kinase, phosphorylating TRPV1 at residue Y200. (2) Synaptic vesicles can also store TRP channels, which are then delivered to the cell surface by a SNARE complex. (3) The TRPV1 channel can be positively regulated by the chaperone sigma-1 receptor (Sig-1R), this interaction upregulates TRPV1 protein levels in the plasma membrane. (Created with BioRender.com.)

24.8 Channelopathies

Most reported TRPV channel channelopathies have been associated with TRPV4 (see later). However, it was suggested that TRPV1 variants (Thr612Met and Asn394del) are associated to malignant hyperthermia (MH), a disorder arising from uncontrolled muscle calcium release in response to an abnormality in the sarcoplasmic reticulum (SR) Ca^{2+}-release mechanism triggered by halogenated inhalational anesthetics. Mutations in and malfunction of TRPV4 have been extensively associated to several diseases. For example, mutations in the N-terminus (ARDs) of TRPV4 are linked to genetic diseases such as Charcot–Marie–Tooth disease type 2C. Gain-of-function mutations (TRPV4-P19S) can lead to increased entrance of Ca^{2+} through the channel and to oversecretion of molecules such as MMP-1 (an interstitial collagenase), leading to chronic obstructive pulmonary disease.

Other mutations in the TRPV4 channel lead to development of several types, as shown in Figure 24.4 (Nilius and Owsianik, 2010; Rosenbaum et al., 2020).

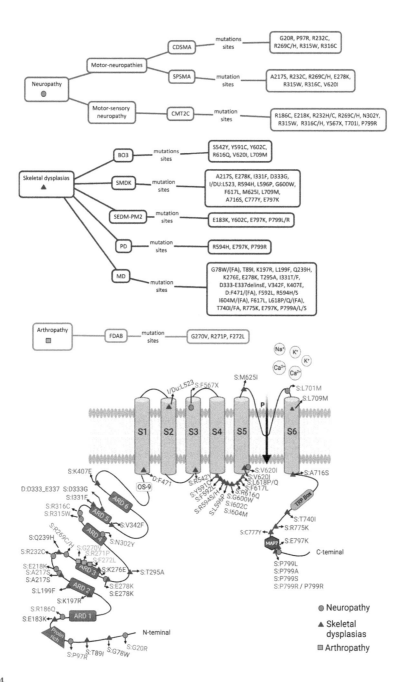

FIGURE 24.4

TRPV4 channelopathies. (Top) Conceptual map of heritable mutants of TRPV4 channel. CDSMA, congenital distal spinal muscle atrophy; SPSMA, scapuloperoneal spinal muscular; CMT2C, Charcot–Marie–Tooth disease type 2C; BO3, autosomal dominant skeleto-dysplasia brachyolmia type 3; SMDK, spondylometaphyseal dysplasia Kozlowski type; SEDM-PM2, spondyloepimetaphyseal dysplasia Maroteaux pseudo-Morquio type 2; PD, parastremmatic dysplasia; MD, metatropic dysplasia, which is combined with fetal akinesia (FA); FDAB, familial digital arthropathy brachydactyly. (Bottom) Localization of mutations in the TRPV4 channel structure. S, substitution; D, deletion; I/Du, insertion/duplication; ARD, ankyrin repeat domain; OS-9, osteosarcoma 9; MAP7, microtubule associated protein 7. (Created with BioRender.com.)

For TRPV5 and TRPV6 it has been found that altered function results in diseases such as milk-alkali syndrome, kidney stone disease, osteoporosis, vitamin D deficiency and rickets (Nilius and Owsianik, 2010).

24.9 Conclusions

Several TRPV channels control Ca^{2+} homeostasis and are involved in several physiological and pathophysiological functions. Much of the information available on the role of these ion channels has come from the use of transgenic animal models. However, extensive work is still necessary to fully understand their function in several tissues and organs.

These channels exhibit complex gating mechanisms and further work is needed to uncover details of the gating of these ion channels by different ligands and temperature. For example, since denaturation of proteins by heat is highly dependent on temperature through perturbations of polar and hydrophobic interactions, partial denaturation of certain regions of the channel structure could be part of the mechanism of temperature gating in temperature-gated TRP channels.

Acknowledgments

I am thankful to Itzel Llorente for figures and table production, and to León D. Islas for suggestions and proofreading. This work was supported by Dirección General de Asuntos del Personal Académico (DGAPA)–Programa de Apoyo a Proyectos de Investigación e Innovación Tecnológica (PAPIIT) grant IN200720, Consejo Nacional de Ciencia y Tecnología (CONACyT) grant A1-S-8760, and Secretaría de Educación, Ciencia, Tecnología e Innovación del Gobierno de la Ciudad de México grant SECTEI/208/2019.

Suggested Reading

Ambudkar, I. S. 2007. "Trafficking of TRP Channels: Determinants of Channel Function." In: *Transient Receptor Potential (TRP) Channels*, edited by Veit Flockerzi and Bernd Nilius, 179:541–57. *Handbook of Experimental Pharmacology*. Berlin, Heidelberg: Springer Berlin Heidelberg. doi:10.1007/978-3-540-34891-7_32.

Blair, Nathaniel T., Ingrid Carvacho, Dipayan Chaudhuri, David E. Clapham, Paul DeCaen, Markus Delling, Julia F. Doerner, et al. 2019. "Transient Receptor Potential Channels (Version 2019.4) in the IUPHAR/BPS Guide to Pharmacology Database." *IUPHAR/BPS Guide to Pharmacology CITE* 2019(4). doi:10.2218/gtopdb/F78/2019.4.

Canul-Sánchez, Jesús Aldair, Ileana Hernández-Araiza, Enrique Hernández-García, Itzel Llorente, Sara L. Morales-Lázaro, León D. Islas, and Tamara Rosenbaum. 2018. "Different Agonists Induce Distinct Single-Channel Conductance States in TRPV1 Channels." *The Journal of General Physiology* 150(12): 1735–46. doi:10.1085/jgp.201812141.

Cao, Erhu. 2020. "Structural Mechanisms of Transient Receptor Potential Ion Channels." *Journal of General Physiology* 152(3): e201811998. doi:10.1085/jgp.201811998.

Caterina, Michael J., and David Julius. 2001. "The Vanilloid Receptor: A Molecular Gateway to the Pain Pathway." *Annual Review of Neuroscience* 24(1): 487–517. doi:10.1146/annurev.neuro.24.1.487.

Denda, Mitsuhiro, Shigeyoshi Fuziwara, Kaori Inoue, Sumiko Denda, Hirohiko Akamatsu, Akiko Tomitaka, and Kayoko Matsunaga. 2001. "Immunoreactivity of VR1 on Epidermal Keratinocyte of Human Skin." *Biochemical and Biophysical Research Communications* 285(5): 1250–52. doi:10.1006/bbrc.2001.5299.

Du, Qian, Qiushi Liao, Changmei Chen, Xiaoxu Yang, Rui Xie, and Jingyu Xu. 2019. "The Role of Transient Receptor Potential Vanilloid 1 in Common Diseases of the Digestive Tract and the Cardiovascular and Respiratory System." *Frontiers in Physiology* 10(August): 1064. doi:10.3389/fphys.2019.01064.

Emir, Tamara Luti Rosenbaum, ed. 2017. *Neurobiology of TRP Channels*, 2nd ed. Frontiers in Neuroscience. Boca Raton (FL): CRC Press/Taylor & Francis. http://www.ncbi.nlm.nih.gov/books/NBK476115/.

Falcón, Debora, Isabel Galeano-Otero, Eva Calderón-Sánchez, Raquel Del Toro, Marta Martín-Bórnez, Juan A. Rosado, Abdelkrim Hmadcha, and Tarik Smani. 2019. "TRP Channels: Current Perspectives in the Adverse Cardiac Remodeling." *Frontiers in Physiology* 10(March): 159. doi:10.3389/fphys.2019.00159.

Ferrandiz-Huertas, Clotilde, Sakthikumar Mathivanan, Christoph Wolf, Isabel Devesa, and Antonio Ferrer-Montiel. 2014. "Trafficking of ThermoTRP Channels." *Membranes* 4(3): 525–64. doi:10.3390/membranes4030525.

Gamper, Nikita, and Tibor Rohacs. 2012. "Phosphoinositide Sensitivity of Ion Channels, a Functional Perspective." In: *Phosphoinositides II: The Diverse Biological Functions*, edited by Tamas Balla, Matthias Wymann, and John D. York, 59:289–333. *Subcellular Biochemistry*. Dordrecht: Springer Netherlands. doi:10.1007/978-94-007-3015-1_10.

Gao, Yuan, Erhu Cao, David Julius, and Yifan Cheng. 2016. "TRPV1 Structures in Nanodiscs Reveal Mechanisms of Ligand and Lipid Action." *Nature* 534(7607): 347–51. doi:10.1038/nature17964.

Holzer, Peter, and Angelo A. Izzo. 2014. "The Pharmacology of TRP Channels: The Pharmacology of TRP Channels." *British Journal of Pharmacology* 171(10): 2469–73. doi:10.1111/bph.12723.

Islas, León D. 2017. "Molecular Mechanisms of Temperature Gating in TRP Channels." In: *Neurobiology of TRP Channels*, edited by Tamara Luti Rosenbaum Emir, 1st ed., 11–25. Boca Raton, FL FCRC Press. doi:10.4324/9781315152837-2.

Jara-Oseguera, Andres, Chanhyung Bae, and Kenton J. Swartz. 2016. "An External Sodium Ion Binding Site Controls Allosteric Gating in TRPV1 Channels." *eLife* 5(February): e13356. doi:10.7554/eLife.13356.

Lee, Bo Hyun, and Jie Zheng. 2015. "Proton Block of Proton-Activated TRPV1 Current." *Journal of General Physiology* 146(2): 147–59. doi:10.1085/jgp.201511386.

Liedtke, Wolfgang. 2005. "TRPV4 Plays an Evolutionary Conserved Role in the Transduction of Osmotic and Mechanical Stimuli in Live Animals: TRPV4 in the Transduction of Osmotic and Mechanical Stimuli." *The Journal of Physiology* 567(1): 53–8. doi:10.1113/jphysiol.2005.088963.

Nieto-Posadas, Andrés, Giovanni Picazo-Juárez, Itzel Llorente, Andrés Jara-Oseguera, Sara Morales-Lázaro, Diana Escalante-Alcalde, León D. Islas, and Tamara Rosenbaum. 2011. "Lysophosphatidic Acid Directly Activates TRPV1 through a C-Terminal Binding Site." *Nature Chemical Biology* 8(1): 78–85. doi:10.1038/nchembio.712.

Nilius, Bernd, Tamás Bíró, and Grzegorz Owsianik. 2014. "TRPV3: Time to Decipher a Poorly Understood Family Member!: TRPV3." *The Journal of Physiology* 592(2): 295–304. doi:10.1113/jphysiol.2013.255968.

Nilius, Bernd, and Grzegorz Owsianik. 2010. "Channelopathies Converge on TRPV4." *Nature Genetics* 42(2): 98–100. doi:10.1038/ng0210-98.

Nilius, Bernd, Grzegorz Owsianik, Thomas Voets, and John A. Peters. 2007. "Transient Receptor Potential Cation Channels in Disease." *Physiological Reviews* 87(1): 165–217. doi:10.1152/physrev.00021.2006.

Ortíz-Rentería, Miguel, Rebeca Juárez-Contreras, Ricardo González-Ramírez, León D. Islas, Félix Sierra-Ramírez, Itzel Llorente, Sidney A. Simon, Marcia Hiriart, Tamara Rosenbaum, and Sara

L. Morales-Lázaro. 2018. "TRPV1 Channels and the Progesterone Receptor Sig-1R Interact to Regulate Pain." *Proceedings of the National Academy of Sciences of the United States of America* 115(7): E1657–66. doi:10.1073/pnas.1715972115.

Owsianik, Grzegorz, Karel Talavera, Thomas Voets, and Bernd Nilius. 2006. "Permeation and Selectivity of TRP Channels." *Annual Review of Physiology* 68(1): 685–717. doi:10.1146/annurev. physiol.68.040204.101406.

Rosenbaum, Tamara, Miguel Benítez-Angeles, Raúl Sánchez-Hernández, Sara Luz Morales-Lázaro, Marcia Hiriart, Luis Eduardo Morales-Buenrostro, and Francisco Torres-Quiroz. 2020. "TRPV4: A Physio and Pathophysiologically Significant Ion Channel." *International Journal of Molecular Sciences* 21(11): 3837.

Sánchez-Moreno, Ana, Eduardo Guevara-Hernández, Ricardo Contreras-Cervera, Gisela Rangel-Yescas, Ernesto Ladrón-de-Guevara, Tamara Rosenbaum, and León D. Islas. 2018. "Irreversible Temperature Gating in Trpv1 Sheds Light on Channel Activation." *eLife* 7: e36372.

Shibasaki, Koji. 2016. "Physiological Significance of TRPV2 as a Mechanosensor, Thermosensor and Lipid Sensor." *The Journal of Physiological Sciences* 66(5): 359–65. doi:10.1007/s12576-016-0434-7.

Yao, Jing, Beiying Liu, and Feng Qin. 2010. "Kinetic and Energetic Analysis of Thermally Activated TRPV1 Channels." *Biophysical Journal* 99(6): 1743–53. doi:10.1016/j.bpj.2010.07.022.

Yao, Xiaoqiang, Hiu-Yee Kwan, and Y. Huang. 2005. "Regulation of TRP Channels by Phosphorylation." *Neurosignals* 14(6): 273–80. doi:10.1159/000093042.

Yelshanskaya, Maria V., Kirill D. Nadezhdin, Maria G. Kurnikova, and Alexander I. Sobolevsky. 2021. "Structure and Function of the Calcium-Selective TRP Channel TRPV6." *The Journal of Physiology* 599(10): 2673–97. doi:10.1113/JP279024.

Zheng, Jie. 2013. "Molecular Mechanism of TRP Channels." In: *Comprehensive Physiology*, edited by Ronald Terjung, c120001. Hoboken, NJ: John Wiley & Sons, Inc. doi:10.1002/cphy.c120001.

25

Store-Operated CRAC Channels

Murali Prakriya

CONTENTS

25.1 Introduction

Ca^{2+} release-activated Ca^{2+} (CRAC) channels are a ubiquitous mechanism for mobilizing Ca^{2+} signals in many animal cells. Once activated, Ca^{2+} influx through CRAC channels drives a wide range of cellular functions, including enzyme activation, gene expression, chemotaxis and gliotransmitter release. Physiologically, CRAC channels are activated by depletion of intracellular Ca^{2+} stores that occurs downstream of cell surface receptors coupled to the generation of inositol trisphosphate (IP3) to evoke store-operated Ca^{2+} entry (SOCE) (Figure 25.1). Clinical studies have shown that mutations in CRAC channel genes cause devastating immunodeficiency, muscle weakness, and abnormalities in the skin and teeth (1). Moreover, animal studies have implicated a growing list of possible diseases, including allergies, cancers, autoimmunity, and neurodegenerative diseases, to loss or gain of SOCE activity (2), highlighting the potential importance of these channels for human health and disease.

DOI: 10.1201/9781003096276-25

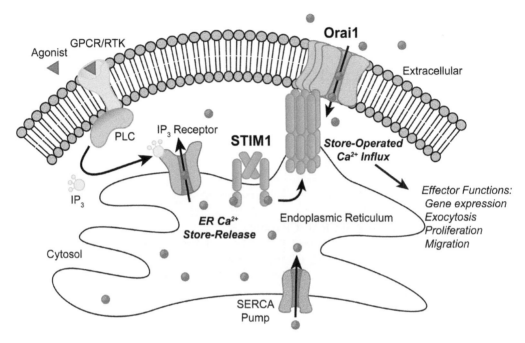

FIGURE 25.1

Store-operated calcium entry. Stimulation of cell surface G protein-coupled receptors (GPCRs) or receptor tyrosine kinases (RTKs) in many cells leads to the production of the diffusible messenger, inositol triphosphate (IP$_3$). Depletion of intracellular Ca^{2+} stores by IP$_3$ is sensed by the ER Ca^{2+} sensor, STIM1. Activated STIM1 molecules migrate to and gather at ER–plasma membrane junctions. In their extended form, STIM1 dimers bind to and gate ORAI1 channels. The resulting Ca^{2+} influx through ORAI1 drives a range of cellular effector functions.

Early studies of SOCE relied on cytosolic Ca^{2+} indicator dyes to examine Ca^{2+} fluxes (3). These investigations were aided by the discovery of thapsigargin, a plant alkaloid that is a potent inhibitor of the SERCA (sarco/endoplasmic reticulum Ca^{2+}-ATPase) pump in the endoplasmic reticulum (ER) membrane. Thapsigargin allowed investigators to directly deplete intracellular Ca^{2+} stores without stimulating surface receptors, thereby eliminating the myriad confounding effects of signaling that results from receptor stimulation (3). Electrophysiological studies that followed these early studies characterized the biophysical and functional features of store-operated currents in T lymphocytes and mast cells (4, 5). The Ca^{2+} currents in these cells, which were termed CRAC currents, exhibit high Ca^{2+} selectivity and could be distinguished from other Ca^{2+}-selective channels based on their low unitary conductance and low permeability to large monovalent cations (reviewed in ref. 2). Despite early characterization of CRAC currents, however, the molecular identity of CRAC channels and the mechanism linking ER Ca^{2+} store depletion to their activation remained unknown for several decades. This changed dramatically following the identification of stromal interaction molecule 1 (STIM1) in 2005 as the ER Ca^{2+} sensor and ORAI1 in 2006 as the prototypic CRAC channel protein (6, 7) (Figure 25.1). We now know that CRAC channels are activated through the binding of the ER Ca^{2+} sensors STIM1 and STIM2 to the CRAC channel ORAI proteins (Figure 25.1). STIM1 and ORAI1 can fully reconstitute SOCE in heterologous expression systems, satisfying both the necessity and sufficiency of these proteins for SOCE.

25.2 Biophysical Properties of CRAC Channels

Long before the identification of STIM and ORAI proteins, several decades of pioneering electrophysiological studies from the Cahalan, Penner, and Lewis groups had revealed many important biophysical and pharmacological features of CRAC channels. Electrophysiological recordings revealed that CRAC currents activate slowly over a period of seconds following ER Ca^{2+} store depletion (Figure 25.2A) and correspondingly deactivate (close) slowly in response to store refilling. The speed of activation–deactivation is dictated mainly by the kinetics of STIM1 activation or its reversal and rate-limited by the migration and accumulation (or its reversal) of STIM1 at ER–plasma membrane junctions.

From a biophysical standpoint, CRAC channels are widely noted for their very high Ca^{2+} selectivity (P_{Ca}/P_{Na} >1000) with an inwardly rectifying current–voltage relationship (Figure 25.2A) and very low conductance, estimated to be only ~10–30 fS for Ca^{2+} and ~1 pS for Na^+ in the absence of extracellular divalent cations (2). Interestingly, high Ca^{2+} selectivity is only manifested in Ca^{2+}-containing solutions. CRAC channels readily conduct a variety of small monovalent ions (Na^+, Li^+ and K^+) in divalent-free solutions (e.g., Figure 25.2B), indicating that high Ca^{2+} selectivity is not an intrinsic feature of the CRAC channel pore but arises due to ion–pore interactions. This is clearly revealed by the blockade of monovalent currents by micromolar concentrations of Ca^{2+} (K_i ~20 μM at –100 mV). Occupancy by a single Ca^{2+} ion appears sufficient to block the large monovalent conductance, and, as expected for a binding site within the pore, Ca^{2+} block is voltage-dependent (2). CRAC channels are also virtually impermeable to large monovalent cations including Cs^+ (P_{Cs}/P_{Na} <0.1) (Figure 25.2B), which is due to the relatively narrow dimensions of the pore that limits the electrodiffusion of larger ions. The channels are not sensitive to common inhibitors of Ca_V or Na^+ channels, but are inhibited by imidazole antimycotics, submicromolar concentrations of lanthanides,

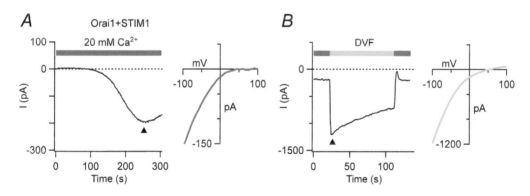

FIGURE 25.2
Electrophysiological characteristics of CRAC currents. (A) Time course of CRAC current activation in a HEK293 cell expressing exogenous ORAI1 and STIM1. CRAC current activates slowly in response to depletion of ER Ca^{2+} stores following dialysis of BAPTA into the cell from the recording electrode after whole-cell break-in at t = 0 s. The current–voltage relationship of the current (blue trace) reveals an inwardly rectifying profile with an extremely positive (>+60 mV) reversal potential, which is indicative of high Ca^{2+} selectivity of CRAC channels. (B) In the absence of extracellular divalent ions (in a divalent-free [DVF] solution), CRAC channels readily conduct large Na^+ currents. The current–voltage relationship of the DVF current (right, green trace) shows an inwardly rectifying profile with a reversal potential around +50 mV, indicating low permeability for Cs^+ ions present in the recording pipette.

and the small molecules 2-APB, BTP2, and CM4620 among others (2). Overall, the high Ca^{2+} selectivity, low conductance and high sensitivity to lanthanides are useful criteria for identifying and distinguishing CRAC currents from other channels in native cells.

25.3 STIM1 Is the ER Ca^{2+} Sensor for CRAC Channel Activation

STIM1 was first discovered in a screen for molecules that bind to pre-B lymphocytes. Initial studies suggested a role for STIM1 in tumor growth, but it was not until 2005 that a clear role was demonstrated for Ca^{2+} signaling through RNAi screens for inhibitors of SOCE (6, 8). Human STIM1 is a 77 KDa single-pass ER membrane protein, with a luminal N-terminal domain containing the signal peptide and a large C-terminal domain in the cytosol (Figure 25.3A). It contains several domains critical to its function including a sterile alpha motif (SAM) and two EF hands in the N-terminus (Figure 25.3A) and three coiled-coil domains, a Ser/Pro-rich region, and a Lys-rich region in the C-terminus. Whereas *Drosophila* has a single STIM gene, mammals have two closely related genes, *STIM1* and *STIM2*, which differ significantly in their C-terminal region. In resting cells, STIM1 is largely localized in the bulk ER. ER Ca^{2+} store depletion triggers unbinding of Ca^{2+} from

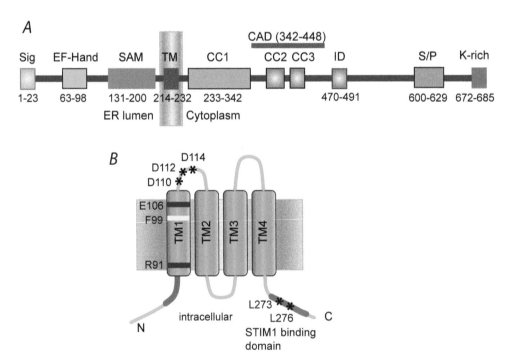

FIGURE 25.3
Topological organization of STIM1 and ORAI1. (A) A topological representation of STIM1 and its key functional domains. These domains include Sig, signal peptide; SAM, sterile alpha motif; TM, transmembrane domain; CC, coiled-coil domain; CAD, CRAC activation domain; ID, inactivation domain of STIM1; S/P, serine proline-rich domain; K-rich, lysine-rich domain. (B) A schematic of a single subunit of ORAI1. The STIM1-binding domains are shaded in red. Key residues labeled include the human disease locus, R91; the glutamate selectivity filter, E106; the channel gate residue, F99; and the acidic residues D110, D112 and D114 located in the outer mouth of the pore.

the luminal EF hand, which ultimately results in the redistribution of STIM1 from the bulk ER into puncta located in close apposition to the plasma membrane (2). EF-hand STIM1 mutants with impaired Ca^{2+} binding form puncta and activate CRAC channels independently of ER Ca^{2+} store depletion (6, 8); the absence of Ca^{2+} binding in these mutants essentially tricks the molecules into responding as if stores are depleted. In this manner, changes in Ca^{2+} binding at the N-terminus are coupled to SOCE initiation through a series of conformational changes and interactions with the Orai1 C-terminus. Thus, STIM1 fulfills two critical roles in the activation process of CRAC channels: it senses the depletion of ER Ca^{2+} stores, as well as communicating the store depletion signal to CRAC channels in the plasma membrane to trigger their opening (Figure 25.1).

25.4 ORAI1 Is the Prototypic CRAC Channel Protein

In 2006, the ORAI proteins (ORAI1-3) were identified using RNAi screens and linkage analysis of a human family with an inherited form of immunodeficiency (7, 9, 10). ORAI1, the best studied member, is a widely expressed 33 kDa cell surface protein with four predicted transmembrane domains, intracellular N- and C-termini (Figure 25.3B), and no significant sequence homology to other previously identified ion channels. Overexpression of ORAI1 together with STIM1 in HEK293 or other heterologous cells evokes large currents with the biophysical characteristics consistent with native CRAC channels. These currents also share the core features of CRAC currents such as Ca^{2+} block of Na^+ current and sensitivity to blockade by 2-APB and La^{3+}. Mutation of a highly conserved acidic residue in the first transmembrane domain of ORAI1 (E106D) (Figure 25.3B) significantly diminishes the Ca^{2+} selectivity of the CRAC channels and alters a wide range of properties intimately associated with the pore, including La^{3+} block, the voltage dependence of Ca^{2+} blockade, and Cs^+ permeation (9–11). Mammalian cells express two other closely related homologues, ORAI2 and ORAI3, that differ primarily in their C-terminal and the 3–4 loop. All three isoforms function similarly in producing store-operated Ca^{2+} entry when coexpressed with STIM1 in HEK293 cells and are widely expressed in mammalian tissues (reviewed in (2)). However, ORAI1 remains the best studied CRAC channel protein and appears to be the predominant isoform mediating SOCE in most cells with strong links to human disease (1, 2). By contrast, there is no direct genetic evidence for a role of ORAI2 or ORAI3 channels for pathology in any cell type yet.

25.5 Activation of STIM1: Oligomerization and Redistribution to ER–Plasma Membrane Junctions

As described earlier, the STIM proteins respond to alterations in ER Ca^{2+} store content through Ca^{2+} unbinding–binding from an EF hand in the luminal domain. The dissociation constant for Ca^{2+} of the luminal EF-hand domain is ~500 μM, which is in line with the range of Ca^{2+} concentrations known to exist in ER lumen. STIM2 has a lower Ca^{2+}-binding affinity in the range of 300 μM compared to the EF hand of STIM1 (~500 μM) (reviewed in (2)). From structural studies of the isolated luminal domain fragments, we now know that Ca^{2+} unbinding from the N-terminal EF hand triggers the unfolding and aggregation of the luminal domain, resulting in the appearance of dimers and higher-order multimers

(12). This process of STIM1 oligomerization is an early step in the channel activation process, occurring well before STIM1 redistribution to the plasma membrane and serves as a critical upstream activation switch that unfolds all subsequent steps of the channel activation process (reviewed in (2)).

Once ER stores are depleted, an intramolecular autoinhibitory brake that keeps STIM1 in a quiescent state in cells with replete ER Ca^{2+} stores is released to trigger STIM1 activation (12, 13). This brake is formed by interactions of the CC1 domain of STIM1 with the catalytic CRAC activation domain (CAD), which results in sequestration of CAD within a folded tertiary structure of the STIM1 C-terminus (13). Dimerization of the luminal EF–SAM domains is a key event in this process, releasing the inhibitory clamp by packing the CC1 domains together to force CAD outward and away from the ER membrane. In this extended form, CAD is able to reach out and interact with ORAI channels in the plasma membrane to trigger ORAI1 activation (Figure 25.1).

Once activated, one of the most striking features of STIM1 is its redistribution from the bulk ER in resting cells with full stores to the plasma membrane where it accumulates into discrete puncta at the junctional sites between the ER and the plasma membrane. STIM1 accumulation at the plasma membrane exhibits the same dependence on ER Ca^{2+} concentration as activation of I_{CRAC} and precedes the development of I_{CRAC} by ~10 s, consistent with the notion that channel activation requires a local interaction between STIM1 and ORAI subunits (reviewed in (2)). Interestingly, truncation of a basic region at the extreme C-terminus of STIM1 attenuates redistribution of the truncated STIM1 to peripheral sites without affecting STIM1 oligomerization (2, 13), indicating that this polybasic domain is important for the interaction of STIM1 with the plasma membrane. Because the polybasic tail is not found in *Caenorhabditis elegans* or *Drosophila* STIM proteins, it is likely a vertebrate adaption that facilitates STIM1 migration to the plasma membrane and efficient gating of ORAI channels. Moreover, it is now increasingly clear that the ER–plasma membrane junctions have, in addition to STIM1, a plethora of other molecules with specialized functions designed to promote and stabilize a host of signaling molecules including proteins involved in lipid exchange, adapter proteins such as the extended synaptotagmins (E-syts), and ion channels (2, 12, 13).

25.6 STIM1 Binds Directly to ORAI1

CRAC channels are operationally defined as store-operated channels. The mechanistic basis for this feature involves direct binding of the ER Ca^{2+} sensor, STIM1 to ORAI1, which leads to CRAC channel activation. The region of STIM1 required for ORAI1 activation encompasses the second and the third coiled-coil (CC) domains, and is variously called the CRAC activation domain (CAD), STIM1–ORAI1 activating region (SOAR), or Ccb9 (2) (Figure 25.3A). The structure of this domain indicates that it adopts an R-shaped configuration as a dimer with the main functional domains (CC2 and CC3), forming a hairpin motif (2, 14). Whether this dimeric module represents the active state structure of CAD and how this differs from its resting state configuration is a matter of active debate.

In ORAI1, the primary interaction site of STIM1 is with the intracellular C-terminus (Figure 25.3B). Here, the point mutations (L273S and L276D) completely disrupt ORAI1–STIM1 interaction. Interestingly, studies using systems of purified components have also shown that STIM1 can bind to the ORAI1 N-terminal fragments (2). However, N-terminal deletion mutations retain significant levels of full-length STIM1–ORAI1 binding, and yet, these mutations abrogate SOCE and I_{CRAC}. Some models have proposed modular roles of the two binding sites, with the C-terminal site mediating STIM1 binding and the N-terminal site thought to regulate channel gating. However, mutations in the putative N-terminal STIM1-binding site (but not the C-terminal site) also abrogate currents in gain-of-function (GOF) ORAI1 mutants independently even in the absence of STIM1 (15). Thus, the N-terminus is required for an aspect of channel function independently of any role in STIM1 binding. Whether the N-terminus truly harbors a STIM1 binding site in the full-length protein remains unknown.

25.7 Pore Architecture and Gating

Soon after the identification of ORAI1, cysteine-scanning analysis revealed that of the four transmembrane domains, only the first transmembrane domain (TM1) forms the pore, and the key pore-facing residues included the glutamate selectivity filter (E106), hydrophobic residues (V102, F99 and L95), and the basic R91 at the base of the pore (16). Subsequent elucidation of the *Drosophila* Orai crystal structure confirmed these findings and revealed a remarkable channel structure with numerous intriguing structural motifs (17) (Figure 25.4A). The structure revealed that the ORAI pore is divided into three regions: an acidic region in the outer pore, a hydrophobic central region and a basic region in the inner pore (Figure 25.4B). The wide outer vestibule of the pore contains 18 aspartate residues (D110/D112/D114 on each subunit) and dramatically narrows near a ring of glutamate residues forming the selectivity filter (E106) at the pore entrance (Figure 25.4B). These negative charges in the outer vestibule and selectivity filter are important for attracting cations to the external mouth and enhancing permeation by funneling ions into the pore (18) with E106 directly forming the high affinity Ca^{2+}-binding site that controls the channel's high Ca^{2+} selectivity.

The central portion of the pore is lined by three rings of hydrophobic residues (V102, F99 and L95) (Figure 25.4B), of which the outer two residues V102 and F99 function as a hydrophobic gate (19, 20). Mutations of either V102 or F99 to more polar residues enhances pore hydration, leading to leaky channels (19, 20). Electrophysiological analysis and molecular dynamics (MD) simulations indicate that pore opening is triggered by a modest rotation of

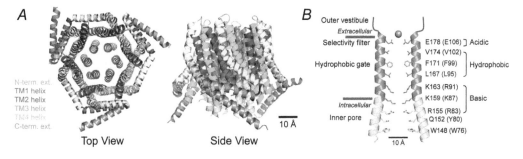

FIGURE 25.4

The structure of the Orai channel. (A) Crystal structure of the *Drosophila* ORAI channel (17). Cross-sectional view of dORAI from the extracellular side showing a hexameric ORAI complex arranged around a central axis. The TM1 helix (blue) forms the pore. The right image shows a side view of the entire channel complex with the primary STIM1-binding domains colored in orange. (B) Architecture of the ORAI pore. Pore-lining residues in two TM1 domains show amino acid side chains projecting into the pore. The channel pore has three major structural regions comprising the acid region, the hydrophobic region and the basic region. Residue numbers correspond to *Drosophila* ORAI, and the corresponding residues for human ORAI1 are shown in parentheses. The ORAI TM1 extension helix that projects into the cytoplasm is highlighted in teal.

the pore helix to move the F99 residues away from the pore axis (Figure 25.4B). This motion solvates the pore and lowers the overall energetic barrier in the hydrophobic stretch to allow ion conduction (20). From computational studies, we know that the orientation of F99 is tightly coupled to pore hydration (15). When F99 is pore facing, the hydrophobic stretch repels water, and when F99 moves away, pore hydration increases. In this manner, F99 and V102 are well designed to regulate the wetting–dewetting transitions in the pore to control ion permeation. More recently, a 3.3 Å cryo-electron microscopy structure of constitutively active mutant H206A dOrai (H134A ORAI1; Protein Data Bank ID: 6BBF) has been solved, and satisfyingly, the open channel configuration of this structure is in very good agreement with previous functional studies. All of the amino acids found to contribute to the pore (hORAI1 R91, L95, G98, F99, V102 and E106) from cysteine accessibility experiments are also observed to do so in the dOrai structures (Figure 25.4B). The new structure also shows slight widening of the V102–F99 hydrophobic stretch, reaffirming the key role for these residues in controlling pore opening.

25.8 Tissue Distribution of STIM and ORAI Proteins

In mammals, the STIM and ORAI proteins are broadly expressed and generally could be considered as ubiquitously found proteins, although levels of expression of particular isoforms differ between different tissues. From mRNA expression analysis, we know that STIM1 and STIM2 are present in skeletal muscle, brain, heart, lung, pancreas and liver. Functional studies have revealed that ORAI1 and STIM1 are the key mediators of store-operated calcium entry in lymphocytes, mast cells, B cells, macrophages, neutrophils and microglia, and expression of these proteins has been demonstrated in these tissues using a variety of approaches including mRNA and protein analysis (see Volume III, Chapter 7 for further description of ion channels regulating immune cell function). Expression of STIM2 in the brain is particularly prominent in structures such as the hippocampus and the cerebellum, and diversity of STIM protein function is further increased by alternative splicing. ORAI1 transcripts and protein expression have also been described in the brain, although the levels of ORAI2 and to some extent, ORAI3, are substantively higher (Allen Brain Atlas). Overlapping expression of multiple ORAI isoforms in the same cells appears to be a recurring theme and has been described in many types of cells including T cells, macrophages and hippocampal neurons. This raises the question of whether there exists heteromultimeric channels. Interestingly, the deletion of ORAI2 paradoxically enhances SOCE in a variety of tissues indicating that the roles of the various ORAI isoforms are not straightforward and that the subunit composition of the CRAC channel may have a profound influence on SOCE. The elucidation of the precise roles of the different Orai isoforms in specific cell types promises to be a significant area of upcoming research.

25.9 Physiological Functions of CRAC Channels

In light of the widespread tissue distribution of ORAI and STIM proteins, many specific roles are to be expected for CRAC channels in different organ systems, a view that is affirmed by knockout studies and human patients with mutated *STIM/ORAI* genes.

One well-characterized physiological role is the generation of long-lasting $[Ca^{2+}]_i$ elevations essential for Ca^{2+}-dependent gene transcription, motility, cytokine production and cell proliferation in many cells. Several studies of human patients have shown that severe immunodeficiencies arise from mutations in CRAC channels that render them inactive (1). The abrogation of CRAC channel function in these cells results in the elimination of Ca^{2+} elevations necessary to drive nuclear translocation of NFAT, an important and widely expressed transcription factor involved in cytokine gene expression. The following sections describe some well-known albeit highly abbreviated list of roles of CRAC channels in some cell types and tissues.

25.9.1 T Cells

As the primary avenue for Ca^{2+} influx in T cells, CRAC channels regulate a wide range of T-cell effector functions including antigen-stimulated cytokine production, clonal expansion and cytolysis (1) (see Volume III, Chapter 7 for a more detailed description of CRAC channels regulating immune cell function). These effector functions are mediated primarily via activation of calcineurin and NFAT, and the downstream induction of cytokines dependent on these mediators, which include such key proinflammatory cytokines like IL-6, IFNγ, and TNFα, and chemokines such as MCP1. CRAC channels also regulate the motility of T cells and by extension, their interaction with antigen-presenting cells (APCs) by functioning as stop signals (1, 21, 22). Other pathways controlled by CRAC that SOCE regulates include the activation of NF-κB, AMPK, ERK1/2 and CREB, which contribute in different ways to shape the T-cell response following antigenic stimulation. However, it is also worth noting that several important aspects of T-cell function are not regulated by CRAC channels. These include T-cell development and positive and negative selection within the thymus, T-cell migration, and proliferation by cytokines that are not connected to Ca^{2+} signaling. The selective control of specific T-cell functions by CRAC channels have clinical implications for human immunity and the development of CRAC channel blockers.

25.9.2 Platelets

In addition to T cells, CRAC channels serve as the primary Ca^{2+} influx route for agonist-evoked Ca^{2+} entry in platelets (23). STIM1- and ORAI1-deficient mice exhibit impaired agonist-stimulated platelet SOCE and impaired thrombus formation resulting in a mild increase in bleeding time following injury. STIM1$^{-/-}$ mice were significantly protected from arterial thrombosis and ischemic brain infarction, and ORAI1-deficient mice exhibited resistance to various measures of thrombus formation including pulmonary thromboembolism, arterial thrombosis and ischemic brain infarction (23). These findings indicate that CRAC channels in platelets are crucial mediators of cerebrovascular thrombus formation, raising the possibility that blockade of channel function might be beneficial for the treatment of this condition.

25.9.3 Skeletal Muscle

In addition to the immune phenotypes, genetic studies in humans and mice reveal defects in skeletal muscle function and development (24, 25). Mice lacking STIM1 show no SOCE in skeletal muscle and this phenotype is accompanied by marked propensity for rapid muscle fatigue during repeated stimulation (24). Interestingly, unlike in non-excitable cells, STIM1 and ORAI1 proteins appear to be prelocalized in close proximity to each other within the triad junction in skeletal muscle under resting conditions, thus permitting extremely fast and efficient trans-sarcolemmal Ca^{2+} influx during store depletion (25). In the most widely accepted model, CRAC channels do not directly regulate the beat-to-beat contractility of the muscle, but are instead necessary to maintain store refilling during fast repetitive contractile events. Loss of CRAC channel function in ORAI1-deficient mice leads to accelerated fatigue, consistent with this role. This is likely directly related to the extreme muscle weakness seen in human patients with loss-of-function CRAC channel mutations. In addition to accelerated fatigue, the weakness in humans with ORAI1 and STIM1 mutations may also arise from loss of fast twitch fibers due to defects in muscle development.

25.9.4 Nervous System

An area of increasing focus in recent years is the role of CRAC channels in the brain. One study found that selective deletion of ORAI1 in excitatory neurons of the forebrain leads to impaired synaptic plasticity and cognitive learning through a mechanism requiring ORAI1-mediated Ca^{2+} signaling in dendritic spines (26). STIM1 has also been implicated in regulating short-term plasticity. In particular, a short isoform of STIM1 termed STIM1B, which is highly expressed in many regions of the brain including the hippocampus, appears critical for synaptic facilitation evoked by high-frequency stimulus (27). In astrocytes, which regulate numerous aspects of brain physiology including inflammation, astrocyte ORAI channels are a key pathway for Ca^{2+} entry and regulate gliotransmitter release to modulate hippocampal circuits (28). Additional studies also indicate a vital role for SOCE- and ORAI–STIM-mediated Ca^{2+} influx in dorsal horn neuron excitability with implications for neuropathic pain (29), raising the possibility that targeting ORAI1 channels in the central nervous system may offer novel avenues for mitigating chronic pain.

25.9.5 Cell Proliferation and Cancer

In numerous cell types ranging from smooth muscle cells to neural stem cells, CRAC channels formed by STIM and ORAI are important for cell proliferation. Genetic or pharmacological suppression of CRAC channel function decreases cell proliferation via suppression of NFAT signaling that plays a key role in driving proliferation in many cells. This phenotype has pathological implications for several types of cancers, including breast, cervical and glioblastoma. In addition, pharmacological blockade or knockdown of ORAI1 or STIM1 by RNA interference decreases tumor metastasis in mice, and human studies have suggested that women with relatively elevated levels of STIM1 fare worse (30). The mechanisms by which Ca^{2+} signaling is altered in breast cancers remain uncertain, but could likely mimic the proliferative states commonly seen in remodeled tissue states such as those found after lung injury or in cell culture. Taken together, these studies establish a critical role of CRAC channels in different disease processes and provide strong impetus for understanding the molecular physiology at many levels.

25.10 Conclusions

The identification of the STIM and ORAI protein families has led to dramatic advances in our understanding of CRAC channel physiology and function. In particular, we now have a firm mechanistic framework for understanding how CRAC channels are activated by depletion of intracellular Ca^{2+} stores, and we know the key structural elements involved in ion permeation and selectivity. However, there remains many broad unresolved issues. One question is the mechanism by which STIM1 binding leads to ORAI1 pore opening. As described earlier, the structures of both closed and open channels are now available, many critical steps of STIM1 activation have been described in considerable detail, and recent studies are beginning to dissect conformational changes that occur during gating. We can therefore expect the coming years to resolve the remaining questions on the structural mechanisms of CRAC channel gating including the structural basis of STIM1–ORAI1 binding and how this binding is transduced to opening of the gate. Another unknown relates to the mechanisms and functions of the "other" ORAI and STIM proteins: ORAI2, ORAI3 and STIM2. The study of these noncanonical CRAC channel proteins is currently constrained by a pressing lack of genetic evidence in both animals and humans. However, recent evidence is revealing fascinating differences in the behavior of these proteins from those of canonical CRAC channels, including but not limited to store-independent regulation, contributions to SOCE, subcellular localization of the proteins and regulation. The development of genetic tools such as specific knockouts and/or transgenic mice should open the door for providing new information on the physiological roles of these less studied isoforms. Finally, although some work has occurred in the development of small molecular drugs against CRAC channels, the mechanism and targets of most CRAC channel inhibitors are unclear. Recent breakthroughs in solving the crystal structures of the CRAC channel proteins should influence structure-based drug design, and hopefully provide a new generation of CRAC channel blockers with therapeutic potential.

Acknowledgments

I thank members of my laboratory and many colleagues and friends for stimulating discussions over the years. I am especially grateful to Priscilla Yeung and Megumi Yamashita for their insights on CRAC channel gating and physiology, and for help with the generation of the figures. The work described in this chapter was supported by grants from the National Institutes of Health.

Suggested Reading

1. Feske, S. 2019. CRAC channels and disease - From human CRAC channelopathies and animal models to novel drugs. *Cell Calcium* 80:112–116.
2. Prakriya, M., and R. S. Lewis. 2015. Store-operated calcium channels. *Physiol Rev* 95(4):1383–1436.
3. Takemura, H., A. R. Hughes, O. Thastrup, and J. W. Putney Jr. 1989. Activation of calcium entry by the tumor promoter thapsigargin in parotid acinar cells. Evidence that an intracellular calcium pool and not an inositol phosphate regulates calcium fluxes at the plasma membrane. *J Biol Chem* 264(21):12266–12271.
4. Hoth, M., and R. Penner. 1992. Depletion of intracellular calcium stores activates a calcium current in mast cells. *Nature* 355(6358):353–356.
5. Zweifach, A., and R. S. Lewis. 1993. Mitogen-regulated Ca²⁺ current of T lymphocytes is activated by depletion of intracellular Ca²⁺ stores. *Proc Natl Acad Sci U S A* 90(13):6295–6299.
6. Zhang, S. L., Y. Yu, J. Roos, J. A. Kozak, T. J. Deerinck, M. H. Ellisman, K. A. Stauderman, and M. D. Cahalan. 2005. STIM1 is a Ca²⁺ sensor that activates CRAC channels and migrates from the Ca²⁺ store to the plasma membrane. *Nature* 437(7060):902–905.

7. Feske, S., Y. Gwack, M. Prakriya, S. Srikanth, S. H. Puppel, B. Tanasa, P. G. Hogan, R. S. Lewis, M. Daly, and A. Rao. 2006. A mutation in Orai1 causes immune deficiency by abrogating CRAC channel function. *Nature* 441(7090):179–185.

8. Liou, J., M. L. Kim, W. D. Heo, J. T. Jones, J. W. Myers, J. E. Ferrell Jr., and T. Meyer. 2005. STIM is a Ca^{2+} sensor essential for Ca^{2+}-store-depletion-triggered Ca^{2+} influx. *Curr Biol* 15(13):1235–1241.

9. Yeromin, A. V., S. L. Zhang, W. Jiang, Y. Yu, O. Safrina, and M. D. Cahalan. 2006. Molecular identification of the CRAC channel by altered ion selectivity in a mutant of Orai. *Nature* 443(7108):226–229.

10. Vig, M., A. Beck, J. M. Billingsley, A. Lis, S. Parvez, C. Peinelt, D. L. Koomoa, J. Soboloff, D. L. Gill, A. Fleig, J. P. Kinet, and R. Penner. 2006. CRACM1 multimers form the ion-selective pore of the CRAC channel. *Curr Biol* 16(20):2073–2079.

11. Prakriya, M., S. Feske, Y. Gwack, S. Srikanth, A. Rao, and P. G. Hogan. 2006. Orai1 is an essential pore subunit of the CRAC channel. *Nature* 443(7108):230–233.

12. Gudlur, A., A. E. Zeraik, N. Hirve, and P. G. Hogan. 2020. STIM calcium sensing and conformational change. *J Physiol* 598(9):1695–1705.

13. Fahrner, M., H. Grabmayr, and C. Romanin. 2020. Mechanism of STIM activation. *Curr Opin Physiol* 17:74–79.

14. Yang, X., H. Jin, X. Cai, S. Li, and Y. Shen. 2012. Structural and mechanistic insights into the activation of stromal interaction molecule 1 (STIM1). *Proc Natl Acad Sci U S A* 109(15):5657–5662.

15. Yeung, P. S., M. Yamashita, C. E. Ing, R. Pomes, D. M. Freymann, and M. Prakriya. 2018. Mapping the functional anatomy of Orai1 transmembrane domains for CRAC channel gating. *Proc Natl Acad Sci U S A* 115(22):E5193–E5202.

16. McNally, B. A., M. Yamashita, A. Engh, and M. Prakriya. 2009. Structural determinants of ion permeation in CRAC channels. *Proc Natl Acad Sci U S A* 106(52):22516–22521.

17. Hou, X., L. Pedi, M. M. Diver, and S. B. Long. 2012. Crystal structure of the calcium release-activated calcium channel Orai. *Science* 338(6112):1308–1313.

18. Frischauf, I., V. Zayats, M. Deix, A. Hochreiter, I. Jardin, B. Muik, B. Lackner, B. Svobodova, T. Pammer, M. Litvinukova, A. A. Sridhar, I. Derler, I. Bogeski, C. Romanin, R. H. Ettrich, and R. Schindl. 2015. A calcium-accumulating region, CAR, in the channel Orai1 enhances Ca(2+) permeation and SOCE-induced gene transcription. *Sci Signal* 8(408):ra131.

19. McNally, B. A., A. Somasundaram, M. Yamashita, and M. Prakriya. 2012. Gated regulation of CRAC channel ion selectivity by STIM1. *Nature* 482(7384):241–245.

20. Yamashita, M., P. S. Yeung, C. E. Ing, B. A. McNally, R. Pomes, and M. Prakriya. 2017. STIM1 activates CRAC channels through rotation of the pore helix to open a hydrophobic gate. *Nat Commun* 8:14512.

21. Negulescu, P. A., T. B. Krasieva, A. Khan, H. H. Kerschbaum, and M. D. Cahalan. 1996. Polarity of T cell shape, motility, and sensitivity to antigen. *Immunity* 4(5):421–430.

22. Bhakta, N. R., D. Y. Oh, and R. S. Lewis. 2005. Calcium oscillations regulate thymocyte motility during positive selection in the three-dimensional thymic environment. *Nat Immunol* 6(2):143–151.

23. Braun, A., J. E. Gessner, D. Varga-Szabo, S. N. Syed, S. Konrad, D. Stegner, T. Vogtle, R. E. Schmidt, and B. Nieswandt. 2009. STIM1 is essential for Fcgamma receptor activation and autoimmune inflammation. *Blood* 113(5):1097–1104.

24. Stiber, J., A. Hawkins, Z. S. Zhang, S. Wang, J. Burch, V. Graham, C. C. Ward, M. Seth, E. Finch, N. Malouf, R. S. Williams, J. P. Eu, and P. Rosenberg. 2008. STIM1 signalling controls store-operated calcium entry required for development and contractile function in skeletal muscle. *Nat Cell Biol* 10(6):688–697.

25. Dirksen, R. T. 2009. Checking your SOCCs and feet: The molecular mechanisms of Ca2+ entry in skeletal muscle. *J Physiol* 587(13):3139–3147.

26. Maneshi, M. M., A. B. Toth, T. Ishii, K. Hori, S. Tsujikawa, A. K. Shum, N. Shrestha, M. Yamashita, R. J. Miller, J. Radulovic, G. T. Swanson, and M. Prakriya. 2020. Orai1 channels are essential for amplification of glutamate-evoked Ca(2+) signals in dendritic spines to regulate working and associative memory. *Cell Rep* 33(9):108464.

27. Ramesh, G., L. Jarzembowski, Y. Schwarz, V. Poth, M. Konrad, M. L. Knapp, G. Schwar, A. A. Lauer, M. O. W. Grimm, D. Alansary, D. Bruns, and B. A. Niemeyer. 2021. A short isoform of STIM1 confers frequency-dependent synaptic enhancement. *Cell Rep* 34(11):108844.
28. Toth, A. B., K. Hori, M. M. Novakovic, N. G. Bernstein, L. Lambot, and M. Prakriya. 2019. CRAC channels regulate astrocyte Ca2+ signaling and gliotransmitter release to modulate hippocampal GABAergic transmission. *Sci Signal* 12(582):eaaw5450:1-15.
29. Dou, Y., J. Xia, R. Gao, X. Gao, F. M. Munoz, D. Wei, Y. Tian, J. E. Barrett, S. Ajit, O. Meucci, J. W. Putney Jr, , Y. Dai, and H. Hu. 2018. Orai1 plays a crucial role in central sensitization by modulating neuronal excitability. *J Neurosci* 38(4):887–900.
30. McAndrew, D., D. M. Grice, A. A. Peters, F. M. Davis, T. Stewart, M. Rice, C. E. Smart, M. A. Brown, P. A. Kenny, S. J. Roberts-Thomson, and G. R. Monteith. 2011. ORAI1-mediated calcium influx in lactation and in breast cancer. *Mol Cancer Ther* 10(3):448–460.

26

Piezo Channels

Jörg Grandl and Bailong Xiao

CONTENTS

26.1 Introduction

In 2010 Bertrand Coste identified in a screen that combined siRNA knockdown, mechanical cell indentation and patch-clamp electrophysiology Piezo1 and Piezo2 proteins as novel components of mechanically activated cation channels (Coste et al., 2010) (see Figure 26.1). The discovered channels are unusual in several aspects: Piezos are extremely large with approximately 2500 amino acids per subunit, which was contrary to the much smaller sizes of known force-gated channels; Piezos control a current carried by cations, which was a novelty for mammalian force-gated channels and implied that they can mediate an excitatory signal; and Piezos show wide expression in different tissues, which immediately suggested a broad range of physiological roles. Subsequent studies have firmly established that Piezo proteins are the bona fide pore-forming subunits by showing the sufficiency of purified Piezo1 in mediating cationic currents upon reconstitution into artificial bilayers, determining the homotrimeric cryo-electron microscopy (cryo-EM) structure of Piezo1 containing a central ion-conducting pore, and functionally identifying the pore-forming domains (Coste et al., 2012; Ge et al., 2015). Today, much about Piezo function and physiology is known: revolutionized cryo-EM enabled capturing the beautiful architecture of Piezos with a total of 114 transmembrane (TM) helices (38 TMs in each subunit), which suggests a unique activation mechanism; functional studies revealed that Piezo1 senses membrane tension with high sensitivity (~1.4 mN/m) that is unmatched by other force-gated channels (Volume 1, Chapter 4); and investigations in other model organisms established that Piezos are not limited to mammals, but are also found in *Arabidopsis*

DOI: 10.1201/9781003096276-26

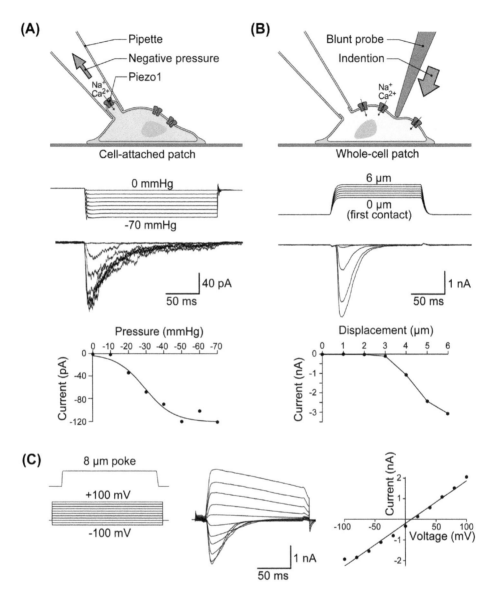

FIGURE 26.1

Discovery of Piezos. (A) Schematic of "stretch" setup, in which negative suction is applied to a cell-attached patch with a high-speed pressure clamp through the patch pipette, stimulating only those channels contained within the patch dome (top). Piezo1 peak current amplitudes initially rise with increasing magnitudes of pressure before reaching saturation (middle). The pressure–response relationship can be fit with a sigmoidal function to measure pressure sensitivity (bottom). (B) Schematic of "poke" setup depicting cell deformation by a blunt probe (typically a fire-polished glass pipette) during a whole-cell recording, which activates a larger population of channels throughout the cell (top). Piezo1 current amplitudes increase with increasing steps of displacement beginning a few micrometers beyond first contact of the probe with the cell membrane. From these experiments, a current–displacement curve can be generated. Typically, currents do not plateau before cell rupture (bottom). (C) Voltage step protocol with a single "poke" displacement during each step (left). A family of currents from a single cell illustrates the voltage dependence of channel inactivation, with severely slowed decay times at positive voltages (middle). An I–V curve plotted from peak current amplitudes reveals a reversal potential near 0 mV, demonstrating nonselective cationic conductance (right). (Reprinted from Jörg Grandl, 2017, "Touch, tension, and transduction: the function and regulation of Piezo ion channels," *Trends Biochem Sci* 42(1):57–71. With permission.)

thaliana, Drosophila melanogaster, Caenorhabditis elegans and *Danio rerio.* However, the largest advances have been made in uncovering the numerous and sometimes unexpected physiological mechanisms that rely on Piezos, and the unique structural features that might enable them to function as versatile mechanotransduction channels. Altogether, these advancements were honored with the shared 2021 Nobel Prize for Physiology or Medicine for Ardem Patapoutian, who discovered Piezos. Still, relatively large gaps of knowledge about Piezo pharmacology, regulation and cell biology exist. Similarly, pathologies related to Piezos have been identified but are largely not understood on a mechanistic level. Consequently, the sections in our chapter describing Piezo physiology, structure and function are able to paint a fairly detailed picture, while sections on Piezo pharmacology, regulation, cell biology and disease are merely sketching out work that remains to be done.

26.2 Physiological Roles

Piezos are fully specialized mechanosensors – no endogenous molecules or physical stimuli other than mechanical force are firmly established to activate Piezo1 or Piezo2. Human *PIEZO1* and *PIEZO2* genes are located in chromosomes 16 and 18 to encode 2521 and 2752 amino acids, respectively. This dedication to one single stimulus modality greatly facilitated and expedited research to uncover Piezo physiology, because gene expression data directly pointed toward a force-sensing function in the respective cells and tissues (Murthy et al., 2017). Consequently, ablating the expression of Piezo2 in neurons of the dorsal root ganglia (DRG) leads to a specific deficit in the detection of light mechanical touch, but not in thermal responses, which are also mediated by these neurons (Ranade et al., 2014). Similarly, specialized mechanotransduction functions have been demonstrated for Piezo2 expressed in Merkel cells (light touch), the vagus nerve (airway stretch and lung inflation), bladder urothelium and innervating sensory neurons (bladder control), and proprioceptive neurons (muscle stretch) (Nonomura et al., 2017; Woo et al., 2015). While not mediating acute mechanical pain responses, Piezo2 contributes to mechanical allodynia or tactile pain under inflammatory conditions (Szczot et al., 2018).

What is more, Piezo1 and Piezo2 have an expression pattern that is largely nonoverlapping, resulting in a clear division of physiological functions: Whereas Piezo2 is predominantly limited to neuronal tissue, Piezo1 is widely expressed. In the cardiovascular system, it senses shear stress, vascular flow, and organ stretch in order to regulate vascular development, blood pressure, arterial remodeling, blood cell volume, and cardiac function. In the skeletal system, it is expressed in osteoblasts and regulates mechanical load-dependent bone formation and remodeling. In the immune system, it regulates both innate and adaptive immunity. In the digestive system, it is involved in gut peristalsis and colitis and pressure-induced pancreatitis (Murthy et al., 2017). The known exceptions to this rule are nodose and petrosal sensory ganglia, where both channels contribute to sensing of blood pressure and the baroreceptor reflex, and gut enterochromaffin cells, where they mediate serotonin release. The genetic regulation for this expression separation is currently unknown, and the nearly identical function of Piezo1 and Piezo2 with their apparently indistinguishable force-sensitivity and response kinetics – only the inactivation kinetics are markedly distinct in Piezo2 (fast) and Piezo1 (slow) – leaves little clues. Despite their generally segregated roles, the in vivo functions of Piezo1 and Piezo2 appear to be exchangeable. For instance, ectopic expression of Piezo1 into primary somatosensory neurons sensitizes and rescues Piezo2-dependent touch and proprioception responses, but unexpectedly suppresses rather than evokes mechanical pain.

26.3 Subunit Diversity and Basic Structural Organization

Mammalian Piezos are large transmembrane proteins without sequence homology to known ion channels. For instance, mouse Piezo1 and Piezo2 have 2547 and 2822 amino acids, respectively, and share around 42% of sequence homology. Alternative splicing of Piezo1 and Piezo2 genes has been shown to occur in a cell-specific manner and can generate channels with altered ion permeation and gating kinetics. For instance, Piezo1 has a splicing variant Piezo1.1, which has wide expression in both mouse and human cell types and tissues. Remarkably, Piezo2 is extensively spliced in both mouse and human sensory neurons in a cell-type-dependent manner. Intriguingly, alternative splicing of a conserved domain containing the key lateral plug gate has been identified in both Piezo1 and Piezo2, demonstrating a common and evolutionarily conserved regulatory mechanism for the Piezo channel family. These Piezo splice variants differ in mechanosensitivity, unitary conductance, ion selectivity and inactivation kinetics. Thus, they may greatly expand and fine-tune the functional and physiological diversity of Piezo1 and Piezo2. Nevertheless, the full extent of functional variation, regulation of splicing and physiological consequences has not yet been explored in detail. For instance, the physiological significance of the splicing variants remains to be determined using mouse genetics models. Our understanding of the structure–function relationship of the complex Piezo channels has been paradigm-shifted by the determination of the cryo-EM structures of the full-length mouse Piezo1 of 2547 residues (Protein Data Bank: 5Z10, 6B3R, 7I28), the full-length mouse Piezo2 of 2822 residues (Protein Data Bank: 6KG7) and the mouse Piezo1.1 variant of 2524 residues (Protein Data Bank: 6LQI) (Wang et al., 2019; Guo and MacKinnon, 2017; Saotome et al., 2018; Zhao et al., 2018). All these structures have revealed that Piezos form homotrimeric architectures, with the extracellular view resembling a three-bladed propeller (Figure 26.2). The three highly curved blades spiral out from the central ion-conducting pore module topped with an extracellular cap, whose structure from mouse Piezo1 (Protein Data Bank: 4RAX) and the single *Caenorhabditis elegans* Piezo homologue (Protein Data Bank: 4PKE) was also independently determined by X-ray crystallography (Figure 26.2A–C).

The structure of Piezo2 has a completely resolved 38 TM topology, while in Piezo1 and Piezo1.1 only the last 26 TMs out of the 38 TMs have been unresolved. Thus, the Piezo1/2 protomer adopts a remarkable 38 TM topological organization, including the N-terminal blade, the C-terminal pore module, and the connecting beam and anchor domains (Figure 26.2D). The blade consists of 36 TM helices (TM1–36), in which every four TMs with a preceding membrane-parallel amphipathic helix are folded into a repetitive structural unit called TM helical unit (THU) or Piezo repeat (Figure 26.2D, E). TM37 and TM38 are termed outer helix (OH) and inner helix (IH), respectively. Together with the extracellular cap (or C-terminal extracellular domain [CED]) and the intracellular C-terminal domain (CTD), the C-terminal OH–Cap–IH–CTD region trimerizes to form the central pore module with the three IHs enclosing the transmembrane pore (Figure 26.2F). Additionally, the 9 nm long intracellular beam extends from TM28 to the bottom of the central pore module, forming layered interfaces with the CTD and the anchor domain that wedges into the space between THU9 and the OH–IH pair, and followed with critical functional domains including the central plug, lateral plug, latch, clasp and unresolved disordered

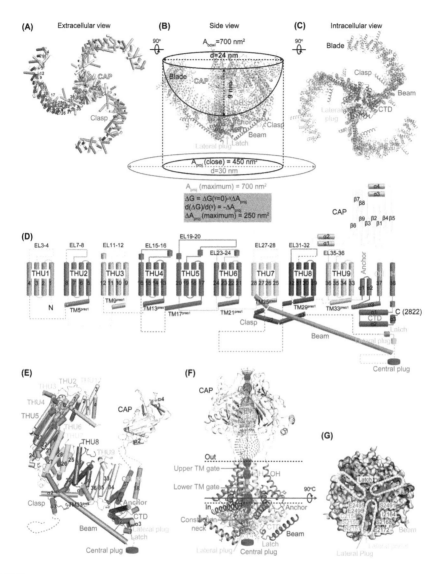

FIGURE 26.2
The homotrimeric structure of Piezos and the 38 TM topological organization. (A–C) The indicated view of illustrated models of Piezo2 with the major structural domains colored and labeled. In panel B, the midplane opening diameter, depth, surface area (A_{bowl}) and projection area (A_{proj}) of the illustrated nanobowl configuration shaped by the highly curved transmembrane regions of the three blades are labeled. When the blades become completely planar, the hypothetical maximal A_{proj} of 700 nm^2 is illustrated. In the gray box, tension (γ)-induced free energy change upon transition from closed to open channel is described by the free energy equation. The tension sensitivity is determined by the change of the projection area (ΔA_{proj}), which might reach a maximal value of 250 nm^2. (D) The 38 TM topological model of a Piezo2 protomer. Dashed lines indicate unresolved regions. (E) The structural model of a Piezo2 protomer. (F) Ribbon diagram showing the OH–Cap–IH–CTD-constituted pore module together with the anchor, lateral plug, latch and beam domains. The central solvent-accessible pathway is marked with a dotted mesh generated by the program HOLE (pore radius: red < 1 Å < green < 2 Å < purple). The extracellular vestibule (EV), membrane vestibule (MV) and intracellular vestibule (IV) are labeled. The constricted sites formed by IHs serve as the upper TM gate and lower TM gate. The cytosolic constriction neck might not serve as a major ion-conducting pathway. (G) Top view of the three lateral portals (dashed rounded rectangles) at the dissection position indicated by the arrow shown in panel F. Residues of Piezo1 lining the lateral portal, the lateral plug and latch are colored and labeled. Collectively mutating the nine labeled residues into positively charged lysine residues converts Piezo1 from a cation-selective channel to an anion-selective channel. (Modified from Bailong Xiao, 2021, "Structural designs and mechanogating mechanisms of the mechanosensitive Piezo channels," *Trends Biochem Sci* 46(6):472–488. With permission.)

region residing in the largest intracellular loop between TM28 (THU7) and TM29 (THU8) (Figure 26.2D, E).

A signature feature of the Piezo1/2 structures is that their TM regions of the three blades adopt a highly curved and nonplanar configuration, which might be facilitated by the intracellular helical layer formed by the membrane-parallel helices preceded in each of the THUs. Based on the complete structure of Piezo2, the three blades of a total of 114 TMs shape a nanobowl configuration of a midplane opening diameter of 24 nm and a depth of 9 nm, producing a midplane bowl surface area of 700 nm^2 and a projected in-plane area of 450 nm^2, which could give rise to a maximal in-plane membrane area expansion of ~250 nm^2 when Piezo2 hypothetically becomes fully planar (Figure 26.2B). The nanobowl shape of the Piezo1/2 structure might drastically curve the residing membrane. Indeed, when reconstituted into small liposomes, Piezo1 proteins locally deformed the lipid bilayer, although the channels were inserted in an inside-out orientation to adopt the membrane curvature of the liposome (Lin et al., 2019). This unusual nanobowl configuration of the Piezo channel–membrane system might serve as the unique structural basis for conferring exquisite mechanosensitivity to Piezo channels (Haselwandter and MacKinnon, 2018).

By possessing a striking 114 TM domains in the trimeric channel complex, Piezo channels are membrane proteins with the largest number of TM domains. Despite their difference in tissue expression, physiological function and biophysical properties, mouse Piezo1 and Piezo2 share a remarkably similar overall architecture, 38 TM topological organization and key structural domains, suggesting that they might possess overall similar mechanogating mechanisms for mediating in vivo mechanotransduction functions.

26.4 Gating

Upon mechanical stimulation Piezos activate rapidly (<1 ms). Several studies have shown that Piezo1 is directly gated by membrane tension with high sensitivity (~1.4 mN/m) (Lewis and Grandl, 2015). The detailed mechanism is unclear, but likely relies on the sensing and transduction of local membrane curvature by the extremely large and unusually curved blade structures. The extreme Piezo curvature induces local bending of the membrane, which has been hypothesized to be essential for its high tension sensitivity (Haselwandter and MacKinnon, 2018).

Under constant mechanical stimulation Piezos inactivate through a channel-intrinsic mechanism that is slowed by depolarization. It is plausible that inactivation is physiologically needed to protect cells from persistent Piezo activation and subsequent calcium overload, because Piezo sensitivity (~1.4 mN/m) is comparable to resting membrane tension (1–2 mN/m) in cells (Opsahl and Webb, 1994). Piezo1 inactivation is slow (τ ~30 ms) in comparison to Piezo2 (τ ~10 ms), and slowed by more positive voltages. Two kinetically distinct inactivated states exist and recovery-from-inactivation is again voltage-dependent (slower at more positive potentials) and slower in Piezo1 as compared to Piezo2 (Lewis et al., 2017; Wu et al., 2017). An emergent property of inactivation is that it modulates the overall mechanical sensitivity of Piezos and endows the channels with the ability to function as frequency filters of repetitive stimuli. Inactivation itself is regulated through many mechanisms, including physical factors such as voltage, pH, temperature, divalent ions, osmotic swelling, and cellular processes such as alternative splicing, protein interactions, membrane lipid composition, and G protein-coupled pathways. The structural determinants of inactivation have been localized to the cap domain and single residues within the inner pore helix, while electrostatic interactions between the cap and the blades and conformational changes within the cap have been shown to be required for gating to occur (Zheng et al., 2019). Removal of the mechanical stimulus results in deactivation, again with kinetics that is distinct between Piezo1 and Piezo2, that is voltage-dependent. Altogether, Piezo gating by membrane tension can be accurately described by a four-state Markov model. In addition to activation by membrane tension, a tether-gating mechanism has been proposed for Piezos, and membrane composition has emerged as an important regulator of gating.

26.5 Ion Permeability

Piezos are nonselective cation channels and their unitary conductances are ~30 pS (Piezo1) and ~22 pS (Piezo2) under physiologically relevant ionic conditions. A central permeation pathway, which is formed by the three IHs (TM38), spans the entire membrane. The Piezo2 structure has a completely sealed transmembrane pore, which is dilated in the Piezo1 structure, suggesting the existence of transmembrane gates that enable force-induced gating (Figure 26.2F). However, defining the entrance and exit points from this central pore has been more complicated: extracellular ions could enter either through the cap, although besides structural cavities there are no functional data that support this pathway exists, or through side fenestrations below the cap, although electron densities of extracellular loops in the blades that may obstruct these fenestrations are currently unresolved. Exiting of ions from the pore into the cell has been shown to occur also through three side fenestrations and three lateral portals, which are gated by the so-called lateral plugs, which follow the beam (Figure 26.2G) (Geng et al., 2020). Collectively mutating the nine residues lining the lateral portal into positively charged lysine residues converts Piezo1 from a cation-selective channel into an anion-selective channel. The lateral plugs likely move cooperatively to gate the channel in an all-or-none fashion through a proposed plug-and-latch mechanism, in which force-induced conformational changes of the blade–beam structure might lead to physically unplug the lateral plug gates. Excitingly, the splice variant Piezo1.1 and mutants lack the lateral plugs, giving rise to increased unitary conductance (~44 pS), mildly reduced calcium permeability and sensitized mechanical sensitivity. What is more, the identical splice variant Piezo2.1 qualitatively mirrors these structural and functional features. Whether ions might also flow through the vertical central constriction neck with a central plug at the bottom requires additional structural and functional evidence.

26.6 Pharmacology

As a novel, distinct and complex class of ion channels, the current pharmacology of Piezo channels remains poor and lacks high potency and specificity. Piezo1- and Piezo2-mediated currents can be blocked by nonspecific blockers, including polycation ruthenium red (RR), gadolinium (Gd^{3+}) and the peptide toxin GsMTx-4. RR and Gd^{3+} are known to block many TRP channels, while GsMTx-4 also inhibits other mechanosensitive channels such as the two-pore potassium channels. RR functions as a pore blocker of mouse Piezo1/2, but not drosophila Piezo, and the species difference of RR sensitivity is determined by the extracellular cap domain. Residue E2133 in the anchor domain and E2495 and E2496 in the lateral portal of Piezo1, in addition to determining unitary conductance, have been shown to be essential for blockage by RR. This function is conserved in Piezo2, here by the equivalent residue E2126, E2769 and E2770. The exact binding site of RR within the ion-conducting pathway of Piezo1 and Piezo2 remains to be structurally determined. GsMTx-4 is a gating modifier of Piezo1 via producing an ~30 mmHg rightward shift in the pressure-gating curve, likely by modulating local membrane tension. GsMTx4 has been proposed to occupy a small surface area in unstressed membranes, but penetrate deeper into the membrane to partially relax the outer monolayer upon tension application to the membrane, thereby reducing the effective magnitude of stimulus acting on the Piezo1 channel.

High-throughput screening of compounds to induce Piezo-dependent Ca^{2+} responses has led to the identification of three Piezo1 chemical activators, including hydrophobic Yoda1 (with a maximal water solubility of about 30 µM) and hydrophilic Jedi1 and Jedi2 (with a maximal water solubility up to 2 mM). Yoda1 activates both mouse and human Piezo1 with an apparent EC_{50} of about 17 µM and 27 µM, respectively, but does not activate Piezo2. Jedi1 and Jedi2 specifically activate Piezo1 but not Piezo2, with an EC_{50} of about 200 µM and 158 µM, respectively. Both Yoda1 and Jedi1/2 can directly bind to Piezo1 in the blade region, resulting in stabilization of its open state via independent mechanisms. Jedi-induced activation of Piezo1 requires extracellular loops between TM15–16 and TM19–20, and two leucine residues L1342 and L1345 in the proximal end of the beam, suggesting a long-distance allosteric gating of the central pore. The relatively higher potency and more robust Ca^{2+} response of Yoda1 as compared to Jedi1 and Jedi2 make it a widely used tool compound to assay the functional expression of Piezo1 in various cell types via Ca^{2+} imaging. The Yoda1 derivative Dooku1 acts as a competitive antagonist to block Yoda1-induced Piezo1 Ca^{2+} responses with an apparent IC_{50} of about 1 µM, suggesting both molecules compete for the same binding site.

26.7 Regulation

Consistent with the unique nanobowl configuration of the Piezo-membrane system, the properties of Piezo channels have been reported to be regulated by the composition of membrane lipids, including phosphoinositides and cholesterols. Pockets filled with lipid densities have been identified in the Piezo1 structure, as well as in molecular dynamics simulation. Piezo activity might also be affected by membrane stiffness. Coexpression of stomatin-like protein 3 (STOML3) potentiates mechanical activation of both Piezo1 and Piezo2. It has been proposed that STOML3 binds and recruits cholesterol to form lipid rafts, leading to an increase in membrane stiffness. By supplementing different dietary fatty acids into cell cultures, it has been found that margaric acid, a saturated fatty acid that increases membrane stiffness, inhibited Piezo1 activation, and polyunsaturated fatty acids (PUFAs) modulated channel inactivation via decreasing membrane stiffness. Notably, in some cell types such as endothelial cells, renal epithelial cells and stem cells, the endogenous Piezo1-mediated current displayed drastically slowed inactivation kinetics.

Endogenous sphingomyelinase SMPD3 has been proposed to slow Piezo1 inactivation by producing ceramide in endothelial cells.

Piezo activities can be modulated by altering cytoskeleton proteins in different cell types, including actin, dynamin, and filamin A, which might indirectly affect membrane stiffness, tension, or curvature. For instance, treatment of cells with cytochalasin D, an actin polymerization inhibitor, inhibited Piezo1-mediated mechanically activated currents. On the other hand, the activity of Piezo1 and Piezo2 was increased when measured during osmotic swelling, which might cause an increase in resting membrane tension.

Additional functional interaction proteins have been identified for Piezo1 and Piezo2, including inhibiting proteins such as polycystin-2, pericentrin, SERCA, Mtmr2, and annexin A6, and potentiating proteins such as the aforementioned STOML3 and TMEM150C/ Tentonin3. Some of them, such as Mtmr2, STOML3 and TMEM150C, appear to indirectly affect Piezo channel activity likely by affecting membrane properties. On the other hand, SERCA2 directly suppresses Piezo1 via binding to the linking region that connects the pore module and the mechanotransduction module.

26.8 Cell Biology

The biogenesis, trafficking and turnover of Piezo channels remain largely unexplored. While much of the research has been focused on Piezo channels localized on the plasma membrane, a large portion of Piezo proteins actually reside in the intracellular membrane system including the endoplasmic reticulum (ER) and nuclear envelope. Intriguingly, a study has proposed that mechanical stretch can cause Piezo1-mediated calcium release from ER, resulting in rapid loss of H3K9 trimethylation (H3K9me3), heterochromatin reduction and nuclear softening, which in turn protects DNA damage from mechanical stress.

Changes of Piezo1 expression have been associated with various disease conditions, including osteoporosis, cardiomyopathy, glioma, gastric cancer, colon cancer, bladder carcinoma and prostate cancer. A positive feedback relationship between the Piezo1 mechanosensor and the mechanical force experienced by the mechanosensitive cells and organs has been illustrated in several biological settings, in which a mechanical load leads to increased expression and activation of Piezo1. For instance, in glioma tissues, activation of Piezo1 by mechanical force upregulated the extracellular matrix and stiffened the tissue, which in turn elevated Piezo1 expression to promote glioma aggression. When subjected to exercise on a treadmill, mice showed increased expression and activation of Piezo1 in their osteoblasts. Such mechanical loading-induced changes of Piezo1 in turn enhanced osteogenesis and bone formation. In reverse, either mechanical unloading (e.g., hind-limb suspension or microgravity treatment) or loss of the Piezo1 mechanosensor (e.g., osteoblast-specific knockout) resulted in impaired osteogenesis and bone loss. Furthermore, an autonomic upregulation of Piezo1 in cardiomyocytes contributes to cardiomyopathy in both cellular and animal models, as well as in human patients, which might have altered mechanical stress conditions. While calcium might be critically involved via activating Ca^{2+}-dependent molecules such as CaMKII and NFAT, the detailed mechanism underlying mechanical force-dependent genetic regulation of Piezo1 expression remains to be fully addressed.

26.9 Channelopathies and Disease

As of today, multiple studies have linked eight human diseases to over 60 genetic mutations in Piezo1 and Piezo2 (Murthy et al., 2017): approximately 40 distinct single-point mutations in Piezo1 have been associated with colorectal adenomatous polyposis, dehydrated hereditary stomatocytosis, generalized lymphatic dysplasia and hemolytic anemia. Paradoxically, there appears to be overlap in symptoms, as some dehydrated hereditary stomatocytosis patients also have generalized lymphatic dysplasia, especially perinatally, while red blood cells in generalized lymphatic dysplasia patients show occasional stomatocytes. Similarly, a family of disorders including distal arthrogryposis type 5, Gordon syndrome, Marden–Walker syndrome and microphthalmia have been linked to approximately 20 single-point mutations in Piezo2. Patients all exhibit congenital joint contractures or abnormal stiffness of joints, but can be distinguished by other specific symptoms. Many of the diseases are not obviously related to the known physiological functions of Piezos. While the effects of most mutations on Piezo function are unknown, the few mutations studied thus far distinctly affect Piezo inactivation. For example, individuals with hereditary xerocytosis have slowed inactivation and develop age-onset iron overload. Strikingly, a relatively mild mutation that is associated with increased plasma iron is present in one-third of individuals of African descent (Ma et al., 2018). Still, our understanding of these diseases and their underlying mechanisms are limited. Humans who are effective knockouts of Piezo2 have severe deficits in proprioception and discriminative touch perception, fail to develop sensitization and painful reactions to touch after skin inflammation, and frequently report compromised bladder function (Szczot et al., 2018). All of these diagnoses are mirrored by phenotypes seen in gene ablation studies in mice. In addition to genetic studies, levels of Piezo1 protein are abnormal in several human disease conditions, such as hypertrophic scar formation, osteoporosis, cancer, and hypertrophic cardiomyopathy, hinting at roles for Piezo1 in wound healing, maintenance of healthy bone structure, cell proliferation and heart function.

26.10 Conclusions

Since the discovery of the Piezo channel family in 2010, the pathophysiological roles of Piezo channels have been demonstrated not only in various animal species, but also in humans in nearly all aspects of the mechanotransduction process. In the meantime, structure–function characterizations have unraveled the unique structural features and elegant gating mechanisms that enable Piezo channels to form sophisticated mechanotransduction machineries to effectively convert mechanical force into cellular signaling. Despite the remarkable research progress, many outstanding questions remain to be addressed, for instance, how Piezo channels might utilize their channel properties to precisely determine the various pathophysiological processes involving mechanotransduction. Given the signature nanobowl configuration of the Piezo-membrane system and the proposed gating mechanisms, it will be essential to understand the dynamic gating mechanisms of Piezo channels in a lipid environment in the presence and absence of the cytoskeleton. The expression control, biogenesis, trafficking and turnover of Piezo channels remain largely unexplored. Last but not least, built on the deep understanding of the molecular mechanisms and disease relevance of Piezo channels, the daunting but ultimate goal might be to develop novel therapeutics for disease treatment such as Piezo2-mediated tactile pain.

Suggested Reading

Coste, B., J. Mathur, M. Schmidt, T. J. Earley, S. Ranade, M. J. Petrus, A. E. Dubin, and A. Patapoutian. 2010. "Piezo1 and Piezo2 are essential components of distinct mechanically activated cation channels." *Science* 330(6000):55–60. doi: 10.1126/science.1193270.

Coste, B., B. Xiao, J. S. Santos, R. Syeda, J. Grandl, K. S. Spencer, S. E. Kim, M. Schmidt, J. Mathur, A. E. Dubin, M. Montal, and A. Patapoutian. 2012. "Piezo proteins are pore-forming subunits of mechanically activated channels." *Nature* 483(7388):176–181. doi: 10.1038/nature10812.

Ge, J., W. Li, Q. Zhao, N. Li, M. Chen, P. Zhi, R. Li, N. Gao, B. Xiao, and M. Yang. 2015. "Architecture of the mammalian mechanosensitive Piezo1 channel." *Nature* 527(7576):64–69. doi: 10.1038/nature15247.

Geng, J., W. Liu, H. Zhou, T. Zhang, L. Wang, M. Zhang, Y. Li, B. Shen, X. Li, and B. Xiao. 2020. "A plug-and-latch mechanism for gating the mechanosensitive piezo channel." *Neuron* 106(3):438-451 e6. doi: 10.1016/j.neuron.2020.02.010.

Guo, Y. R., and R. MacKinnon. 2017. "Structure-based membrane dome mechanism for Piezo mechanosensitivity." *eLife* 6:e33660. doi: 10.7554/eLife.33660.

Haselwandter, C. A., and R. MacKinnon. 2018. "Piezo's membrane footprint and its contribution to mechanosensitivity." *eLife* 7:: e41968. doi: 10.7554/eLife.41968.

Lewis, A. H., A. F. Cui, M. F. McDonald, and J. Grandl. 2017. "Transduction of repetitive mechanical stimuli by Piezo1 and Piezo2 ion channels." *Cell Rep* 19(12):2572–2585. doi: 10.1016/j.celrep.2017.05.079.

Lewis, A. H., and J. Grandl. 2015. "Mechanical sensitivity of Piezo1 ion channels can be tuned by cellular membrane tension." *eLife* 4::e12088. doi: 10.7554/eLife.12088.

Lin, Y. C., Y. R. Guo, A. Miyagi, J. Levring, R. MacKinnon, and S. Scheuring. 2019. "Force-induced conformational changes in PIEZO1." *Nature* 573(7773):230–234. doi: 10.1038/s41586-019-1499-2.

Ma, S., S. Cahalan, G. LaMonte, N. D. Grubaugh, W. Zeng, S. E. Murthy, E. Paytas, R. Gamini, V. Lukacs, T. Whitwam, M. Loud, R. Lohia, L. Berry, S. M. Khan, C. J. Janse, M. Bandell, C. Schmedt, K. Wengelnik, A. I. Su, E. Honore, E. A. Winzeler, K. G. Andersen, and A. Patapoutian. 2018. "Common PIEZO1 allele in African populations causes RBC dehydration and attenuates plasmodium infection." *Cell* 173(2):443-455.e12. doi: 10.1016/j.cell.2018.02.047.

Murthy, S. E., A. E. Dubin, and A. Patapoutian. 2017. "Piezos thrive under pressure: Mechanically activated ion channels in health and disease." *Nat Rev Mol Cell Biol* 18(12):771–783. doi: 10.1038/nrm.2017.92.

Nonomura, K., S. H. Woo, R. B. Chang, A. Gillich, Z. Qiu, A. G. Francisco, S. S. Ranade, S. D. Liberles, and A. Patapoutian. 2017. "Piezo2 senses airway stretch and mediates lung inflation-induced apnoea." *Nature* 541(7636):176–181. doi: 10.1038/nature20793.

Opsahl, L. R., and W. W. Webb. 1994. "Lipid-glass adhesion in giga-sealed patch-clamped membranes." *Biophys J* 66(1):75–79. doi: 10.1016/S0006-3495(94)80752-0.

Ranade, S. S., S. H. Woo, A. E. Dubin, R. A. Moshourab, C. Wetzel, M. Petrus, J. Mathur, V. Begay, B. Coste, J. Mainquist, A. J. Wilson, A. G. Francisco, K. Reddy, Z. Qiu, J. N. Wood, G. R. Lewin, and A. Patapoutian. 2014. "Piezo2 is the major transducer of mechanical forces for touch sensation in mice." *Nature* 516(7529):121–125. doi: 10.1038/nature13980.

Saotome, K., S. E. Murthy, J. M. Kefauver, T. Whitwam, A. Patapoutian, and A. B. Ward. 2018. "Structure of the mechanically activated ion channel Piezo1." *Nature* 554(7693):481–486. doi: 10.1038/nature25453.

Szczot, M., J. Liljencrantz, N. Ghitani, A. Barik, R. Lam, J. H. Thompson, D. Bharucha-Goebel, D. Saade, A. Necaise, S. Donkervoort, A. R. Foley, T. Gordon, L. Case, M. C. Bushnell, C. G. Bonnemann, and A. T. Chesler. 2018. "PIEZO2 mediates injury-induced tactile pain in mice and humans." *Sci Transl Med* 10(462). doi: 10.1126/scitranslmed.aat9892.

Wang, L., H. Zhou, M. Zhang, W. Liu, T. Deng, Q. Zhao, Y. Li, J. Lei, X. Li, and B. Xiao. 2019. "Structure and mechanogating of the mammalian tactile channel PIEZO2." *Nature* 573(7773):225–229. doi: 10.1038/s41586-019-1505-8.

Woo, S. H., V. Lukacs, J. C. de Nooij, D. Zaytseva, C. R. Criddle, A. Francisco, T. M. Jessell, K. A. Wilkinson, and A. Patapoutian. 2015. "Piezo2 is the principal mechanotransduction channel for proprioception." *Nat Neurosci* 18(12):1756–1762. doi: 10.1038/nn.4162.

Wu, J., M. Young, A. H. Lewis, A. N. Martfeld, B. Kalmeta, and J. Grandl. 2017. "Inactivation of mechanically activated Piezo1 ion channels is determined by the C-terminal extra-cellular domain and the inner pore helix." *Cell Rep* 21(9):2357–2366. doi: 10.1016/j.celrep.2017.10.120.

Zhao, Q., H. Zhou, S. Chi, Y. Wang, J. Wang, J. Geng, K. Wu, W. Liu, T. Zhang, M. Q. Dong, J. Wang, X. Li, and B. Xiao. 2018. "Structure and mechanogating mechanism of the Piezo1 channel." *Nature* 554(7693):487–492. doi: 10.1038/nature25743.

Zheng, W., E. O. Gracheva, and S. N. Bagriantsev. 2019. "A hydrophobic gate in the inner pore helix is the major determinant of inactivation in mechanosensitive Piezo channels." *eLife* 8:e44003. doi: 10.7554/eLife.44003.

27

Ryanodine Receptors

Jean-Pierre Benitah, Laetitia Pereira, Liheng Yin, Jean-Jacques Mercadier,
Marine Gandon-Renard, Almudena Val-Blasco, Romain Perrier and Ana Maria Gomez

CONTENTS

27.1 Introduction

The ryanodine receptor (RyR) is the largest known channel (more than 2.2 MDa). With a large conductance, and monovalent and divalent cation selectivity, the RyR physiologically works as a Ca^{2+} release channel. It sits in the membrane of an internal organelle, the sarco/endoplasmic reticulum, releasing Ca^{2+} to the cytosol when activated. It was identified in 1985 in sarcoplasmic reticulum (SR) preparations from rabbit skeletal and cardiac muscles by a strong Ca^{2+}-dependently binding to ryanodine, the alkaloid toxin extracted from the plant *Ryania speciosa*, which irreversibly binds the open RyR, giving its name (Pessah, Waterhouse, and Casida 1985). There are three isoforms of RyR in mammals: RyR1, which is mostly found in skeletal muscle; RyR2, which is mostly found in heart and brain; and RyR3, which is more ubiquitous but with lower levels of expression, in smooth muscle, brain and slow skeletal muscle.

RyR is a homotetramer; each monomer comprises about 5000 amino acids, with six transmembrane domains near the small cytosol-facing C-terminus, and a large cytosolic central and N-terminal domain that are visible in electron micrographs as dense structures, named "feet" (Figure 27.1). Due to its large size, RyR is difficult to crystallize, but recent studies have allowed visualization of the 3D structure and resolved some parts by cryo-EM and X-ray, mostly from the RyR1, although some advances are being made with RyR2.

SKELETAL **CARDIAC**

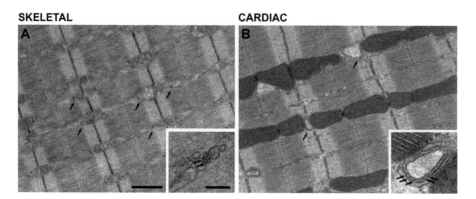

FIGURE 27.1
Position and morphology of calcium release units (CRUs) in skeletal fibers and cardiac muscle. (A) In adult skeletal muscle fibers, Ca^{2+} release units (CRUs, pointed by empty arrows) are located in proximity of the transition between the A and I band of sarcomere, on both sides of the Z-line. Mitochondria (M), when present, are positioned adjacent to the CRUs, closer to Z-lines. Inset: In skeletal fibers, CRUs are called triads, because they are formed by three elements, one central transverse tubule (TT) flanked by two terminal cisternae of the sarcoplasmic reticulum (SR). Under higher magnification in thin sections ryanodine receptors (RyRs) are visible as electron densities that span the gap between TT and SR membranes (pointed by small black arrows). (B) In adult cardiac muscle cells, CRUs (pointed by empty arrows) are in the I band of sarcomere, approximately in correspondence of Z-lines. Mitochondria (M) are bigger in size, darker and disposed in longitudinal columns between myofibrils. Inset: CRUs are called dyads, because they are formed by the association of a TT, presenting a wide profile, with a narrow and flat SR terminal cisterna, that is wrapped around the TT. Under higher magnification RyRs are visible as electron densities (named feet, pointed small black arrows) placed in the junctional gap between TT and SR membranes. Scale bars: Panels A and B, 1 μm; insets, 0.2 μm. (Figure courtesy of L. Pietrangelo and F. Protasi, University G. d'Annunzio of Chieti Pescara, Italy.)

27.2 Physiological Role

The RyR is expressed in excitable cells where its physiological function is to release enough Ca^{2+} to transform an electrical signal (action potential) into a functional outcome (contraction or secretion, depending on cell type). In muscle cells, RyRs are key elements of the excitation–contraction coupling mechanism, and in other excitable cells such as neurons and secretory cells (e.g., chromaffin cells and β-cells from islets of Langerhans) they are key in excitation–secretion coupling, initiating exocytosis to release neurotransmitters or hormones. In addition to these physiological functions, RyRs can participate in the regulation of gene transcription and expression activated through the excitation-transcription coupling (for review, see Bers 2008).

The RyRs are activated by Ca^{2+} through the Ca^{2+}-induced Ca^{2+} release mechanism. A small amount of Ca^{2+} entry through voltage-sensitive Ca^{2+} channels activates RyR, which in turn markedly amplifies this signal by releasing Ca^{2+} from the SR. However, in skeletal muscle, RyR1s can open in the absence of Ca^{2+}. In fact, the opening of the skeletal L-type Ca^{2+} channel (Cav1.1) is directly coupled to the opening of the RyR1. A change in conformation of the voltage sensor of Cav1.1 is transmitted to the RyR1 via direct protein–protein interaction.

In the cardiac and skeletal muscles, RyRs are mostly distributed at the SR membrane directly in front of sarcolemmal voltage-dependent L-type Ca^{2+} channels (LTCC) located on T-tubular invaginations, being part of the dyad (heart) or triad (skeletal muscle) (Figure

27.1). This specific spatial organization renders RyRs essential for the initiation of contraction. In the heart, upon depolarization, the opening of the LTCC (Cav1.2) generates a small inward Ca^{2+} influx insufficient to produce cardiomyocyte contraction. However, this Ca^{2+} influx activates the RyR2s. As a result, RyR2s massively release the Ca^{2+} stored in the SR into the cytoplasm. This RyR2-mediated Ca^{2+} release is synchronized by action potential and takes place at all T-tubule–SR junctions in ventricular cells. The Ca^{2+} release by RyR2s is massive (~10 μM) and the resulting increase of Ca^{2+} concentration into the cytosol is sufficient to generate cardiomyocyte contraction by Ca^{2+} binding to the myofilaments. This amplification process, first described by A. Fabiato and F. Fabiato, is known as Ca^{2+}-induced Ca^{2+} release. The strength of cardiac contraction depends on the amount of Ca^{2+} released by the RyR2 during the excitation–contraction coupling and is directly related to the amplitude of Ca^{2+} influx through the LTCCs and the amount of Ca^{2+} available inside the SR, also referred to as the SR Ca^{2+} load. In other cardiac myocytes, such as pacemaker cells, RyR2 is also involved in action potential generation, thus regulating automaticity (Lakatta and DiFrancesco 2009). RyRs are multimolecular complexes scaffolding a number of regulatory proteins that participate in the regulation of cardiac contraction such as calmodulin (CaM), Ca^{2+}/CaM kinase II (CaMKII), protein kinase A (PKA), FK506 binding protein (FKBP12.6), sorcin, junctin, triadin and calsequestrin. This regulation of the RyRs will be described later.

27.3 Subunit Diversity and Basic Structural Organization

The RyR is a homotetramer with hydrophobic segments of the four identical subunits forming a central Ca^{2+} pore. The three proteins RyR1, RyR2 and RyR3 share about 70% identity. The genes coding for each isoform were identified from 1989 to 1990. In humans, they are located on chromosome 19q13.2 and span 104 exons (RyR1), on chromosome 1q43 and spans 10 exons (RyR2), and on chromosome 15q13.3-14 and span 103 exons (RyR3).

Single-particle electron microscopy has provided the best resolution structure of the channel, with a resolution of 3.8 Å of about 70% of the rabbit and the pig RyR1 (Yan et al. 2015; Zalk et al. 2015; Efremov et al. 2015). Portions of the RyR2 structure have also been crystalized and are superimposed on the RyR1 structure to get insights into the RyR2 structure. As seen in Figure 27.2A, the side view of the RyR has a mushroom-like shape with the stalk in the SR membrane and a large cytoplasmic portion. The EF hand is the putative Ca^{2+} binding site. The top view is square-shaped (Figure 27.2C). The C-terminal domain (six transmembrane domains) forms the pore and has similarities with voltage-gated channels. The cytoplasmic, regulatory portion has three core solenoids (long helical domains), two in the central domain and one in the N-terminal domain. Schematic views are also shown in Figure 27.2E, F, with different color codes for each portion.

The phosphorylation hotspot is located in repeat 3–4 (it contains the sites equivalent to 2808 and 2814, which are the most studied sites in RyR2). These serines are phosphorylated by PKA and CaMKII (see Section 27.6). Recent studies have shown that the catalytic subunit of PKA embraces the whole repeat 3–4 of the RyR2, which is quite flexible. Phosphorylation by CaMKII at the S2814 equivalent site exposes the S2808, which becomes more phosphorylatable by PKA (Haji-Ghassemi, Yuchi, and Van Petegem 2019), thus

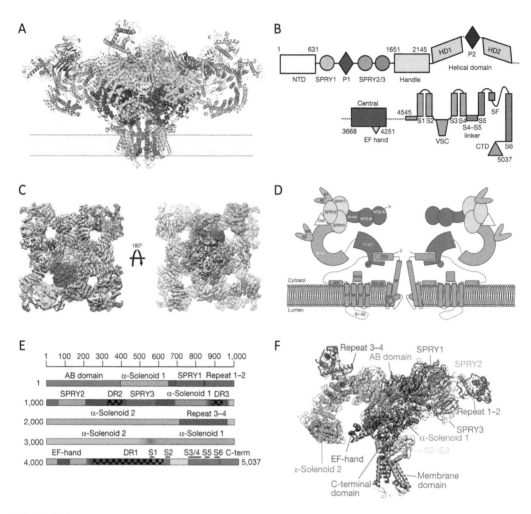

FIGURE 27.2

RyR structure. (A) Structure of the tetrameric RyR1 in complex with FKBP12 in side view. (B) Schematic illustration of domain organization in one RyR1 protomer. (C) RyR1 at 5.0 Å resolution, with one protomer segmented according to the domains assigned in the model, colored as follows: blue, N-terminal domain; cyan, SPRY1, SPRY2 and SPRY3; salmon, clamp region (RY12 repeats) and phosphorylation domain (RY34 repeats); yellow, FKBP12; green, the bridge solenoid scaffold; red, the core solenoid; orange, transmembrane and C-terminal domains; purple, putative Ca^{2+}-binding domain (EF). (D) Schematic representation of two monomers, with the color code as in panel C. B-sol, bridge solenoid; C-sol, core solenoid; N-sol, N-terminus solenoid. (E) Primary structure of one RyR protomer with positions of each region. (F) Side view of molecular density of one RyR protomer, with the color code as in panel E. The α-solenoid 2 and repeat 3–4 domains are omitted for clarity. For equivalence among panels: NTD in panels A and B is subdivided in NTD-A, B, and N-sol in panel C and D, and AB domain plus α-solenoid1 in panel E. SPRY are named the same in all panels, repeats 1–4 in panel F are named R1–4 in D, and P1 and P2 in B. HD1 and HD2 in B correspond to B-sol in panel D and α-solenoid 2 in panel F. Central in panel B corresponds to C-sol in panel D and α-solenoid 1 in panel F. CTD in panels B and D is C-terminal domain in panel F. (Panels A and B from Yan et al. 2015; panels C and D from Zalk et al. 2015; panels E and F from Efremov et al. 2015.)

showing that these sites in the phosphorylation hotspots are not independent. The FKBP12 and 12.6 kDa (also named Calstabin 1/2) bind to the RyR between the solenoids (helical domains) 1 and 2 (marked in yellow in Figure 27.2C, D), stabilizing its closed state. Around amino acids 3500–3600 is the region that binds to calmodulin, which also interacts with channel gating by stabilizing its closed state.

27.4 Gating and Ion Permeability

Mutagenesis analysis, cryogenic electron microscopy and molecular dynamics simulations revealed that the transmembrane pore has an overall architecture similar to that of a voltage-gated Na^+ or K^+ channel. The first four transmembrane helices (S1–S4) form a compact bundle termed the pseudo voltage-sensor domain, while the last two helices (S5 and S6) from each protomer form the pore domain, with S6 helices lining the ion conduction pathway and a hemipenetrant pore helix forming the luminal mouth of the channel, analogous to the selectivity filters of K^+ and Na^+ channels (Meissner 2017). The RyR channels conduct both monovalent and divalent cations, while being repulsive for anions due to the negative net charge of their pore. Indeed, a GGGIGDE motif on each protomer at the selectivity filter is essential for efficient ion permeation. The RyR channel is thus cation-selective and has an unusually large ion conductance for both monovalent cations ($\sim$800 pS with 250 mM K^+ as conducting ion) and divalent cations ($\sim$150 pS with 50 mM Ca^{2+}), while maintaining modest divalent versus monovalent cation selectivity (PCa/PK $\sim$7). This modest selectivity between monovalent and divalent cations with the following relative specificity $K^+ > Rb^+ > Na^+ \approx Cs^+ > Li^+$, and similar selectivity for divalent cation ($Ba^{2+} \approx Sr^{2+} \approx Ca^{2+} > Mg^{2+}$), is unusual for Ca^{2+} channel. For comparison, the LTCC displays a much higher degree of discrimination between monovalent and divalent cations (PCa/PK >20). The apparent deficiency in Ca^{2+} selectivity may be related to its high conductance. A distinct permeation feature of RyR compared to the highly selective K^+, Na^+ and Ca^{2+} channels is that Ca^{2+} ions are nearly fully hydrated with the first solvation shell intact during the translocation that leads to the weaker ion selectivity of RyRs (Zhang et al. 2020). Indeed, the transmembrane pore region is wide enough to allow hydrated cations to permeate and therefore facilitate a large ion conductance. Nevertheless, the selectivity among cations is mainly achieved by the charge–space competition, as Ca^{2+} could accommodate the most charge in the least space compared with K^+ and show higher affinity (the presence of Ca^{2+} in the solution significantly reduces the permeation of K^+). Another feature is that the selectivity filter of RyR is highly dynamic; its conformation changes during the state transition. The mobile nature of the selectivity filter would allow all cations smaller than the flexible aperture of the selectivity filter to pass through upon activation, hence making the RyR a broad-spectrum cation-conduction channel. The relatively poor Ca^{2+} selectivity of RyR channels does have important physiological implications. Notably, RyRs must mediate their own K^+ countercurrents to minimize the formation of a membrane potential during SR Ca^{2+} release that would otherwise rapidly impede further release of Ca^{2+}.

27.5 Pharmacology

Many pharmacological agents have been described for both experimental and therapeutic regulation of RyRs (for review, see Dulhunty et al. 2017). Among these agents, caffeine, ryanodine and tetracaine are commonly experimentally used to assess RyRs role and function in SR Ca^{2+} release. Caffeine regulates the RyRs by lowering the threshold of Ca^{2+} activation. Sustained exposure to caffeine keeps the RyRs in an open state, provoking an abrupt release of the Ca^{2+} stored into the SR while preventing SR Ca^{2+} filling. As a result, caffeine empties the SR allowing the estimation of SR Ca^{2+} content. As for ryanodine, its binding is highly specific, but concentration-dependent. It locks the open channel maintaining it in a lower subconductance state at low concentrations (nanomolar), but fully closes them under high concentrations (millimolar range). Pioneering analyses of RyR1 and RyR2 proposed four binding sites with decreasing affinity and opposed effects. The higher affinity (1–4 nM) activates, and the lowest affinity (2-4 µM) inhibits it, with negative cooperativity among sites (Pessah and Zimanyi 1991). The binding to the high affinity site reduces the single-channel conductance while increasing open probability. The transmembrane site Q4863 in RyR2 (equivalent to Q4933 in RyR1) is important in ryanodine binding (Wang et al. 2003; Ranatunga et al. 2007). Once bound to ryanodine, the channel becomes insensitive to Ca^{2+}, Mg^{2+} and ATP. Ryanodine is also used in a radiolabeled form ([^{3}H] ryanodine) to quantify RyRs and study their structure and function. [^{3}H] ryanodine binding analyses are also used to assess the dependence of the channel on Ca^{2+}. As ryanodine only binds to the open channel, binding depends on the open probability of the channel, which highly depends on [Ca^{2+}] Tetracaine is an anesthetic agent that inhibits the RyRs by keeping them in a closed state.

Several pharmacological compounds able to regulate the RyRs are also considered as therapeutic agents of great potential to prevent cardiac dysfunction, arrhythmias or hypertrophy associated with altered SR Ca^{2+} release. Among them, flecainide, a class I anti-arrhythmic agent (a Na^+ channel blocker), has been also identified as a RyRs inhibitor using RyR channels incorporated into planar phospholipid bilayers. Flecainide blocks the open state of the RyRs preventing the increased diastolic Ca^{2+} release underlying arrhythmogenic events (Kryshtal et al. 2021). In addition, dantrolene has been shown to inhibit RyR1 and is used for the treatment of malignant hyperthermia as a skeletal muscle relaxant (Zhao et al. 2001). In cardiomyocytes, dantrolene inhibits RyR2 by enhancing the calmodulin-binding affinity to RyR2, which lowers the pro-arrhythmic diastolic Ca^{2+} release (measured as sparks and waves) notably in catecholaminergic polymorphic ventricular tachycardia (CPVT), rendering it as a promising therapeutic anti-arrhythmic agent. Another strategy used to decrease the SR Ca^{2+} leak in order to prevent arrhythmias has been to stabilize the RyR2 binding to FKBP12.6 using JTV519 (or K201). JTV519 is a 1,4 benzothiazepine derivative that prevents the dissociation of FKBP12.6 from the RyR2, thus reducing SR Ca^{2+} leak. It has been suggested that JTV519 stabilizes RyR2 by restoring normal RyR2 interdomain interactions (Yamamoto et al. 2008). Current research efforts are directed toward molecules that can stabilize the RyR closed states and might be useful to treat neuromuscular disorders and cardiovascular diseases, such as arrhythmogenic genetic or acquired diseases and heart failure.

27.6 Regulation

RyR is a channel regulated by multiple factors on both sides of the SR membrane (cytosolic and luminal), such as ions, molecules, proteins and posttranslational modifications. The main RyR regulator is Ca^{2+}, which activates it but can also inhibit it at higher concentrations. In the presence of ATP, Ca^{2+} activates the channel from concentrations less than 10 µM, and inhibits it at concentrations higher than 100 µM (see Figure 27.3) (Gyorke and Gyorke 1998). It has been recognized that the RyR2 contains a Ca^{2+}-binding site of high affinity in the cytosolic portion that activates it (Ikemoto and Yamamoto 2002), with E3987 (corresponding to E4032 in RyR1 and E3885 in RyR3) being important (Li and Chen 2001). Thus, at the cytosolic side, Ca^{2+} is the physiological activator (although RyR1 can open in

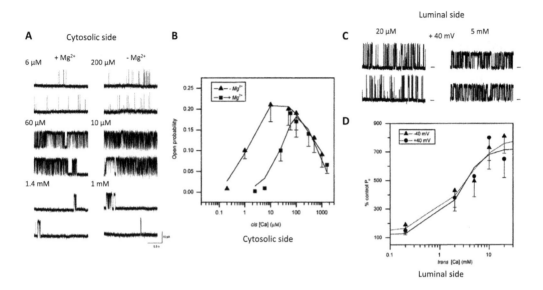

FIGURE 27.3

RyR single-channel currents and Ca^{2+} sensitivity. (A) Effect of varying cytosolic $[Ca^{2+}]$ on the activity of RyR2. Single-channel currents (upward deflections) activated by Ca^{2+} and ATP with (left) and without (right) Mg^{2+} at 40 mV holding potential. The trans chamber contained 20 µM Ca^{2+}. (B) Averaged data on the dependence of RyR2 open probability on cytosolic $[Ca^{2+}]$. (C) Effect on channel activity of increasing luminal $[Ca^{2+}]$. Channel recordings at 40 mV before and after increasing luminal $[Ca^{2+}]$ from 20 µM to 5 mM. Closed levels marked with dashes. The cis chamber contained 3 mM Na 2ATP and 1 µM free $[Ca^{2+}]$. (D) Averaged data of relative open probability (Po) as a function of luminal $[Ca^{2+}]$ at –40 and +40 mV. (From Gyorke and Gyorke 1998.)

the absence of Ca^{2+} through direct activation by the Cav1.1). Other ions, such as Mg^{2+}, bind to the same site as Ca^{2+}, but fail to activate it, thus reducing RyR activity. Among the molecules that regulate RyR, there are nucleotides, with ATP being the most active, enhancing its activity in the presence of micromolar Ca^{2+} concentrations, the RyR1 being more sensitive than the RyR2 (Laver 2006), whereas ATP in itself is a poor activator as shown with $[Ca^{2+}]$ <50 nM. Since most ATP in cells is bound to Mg^{2+}, it is likely that MgATP rather than free ATP is the physiological effector on RyR channels, the effect being more pronounced on RyR1 than on RyR2.

Reactive oxygen species and antioxidant enzymes modulate excitation–contraction coupling, in particular through their effects on RyRs. Chronic oxidation activates RyR2, which increases SR Ca^{2+} leak and decreases SR Ca^{2+} load (Terentyev et al. 2008). Nitric oxide (NO) modulates the activity of RyR2 through nitrosylation of a number of serine residues. Increasing NO synthase (NOS) activity favors RyR2 nitrosylation, which increases SR Ca^{2+} leak (Carnicer et al. 2012). Conversely, mice with NOS deletion exhibit decreased RyR2 open probability and spark frequency, associated with increased SR Ca^{2+} load (Wang et al. 2010).

Besides ions, other molecules, and their redox and nitrosylated analogues, RyR channels are regulated by several associated proteins whose function is to modulate the channel permeability and sensitivity to Ca^{2+} on both sides of the SR membrane.

On the cytosolic side, these proteins comprise CaM, sorcin, and FKBP12 and FKBP12.6 plus the kinases and phosphatases involved in the phosphorylation/dephosphorylation balance of the tetramer that plays an important role in its function.

Each RyR monomer binds one CaM molecule with a high affinity, irrespective of the number of Ca^{2+} ions bound to it (1 to 4), although high $[Ca^{2+}]$ favors the binding. The effect of CaM binding is an alteration of the RyR Ca^{2+}-dependent activation, resulting in channel opening inhibition. This favors channel closure during diastole (Meissner 2017). During systole, the increase in $[Ca^{2+}]$ leads to the binding of four Ca^{2+} ions to CaM, which results in CaMKII activation, an activation that may persist even when $[Ca^{2+}]$ decreases due to CaMKII autophosphorylation. Sorcin is a 22 kDa protein that binds to both RyRs and LTCC. It regulates RyR opening interacting with the channel when $[Ca^{2+}]$ increases, thus facilitating its closure (Farrell et al. 2003).

Among the RyR regulatory proteins, FKBPs play an important role. In cardiomyocytes, two FKBP isoforms bind and modulate RyR2 function: FKBP12 of 12 kDa, and FKBP12.6 of 12.6 kDa. FKBP12.6 has a much lower abundance but a much higher affinity for RyR2 than FKBP12 (Zissimopoulos et al. 2012). However, the relative importance and the mechanism of action of each isoform is still a matter of debate. There is a general agreement on the fact that FKBP12.6 promotes the coupled gating between RyR2 channels during systole (Marx et al. 2000) and stabilizes RyR2 in its closed state, therefore limiting SR Ca^{2+} release during diastole (Gellen et al. 2008). Because SR Ca^{2+} release during diastole may lead to arrhythmias due to the activation of the NCX-mediated depolarizing current, it has been proposed that FKBP12.6 has anti-arrhythmic properties. Indeed, promoting FKBP12.6–RyR2 binding with JTV519 prevents stress-induced arrhythmias and sudden death observed in cardiac heterozygous FKBP12.6 KO mice (FKBP12.6 KO$^{+/-}$ mice) (Wehrens et al. 2004). The role of FKBP12 in the heart is less documented. FKBP12 seems to promote pro-arrhythmogenic events from RyR2, an effect that is antagonized by FKBP12.6 (Galfre et al. 2012). In skeletal muscle cells, the two FKBP 12 and 12.6 isoforms are also expressed but in contrast to cardiomyocytes FKBP12 has a much higher affinity for RyR1 than FKBP12.6.

As described earlier, RyR phosphorylation status (phosphorylation–dephosphorylation balance), resulting from the synergic action of a number of kinases (PKA, CaMKII), phosphatases (PP1, PP2A), and phosphodiesterases (PDE4D) linked to the channel by anchoring proteins, plays a major role in RyR2 function. The three main phosphorylation sites of RyR2 are S2808, S2814 and S2030. There is an overall increase in RyR2 open probability with increased phosphorylation, resulting in an increased SR Ca^{2+} release during sympathetic stimulation. The analytical approach has attempted to assign each phosphorylation site with specific kinases and phosphatases, and a specific effect on the channel function. This has led to a considerable number of controversies still ongoing, although there seems to now be a consensus on the fact that Ser2814 hyperphosphorylation participates in the diastolic SR Ca^{2+} leak responsible for arrhythmias observed in several pathological conditions (Camors and Valdivia 2014). As mentioned earlier, the specificity of the various kinases for a given phosphorylation site has been questioned and there is now a more holistic view regarding the effects of the phosphorylation of each individual site on RyR2 function with various degrees of cooperativity between the sites. In this context, recent description of a Ser2367 residue phosphorylated by a striated muscle preferentially expressing protein kinase SPEG is of particular interest (Campbell et al. 2020). Unlike phosphorylations mediated by PKA and CaMKII that increase RyR2 activity, SPEG phosphorylation at Ser2367 reduces RyR2-mediated SR Ca^{2+} release. Therefore, the resulting effect of the various phosphorylations on RyR2 function is not only the result of a subtle balance between kinases and phosphatases, but also between those that favor and those that mitigate SR Ca^{2+} release.

RyRs are also regulated on their luminal side. Contrary to the cytosolic side, Ca^{2+} activates the RyR in a dose-dependent manner (Figure **27.3**). Some membrane-embedded proteins (junctin, triadin) or proteins in the SR lumen, controlling free $[Ca^{2+}]$ (calsequestrin, histidine-rich Ca^{2+}-binding protein), regulate Ca^{2+} sensitivity of the inner mouth of the RyR channel (Gyorke and Terentyev 2008). Moreover, triadin plays an important role in the structural and functional integrity of the LTCC–RyR junction and in the SR Ca^{2+} release (Chopra and Knollmann 2013).

27.7 Cell Biology (Biogenesis, Trafficking, Turnover)

Like most proteins, ryanodine receptors are synthesized in the cytosol and are then targeted to their final destinations. RyRs expression requires transcription factors binding, for example, binding of specificity protein 1 on RyR1 and RyR2 promoters is required for their expressions (Schmoelzl et al. 1996; Mackrill 2010). Tissue-specific expression of RyR1 results from a transcriptional repression in non-muscle cells mediated by the first intron of the gene (Schmoelzl et al. 1996). RyR2 expression is also regulated by circadian rhythm in neurons, as its expression is increased by BMAL1 and repressed by mCRY1 (Pfeffer et al. 2009). The three RyR isoforms are mainly present at the endoplasmic and sarcoplasmic reticulum (ER/SR) membrane, but can be also on the plasma membrane, secretory vesicles and endosomes. However, little is known about how RyRs are targeted to ER/SR. It has been shown that the first transmembrane domain of RyR1 provides a strong ER retention signal and is sufficient for RyR1 targeting in the ER (Meur et al. 2007), but the specific signal for the localization of RyR at the junctional SR is not known. In young rats, it has been shown that the half-life of RyRs is about eight days, whereas it increases with age thereby decreasing RyR turnover in skeletal muscle (Ferrington, Krainev, and Bigelow 1998; Denniss, Dulhunty, and Beard 2018).

27.8 Channelopathies and Disease

To date, more than 300 RyR gene mutations have been identified and linked to different pathologies. In RyR1, most of them are associated with life-threatening myopathies, and over 200 mutations in RyR2 isoform are related to cardiac diseases. Regarding RyR3, much less is known due to its low expression levels, however, it has been related recently to Alzheimer's disease and some cases of hypertension and diabetes.

Regarding RyR1-related myopathies, malignant hyperthermia (MH) is an autosomal dominant pharmacogenetic disease that manifests itself as a sudden increase in body temperature and muscle stiffness triggered by anesthesia or muscle relaxants. The first identified RyR1 human mutation related to MH was R614C (Hogan et al. 1992). RyR1 mutations in MH produce increase SR Ca^{2+} release and intracellular Ca^{2+} overload leading to enhanced Ca^{2+} binding to the myofilaments dysregulating the cell contraction. It generally develops in muscle rigidity and, consequently, a hypermetabolic state that leads to body temperature rise. The only treatment approved for MH is dantrolene. Another RyR1-related disease is central core disease (CCD), which is a congenital myopathy (I4898T, G4890R, R4892W, G4898R, A4905V) commonly developed in childhood, typically manifesting in muscle weakness and delay in motor development (Ogasawara and Nishino 2021). Typical findings include reduction or absence of mitochondria, variable degrees of myofibrillar disorganization, and the absence of oxidative and glycolytic enzymatic activity from central core regions. Although there is no consensus on the pathophysiological mechanism of RyR1 mutation, it has been suggested that the coupling between depolarization and RyR1 is altered, therefore altering Ca^{2+} release and leading to decreased $[Ca^{2+}]_i$ and defective muscle contraction. Nevertheless, the specific mechanisms by which RyR1 mutations lead to pathology remain unclear.

Given the fundamental role of RyR2 in excitation–contraction coupling, the malfunction of RyR2 leads to more or less severe cardiac dysfunction. Specific mutations in RyR2 cause pathological situations especially under stress conditions. In this regard, CPVT is an inherited channelopathy that causes syncope and sudden death in children and young adults in a context of exercise or emotional stress. Until now, more than 150 variants associated with CPVT have been identified in the RyR2 gene (CPVT1). Most of them have been located at three "hot spots" located at the N-terminal domain (residues 44–466) (32%), the central domain (residues 2246–2534) (30%), and the C-terminal domain (residues 3778–4959) (38%), since Priori et al. first reported a series of RyR2 mutations present in CPVT patients. Most of the analyzed RyR2 mutations found in CPVT have been classified as gain of function, although few have shown a loss-of-function alteration. The gain-of-function mutations increase the diastolic SR Ca^{2+} leak in ventricular myocytes especially under stress conditions, which can be detected as an increase in the frequency or amount of elementary Ca^{2+} release events, such as Ca^{2+} sparks and induces DADs (delayed afterdepolarizations) via the Na^+–Ca^{2+} exchanger, leading to triggered action potential. Analyses in cardiomyocytes from mice with different RyR2 mutations have found several alterations in the mutated channel, which are proposed as pro-arrhythmogenic mechanisms. These alterations are, to date, an increase in cytosolic or luminal Ca^{2+} sensitivity (Fernandez-Velasco et al. 2009), FKBP12.6 unbinding (Lehnart et al. 2008), alteration of the zipping–unzipping interaction between N-terminal and central domains (Uchinoumi et al. 2010), decreased CaM interaction after phosphorylation (Xu et al. 2010), or destabilization of the binding to junctophilin 2 (Yin et al. 2021). Findings in different RyR2 CPVT mutations are summarized in Table 27.1. The gain-of-function mutations show an increase in diastolic SR Ca^{2+} release events such as Ca^{2+} sparks or waves, which in conditions of sympathetic stimulation trigger Ca^{2+} waves and DADs (Figure 27.4).

While most CPVT1 mutations are gain-of-function, loss-of-function mutations have also been demonstrated, such as RyR2^{L433P} in the HEK293 cell line and RyR2^{A4860G} in the HEK293 cell line and transgenic mouse model, thus complicating the arrhythmogenic mechanism in CPVT. Functional study of RyR2^{A4860G} mutation in ventricular cardiomyocytes from A4860G knock-in mice revealed the novel mechanism that the mutation decreased the peak of Ca^{2+} release during systole, gradually overloading the SR with Ca^{2+}, which in turn caused random bursts of prolonged Ca^{2+} release, activating electrogenic Na^+–Ca^{2+} exchanger activity and triggering early afterdepolarizations (EADs) (Zhao et al. 2015) (Figure 27.4).

Most CPVT studies have focused on the ventricles, although bradycardia is common in CPVT patients. Accordingly, new studies have shifted to pacemaker cells or Purkinje fibers as a new hub to understand CPVT. In this sense, the RYR2^{R4496C} mutation was associated in pacemaker cells with a high occurrence of Ca^{2+} sparks and Ca^{2+} waves in diastole under β-adrenergic stimulation, which unloaded the SR Ca^{2+}, provoking pauses in action potential generation (Neco et al. 2012). Similarly, a slower heart rhythm and high diastolic SR Ca^{2+} leak were observed in pacemaker tissues dissected from RyR2^{R420Q} knock-in mice. In this case, the enhanced Ca^{2+} leak from the SR during diastole partially inactivated the L-type Ca^{2+} channel, which is also involved in automaticity, slowing the rate of action potential generation (Wang et al. 2017).

TABLE 27.1

Main Findings in Some RyR2 CPVT Mutations

Mutation	Location	Model	Finding/Mechanism	Reference
		HEK293	$\uparrow Ca^{2+}$ release at diastolic cytosolic $[Ca^{2+}]$	Domingo et al. 2015
		Mice	Bradycardia by disturbing the coupled clock pacemaker	Wang et al. 2017
R420Q	N-T	hiPS-CM	ISO was either ineffective, $\uparrow$ arrhythmias, $\uparrow$ diastolic $[Ca^{2+}]$	Novak et al. 2015
		Mice, hiPS-CM	$\uparrow Ca^{2+}$ spark duration, disruption of JPH2 binding to RyR2, ultrastructure alteration (smaller RyR2 clusters and widened jSR), $\uparrow$ NT-core solenoid interaction	Yin et al. 2021
		HEK293	$\downarrow$ sensitivity to channel activation	Thomas et al. 2004
		HL-1$_r$, HEK293	$\uparrow$ propensity for SOICR	Jiang et al. 2005
L433P	N-T		$\downarrow$ threshold for Ca^{2+} release termination and $\uparrow$ fractional release	Tang et al. 2012
		Mice	$\uparrow$ diastolic SR Ca^{2+} leak in atrial myocytes	Shan et al. 2012
		Yeast	Disruption of RyR2 N-terminal self-association	Seidel et al. 2015
		HL-1, HEK293	$\uparrow$ propensity for SOICR	Jiang et al. 2006
		HEK293	Dissociation of FKBP12.6 from RyR2	Wehrens et al. 2008
			$\downarrow$ binding of FKBP12.6 subunit	Lehnart et al. 2008
			$\downarrow$ RyR2 interdomain interaction, $\downarrow$ CaM binding, $\uparrow$ spontaneous Ca^{2+} leak	Xu et al. 2010
			$\uparrow$ diastolic SR Ca^{2+} leak in atrial myocytes and $\downarrow$ FKBP12.6 from RyR2	Shan et al. 2012
R2474S	Ctr	Mice	$\uparrow$ SR Ca^{2+} leak, $\uparrow Ca^{2+}$ alternans and Ca^{2+} waves $\uparrow$ atrial arrhythmias	Xie et al. 2013
			$\uparrow$ spontaneous Ca^{2+} leak, DADs, triggered activity and Ca^{2+} spark frequency $\rightarrow$ correction by $\uparrow$ CaM binding to the RyR2	Fukuda et al. 2014
			Exercise training $\downarrow$ ventricular tachycardia in CPVT through $\downarrow$ CaMKII-dependent phosphorylation of RyR2	Manotheepan et al. 2016
			$\uparrow$ cytosolic RyR2 Ca^{2+} sensitivity and $\uparrow$ pro-arrhythmic events upon sympathetic stimulation	Danielsen et al., 2018
		HL-1	$\uparrow Ca^{2+}$ release and RyR2: FKBP12.6 dissociation	George, Higgs and Lai 2008
		HEK293	$\uparrow$ RyR2 activity at low $[Ca^{2+}]$, RyR2: FKBP12.6 dissociation	Jiang et al., 2002 Wehrens et al. 2008

(Continued)

TABLE 27.1 (CONTINUED)

Main Findings in Some RyR2 CPVT Mutations

Mutation	Location	Model	Finding/Mechanism	Reference
			↑ RyR2 luminal Ca^{2+} activation, ↓ threshold for SOICR	Jiang et al. 2004
			Predisposition to VT and VF in response caffeine and/or adrenergic stimulation of the murine heart	Cerrone et al. 2005; Cerrone et al. 2007
			Abnormal Ca^{2+} transients, RyR2:FKBP12.6 interaction is not involved	Liu et al. 2006
			↑ Diastolic spontaneous Ca^{2+} release, due to ↑ in Ca^{2+} sensitivity	Fernandez-Velasco et al. 2009
			↑ RyR2 leak ↓ SR Ca^{2+} load and ↓ Ca^{2+} waves threshold. β-AR stimulation → ↑ SR Ca^{2+} load and ↑ Ca^{2+} waves	Kashimura et al. 2010
			↑ Purkinje cells abnormalities in intracellular Ca^{2+} handling	Kang et al., 2010
			His-Purkinje system focally activated arrhythmias	Herron et al. 2010
			Na*-dependent SRCa^{2+} overload in the absence of β-AR stimulation ↑ triggered arrhythmias	Sedej et al. 2010
R4497C / M496C	C-T		Overexpression of CaMKII in RyR2 R4496CT$^{+/-}$ knock-in mice → altered intracellular Ca^{2+} handling and ↑ mortality	Dybkova et al. 2011
			CaMKII inhibition ↓ catecholamine-induced sustained VT in vivo,	
		Mice	↓ triggered activity and ↓ transient inward currents induced by ISO in vitro, mechanistically by ↑ SR Ca^{2+} leak and ↓ SERCA activation	Liu et al. 2011
			↓ SAN automaticity by ↓ l$_{csi}$ and SR Ca^{2+} depletion during diastole	Neco et al. 2012
			Mutated RyR2s are functionally normal at rest but display a ↑ CRV on intense β-AR stimulation	Chen et al. 2012
			↑ RyR2 activity,↑ Ca^{2}spark and ↑ Ca^2 waves	Savio-Galmberti and Knollmann 2015
			[Na$^+$] excess of Purkinje cells →↑ triggered activity and arrhythmogenesis at lower levels of stress than ventricular myocytes	Willis et al. 2016
			↓ positive inotropic responses due to ↑ SR Ca^{2+} leak resulting from faster recovery from inactivation of the RyR2 R4496C channels	Ferrantini et al., 2015

(Continued)

TABLE 27.1 (CONTINUED)

Main Findings in Some RyR2 CPVT Mutations

Mutation	Location	Model	Finding/Mechanism	Reference
			Selectively silenced mutant RYR2 R4496C mRNA reduces ISO-induced DADs and triggered activity, adrenergically mediated VT, ultrastructural abnormalities and mitochondrial abnormalities	Bongianino et al., 2017
		HEK.293,HL-1	↓ RyR2 activation by luminal Ca^{2+}, and =RyR2 activation by cytosolic Ca^{2+}	Jiang et al., 2007
A4860Q	C-T	HEK293	↑ SOICR threshold	Jones et al., 2008
		Mice	↓Ca^{2+} transient peak during systole, gradually overloading the SR, activating electrogenic Na$^+$– Ca^{2+} exchanger and triggering EADs	Zhao et al., 2015a

Abbreviations: N-Terminal (N-T), central (Ctr), C-terminal domain (C-T), isoproterenol (ISO), store-overload induced calcium release (SOICR), sarcoplasmic reticulum (SR), β-adrenergic (β-AR), calmodulin (CaM), delayed afterdepolarization (DAD), early afterdepolarization (EAD), Ca^{2+}/calmodulin kinase II (CaMKII), ventricular tachycardia (VT), ventricular fibrillation (VF), sarcoplasmic reticulum calcium ATPase (SERCA), sino-atrial node (SAN), L-type calcium current (I_{CaL}), Ca^{2+} release variability (CRV), junctophilin2 (JPH2), junctional sarcoplasmic reticulum (jSR).

Robust evidence links RyR2 mutations with CPVT1, but, interestingly, rare mutations in RyR2 have been reported lately relating to other cardiac pathologies. For instance, the RyR2^{P1124L} mutation was identified in a patient with sarcomere mutation-negative hypertrophic cardiomyopathy. Valdivia and colleagues showed that the RyR2^{P1124L} mutation is a cytosolic loss-of-function and luminal gain-of-function mutation triggering diastolic Ca^{2+} release and cardiac arrhythmias, and induces a decrease in cardiac contractility that would lead to a hypertrophic phenotype (Alvarado et al. 2019).

Furthermore, another genetic arrhythmia, arrhythmogenic right ventricular cardiomyopathy (ARVC), also known as arrhythmogenic cardiomyopathy (ACM), has been reported to be associated with mutations in RyR2 (Te Riele et al. 2015). However, whether RyR2 mutations cause ARVC is controversial, since the identification of RyR2 mutations in ARVC might reflect clinical misdiagnosis (Roux-Buisson et al. 2014).

Besides specific channelopathies due to RyR2 mutations, other cardiovascular diseases characterized by structural changes and Ca^{2+} mishandling show modifications in the structure or function of RyR2, such as heart failure (HF). HF is associated with enhanced phosphorylation and redox modifications of RyR2s, leading to increased diastolic Ca^{2+} leak and, consequently, to arrhythmias (Table 27.1). During HF progression, maladaptive chronic β-adrenergic activation results in posttranslational changes in RyR2 inducing Ca^{2+} handling dysregulation together with later alteration of cardiac contractility, arrhythmogenic events and cardiac failure. The general consensus is that RyR2 sites are hyperphosphorylated by PKA (Ser 2808 and Ser 2030 in mice) and by CaMKII (Ser2808 and Ser2814 in mice) during HF, the latter being involved in favoring Ca^{2+} leak and DADs. Moreover, oxidation can also affect RyR2 intersubunit interaction, modifying its function and increasing SR Ca^{2+} release during diastole (Val-Blasco et al. 2021). In addition, the same posttranslational

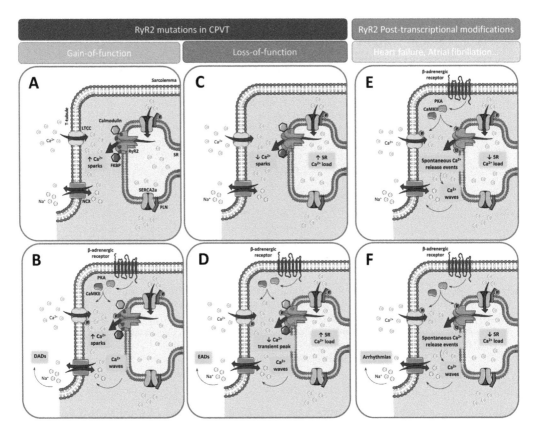

FIGURE 27.4
Cardiac Ca^{2+} mishandling caused by RyR2 modifications. (A and B) Gain-of-function mutations in CPVT1. During diastole, RyR2 mutations favor the increase of spontaneous Ca^{2+} leak in form of small Ca^{2+} sparks (A). Under β-adrenergic stimulation, Ca^{2+} waves are produced, activating Ca^{2+} extrusion through the Na^+–Ca^{2+} exchanger (NCX) that depolarizes the membrane producing delayed-afterdepolarizations (DADs) (B) that are pro-arrhythmogenic. (C and D) Loss-of-function mutations in CPVT1. In loss-of-function mutations there is a decrease in diastolic Ca^{2+} leak, which progressively overloads the SR with Ca^{2+} (C), which is enhanced by β-adrenergic stimulation, triggering spontaneous Ca^{2+} waves during diastole. The decrease in systolic Ca^{2+} transient induces less I_{Ca} inactivation and early afterdepolarizations (EADs) (D). (E and F) Posttranscriptional modifications of RyR2. In heart failure, the RyR2 hyperphosphorylation due to enhanced sympathetic tone, and other posttranslational modifications increase RyR2 open probability. This leads to enhanced SR Ca^{2+} leak and Ca^{2+} waves despite decreased SR Ca^{2+} load due to decreased SERCA2a function, leading to DADs. Moreover, alteration in other ionic currents that prolongs action potential duration also promotes EADs.

modifications of RyR2 have been related to atrial fibrillation (AF), which is the most common form of cardiac arrhythmias. In paroxysmal AF, RyR2 protein levels are upregulated, resulting in an increase in the SR Ca^{2+} leak, while in chronic AF, hyperactivity of RyR2 is mediated by an increase in phosphorylation by CaMKII and PKA (Heijman et al. 2014).

Besides muscles, RyR malfunctioning is also associated with some neurological diseases such as Alzheimer's disease, Parkinson's disease and dementia by hypoperfusion (Abu-Omar et al. 2018).

27.9 Conclusions

Ryanodine receptors are the largest ion channels, located in the membrane of the sarcoplasmic or endoplasmic reticulum. Despite their poor ion selectivity, they function as a Ca^{2+} release channel, and despite their similarity with voltage-gated channels, their opening is triggered by Ca^{2+} or by the direct opening of the L-type voltage channels located in the sarcolemma of skeletal muscle fibers. They are key elements in activating contraction in cardiac and skeletal muscles, controlling arterial tone, and activation of secretion in secretory cells and neurons. They serve as a scaffold for multiple proteins, which regulates their function, and can be modulated by several posttranslational modifications such as phosphorylation, oxidation and nitrosylation. Alterations in these modulatory mechanisms or genetic mutations are involved in many pathological states. Thus, current research is being directed toward finding pharmacological modulators for these channels that can stabilize the closed state without interfering with the opening during the action potential stimulus, which is essential for cellular function.

Suggested Reading

This chapter includes additional bibliographical references hosted only online. Please visit https://www.routledge.com/9780367538163 to access the additional references for this chapter, found under "Support Material" at the bottom of the page.

Abu-Omar, N., J. Das, V. Szeto, and Z. P. Feng. 2018. "Neuronal ryanodine receptors in development and aging." *Mol Neurobiol* 55(2):1183–92. doi: 10.1007/s12035-016-0375-4.

Alvarado, F. J., J. M. Bos, Z. Yuchi, C. R. Valdivia, J. J. Hernandez, Y. T. Zhao, D. S. Henderlong, Y. Chen, T. R. Booher, C. A. Marcou, F. Van Petegem, M. J. Ackerman, and H. H. Valdivia. 2019. "Cardiac hypertrophy and arrhythmia in mice induced by a mutation in ryanodine receptor 2." *JCI Insight* 5(7):e126544. doi: 10.1172/jci.insight.126544.

Camors, E., and H. H. Valdivia. 2014. "CaMKII regulation of cardiac ryanodine receptors and inositol triphosphate receptors." *Front Pharmacol* 5:101. doi: 10.3389/fphar.2014.00101.

Campbell, H. M., A. P. Quick, I. Abu-Taha, D. Y. Chiang, C. F. Kramm, T. A. Word, S. Brandenburg, M. Hulsurkar, K. M. Alsina, H. B. Liu, B. Martin, D. Uhlenkamp, O. M. Moore, S. K. Lahiri, E. Corradini, M. Kamler, A. J. R. Heck, S. E. Lehnart, D. Dobrev, and X. H. T. Wehrens. 2020. "Loss of SPEG inhibitory phosphorylation of ryanodine receptor Type-2 promotes atrial fibrillation." *Circulation* 142(12):1159–72. doi: 10.1161/CIRCULATIONAHA.120.045791.

Carnicer, R., A. B. Hale, S. Suffredini, X. Liu, S. Reilly, M. H. Zhang, N. C. Surdo, J. K. Bendall, M. J. Crabtree, G. B. Lim, N. J. Alp, K. M. Channon, and B. Casadei. 2012. "Cardiomyocyte GTP cyclohydrolase 1 and tetrahydrobiopterin increase NOS1 activity and accelerate myocardial relaxation." *Circ Res* 111(6):718–27. doi: 10.1161/CIRCRESAHA.112.274464.

Efremov, R. G., A. Leitner, R. Aebersold, and S. Raunser. 2015. "Architecture and conformational switch mechanism of the ryanodine receptor." *Nature* 517(7532):39–43. doi: 10.1038/nature13916.

Farrell, E. F., A. Antaramian, A. Rueda, A. M. Gomez, and H. H. Valdivia. 2003. "Sorcin inhibits calcium release and modulates excitation-contraction coupling in the heart." *J Biol Chem* 278(36):34660–6. doi: 10.1074/jbc.M305931200.

Galfre, E., S. J. Pitt, E. Venturi, M. Sitsapesan, N. R. Zaccai, K. Tsaneva-Atanasova, S. O'Neill, and R. Sitsapesan. 2012. "FKBP12 activates the cardiac ryanodine receptor Ca2+-release channel and is antagonised by FKBP12.6." *PLOS ONE* 7(2):e31956. doi: 10.1371/journal.pone.0031956.

Gellen, Barnabas, Maria Fernandez-Velasco, Francois Briec, Laurent Vinet, Khai LeQuang, Patricia Rouet-Benzineb, Jean-Pierre Benitah, Mylene Pezet, Gael Palais, Noemie Pellegrin, Andy Zhang, Romain Perrier, Brigitte Escoubet, Xavier Marniquet, Sylvain Richard, Frederic Jaisser, Ana Maria Gomez, Flavien Charpentier, and Jean-Jacques Mercadier. 2008. "Conditional FKBP12.6 overexpression in mouse cardiac myocytes prevents triggered ventricular tachycardia through specific alterations in excitation-contraction coupling." *Circulation* 117(14):1778–86. doi: 10.1161/circulationaha.107.731893.

Gyorke, I., and S. Gyorke. 1998. "Regulation of the cardiac ryanodine receptor channel by luminal Ca2+ involves luminal Ca2+ sensing sites." *Biophys J* 75(6):2801–10. doi: 10.1016/S0006-3495(98)77723-9.

Gyorke, S., and D. Terentyev. 2008. "Modulation of ryanodine receptor by luminal calcium and accessory proteins in health and cardiac disease." *Cardiovasc Res* 77(2):245–55.

Heijman, J., N. Voigt, S. Nattel, and D. Dobrev. 2014. "Cellular and molecular electrophysiology of atrial fibrillation initiation, maintenance, and progression." *Circ Res* 114(9):1483–99. doi: 10.1161/CIRCRESAHA.114.302226.

Hogan, K., F. Couch, P. A. Powers, and R. G. Gregg. 1992. "A cysteine-for-arginine substitution (R614C) in the human skeletal muscle calcium release channel cosegregates with malignant hyperthermia." *Anesth Analg* 75(3):441–8. doi: 10.1213/00000539-199209000-00022.

Kryshtal, D. O., D. J. Blackwell, C. L. Egly, A. N. Smith, S. M. Batiste, J. N. Johnston, D. R. Laver, and B. C. Knollmann. 2021. "RYR2 channel inhibition is the principal mechanism of flecainide action in CPVT." *Circ Res* 128(3):321–31. doi: 10.1161/CIRCRESAHA.120.316819.

Li, P., and S. R. Chen. 2001. "Molecular basis of Ca(2)+ activation of the mouse cardiac Ca(2)+ release channel (ryanodine receptor)." *J Gen Physiol* 118(1):33–44.

Mackrill, J. J. 2010. "Ryanodine receptor calcium channels and their partners as drug targets." *Biochem Pharmacol* 79(11):1535–43. doi: 10.1016/j.bcp.2010.01.014.

Meur, G., A. K. Parker, F. V. Gergely, and C. W. Taylor. 2007. "Targeting and retention of type 1 ryanodine receptors to the endoplasmic reticulum." *J Biol Chem* 282(32):23096–103. doi: 10.1074/jbc.M702457200.

Ogasawara, M., and I. Nishino. 2021. "A review of core myopathy: Central core disease, multiminicore disease, dusty core disease, and core-rod myopathy." *Neuromuscul Disord* 31(10):968–77. doi: 10.1016/j.nmd.2021.08.015.

Pessah, I. N., A. L. Waterhouse, and J. E. Casida. 1985. "The calcium-ryanodine receptor complex of skeletal and cardiac muscle." *Biochem Biophys Res Commun* 128(1):449–56.

Pessah, I. N., and I. Zimanyi. 1991. "Characterization of multiple [3H]ryanodine binding sites on the Ca2+ release channel of sarcoplasmic reticulum from skeletal and cardiac muscle: Evidence for a sequential mechanism in ryanodine action." *Mol Pharmacol* 39(5):679–89.

Pfeffer, M., C. M. Muller, J. Mordel, H. Meissl, N. Ansari, T. Deller, H. W. Korf, and C. von Gall. 2009. "The mammalian molecular clockwork controls rhythmic expression of its own input pathway components." *J Neurosci* 29(19):6114–23. doi: 10.1523/JNEUROSCI.0275-09.2009.

Ranatunga, K. M., S. R. Chen, L. Ruest, W. Welch, and A. J. Williams. 2007. "Quantification of the effects of a ryanodine receptor channel mutation on interaction with a ryanoid." *Mol Membr Biol* 24(3):185–93. doi: 10.1080/09687860601076522.

Wang, H., S. Viatchenko-Karpinski, J. Sun, I. Gyorke, N. A. Benkusky, M. J. Kohr, H. H. Valdivia, E. Murphy, S. Gyorke, and M. T. Ziolo. 2010. "Regulation of myocyte contraction via neuronal nitric oxide synthase: Role of ryanodine receptor S-nitrosylation." *J Physiol* 588(15):2905–17. doi: 10.1113/jphysiol.2010.192617.

Wang, R., L. Zhang, J. Bolstad, N. Diao, C. Brown, L. Ruest, W. Welch, A. J. Williams, and S. R. Chen. 2003. "Residue Gln4863 within a predicted transmembrane sequence of the Ca2+ release channel (ryanodine receptor) is critical for ryanodine interaction." *J Biol Chem* 278(51):51557–65. doi: 10.1074/jbc.M306788200.

Wehrens, X. H., S. E. Lehnart, S. R. Reiken, S. X. Deng, J. A. Vest, D. Cervantes, J. Coromilas, D. W. Landry, and A. R. Marks. 2004. "Protection from cardiac arrhythmia through ryanodine receptor-stabilizing protein calstabin2." *Science* 304(5668):292–6.

Yamamoto, T., M. Yano, X. Xu, H. Uchinoumi, H. Tateishi, M. Mochizuki, T. Oda, S. Kobayashi, N. Ikemoto, and M. Matsuzaki. 2008. "Identification of target domains of the cardiac ryanodine receptor to correct channel disorder in failing hearts." *Circulation* 117(6):762–72. doi: 10.1161/CIRCULATIONAHA.107.718957.

Yan, Z., X. Bai, C. Yan, J. Wu, Z. Li, T. Xie, W. Peng, C. Yin, X. Li, S. H. W. Scheres, Y. Shi, and N. Yan. 2015. "Structure of the rabbit ryanodine receptor RyR1 at near-atomic resolution." *Nature* 517(7532):50–5. doi: 10.1038/nature14063.

Zalk, R., O. B. Clarke, A. des Georges, R. A. Grassucci, S. Reiken, F. Mancia, W. A. Hendrickson, J. Frank, and A. R. Marks. 2015. "Structure of a mammalian ryanodine receptor." *Nature* 517(7532):44–9. doi: 10.1038/nature13950.

Zhao, F., P. Li, S. R. Chen, C. F. Louis, and B. R. Fruen. 2001. "Dantrolene inhibition of ryanodine receptor Ca2+ release channels. Molecular mechanism and isoform selectivity." *J Biol Chem* 276(17):13810–6. doi: 10.1074/jbc.M006104200.

Zissimopoulos, S., S. Seifan, C. Maxwell, A. J. Williams, and F. A. Lai. 2012. "Disparities in the association of the ryanodine receptor and the FK506-binding proteins in mammalian heart." *J Cell Sci* 125(7):1759–69. doi: 10.1242/jcs.098012.

28

Proton Channels

Emily R. Liman and I. Scott Ramsey

CONTENTS

28.1 Introduction

Control of intracellular pH (pH_i) is biochemically fundamental, and transmembrane proton movement is precisely regulated in most if not all cells. Examples of H^+-selective channels in metazoans include the voltage-gated proton channel HVCN1 (Hv1) and the recently identified family of otopetrin (OTOP) proteins. Proton channels must operate under a unique set of chemical and biophysical constraints, and specialized mechanisms of H^+ transfer and selectivity appear to have evolved independently in Hv1 and OTOP proteins. Under physiological conditions (pH 7.4 or 400 nM [H^+]), the concentration of protons is ~0.4 million times lower than that of extracellular Na^+ (~140 mM), meaning that H^+ currents mediated by Hv1 and OTOP channels, which are on the same order of magnitude as most others in this book, would be expected to be gargantuan if measured at concentrations similar to physiologically important inorganic ions (i.e., 100 mM H^+ or pH 1.0). Counterintuitively, Hv1 and OTOP channels are also exquisitely proton-selective, and at least Hv1 channels appear to utilize unconventional -type hydrogen-bonded chain and/or water-wire mechanisms for proton transfer.

DOI: 10.1201/9781003096276-28

Due to differences in their biophysical properties, Hv1 and OTOP channels generally support steady-state proton fluxes in opposite directions: Hv1 channels primarily mediate proton efflux while OTOP channels mediate influx or bidirectional flux depending on isoform. From the perspective of intracellular pH (pH_i) control, the direction of proton flux is crucial: outward H^+ currents through Hv1 will alkalize the cytoplasm and can facilitate recovery from acid loads, whereas inwardly directed H^+ currents will acidify the cell and may meaningfully alter pH_i locally or globally. Changes in pH_i can have important consequences for biochemical homeostasis and cellular signaling. For instance, H^+ competes with the pleiotropic second messenger ion Ca^{2+} for binding to acidic side chains, and lowering pH_i (i.e., increasing proton concentration) will therefore raise the concentration of intracellular free (i.e., unbound) Ca^{2+} ($[Ca^{2+}]_i$). Membrane potential, which determines the driving force of movement of all other ions and electrolytes, is also reciprocally controlled by inward (which depolarize) versus outward H^+ currents (which hyperpolarize). Although the cellular ramifications of altering the size and direction of proton currents are potentially ubiquitous, our knowledge of proton channel physiology is probably still in its infancy.

Although the experimental and model structures of Hv1 and OTOP channels have been reported, efforts to decipher the structural bases of H^+ permeation, ion selectivity, and gating mechanisms in OTOP and Hv1 channels have so far been stymied by an unusual set of constraints. The low electron density of H^+ renders it invisible to mainstay techniques (X-ray crystallography and cryo-electron microscopy [cyro-EM]) for experimental determination of ion channel structure. Other promising methods, such as nuclear magnetic resonance (NMR), Raman, and time-resolved Fourier transform infrared spectroscopy have not yet been adapted for use under the near-physiological conditions (i.e., sparsely packed proteins in membrane bilayers flanked by intact ion concentration gradients) that are typical for voltage-clamp electrophysiology, potentially confounding efforts to correlate atomic-scale conformational changes with measured channel activity.

In yet another cruel twist of fate, the proton equivalents H^+, H_3O^+ and OH^- share the same Nernst equilibrium potential (i.e., $E_H{}^+$), which severely limits the utility of conventional electrophysiological approaches to discriminate the identity of current-carrying ions in eponymous H^+ channels. The anomalously high mobility of H^+ in aqueous solution (proton diffusion is ~5 times faster than K^+ and ~12 times faster than Tris or other commonly used pH buffer molecules[1]), which is attributed to Grotthuss-type proton transfer (Cukierman, 2006), means that currents carried by H^+-selective channels can easily overwhelm the ability of pH buffers to accurately control both pH_i and pH_o when channels are open. Incomplete control of the pH gradient ($\Delta pH = pH_o - pH_i$) during a voltage-clamp experiment with H^+ channels directly challenges the experimenter's ability to control the driving force for ion movement and thus accurately measure current amplitude and current reversal potentials. The aforementioned and other limitations have impeded progress in the field, and the structural bases for H^+ conduction and selectivity in Hv1 and OTOP channels remain enigmatic. In consequence, exciting scientific challenges await the intrepid investigator of proton channel mechanisms.

28.2 OTOP Channel Introduction

The family of otopetrin-like proteins is evolutionarily ancient, with members found in divergent metazoans from *Caenorhabditis elegans* to humans (Hughes et al., 2008). Mice with mutations in the otopetrin1 gene (*Otop1*) lack calcium-carbonate-based otoconia that are required for detection of head acceleration and gravitational force, hence the name: *otós* (ear) and *pétrā* (stone) (Hurle et al., 2003). The first evidence that members of the OTOP protein family formed ion channels came from efforts to identify a sour taste receptor (Tu et al., 2018). Heterologous expression of candidate taste receptors showed that both vertebrate and invertebrate OTOP channels are sufficient to form novel proton-selective conductances (Tu et al., 2018). The structures of several OTOP channels were soon resolved by cryo-EM (Chen et al., 2019; Saotome et al., 2019). Being a relative newcomer to the ion channel superfamily, progress in the OTOP channel field is evolving rapidly, and the newest findings may not be described in this review.

28.2.1 Physiological Roles of OTOP1 in Vestibular and Taste Systems

H^+ currents subsequently shown to be mediated by OTOP1 were first described in patch-clamp recordings from mouse taste cells that transduce sour taste (Chang et al., 2010; Bushman et al., 2015). These experiments established some of the basic properties of OTOP channels, such as inhibition by extracellular Zn^{2+} and an absence of voltage-dependent gating. The latter immediately distinguished H^+ currents in sour taste cells from those carried by Hv1, the only other known proton-selective channel in mammals and suggested that they were carried by a novel membrane protein that also functions as a sour taste receptor.

Vestibular deficits in *Otop1*-mutant mice and zebrafish were described separately (Hurle et al., 2003; Hughes et al., 2004; Sollner et al., 2004). Otoconia are calcium carbonate-based structures which transduce mechanical force into an electrical response in the vestibular "hair" cells (Hurle et al., 2003). In mice, *Otop1* gene products are expressed in supporting cells that line the epithelium of two vestibular structures – the utricle and sacculus – and mice carrying mutations in *Otop1* (*tlt* or *mlg*) cannot right themselves in a forced swim task (Hurle et al., 2003). How the OTOP1 proton channel participates in the formation of otoconia is still not known. Although mutations in murine (*Mm*) OTOP1 were originally thought to perturb purinergic signaling (Hughes et al., 2007), a role for OTOP1 in regulating pH to promote biomineralization now seems more likely.

Given its established role in the vestibular system, it came as a surprise when *Otop1* was identified in a differential RNAseq screen of genes that are enriched in sour taste cells and the *Mm* OTOP1 protein was found to exhibit proton channel activity and putatively function as a sour taste receptor (Tu et al., 2018). Expression of either *Mm* or *Hs* OTOP1 is sufficient to confer robust acid-evoked H^+ currents in heterologous cells, and mutations that reduce or disrupt OTOP1 channel function lead to reductions in native sour taste receptor cell proton currents (Tu et al., 2018; Teng et al., 2019). Subsequent studies with *Otop1*$^{-/-}$ mice showed that the *Mm* OTOP1 channel is required for responses to acid stimuli measured in taste receptor cells, gustatory nerve fibers and gustatory brain regions (Teng et al., 2019; Zhang et al., 2019). However, the behavioral aversion to ingestion of acids appears to be unaltered in *Otop1*$^{-/-}$ mice, suggesting that OTOP1 may be one of several molecules involved in sensing acids in the oral cavity; the identities of other putative receptors/ion channels remain unknown.

28.2.2 Gene Structure, Subunit Diversity, and Structure of OTOP Channels

The otopetrin gene family, which is also called the otopetrin domain protein (ODP) gene family, is highly conserved from *C. elegans* to humans (Hughes et al., 2008) (Figure 28.1A). In most vertebrates, three genes (*Otop1*, *Otop2* and *Otop3*) encode approximately 600 amino acid (aa) proteins with no known similarity to other ion channels or transporters. Mouse

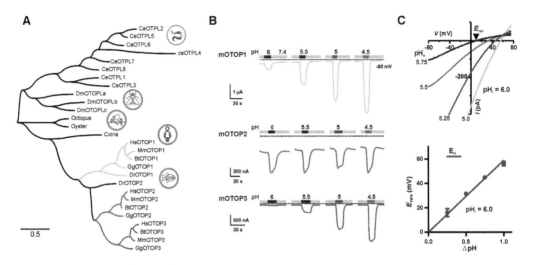

FIGURE 28.1
Otopetrin gene family encodes proton-selective ion channels. (A) Phylogenetic tree of representative OTOP proteins. One ancestral Otop gene likely gave rise to extant Otop gene, which underwent independent expansions multiple times in evolution. (B) Currents mediated by each of the murine OTOP channels expressed in *Xenopus* oocytes, evoked in response to lowering extracellular pH as indicated. (C) Current–voltage relationship for *Mm* OTOP1 expressed in HEK-293 cells in response to varying extracellular pH, as indicated. The current reversal potential (E_{rev}) shifts ~59 mV/ΔpH, similar to the shift predicted by the Nernst equation for a proton-selective ion channel.

Otop2 and *Otop3* are located in a tandem array on chromosome 11 (17 in humans), while *Otop1* is on chromosome 5 (4 in humans). Invertebrate orthologs of OTOP channels have been designated as the domain of unknown function 270 (DUM270) family. There are three known OTOP/DUM270 genes in *Drosophila melanogaster* and other arthropods, but these are not direct homologs of the three vertebrate genes. Similarly, the eight *C. elegans* genes that encode OTOP-like proteins evidently diverged since the last common ancestor with vertebrates and arthropods, suggesting that the ancestral genome probably contained a single Otop-like gene (Hughes et al., 2008). Amino acid sequence identity is highest in transmembrane segments 5, 6, 11 and 12, and all OTOP-like proteins contain a C-terminal motif termed the Phe-Tyr-Arg (FYR) box. All three vertebrate OTOP proteins and a sampling of invertebrate OTOP proteins have been shown to function as proton channels (Figure 28.1B, C), and it seems likely that other species orthologs will function similarly (Tu et al., 2018).

The structures of zebrafish (*Dr*), chicken (*Gg*) and *Xenopus* (*Xl*) OTOP proteins were reported at near-atomic resolution in 2019 (3.0 Å and 3.3 ÅÅ for *Dr* OTOP1 [Protein Data Bank: 6NF4] and *Gg* OTOP3 [Protein Data Bank: 6NF6], respectively, and 3.7 ÅÅ for *Xl* OTOP3 [Protein Data Bank: 6O84]) (Chen et al., 2019; Saotome et al., 2019), and show the channels to be cuboid-shaped homodimers with dimensions of roughly 70 Å 50 Å 50 Å with cytosolic N- and C-termini. The 12 alpha-helical segments in each protomer can be further divided into two bundles of six transmembrane (TM) helices (TM 1–6 and TM 7–12), each of which forms homologous barrel structures termed the "N" and "C" domains. The quaternary structure of dimeric OTOP channel complexes is pseudo-tetrameric and contains a total of 24 TM helices like voltage-gated Na^+, K^+ and Ca^{2+} channels (VGCs) (Figure 28.2 A, B).

Unlike VGCs, the central axis of the OTOP channel complex is occupied by cholesterol or cholesterol-like lipids, and therefore does not seem likely to form an ion-permeable pore (Saotome et al., 2019). Molecular dynamics (MD) simulations show that water molecules can invade other solvent-exposed cavities that potentially could contribute to the

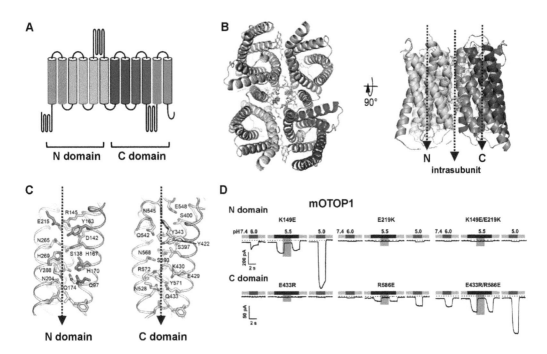

FIGURE 28.2

Structure of OTOP1 and mechanism of ion permeation. (A) OTOP1 contains 12 membrane-spanning helices, arranged in two domains that bear structural similarity. (B) Structure of *Dr* OTOP1 (Protein Data Bank: 6NF4), shows that the channel assembles as a dimer, with a pseudo-tetrameric stoichiometry. The central fenestration is filled with lipids and thus unlikely to permeate ions. Instead, three possible permeation pathways have been identified that are mostly hydrated in MD simulations. One is found in each the centers of the N and C domains, and the third is at the intrasubunit interface (between and N and C domains). (C) The putative pores within the N and C domains share several features in common, including the presence of charged residues, and a hydrophobic NPY triad may impede ion conduction in the closed state. (D) Mutation of charged residues in the N and C domains of *Mm* OTOP1 show that negatively charged residues E219 (E215 in *Dr* OTOP1) and E433 (E429 in *Dr* OTOP1) are indispensable for channel function or assembly, while positively charged residues K149 (R145 in *Dr* OTOP1) and R586 (R572 in *Dr* OTOP1) are dispensable. A double mutant in the C domain rescues channel function, consistent with the formation of a salt bridge between E433 and R586.

formation of proton permeation pathways (Saotome et al., 2019). Each N and C domain contains one putative pore, and the third is found at the intrasubunit interface between N and C domains (Figure 28.2B). The intersubunit interface is poorly conserved among OTOP proteins and therefore seems less likely to be required for shared H⁺ channel function. However, tryptophan residues located at the intersubunit interface (W394 and W398 in *Dr* OTOP1) are required for dimer formation, and mutations either disrupt dimerization or plasma membrane targeting (Saotome et al., 2019).

28.2.3 Ion Selectivity, Pharmacology, and Gating of OTOP Channels

The functional properties of murine OTOP channels have been described in detail. *Mm* OTOP1 is extraordinarily selective for H⁺ over Na⁺: $P_{H}^{+}/P_{Na}^{+} > 10^5$, as determined from the shift of the current reversal potential (E_{rev}) when H⁺ and Na⁺ concentrations are varied (Tu et al., 2018). The exquisite selectivity for H⁺ over more abundant monovalent cations has given rise to the hypothesis that OTOP channels (like Hv1, discussed later), may be perfectly selective (i.e., specific) for proton permeation.

Although the mechanism(s) of ion selectivity and the location(s) of the ion permeation pathway are not presently known, OTOP channel structures do provide some hints. A close examination of solvent-exposed vestibules in the N and C domains reveals the presence of crevices that narrow toward the middle of the membrane and evidently prevent waters from crossing the membrane. N and C crevices may therefore allow H^+ in the bulk to access a selectivity filter region, similar to what has been reported for Hv1 (see later). In OTOP channels, the central constriction to solvent access is located near well-conserved Gln-Asn-Tyr (QNY) triads (Figure 28.2C), and mutations in either the N or C domain triad abolish proton currents (Chen et al., 2019; Saotome et al., 2019). Extracellular to the QNY triad, both putative pores contain acidic and basic residues that could contribute to proton conduction and/or stabilize protein structure.

Experimental evidence strongly supports the existence of a C domain salt bridge: charge-reversing Glu → Arg mutation (E433 in *Mm* OTOP1) of a conserved acidic residue abolishes H^+ currents, but currents were subsequently restored when the putative countercharge was reciprocally mutated to Glu (R586E) (Saotome et al., 2019). However, mutations of conserved Glu and Lys residues in the N domain do not exhibit similar functional complementation, suggesting functional asymmetry between N and C domain crevices (Saotome et al., 2019) (Figure 28.2D). Charge-neutralizing mutations of highly conserved charged and polar residues located in the crevice at the intrasubunit also disrupt H^+ currents (Saotome et al., 2019). Although conserved residues in OTOPs are required for channel function, it is unclear whether mutations disrupt gating or ion permeation, and mechanistic causality remains to be established.

Native proton currents in sour taste cells and heterologously expressed OTOP channel currents are inhibited by extracellular Zn^{2+} (Chang et al., 2010; Tu et al., 2018). As expected for a discrete Zn^{2+}-binding site formed by ionizable side chains (De La Rosa et al., 2018), inhibition of OTOP channels by $ZnCl_2$ is strongly pH-dependent, exhibiting lower (millimolar) potency at acidic extracellular pH (pH_o), and higher (near-micromolar) potency at more alkaline pH_o (Bushman et al., 2015; Teng et al., 2019). Unlike for Hv1 (Ramsey et al., 2006; De La Rosa et al., 2018; Qiu et al., 2016), residues that compose the putative Zn^{2+}-binding site in OTOP proteins have not yet been identified.

Gating mechanisms of OTOP channels are still largely unknown, but recent evidence shows that some are acid-activated (Teng et al., 2022). Notably, murine OTOP1 and OTOP3 exhibit a sharp increase in conductance between pH_o 5.5 and 6.0, whereas OTOP2 currents can be elicited over a wide range of pH_o including in response to alkaline pH (Figure 28.1B). The hypothesis that OTOP1 and OTOP3, but not OTOP2, channels are positively gated by protons is further substantiated by the observation that in response to acid stimuli, the activation kinetics of OTOP1 and OTOP3 currents, but not OTOP2, are pH-dependent. Whether intracellular protons that accumulate upon influx through OTOP channels could modulate the gating of OTOP channels is not yet known, and could vary among the different subtypes. One study showed that during sustained acid stimuli, the rate of decay of proton currents in taste cells, which express OTOP1, was well correlated to current amplitude (Bushman et al., 2015). However, understanding the relative contributions of altered driving force for H^+ entry and the effects of pH_i changes on current decay kinetics (i.e., via desensitization [Chen et al., 2019] or inactivation) in various OTOP channels requires additional study.

28.2.4 OTOP Channels Distribution and Related Diseases

OTOP channels are expressed in a wide array of tissues and cell types (Tu et al., 2018), where they can specifically and rapidly transport protons across the cell membrane.

OTOP1: In addition to the taste and vestibular systems, *Mm* OTOP1 is highly expressed in brown adipose tissue and at lower levels in white adipose tissue (Tu et al., 2018). Interestingly, Otop1 gene expression is upregulated in white adipose tissue in obese mice and is protective against adipose tissue inflammation (Wang et al., 2014). How or whether OTOP1 proton channel activity contributes to adipose tissue physiology is not known. Brown adipose tissue generates heat by engaging a futile mitochondrial cycle whereby proton accumulation that generally powers the synthesis of ATP is shunted through the mitochondrial proton channel UCP1. Thus, pH regulation by OTOP1 may be significant in brown adipose physiology. Despite high to moderate levels of *Otop1* expression in mammillary and adrenal glands, uterus, and mast cells, the roles of OTOP channels in these tissues remain uninvestigated (Tu et al., 2018).

OTOP2 and OTOP3: *Otop2* gene expression has been detected in the olfactory bulb, stomach and colon (Tu et al., 2018). In the human colon, *Otop2* gene expression is found in a subset of cells that coexpress the calcium-sensitive chloride channel BEST4 and the satiety peptide uroguanylin (Parikh et al., 2019). The cells that express OTOP2 are depleted in colorectal tumors and inflammatory bowel syndrome, suggesting that OTOP2 may be protective against colorectal disease (Parikh et al., 2019). Consistent with this possibility, the *Otop2* gene was shown to be under the control of p53 and to suppress cancer cell proliferation in vitro (Qu et al., 2019). The OTOP2 protein is expressed in the chicken uterus, where it may play a role in the biomineralization and the formation of the eggshell (Sah et al., 2018). The *Otop3* gene is expressed in the epidermis and the gastrointestinal tract; interestingly, higher expression levels appear to be part of a signature for disease progression in colon cancer (Yang et al., 2019).

The distribution and functions of invertebrate OTOP channels are just beginning to be described. Two recent reports show that in *Drosophila melanogaster* (*Dm*), an OTOP-like channel (OtopLa) is involved in acid sensing by the gustatory system (Ganguly et al., 2021; Mi et al., 2021), and in sea urchins, an OTOP channel is involved in biomineralization, analogous to its role in the vertebrate inner ear (Chang et al., 2021). The role of OTOP channels in diverse species and other tissues remains to be uncovered but based on the few systems in which they have been studied, it can be expected that they play essential roles in acid sensing, pH homeostasis and biomineralization.

28.3 Hv1 Channel Introduction

Voltage-gated proton currents were measured in a variety of cell types (including snail neurons, alveolar epithelial cells and neutrophils) before genes encoding the Hv1 protein were identified (DeCoursey, 2003). In 2006, genes encoding Hv1 (Ramsey et al., 2006) and VSOP (Sasaki et al., 2006) were identified in human, mouse and *Ciona intestinalis* (*Ci*) genomes based on sequence similarity to other voltage sensor (VS) domain-containing proteins (Ramsey et al., 2006; Sasaki et al., 2006). A standardized nomenclature (HVCN1) was subsequently adopted, but we prefer Hv1 because of its obvious relationship to this historical VGC nomenclature (i.e., Kv1.2). Although other candidate voltage-gated H^+ channel proteins had been proposed prior to the identification of Hv1, the current consensus is that Hv1 remains the only authentic mammalian voltage-gated H^+ channel protein (Decoursey, 2012).

Heterologous expression of mammalian Hv1 proteins is sufficient to reconstitute the hallmark biophysical properties of native proton-selective and voltage-dependent and pH-dependent conductances (G_{vH}^+) (Ramsey et al., 2006). Targeted disruption of the mouse *Hvcn1* allele abolishes Hv1 protein expression and G_{vH}^+, indicating that Hv1 is also necessary. Consistent with early reports (DeCoursey, 2003), Hv1 protein and mRNA are highly expressed in human and mouse immune cells and tissues (Ramsey et al., 2006). More recent work shows that Hv1 expression is more widespread, and now includes mammalian microglia, breast cancer cell lines, pancreatic β-cells, spermatozoa and other cells (Wu et al., 2012; Pang et al., 2020; Lishko et al., 2010; Decoursey, 2012; Bare et al., 2020).

28.3.1 Physiological Roles of Hv1 Channels

The primary physiological role for Hv1 channels in most eukaryotic cells is efflux of acid equivalents from the cytoplasm (DeCoursey, 2003, 2012). This is a direct consequence of the characteristic coupling between voltage- and pH-dependent gating that results in the channels being open mainly when the driving force for H^+ movement is outwardly directed (Cherny et al., 1995; DeCoursey, 2003). Perhaps the best understood role of Hv1 channels is during the "respiratory burst" of phagocytic leukocytes, where the sustained, high-level production of reactive oxygen species (ROS) is bactericidal. ROS production by NADPH oxidase causes pH_i to fall precipitously, and the ensuing normalization of pH_i requires H^+ efflux through Zn^{2+}-sensitive proton channels (DeCoursey, 2003). As expected, if Hv1 channel activity supports NADPH oxidase activity, Hv1-null mice exhibit lower ROS production and decreased bacterial killing (Ramsey et al., 2009; Okochi et al., 2009; El Chemaly et al., 2010).

In brain-resident microglia, Hv1 channels augment inflammatory responses in murine models of cerebral ischemia and neurotoxicity (Wu et al., 2012; Tian et al., 2016). Microglia are functionally similar to peripheral macrophages, and the role of Hv1 to augment neurotoxicity is consistent with its role to facilitate NADPH oxidase-dependent ROS production (Wu et al., 2012; Tian et al., 2016; Ramsey et al., 2009). Although key biophysical properties of G_{vH}^+ were first described in airway epithelial cells (DeCoursey, 2003), the role of Hv1 in the lung remains poorly understood. Recently, extracellular alkalization of cultured bronchial epithelial cells was shown to alter pH_i in a Zn^{2+}-dependent fashion and effectively prevent SARS-CoV-2 replication (Davis et al., 2021), suggesting that modulating Hv1 activity in the lung could be therapeutically beneficial in both viral and nonviral inflammatory airway diseases.

Emerging evidence points to a role for Hv1 in the control of Ca^{2+} signaling. For example, H^+ currents decrease the driving force for Ca^{2+} entry and modulate neutrophil chemotaxis (El Chemaly et al., 2010). In spermatozoa, Hv1 channels cause pH_i alkalization that is required for hyperactivation and fertilization (Lishko and Kirichok, 2010). Hv1 is also required for Ca^{2+}-dependent glucose-dependent insulin secretion from pancreatic β-cells, and targeted deletion causes hyperglycemia in rodents (Pang et al., 2020).

Recent studies of Hv channels from protists, fungi, corals, and molluscs show intriguing differences from their mammalian counterparts that suggest specialized roles in the physiology of these organisms, but this work is outside the scope of this review. In summary, Hv1 channels are widespread in eukaryotic cells, and Hv1 appears to be crucial for controlling pH_i, membrane potential and Ca^{2+} signaling, but future discoveries are likely to expand the palette of physiological processes that depend on Hv1 channels.

28.3.2 Hv1 Structure

Hv1 is encoded by a single gene, and the sufficiency of Hv1 to reconstitute most biophysical features of G_{vH}^+ suggests that auxiliary proteins or other cofactors are not required

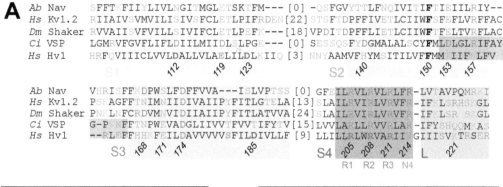

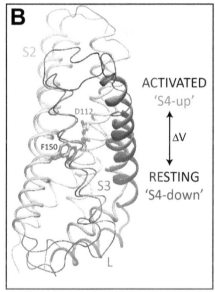 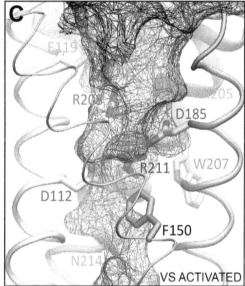

FIGURE 28.3

Amino acid sequence alignments of S4 helical segments in VS domain proteins. (A) Multiple sequence alignments of amino acids in the S3 and S4 helices from the *Arcobacter nitrofigilis* DSM 7299 sodium channel (*Ab* Nav; ADG92959), *Hs* Kv1.2 (NP_004965), *Dm* Shaker (CAA29917), *Ci* VSP (BAD98733), and *Hs* Hv1 (NP_115745) show the positions of conserved residues and S1–S4 helices (colored boxes); L is the S4–S5 linker helix in VGCs or putative S4-CC linker in Hv1, and the gray box indicates residues that were swapped in the mHv1cc chimera (Protein Data Bank: 3WKV). Diagonal numbers indicate amino acid position in *Hs* Hv1. Charged and polar residues are colored and conserved S4 gating charges are highlighted below S4. (B) Putative resting-state (Hv1 F; Randolph et al., 2016) and activated-state (Hv1 J) Hv1 VS domain model structures are overlaid to illustrate differences in S4 position. (C) A snapshot taken from an Hv1 J MD trajectory shows that water molecules penetrate deeply into the Hv1 VS central crevice. In panels B and C, helices are shown as colored ribbons and selected side chains are shown in licorice representations and labeled.

for H[+] channel activity (Sasaki et al., 2006; Ramsey et al., 2006). Human (*Hs*) Hv1 is 273 aa in length: the ~100 aa N-terminal structure is unknown, the middle ~120 aa form a canonical VS domain that is structurally and functionally homologous to those in VGCs and voltage-sensitive phosphatases (VSPs) (Figure 28.3A,B), and a C-terminal coiled-coil (CC) motif self-associates to nucleate the channel's dimeric quaternary architecture. Hv1 protein lacks the canonical pore domain of VGCs and, as discussed later, protons permeate through the VS.

Although biochemical studies indicate that functional Hv1 channels are most likely to form a homodimeric quaternary structure (Koch et al., 2008; Lee et al., 2008; Tombola et al., 2008), structural studies have not yet fully resolved the structure of the intersubunit interface (Li et al., 2015). CC truncation (Hv1-ΔC) attenuates dimerization but does not impair function, suggesting that Hv1 channels could function as monomers (Koch et al., 2008; Lee et al., 2008; Tombola et al., 2008). However, Hv1-ΔC proteins do dimerize ($K_d \approx$ 3 μM) in proteoliposomes (Li et al., 2015) and may also self-associate in cell membranes, raising the possibility that Hv1-ΔC dimers account for part or all of the H^+ current that is commonly attributed to monomeric channels (Koch et al., 2008; Gonzalez et al., 2010; Qiu et al., 2013; Gonzalez et al., 2013; Chamberlin et al., 2014; Carmona et al., 2018; Tombola et al., 2008, Musset et al., 2010).

Although high-resolution structures of full-length (dimeric) Hv1 proteins remain elusive, the experimentally determined VS (Protein Data Bank: 3WKV, 5OQK) and CC domain (Protein Data Bank: 3A2A, 3VMX, 3VMY, 3VYI, 3VN0) structures represent a starting point. The sole X-ray structure (Protein Data Bank: 3WKV) was solved for a chimeric protein that contains VSP (S2–S3 loop; Figure 28.3A) sequence and the CC from an unrelated protein (GCN4; Protein Data Bank: 3WKV) (Takeshita et al., 2014); an NMR-determined ensemble of putative Hv1 VS structures (Protein Data Bank: 5OQK) was also reported (Bayrhuber et al., 2019). However, the physiological relevance of these structures is not yet firmly established and certain features (i.e., functionally critical ionizable side chains located within solvent-free pockets) are difficult to reconcile with experimental evidence; additional work to elucidate *in vivo* protein conformations is therefore needed.

Combinations of molecular modeling, MD simulations, mutagenesis, and voltage-clamp electrophysiology have driven most of the progress toward understanding the structural underpinnings of mechanisms in Hv1. Most in silico Hv1 VS domain model structures are based on related VGC or VSP structures (Figure 28.3B, C) (Ramsey et al., 2010; Li et al., 2015; Geragotelis et al., 2020; Randolph et al., 2016; De La Rosa et al., 2018), but the *in silico* models lack the N-terminal ~100 aa and are necessarily incomplete. Long time-scale simulations (Geragotelis et al., 2020), together with experimental testing of model structure-based hypotheses, should help drive progress even in the absence of definitive experimental structures. An encouraging trend is the development of Hv1 models that *a priori* predict experimental outcomes (De La Rosa et al., 2018).

Despite the current lack of bulletproof structural support, convergent lines of evidence support key structural features of Hv1 channels: (1) Hv1 contains an hourglass-shaped solvent-accessible central crevice that is most constricted near an S1 Asp (D112 in *Hs* Hv1) and S2 Phe (F150 in *Hs* Hv1) in both VS resting-state (i.e., closed H^+ channel) and activated-state (i.e., open H^+ channel) conformations (Ramsey et al., 2010; Kulleperuma et al., 2013; Geragotelis et al., 2020) (Figure 28.3B). (2) D112 and one or more S4 Arg (R208/R2 and or R211/R3 in *Hs* Hv1) side chains, together with protein-associated water molecules, are necessary for the exquisite H^+ selectively that is characteristic of Hv1 channels (Musset et al., 2011; Berger and Isacoff, 2011; Randolph et al., 2016). (3) In contrast to OTOPs (see earlier), neutralizing mutations of conserved ionizable residues have large effects on voltage-dependent gating but do not abolish currents in Hv1 channels (Musset et al., 2011; Berger and Isacoff, 2011; Ramsey et al., 2010). Together, the available data appear to be most parsimonious with the hypothesis that the solvent-filled central crevice in Hv1 mediates H^+ transfer, but the details remain a matter of active investigation and debate (Bennett and Ramsey, 2017; DeCoursey, 2017).

28.3.3 Hv1 Channel Gating

Like VGCs, the amplitude and kinetics of Hv1 current activation are strongly voltage-dependent, but Hv1 channels tend to be relatively slow to open (activation time constants, τ_{ACT}, are typically in the hundreds of milliseconds range for *Hs* Hv1) (Vargas et al., 2012; DeCoursey, 2003; Cherny et al., 1995; Ramsey et al., 2006; Sasaki et al., 2006) (Figure 28.4A). Also, like other VGCs, Hv1 generates transient "gating currents" that are associated with the movement of three highly conserved S4 Arg "gating charge" residues (R208–R211 in *Hs* Hv1) most of the way through the transmembrane electric field (Vargas et al., 2012; De La Rosa and Ramsey, 2018; Carmona et al., 2018) (Figure 28.5D). Neutralizing mutations at S4 gating charge and predicted acidic "countercharge" residue generally result in P_{OPEN}-V shifts in the expected direction, indicating that as in VGCs, S4 gating charge chains in Hv1 are likely to make state-dependent salt bridges with countercharge side chains that are organized into extracellular and intracellular Coulombic networks (De La Rosa et al., 2018; Randolph et al., 2016; Ramsey et al., 2010; Geragotelis et al., 2020; Vargas et al., 2012). Consistent with the hypothesis that VS structure and function are highly homologous, R1H mutations also confer similar resting-state H^+ currents in Hv1 and VGCs (Figure 28.5A–C) (Randolph et al., 2016).

Hv1 also manifests functional and structural peculiarities that are not shared by most other VS domain proteins. First, Hv1 is unusual in lacking a basic side chain at the R4 position (N214 in *Hs* Hv1; Figure 28.3A); introducing either Arg or Lys here produces mutant channels that generate smaller outward currents (I_{STEP}) but retain the robust inwardly directed instantaneous "tail" currents (I_{TAIL}) that are seen in wild-type (WT) Hv1 (Ramsey et al., 2010; Randolph et al., 2016; Sakata et al., 2010) (Figure 28.4A). The behavior of Hv1 N214R/K mutant channels has been interpreted to reflect voltage-dependent block of the H^+ permeation pathway by the cationic charge associated with introduced Arg or Lys, suggesting that R4 in VGCs and VSPs helps to prevent H^+ "leakage" through the VS in its activated-state conformation (Randolph et al., 2016). In the background of *Hs* Hv1 R1H, N214R selectively abrogates the intrinsic, activated-state H^+ current without affecting the resting-state current (Figure 28.5B, C) (Randolph et al., 2016). The N214 side chain is thus likely to be located close to F150 and the hydrophobic constriction in the Hv1 VS activated-state conformation (Figure 28.3C) (Randolph et al., 2016).

A defining feature of Hv1 channels is their sensitivity to changes in the pH gradient ($\Delta pH = pH_o - pH_i$), which causes dramatic (~40 mV/pH unit) shifts in the position of the apparent open probability (P_{OPEN}) curve along the voltage axis (Figure 28.4B, D) (Cherny et al., 1995; Ramsey et al., 2006; Sasaki et al., 2006; Ramsey et al., 2010; De La Rosa and Ramsey, 2018). Intriguingly, pH_i acidification produces V_{THR} shifts of equal magnitude but in the opposite direction as pH_o acidification, leading to the hypothesis that the pH gradient ($\Delta pH = pH_o - pH_i$), rather than absolute intra- or extracellular pH, is the key energetic determinant of Hv1 gating (Cherny et al., 1995; DeCoursey, 2003). Alternatively, separate pH_i and pH_o sensors could allosterically modulate a common voltage-sensitive transition, with apparent ΔpH dependence being an emergent property of the system (Rangel-Yescas et al., 2021). pH-dependent shifts in gating charge movement (Figure 28.5F) argue that ΔpH sensitivity is intrinsic to the Hv1 VS domain (De La Rosa and Ramsey, 2018), but a neutralizing mutagenesis screen of candidate pH-sensing residues showed that pH_o sensitivity is intact in mutant Hv1 channels (Figure 28.4D) (Ramsey et al., 2010). The absence of Hv1-like ΔpH sensitivity in VGCs and VSPs (Vargas et al., 2012) remains mysterious.

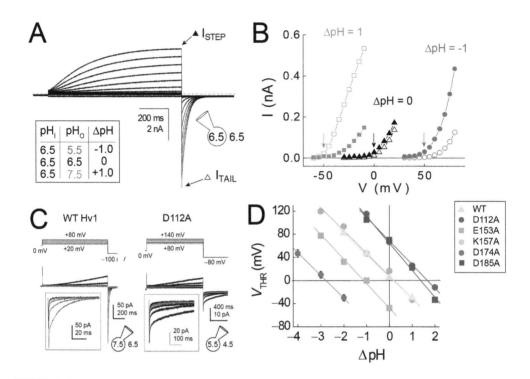

FIGURE 28.4

Voltage and pH-dependent gating of ionic currents in WT and mutant Hv1 channels. (A) Whole-cell currents in a HEK-293 cell expressing *Hs* Hv1 were evoked by successive 10 mV depolarizations at symmetrical pH 6.5. Steady-state currents at the end of the applied voltage step (I_{STEP}) are distinct from transient inward "tail" currents measured at −80 mV after successive voltage steps (I_{TAIL}). (B) Current activation (I_{STEP}, open symbols; −I_{TAIL}, filled symbols) measured at pH_o 7.5 (blue), 6.5 (black) or 5.5 (red) are shifted ~40 mV/pH unit. (C) The apparent threshold (V_{THR}) for activation of Hv1 channels is empirically determined as the smallest depolarization that is sufficient to elicit a tail current (magnified in insets) that is larger than the background (red traces). Neutralization of the "selectivity filter" residue D112 does not prevent ion conduction, but current amplitudes are smaller than WT Hv1. (D) Mutant channels in which candidate pH-sensing residues are neutralized retain WT-like sensitivity to changes in pH_o. (Data in panels A and B are reproduced from Ramsey et al., 2006, with changes. Data in C and D are reproduced from Ramsey et al., 2010, with changes.)

Although the underlying pH-sensing mechanism(s) in Hv1 remain poorly understood, recent studies reflect progress toward understanding this fundamental mystery. Extracellular Zn^{2+} and H^+ both operate as allosteric modulators of voltage-dependent gating in Hv1, but they have opposite effects on the position of the P_{OPEN}-V and Zn^{2+} potency is lower at acidic pH_o, suggesting that Zn^{2+} and H^+ may compete for a common binding site (Cherny and DeCoursey, 1999; De La Rosa et al., 2018, DeCoursey, 2003). Neutralizing two extracellular His residues (H140 and H193; Figure 28.3A) in *Hs* Hv1 is sufficient to abolish the potent Zn^{2+}-dependent modulation of Hv1 gating, but H140A-H193A channels are, like WT Hv1, strongly pH_o-dependent (Ramsey et al., 2006, De La Rosa et al., 2018). Other residues are evidently required to form the extracellular pH sensor, and the recent discovery of that Hv1 orthologs in reef-building corals exhibit diminished pH_o sensitivity (Rangel-Yescas et al., 2021) suggests an avenue to better understanding how the putative extracellular pH sensor is constructed. The nature of the putative intracellular pH sensor is similarly instructed by the reduced pH_i sensitivity reported in *Hs* Hv1 H168A, which was inspired by a snail (*Helisoma trivolvis, Ht*) Hv1 sequence difference (Cherny et al., 2018).

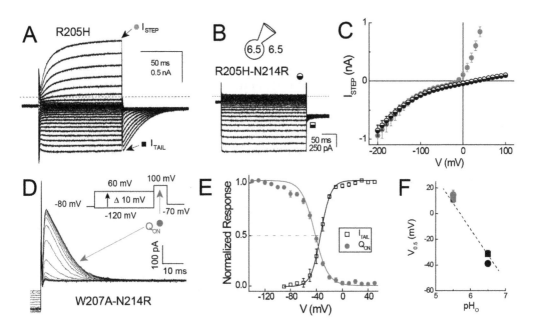

FIGURE 28.5

Resting-state and gating currents in Hv1 channels. (A) R205H (R1H) in *Hs* Hv1 confers a resting-state current at negative potentials that is not observed in WT Hv1. (B) Incorporating N214R into the background of R205H abolishes the intrinsic, activated-state currents seen in WT Hv1 but does not alter resting-state currents. (C) I_{STEP}–V relations for R205H (gray circles) and R205-N214R (half-filled circles) show that R205H exhibits double-rectification, whereas R205-N214R is a pure inward rectifier. (D) Transient "on" gating currents in a double mutant that exhibits block of outward current (N214R) and fast activation kinetics (W207A) and are measured at +100 mV after voltage steps to a range of potentials; integration of the gating current yields the gating charge (Q_{ON}). (E) Normalized Q_{ON}–V and I_{TAIL}–V relations illustrate that gating charge movement and channel opening are separated on the voltage axis. (F) Altering pH_o from 5.5 (red symbols) to 6.5 (black symbols) shifts the mean midpoint of both the Q_{ON}–V (circles) and I_{TAIL}–V (squares) relations ~40 mV/pH unit (dashed line). (Data in panels A–C are reproduced from Randolph et al., 2016. Data in panels D–F are reproduced from De La Rosa and Ramsey, 2018, with changes.)

Unfortunately, transplanting H168 from *Hs* Hv1 into the equivalent position in *Ht* Hv1 failed to restore the pH_i sensitivity of the human channel, indicating that additional determinants of the pH_i-sensing mechanism remain to be identified (Cherny et al., 2018). Taken together, the available data suggest that the hypothesized pH_i and pH_o sensors in Hv1 are unlikely to be monoprotic H^+ binding sites and may instead be formed by networks of ionizable side chains that harbor a "delocalized" proton(s).

28.3.4 Ion Permeation and Selectivity in Hv1

Like OTOP channels, native and expressed Hv1 channels appear to be strongly selective for H^+ over other monovalent cations (DeCoursey, 2003; Ramsey et al., 2006; Sasaki et al., 2006; Musset et al., 2011; Berger and Isacoff, 2011). Two competing (but not mutually exclusive) hypotheses for proton selectivity have been articulated: (1) Intraprotein "water wires" create an "aqueous" pathway for H^+ to migrate through the VS domain (Figure 28.3C) via a Grotthuss-type H^+-hopping mechanism that does not require explicit ionization of side chains (Ramsey et al., 2010; Randolph et al., 2016; Bennett and Ramsey, 2017). (2) Permeating protons are "shuttled" across the central crevice constriction by sequential cycles of ionization/deionization at a conserved S1 Asp (D112 in *Hs* Hv1; Figure 28.3A) (Musset et al., 2011; Dudev et al., 2015; DeCoursey, 2017). Direct arguments for and against each hypothesis are summarized elsewhere (Bennett and Ramsey, 2017; DeCoursey, 2017). In short, both mechanisms are theoretically H^+-specific because proton movement requires the formation of hydrogen bonds (H-bonds), and only protons can be transferred via H-bond rearrangements. The absence of a continuous hydrated pathway for ion movement in OTOP structures (see earlier) suggests that, despite their divergent sequence and structure, similar mechanisms could be in play.

An experimental test of hypothesis 2 showed that neutralizing mutations of ionizable side chains in Hv1 fails to abrogate the measured current as expected (Figure 28.4C), suggesting that an alternative (i.e., "aqueous") mechanism of H^+ transfer is a more parsimonious explanation for the data (Ramsey et al., 2010). It was subsequently shown that H^+ selectivity (but not ΔpH sensitivity; Figure 28.4D; Ramsey et al., 2010) is eroded in D112-mutant *Hs* Hv1 channels, indicating that this residue is part of the Hv1 "selectivity filter" structure (Musset et al., 2011). Unlike WT Hv1 channels, D112X mutants are permeable to anions, raising the possibility that OH^- (rather than H^+) and/or other anions carry the measured current (Musset et al., 2011). Unfortunately, this hypothesis is difficult to test further using voltage-clamp methods because various H^+-equivalent charge carriers (i.e., H^+, OH^- and H_3O^+) each share the same Nernst equilibrium potential.

A parsimonious explanation that is consistent with all of the available experimental data is that D112 mutants, like WT Hv1 channels, retain the capacity to mediate H^+ transfer in a water wire; however, unlike WT channels, D112X mutants also allow solution ions like Cl^- (and potentially OH^- and/or H_3O^+) to diffuse through the altered central crevice structure (Bennett and Ramsey, 2017). Nonselective diffusion through the mutated Hv1 pore makes sense if (a) waters take the place of the mutated Asp side chain, and (b) D112 has a low enough pK_a in vivo that it is nearly always anionic and electrostatically repels anions that find their way into the central crevice, such that removing this electrostatic barrier enables anion permeation. Neutralization of D185, which is located near D112 and selectively conserved in Hv1 VS domains (Figure 28.3C), evidently does not erode H^+ selectivity like D112 mutants (Musset et al., 2011), suggesting that despite structural differences conferred by mutations, an aqueous mechanism might be malleable enough to support rapid and selective proton transfer, consistent with experimental data showing that a wide variety of neutralizing mutations in Hv1 remain functional (Figure 28.4D) (Ramsey et al., 2010). Considering that thermal fluctuations in Hv1 protein structure could produce dynamic changes in electronic structure on the picosecond time scale (where H-bonds are broken and reformed but many orders of magnitude faster than voltage-dependent gating events), it seems reasonable to hypothesize that WT Hv1 may utilize an ensemble of molecularly similar but electronically distinct water structures for H^+ transfer.

Several groups have estimated the unitary conductances for voltage-gated H^+ channels (γ_{H^+}) using indirect (i.e., current fluctuation analysis) methods; although the reported γ_{H^+} values differ, studies agree that unitary H^+ currents are among the tiniest of any known channel (Cherny et al., 2003). The inability to directly measure single-channel currents is complicated by the fact that even at pH values that commonly elicit experimental failure in voltage-clamp experiments with intact cells (i.e., pH 4.0 or $[H^+] = 0.1$ mM), the concentration of the permeant ion is still relatively small compared to the conditions used for single-channel recording in most VGCs. Given the exotic chemical properties of H^+ in the "universal solvent," it is intriguing to speculate that nature may have evolved distinct and specific mechanisms for H^+-selective transmembrane transfer that are poorly informed by our existing knowledge of other ion channel mechanisms.

28.3.5 Hv1 Pharmacology

Zn^{2+} is the best known ligand for Hv1 channels. As previously mentioned, extracellular $ZnCl_2$ causes concentration- and pH_o-dependent shifts in the apparent P_{OPEN}-V relation toward positive voltages; at neutral pH_o, Zn^{2+} potency is high (micromolar) (Cherny and DeCoursey, 1999; Ramsey et al., 2006; De La Rosa et al., 2018). Zn^{2+} potency is reduced ~3 orders of magnitude to the low millimolar range in *Hs* Hv1 H140A-H193 (Ramsey et al., 2006). Bound Zn^{2+} is most likely also liganded by at least one acidic side chain (i.e., E119 or D185) and a water oxygen atom (De La Rosa et al., 2018). For example, Zn^{2+} potency is decreased in E119A, enhanced in E119H, and similar to WT Hv1 in E119H-H140A-H193A, suggesting that (a) E119 is in an ideal location to interact with Zn^{2+} and (b) alternative divalent metal coordination geometries (i.e., in E119H) are also apparently capable of binding Zn^{2+} with high apparent affinity.

The mechanism by which Zn^{2+} binding shifts voltage-dependent gating in Hv1 is still not fully understood. Apparent Zn^{2+} potency is known to be modulated by mutations at both extracellular (E119, D123, H140, D185 and H193) and intracellular (E153) residues (De La Rosa et al., 2018; Ramsey et al., 2006; Qiu et al., 2016); C-terminal truncation produces additional kinetic complexity that is incompletely understood (Musset, Smith, et al., 2010), suggesting the existence of long-range allosteric coupling mechanisms in Hv1. And similar to ΔpH changes, intracellular Zn^{2+} causes P_{OPEN}-V relations to shift in the opposite direction (negatively, making it a positive allosteric modulator or PAM) as extracellular Zn^{2+} (Cherny and DeCoursey, 1999). Little is known about the structural basis of intracellular Zn^{2+} effects. Albumin was also recently reported to function as an Hv1 PAM (Zhao, Dai, et al., 2021), and PAMs could be useful for elucidation of structure and selective ligands that compete with albumin might effectively function as NAMs.

In addition to extracellular Zn^{2+}, Hv1 NAMs include several small organic molecules, some of which appear to mimic the guanidinium ion of Arg to bind in and block the Hv1 pore (Hong et al., 2014; Geragotelis et al., 2020; Zhao, Hong, et al., 2021) and the peptide toxin Corza-6 (C6) (Zhao et al., 2018). Given the potential involvement of Hv1 in human disease-relevant pathologies, there is substantial interest in developing potent, Hv1-selective and drug-like NAMs. However, the structural and functional similarity between Hv1 and other VS domain proteins suggests that caution is warranted. In summary, much work remains to be done before we will know whether Hv1 ligands can fulfill their ballyhooed potential as therapeutics for neuroinflammation, asthma and other major human diseases.

28.3.6 Hv1 Regulation

A single study reported that protein kinase A (PKA)-dependent phosphorylation of *Hs* Hv1 at T29 causes a negative shift in V_{THR} and an increase in the speed of activation kinetics (Musset, Capasso, et al., 2010). Because the structure of the Hv1 N-terminus remains unknown, the mechanism by which T29 phosphorylation modulates gating remains unknown. Nonetheless, negative shifts in the apparent P_{OPEN}-V relation resulting from T29 phosphorylation (Musset, Capasso, et al., 2010), intracellular acidification (and/or extracellular alkalization) (Cherny et al., 1995; DeCoursey, 2003), or extracellular albumin (Zhao, Dai, et al., 2021) could be considered to be physiologically relevant mechanisms of Hv1 channel regulation because they are likely to increase channel activity at the cell's resting membrane potential. Future work to identify the structural bases for pH_i and pH_o sensing and the mechanisms of allostery between ligand binding and voltage-sensor activation could help open the door to understanding the mechanisms that control Hv1 channel activity *in vivo*.

Note

1. doi: https://medicalsciences.med.unsw.edu.au/sites/default/files/soms/page/ElectroPhysSW /JPCalcWin-Demo%20Manual.pdf (see Appendix A).

Suggested Reading

This chapter includes additional bibliographical references hosted only online as indicated by citations in blue color font in the text. Please visit https://www.routledge.com/9780367538163 to access the additional references for this chapter, found under "Support Material" at the bottom of the page.

OTOP Channels

Chang, R. B., H. Waters, and E. R. Liman. 2010. "A proton current drives action potentials in genetically identified sour taste cells." *Proc Natl Acad Sci U S A* 107(51):22320–5. doi: 10.1073/ pnas.1013664107.

Chang, W. W., A. S. Matt, M. Schewe, M. Musinszki, S. Grussel, J. Brandenburg, D. Garfield, M. Bleich, T. Baukrowitz, and M. Y. Hu. 2021. "An otopetrin family proton channel promotes cellular acid efflux critical for biomineralization in a marine calcifier." *Proc Natl Acad Sci U S A* 118(30):e2101378118. doi: 10.1073/pnas.2101378118.

Chen, Q., W. Zeng, J. She, X. C. Bai, and Y. Jiang. 2019. "Structural and functional characterization of an otopetrin family proton channel." *eLife* 8:e46710. doi: 10.7554/eLife.46710.

Ganguly, A., A. Chandel, H. Turner, S. Wang, E. R. Liman, and C. Montell. 2021. "Requirement for an Otopetrin-like protein for acid taste in Drosophila." *Proc Natl Acad Sci U S A* 118(51):e2110641118. doi: 10.1073/pnas.2110641118.

Hughes, I., J. Binkley, B. Hurle, E. D. Green, Nisc, A. Sidow, and D. M. Ornitz. 2008. "Identification of the otopetrin domain, a conserved domain in vertebrate otopetrins and invertebrate otopetrin-like family members." *BMC Evol Biol* 8:41. doi: 10.1186/1471-2148-8-41.

Hughes, I., M. Saito, P. H. Schlesinger, and D. M. Ornitz. 2007. "Otopetrin 1 activation by purinergic nucleotides regulates intracellular calcium." *Proc Natl Acad Sci U S A* 104(29):12023–8. doi: 10.1073/pnas.0705182104.

Hurle, B., E. Ignatova, S. M. Massironi, T. Mashimo, X. Rios, I. Thalmann, R. Thalmann, and D. M. Ornitz. 2003. "Non-syndromic vestibular disorder with otoconial agenesis in tilted/mergulhador mice caused by mutations in otopetrin 1." *Hum Mol Genet* 12(7):777–89. doi: 10.1093/hmg/ ddg087.

Mi, T., J. O. Mack, C. M. Lee, and Y. V. Zhang. 2021. "Molecular and cellular basis of acid taste sensation in Drosophila." *Nat Commun* 12(1):3730. doi: 10.1038/s41467-021-23490-5.

Qu, H., Y. Su, L. Yu, H. Zhao, and C. Xin. 2019. "Wild-type p53 regulates OTOP2 transcription through DNA loop alteration of the promoter in colorectal cancer." *FEBS Open Bio* 9(1):26–34. doi: 10.1002/2211-5463.12554.

Saotome, K., B. Teng, C. C. A. Tsui, W. H. Lee, Y. H. Tu, J. P. Kaplan, M. S. P. Sansom, E. R. Liman, and A. B. Ward. 2019. "Structures of the otopetrin proton channels Otop1 and Otop3." *Nat Struct Mol Biol* 26(6):518–25. doi: 10.1038/s41594-019-0235-9.

Teng, B., C. E. Wilson, Y. H. Tu, N. R. Joshi, S. C. Kinnamon, and E. R. Liman. 2019. "Cellular and neural responses to sour stimuli require the proton channel Otop1." *Curr Biol* 29(21):3647–3656. e5. doi: 10.1016/j.cub.2019.08.077.

Teng, Bochuan, Joshua P. Kaplan, Ziyu Liang, Zachary Kreiger, Yu-Hsiang Tu, Batuujin Burendei, Andrew Ward, and Emily R. Liman. 2022. "Structural motifs for subtype-specific pH-sensitive gating of vertebrate otopetrin proton channels." *eLife* 11:e77946. doi: 10.7554/eLife.77946. PMID: 35920807.

Tu, Y. H., A. J. Cooper, B. Teng, R. B. Chang, D. J. Artiga, H. N. Turner, E. M. Mulhall, W. Ye, A. D. Smith, and E. R. Liman. 2018. "An evolutionarily conserved gene family encodes proton-selective ion channels." *Science* 359(6379):1047–50. doi: 10.1126/science.aao3264.

Wang, G. X., K. W. Cho, M. Uhm, C. R. Hu, S. Li, Z. Cozacov, A. E. Xu, J. X. Cheng, A. R. Saltiel, C. N. Lumeng, and J. D. Lin. 2014. Otopetrin 1 protects mice from obesity-associated metabolic dysfunction through attenuating adipose tissue inflammation." *Diabetes* 63(4):1340–52. doi: 10.2337/db13-1139.

Hv1 Channels

Berger, T. K., and E. Y. Isacoff. 2011. "The pore of the voltage-gated proton channel." *Neuron* 72(6):991–1000. doi: 10.1016/j.neuron.2011.11.014.

De La Rosa, V., A. L. Bennett, and I. S. Ramsey. 2018. "Coupling between an electrostatic network and the Zn(2+) binding site modulates Hv1 activation." *J Gen Physiol* 150(6):863–81. doi: 10.1085/jgp.201711822.

De La Rosa, V., and I. S. Ramsey. 2018. "Gating currents in the Hv1 proton channel." *Biophys J* 114(12):2844–54. doi: 10.1016/j.bpj.2018.04.049.

DeCoursey, T. E. 2003. "Voltage-gated proton channels and other proton transfer pathways." *Physiol Rev* 83(2):475–579. doi: 10.1152/physrev.00028.2002.

Koch, H. P., T. Kurokawa, Y. Okochi, M. Sasaki, Y. Okamura, and H. P. Larsson. 2008. "Multimeric nature of voltage-gated proton channels." *Proc Natl Acad Sci U S A* 105(26):9111–6. doi: 10.1073/pnas.0801553105.

Li, Q., R. Shen, J. S. Treger, S. S. Wanderling, W. Milewski, K. Siwowska, F. Bezanilla, and E. Perozo. 2015. "Resting state of the human proton channel dimer in a lipid bilayer." *Proc Natl Acad Sci U S A* 112(44):E5926–35. doi: 10.1073/pnas.1515043112.

Mackinnon, J. A. Letts, and R. Mackinnon. 2008. "Dimeric subunit stoichiometry of the human voltage-dependent proton channel Hv1." *Proc Natl Acad Sci U S A* 105(22):7692–5. doi: 10.1073/pnas.0803277105.

Musset, B., S. M. Smith, S. Rajan, D. Morgan, V. V. Cherny, and T. E. Decoursey. 2011. "Aspartate 112 is the selectivity filter of the human voltage-gated proton channel." *Nature* 480(7376):273–7. doi: 10.1038/nature10557.

Ramsey, I. S., Y. Mokrab, I. Carvacho, Z. A. Sands, M. S. Sansom, and D. E. Clapham. 2010. "An aqueous H+ permeation pathway in the voltage-gated proton channel Hv1." *Nat Struct Mol Biol* 17(7):869–75. doi: 10.1038/nsmb.1826.

Ramsey, I. S., M. M. Moran, J. A. Chong, and D. E. Clapham. 2006. "A voltage-gated proton-selective channel lacking the pore domain." *Nature* 440(7088):1213–6. doi: 10.1038/nature04700.

Ramsey, I. S., E. Ruchti, J. S. Kaczmarek, and D. E. Clapham. 2009. "Hv1 proton channels are required for high-level NADPH oxidase-dependent superoxide production during the phagocyte respiratory burst." *Proc Natl Acad Sci U S A* 106(18):7642–7. doi: 10.1073/pnas.0902761106.

Randolph, A. L., Y. Mokrab, A. L. Bennett, M. S. Sansom, and I. S. Ramsey. 2016. "Proton currents constrain structural models of voltage sensor activation." *eLife* 5:e18017. doi: 10.7554/eLife.18017.

Sasaki, M., M. Takagi, and Y. Okamura. 2006. "A voltage sensor-domain protein is a voltage-gated proton channel." *Science* 312(5773):589–92. doi: 10.1126/science.1122352.

Takeshita, K., S. Sakata, E. Yamashita, Y. Fujiwara, A. Kawanabe, T. Kurokawa, Y. Okochi, M. Matsuda, H. Narita, Y. Okamura, and A. Nakagawa. 2014. "X-ray crystal structure of voltage-gated proton channel." *Nat Struct Mol Biol* 21(4):352–7. doi: 10.1038/nsmb.2783.

Tombola, F., M. H. Ulbrich, and E. Y. Isacoff. 2008. "The voltage-gated proton channel Hv1 has two pores, each controlled by one voltage sensor." *Neuron* 58(4):546–56. doi: 10.1016/j. neuron.2008.03.026.

Zhao, C., L. Hong, S. Riahi, V. T. Lim, D. J. Tobias, and F. Tombola. 2021. "A novel Hv1 inhibitor reveals a new mechanism of inhibition of a voltage-sensing domain." *J Gen Physiol* 153(9). doi: 10.1085/jgp.202012833.

29

P2X Receptors

Kate Dunning and Thomas Grutter

CONTENTS

29.1 Introduction

P2X receptors are a family of trimeric ligand-gated ion channels (LGICs), forming a non-selective cation-permeable pore upon activation by extracellular ATP. Compared to other LGIC, such as the cys-loop or ionotropic glutamate receptor family, P2X receptors have a relatively recent history, with the concept of purinergic signaling first evoked by Burnstock in the 1970s, and their molecular identification occurring only in the 1990s (North, 2002). In quick succession, all seven members of the P2X family, P2X1–P2X7, had been identified in mammals. P2X receptors are also found in other eukaryotes including invertebrates and amoeba such as *Dictyostelium discoideum*. Fifteen years after the first cloning, the X-ray structure of a P2X member was obtained, revealing a unique molecular architecture (Kawate et al., 2009). Despite a certain level of similarity in terms of amino acid sequence and structure, the functional characteristics, pharmacology, and physiological roles of these subtypes vary considerably, as will be seen throughout this chapter.

29.2 Physiological Roles

Reflective of their wide-ranging cellular distribution, the physiological roles of P2X receptors are equally as varied, and differ between subtype and cellular localization.

DOI: 10.1201/9781003096276-29

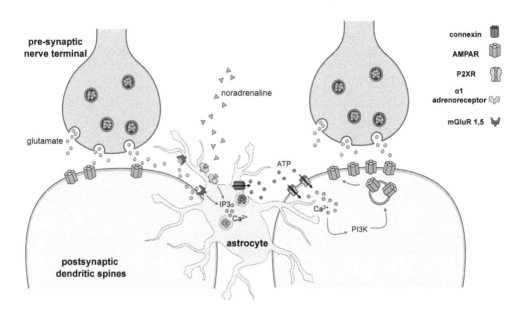

FIGURE 29.1

Neuromodulation by P2X receptors. Release of noradrenaline (NA) and glutamate from presynaptic sources causes an inositol 1,4,5-trisphosphate (IP3)-mediated release of Ca^{2+} in astrocytes, following recognition at mGluR (glutamate) and α1-adrenergic (NA) receptors. In turn, this triggers glial release of ATP, which acts upon P2X receptors in postsynaptic dendritic spines, causing an influx of calcium, activation of phosphatidylinositol 3-kinase (PI3K) and subsequent promotion of postsynaptic AMPA insertion. (Recreated from Khakh and North, 2012. Copyright 2012, with permission from Elsevier. Parts of the figure were drawn by using pictures from Servier Medical Art. Servier Medical Art by Servier is licensed under a Creative Commons Attribution 3.0 Unported License. https://creativecommons.org/licenses/by/3.0/.)

One of the most well-studied physiological roles of P2X receptors is neuromodulation. For example, P2X2, P2X4 and P2X7 are implicated in neuromodulatory processes in magnocellular neurons within the hypothalamus. Through a myriad of cellular signaling processes, extrasynaptic P2X receptors are activated by ATP, leading to an influx of Ca^{2+}, which in turn causes activation of phosphatidylinositol 3-kinase (Figure 29.1). There are multiple sources of extracellular ATP, including ATP release from glial cells. For astrocytes, a variety of mechanisms has been suggested, including vesicular and nonvesicular release of ATP through the participation of ion channels, such as anion channels and connexins (Boué-Grabot and Pankratov, 2017). The overall consequence of this chain of events is the promotion of AMPA receptor insertion into the postsynaptic dendritic spine. This P2X-dependent modulatory network therefore acts to reinforce synaptic strength (Boué-Grabot and Pankratov, 2017).

A further physiological role linked to Ca^{2+} signaling is the implication of P2X1 in the contraction of smooth muscle cells, where its activation leads to rapid calcium influx and membrane depolarization (Ralevic, 2012). In these conditions, voltage-gated Ca^{2+} channels become activated, further increasing intracellular calcium content and ultimately resulting in the calmodulin-dependent mobilization of myosin light-chain kinase, a key component of the myosin-actin ratchet mechanism required for muscle contraction.

Recently, a role for P2X4 in keratinocyte-derived mechanosensory transduction has been elucidated (Moehring et al., 2018). Keratinocytes, skin cells located on the outermost layer of the epidermis, were shown to release ATP upon mechanostimulation, leading to the subsequent activation of P2X4 receptor on nearby neurons, and the transmission of both innocuous and noxious touch to the spinal cord.

29.3 Subunit Diversity and Basic Structural Organization

In terms of multimeric organization, the seven P2X subtypes vary depending on the specific subtype in question (North, 2002). P2X6, for example, is the only variant unable to form homotrimeric structures in recombinant conditions, requiring heteromerization with P2X2 or P2X4 for functional expression. A number of heteromeric P2X receptors have also been shown to form in heterologous systems, but in native tissue, P2X1/5 and P2X2/3 are those for which the most convincing evidence exists (Saul et al., 2013).

Although a number of studies using complementary biochemical, fluorescence and microscopy techniques had provided key structural insights, definitive confirmation of P2X structural organization came with the resolution of the first crystallographic structure in 2009, a truncated zebrafish P2X4 construct in the *apo* state (Kawate et al., 2009). Throughout the decade following this landmark feat, a suite of P2X structures has been resolved, allowing insight into the structural details of several subtypes, in differing allosteric conformational states, and in complex with agonists, antagonists, and modulators.

These structures confirm a trimeric organization, formed by three highly intertwined monomeric units in a chalice-like arrangement (Figure 29.2A). The N- and C-termini are located intracellularly, and, in the ATP-bound open state, form the cytoplasmic cap by way of extensive domain-swapping interactions between all three monomers. This beta-sheet structure is suggested to stabilize the open state, effectively locking it into place. In contrast, it appears that the cap is not present in the *apo* state: this domain was not observed in the *apo* structure of human P2X3 (hP2X3), suggesting these regions to be disordered in the resting state. Linked to these termini are two transmembrane (TM) helical domains, named TM1 and TM2. The ion-conducting pore is lined by the TM2 helices from the three subunits. Finally, the ectodomain, which links TM1 and TM2, is an extensive structure held in place by various disulfide bridges and features the ATP binding site, which is located at the interface of two monomers, forming a "jaw"-like structure (Figure 29.2A). Each monomeric subunit itself has been likened to a leaping dolphin, with the flippers, head and dorsal fins located in the extracellular domain, while the transmembrane domains represent the fluke (Figure 29.2B).

This architecture remains, for the most part, highly similar among P2X family members, with the exception of subtle structural differences resulting from variations in amino acid sequence, giving rise to altered or unique pharmacological binding pockets. Examples of this include the ATP binding site itself, and the unique allosteric binding site found in the extracellular domain of P2X7.

The most significant variation from this generic architecture is the P2X7 subtype, for which cryo-EM structures revealed a number of key differences within the intracellular domains (Figure 29.2C) (McCarthy et al., 2019). While the cytoplasmic cap is visible in the open state structure, similarly to hP2X3, it is also visible in the *apo* state, thus appearing to be present as a permanent structural scaffold in this subtype. Within this cap, a unique motif of 18 amino acids rich in cysteine residues, named the Ccys anchor, is present. The Ccys anchor features at least six palmitoylated residues, whose palmitoyl groups extend into the membrane bilayer, anchoring the cytoplasmic cap into place, a mechanism that is thought to be at the origin of the nondesensitizing behavior of P2X7. The final major structural difference of the P2X7 subtype resides in its elongated C-terminal domain, which forms a uniquely folded "ballast" domain, hanging below the cytoplasmic cap. Within the ballast fold, for which no structural homolog has been identified, two binding motifs exist: a dinuclear zinc site, and a GDP binding site (Figure 29.2C). The physiological role of the ballast remains to be elucidated and will no doubt be the focus of future research efforts.

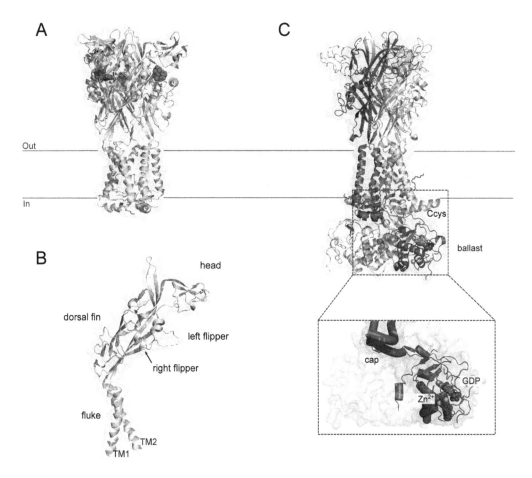

FIGURE 29.2

Structure of P2X receptors. (A) Crystallographic structure of hP2X3 (Mansoor et al., 2016) in its trimeric form in the ATP-bound open state. ATP is shown in its binding site in red spheres. Protein Data Bank: 5SVK. (B) Monomer subunit of hP2X3 in the *apo*, closed state, labeled according to the leaping dolphin analogy. The two transmembrane domains, TM1 and TM2, are also labeled. (C) Cryo-EM resolved structure of full-length rP2X7 (McCarthy et al., 2019), here shown in the *apo*, closed state. Palmitoyl groups within the Ccys anchor are shown as red sticks. Within the dashed rectangle is shown a zoomed image of the ballast region, with the GDP molecule and two Zn²⁺ ions visible. Protein Data Bank: 6U9V. The gray lines indicate the boundaries of the outer (out) and inner (in) leaflets of the membrane bilayer.

29.4 Gating

Currents elicited by the application of agonist differ among P2X subtypes (Figure 29.3A). To account for these differences, it has been postulated that P2X receptors are able to access a number of allosteric conformational states in a dynamic, reversible manner (Figure 29.3B).

The gating transitions from a resting *apo* state to the ATP-bound open state have been extensively studied at the molecular level by a number of techniques. Comparison of resolved structures in the closed, *apo* state with those in an ATP-bound open state has also greatly facilitated deduction of the molecular motions involved in the transition from a closed to open channel state. These motions can broadly be described by a four-step

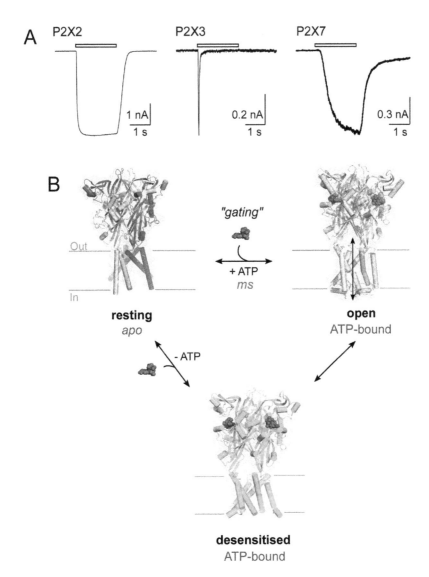

FIGURE 29.3
Gating of P2X receptors. (A) Whole-cell electrophysiological traces of HEK293 cells expressing rP2X2 (left), rP2X3 (middle), or rP2X7 (right) receptors, stimulated with agonist (100 μM ATP in the case of P2X2 and P2X3, and 10 μM BzATP in the case of P2X7). Note that longer applications of ATP (e.g., 20s) result in desensitization of P2X2, and facilitation/sensitization of P2X7 currents. (B) The P2X gating cycle, illustrated by crystallographic structures of hP2X3, in the *apo*, resting state (blue; Protein Data Bank: 5SVJ), ATP-bound open state (yellow; Protein Data Bank: 5SVK), and ATP-bound desensitized state (green; Protein Data Bank: 5SVL) (Mansoor et al., 2016). Upon binding of ATP (shown as red spheres), the gating process takes place: conformational changes lead to the opening of a cation-permeable pore and the formation of the cytoplasmic cap. From this open state, the desensitized state may be reached by certain subtypes, whereby, despite the presence of bound ATP, the channel is closed. Cycling between these allosteric conformational states is dynamic.

mechanism (Chataigneau et al., 2013): (1) binding of ATP induces a tightening of the binding "jaw," bringing the head domain and dorsal fin of the neighboring subunit into proximity; (2) jaw closure results in the upward motion of the dorsal fin; (3) motion of the dorsal fin, in turn, produces an outward flexing motion in the beta-sheets of the lower body; and

(4) being coupled to the TM domains, this outward flex of the lower body causes the TM helices to rotate in what has been described as an "iris"-like motion, leading to the opening of an ion-permeable pore of approximately 5 Å. This cascade of movements occurs on the millisecond time scale, and is common to all P2X receptors. Depending on the subtype in question, the concentration of ATP required for activation varies from the nanomolar to millimolar scale. In 2016, the resolution of the structure of hP2X3 in the open state, with more of the intracellular domains visible than had previously been achieved, revealed that the formation of the cytoplasmic cap also occurs during, or shortly after, the gating process (Mansoor et al., 2016). Further studies on the precise molecular movements and the dynamics of cap formation will allow for an even more complete view of the P2X gating cycle.

In addition to the gating transitions between closed and open channel states, several subtype variants are able to access a third allosteric conformational state, the desensitized state, whereby the channel is closed despite the continued fixation of ATP. Desensitization varies greatly between subtype variants, with P2X1 and P2X3 exhibiting fast desensitization kinetics (within less than a second), while P2X2 and P2X4 desensitize at a much slower rate (several tens of seconds) (North, 2002). In sharp contrast, the P2X7 subtype shows the opposite behavior, in the form of "sensitization" or facilitation – a cholesterol-dependent, progressive increase in ATP-evoked currents (Robinson et al., 2014). Figure 29.3A shows representative traces from P2X2, P2X3 and P2X7 subtypes, illustrating the differences in their response to agonists.

29.5 Ion Permeability

P2X receptors are nonselective cation channels, allowing the passage of Na^+ and Ca^{2+} into the cell, and K^+ efflux along their electrochemical gradients (Samways et al., 2014). The P2X5 subtype is known to be permeable, in addition to cations, to chloride ions (relative permeability of Cl^- to Na^+, $P_{Cl^-}/P_{Na^+} = 0.5$ in human P2X5 [hP2X5]).

In the closed state, the channel gate of hP2X3 is formed by Ile-323, Val-326 and Thr-330, which form a constriction that is too narrow to allow the passage of hydrated Na^+ (Mansoor et al., 2016) (Figure 29.4A). In the open state, however, this constriction vanishes due to an outward rotation of Ile-323 and Val-326, resulting in the formation of a continuous ion-conducting pathway wide enough (~3 Å) to allow the passage of partially hydrated cations (Figure 29.4B). Thr-330 occupies the narrowest point in the conduction pathway and plays a key role in the ion selectivity filter. Mutating this residue to lysine at the homologous site in the rat P2X2 (rP2X2) converts ion selectivity from cationic to anionic (Browne et al., 2011). However, despite featuring an alanine residue at this homologous position, hP2X5 displays significant Cl^- permeability. Thus, the molecular basis of the preference for cations is not yet fully understood.

These ions enter the channel in a lateral manner via three fenestrations located above the transmembrane domains before transiting down the threefold symmetry axis of the open channel (Figure 29.4B). The structure of hP2X3, with cytoplasmic cap visible, provided further information on the ion permeation pathway, demonstrating that the channel orifice directly below the threefold symmetry axis is too narrow for ion exit into the cytosol. Instead, this most likely occurs through three cytoplasmic fenestrations located just above the cytoplasmic cap (Mansoor et al., 2016) (Figure 29.4B).

In addition to small metal cation permeability, some P2X receptors (P2X2, P2X4 and P2X7) are permeable to organic cations of greater size, including N-methyl-D-glucamine (NMDG+), spermidine, and fluorescent organic dyes such as YO-PRO-1. In the case of P2X7,

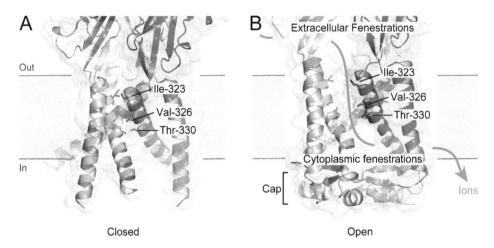

FIGURE 29.4
Ion permeation of P2X receptors. (A) Crystal structure of hP2X3 in the *apo*, closed state (Protein Data Bank: 5SVJ). (B) Crystal structure of hP2X3 in the ATP-bound open state (Protein Data Bank: 5SVK). Pore-lining residues Ile-323, Val-326 and Thr-330; extracellular and intracellular fenestrations; cap; and the putative ion pathway are indicated. The surface of the protein is illustrated, and subunits are shown in light blue with one subunit highlighted in dark blue.

this large molecule permeability extends to species reaching <900 Da in molecular weight, of cationic and anionic natures, with high variability observed between different cell types and experimental conditions (Ugur and Ugur, 2019). This property, unique to P2X7, has been defined as the so-called P2X7 "macropore". The molecular mechanisms behind the large molecular permeability of P2X have remained elusive and largely disputed in the literature (Peverini et al., 2018). For a long time, the favored hypothesis as to the origin of this large cation flow postulated the implication of a further allosteric conformational state, featuring a progressively dilated P2X pore. This was primarily supported by electrophysiological experiments showing a progressive increase in large cation permeability over time. However, a key study by Li et al. (2015) demonstrated that this progressive change in permeability is likely artifactual, as a result of ionic depletion effects occurring in bi-ionic experimental conditions. This study, and others, have determined that flow of some larger organic cations, such as NMDG$^+$ and spermidine, likely occurs on the same millisecond timescale as small metallic cations, but at a reduced rate due to significant conformational and orientational limitations (Harkat et al., 2017).

Another historically prominent hypothesis postulates the involvement of a secondary "annexe" channel. In this scenario, smaller species such as metallic cations would pass through the P2X pore, while larger molecular weight species would transit through a secondary channel that is functionally coupled to P2X. This scenario is particularly appealing in the case of the P2X7 macropore, whose open channel diameter of 5 Å seems highly incompatible with those large molecular weight species permeating the macropore. Although Pannexin-1 was initially identified as a candidate, subsequent literature disputed its implication (Kaczmarek-Hájek et al., 2012). Recent work using a truncated form of panda P2X7 receptor reconstituted into liposomes has also demonstrated that YO-PRO-1, a cationic dye of 376 Da, is able to transit the intrinsic P2X7 pore (Karasawa et al., 2017). However, the permeation pathway for molecules larger than this remains unclear, although a role for the anoctamin family of ion channel/scramblases has been suggested (Kopp et al., 2019; Dunning et al., 2021). This question will undoubtedly remain a focus of P2X research.

29.6 Pharmacology

The endogenous agonist of P2X receptors is extracellular ATP, which has varying EC_{50} values according to subtype (Coddou et al., 2011). These values range from several hundred nanomolars for P2X1 and P2X3, to micromolars for P2X2 and P2X4. In marked contrast to the other P2X subtypes, P2X7 requires unusually high concentrations of ATP for activation (millimolar range). The recently resolved rat P2X7 structure provides a possible explanation for this difference in the form of a much narrower solvent-accessible entrance leading to the ATP binding site (McCarthy et al., 2019). This restricted entrance likely imposes constraints on the entry of ATP, which in turn, may affect ATP binding. 2,3-O-(4-benzoylbenzoyl)-ATP (BzATP), an ATP analog, is frequently used to study P2X7 owing to its increased potency (Donnelly-Roberts et al., 2009). A number of other ATP analogs have varying activity at different subtypes. For example, the use of α,β-meATP allows to distinguish fast-desensitizing subtypes P2X1 and P2X3, where it is highly potent, from the remaining subtype variants who are largely insensitive to this ATP analog (Coddou et al., 2011). In keeping with its unique agonist pharmacology, P2X7 also has the particularity of activation by the nucleotide cofactor NAD^+ (in the case of mouse P2X7), and even non-nucleotide agonists such as lipopolysaccharide and amyloid-β (Di Virgilio et al., 2018).

In terms of antagonists, a variety of P2X antagonists exist, both competitive and non-competitive, with differing efficacy, and differing selectivity according to the subtype in question. Some common antagonists include suramin, a polysulfonated urea derivative; PPADS, a pyroxidal-based disulfonic acid; and TNP-ATP, a competitive antagonist (Syed and Kennedy, 2012). Several synthetic antagonists have also been developed in an industrial setting, with the aim of providing greater subtype selectivity and potency. Given its strong therapeutic potential, many of these compounds target the P2X7 subtype. Some of these have been crystallized in complex with P2X7 and shown to occupy a unique allosteric binding pocket (Karasawa and Kawate, 2016).

Allosteric modulators are, again, highly variable between different P2X subtypes. An example of this is Zn^{2+}, which positively modulates P2X2-5 responses, while acting as a negative modulator at P2X1 and P2X7 subtypes (Coddou et al., 2011). Other modulators of P2X activity include small metal cations such as Cd^{2+} and Cu^{2+}, protons, ivermectin (P2X4, 7), and lipids such as phosphatidylinositols (Coddou et al., 2011). Aside from this pharmacological variation between subtypes, P2X7 pharmacology differs even between orthologs – KN-62, for example, is a potent antagonist at human and mouse P2X7, but inactive at the rat counterpart (Donnelly-Roberts et al., 2009).

29.7 Regulation

A number of different mechanisms intervene in the regulation of P2X receptor function (Figure 29.5), often with specificities for each subtype.

Posttranslational modifications represent one such regulatory mechanism. Phosphorylation of a protein kinase C (PKC) site in the N-terminal domain of P2X1, for example, has been shown to exert a regulatory effect on channel function, altering current amplitudes and the time course of desensitization (Kaczmarek-Hájek et al., 2012). This PKC site is highly conserved, and phosphorylation within this domain has been shown to regulate

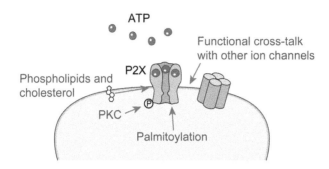

FIGURE 29.5
Functional regulation of P2X receptors. ATP-gated P2X receptors are regulated by different mechanisms. See text for details.

P2X2 receptor desensitization (Boué-Grabot et al., 2000). Palmitoylation, another post-translational modification, is particularly important for the functional regulation of P2X7. The intracellular Ccys anchor contains at least six residues susceptible to palmitoylation. Alanine mutagenesis of the residues that undergo palmitoylation or removal of the motif showed that palmitoylation within this region is a critical determinant of the nondesensitizing behavior of P2X7 (Robinson et al., 2014; McCarthy et al., 2019). It has been suggested that these palmitoyl groups, which can be seen extending into the membrane in cryo-EM structures, serve to lock the cytoplasmic cap into place, thus providing stability for an open channel state (McCarthy et al., 2019).

A number of interactions between P2X receptors and various protein partners have also been identified to be of regulatory importance. Functional cross-talk has been demonstrated between P2X receptors and other LGICs of the cys-loop family, including nicotinic, GABA and 5HT₃ (Kaczmarek-Hájek et al., 2012). P2X7 has a vast number of interacting proteins, many of which interact via its elongated C-terminus, which harbors a wealth of binding domains. More than 50 interacting proteins have been identified, a significant number of which are involved in the immune response, including CD14, MyD88 and NLRP3 (Kopp et al., 2019). Functional interactions with the pannexin-1 hemichannel, the anoctamin family of Ca^{2+}-activated chloride channels and calmodulin have also been described (Kopp et al., 2019; Dunning et al., 2021).

A final important source of P2X regulation is by elements of the cell membrane, including phospholipids and cholesterol. Phosphoinositides (PIPn), which are anionic phospholipids present in the membrane, are known to positively modulate P2X activity in all subtypes, with the exception of P2X5 where no effect is observed (Bernier et al., 2013). While evidence exists for a direct binding of PIPn to a polybasic domain in the P2X C-terminus, some subtypes, such as P2X3, are suggested to undergo regulation by PIPn in an indirect manner.

P2X1, 2, 3, 4 and 7 are found in lipid raft fractions, which are microenvironments within the cell membrane, enriched in cholesterol (Murrell-Lagnado, 2017). Cholesterol has equally been shown to regulate P2X behavior, although the nature of this regulation is dependent on the subtype in question. While P2X2, 3 and 4 have been shown to be largely insensitive to the removal of cholesterol, P2X1 is inhibited by its removal, exhibiting currents reduced by up to 90% (Murrell-Lagnado, 2017). This regulation is indirect and occurs via an interaction between cholesterol and the actin cytoskeleton rather than P2X1 itself. In contrast, cholesterol has an inhibitory effect on the P2X7 subtype, which, in cholesterol-depleted conditions, exhibits greatly potentiated currents and a striking acceleration of sensitization, due to an increase of the open channel probability (Dunning et al., 2021).

Several lines of evidence indicate that this modulation is by way of a direct interaction with the P2X7 channel, including the identification of four cholesterol recognition amino acid sequences in the proximal C-terminus (Robinson et al., 2014), and the use of a reconstituted, truncated panda P2X7 construct, which retains cholesterol sensitivity, and suggests an interaction site located in the transmembrane domains (Karasawa et al., 2017).

29.8 Cell Biology (Biogenesis, Trafficking, Turnover)

P2X receptors are synthesized, assembled and core glycosylated in the endoplasmic reticulum (ER) before undergoing complex glycosylation in the trans Golgi apparatus (Kaczmarek-Hájek et al., 2012). Most P2X receptors are predominantly localized at the plasma membrane, with the exception of homomeric P2X6, which is retained in the ER. While full-length human P2X5 is trafficked to the cell membrane, the predominant allele in humans produces a variant containing an exon deletion, which is also retained in the ER (Robinson and Murrell-Lagnado, 2013).

Subsequent trafficking and stability of P2X receptors at the plasma membrane is variable according to the subtype in question and depending on cell type (Kaczmarek-Hájek et al., 2012). One motif that is common to all P2X subtypes, however, is the YXXXK motif, located in the C-terminal region. This motif regulates the surface expression of P2X and is suggested to stabilize receptors once at the cell surface membrane rather than preventing their transit from the ER (Chaumont et al., 2004). Disulfide bond formation in the ectodomain has been determined necessary for the proper plasma membrane expression of certain subtypes, as has the presence of at least two N-glycosylated residues (Kaczmarek-Hájek et al., 2012).

The turnover of P2X receptors is, once again, highly dependent on subtype. P2X1, for example, undergoes rapid cycling between the membrane and recycling endosomes, whereas P2X2 is relatively stable at the cell membrane, undergoing little constitutive internalization (Robinson and Murrell-Lagnado, 2013). Rapid constitutive internalization is observed for P2X3 and P2X4 receptors, although this occurs through lysosomes and late endosomes, in contrast to the recycling endosomes implicated in P2X1 mobility (Robinson and Murrell-Lagnado, 2013). Internalization may also be dependent on additional factors, such as cholesterol in the case of P2X3 and dynamin for P2X4 (Vacca et al., 2009; Kaczmarek-Hájek et al., 2012).

29.9 Channelopathies and Disease

P2X receptors have been shown to intervene in numerous pathologies. P2X4, for example, has a well-characterized role in the brain-derived neurotrophic factor (BDNF) pathway of neuropathic pain, P2X1 is involved in cardiac and bladder disease, and P2X3 and P2X2/3 heteromers are thought to play a role in dermatitis and visceral pain (Burnstock and Kennedy, 2011).

The P2X7 subtype, however, once again stands apart in that it is involved in a vast range of different diseases, thus making it a key therapeutic target. Pathologies in which the implication of P2X7 has been suggested include inflammatory pain, neuropathic pain, cancer, neurodegenerative conditions such as Alzheimer's and Huntington's disease, Crohn's disease, skin disorders such as dermatitis and psoriasis, as well as psychiatric disorders including depression and schizophrenia (Burnstock and Knight, 2018). A number of single-nucleotide polymorphisms (SNPs) with gain-of-function or loss-of-function effects on P2X7 have been identified, several of which have been shown to correlate with various pathologies (Di Virgilio et al., 2017).

P2X7 is highly expressed in immune cells and is a key component of the NLRP3 inflammasome, with implications for the release of interleukin-1β (IL-1β), a pro-inflammatory cytokine. Accordingly, many of these pathologies are the result of an immune system dysregulation, in particular with regard to the inflammatory response. While a number of clinical trials have focused on P2X7 as a therapeutic target, thus far, no P2X7-directed therapy is in routine use. Recent work, however, has focused on developing biological P2X7-targeted agents, including specific antibodies and nanobodies, which may open up new therapeutic avenues (Di Virgilio et al., 2017).

29.10 Conclusion

Over the last 30 years, collective research efforts have provided an insight into the physiological roles, pathological implications, and molecular structure and function of the P2X family. Many further details, however, remain to be elucidated, and with new advances come new questions. Future research will be needed to unravel areas that remain poorly understood, such as the desensitization processes and its determinants, the role of the novel P2X7 ballast domain, as well as precise and comprehensive mechanistic details of the macropore phenomenon. Another key focal point for future P2X research will be the development of P2X-targeted therapeutics. While a number of clinical trials involving P2X receptors have been registered, no P2X-directed therapeutic agents are in current clinical use.

Suggested Reading

Bernier, Louis-Philippe, Ariel Ase, and Philippe Seguela. 2013. "Post-Translational Regulation of P2X Receptor Channels: Modulation by Phospholipids." *Frontiers in Cellular Neuroscience* 7:226. doi:10.3389/fncel.2013.00226.

Boué-Grabot, Éric, Vincent Archambault, and Philippe Séguéla. 2000. "A Protein Kinase C Site Highly Conserved in P2X Subunits Controls the Desensitization Kinetics of P2X2 ATP-Gated Channels." *Journal of Biological Chemistry* 275(14): 10190–95. doi:10.1074/jbc.275.14.10190.

Boué-Grabot, Eric, and Yuriy Pankratov. 2017. "Modulation of Central Synapses by Astrocyte-Released ATP and Postsynaptic P2X Receptors." *Neural Plasticity* 2017: 1–11. doi:10.1155/2017/9454275.

Browne, Liam E., Lishuang Cao, Helen E. Broomhead, Laricia Bragg, William J. Wilkinson, and R. Alan North. 2011. "P2X Receptor Channels Show Threefold Symmetry in Ionic Charge Selectivity and Unitary Conductance." *Nature Neuroscience* 14(1): 17–8. doi:10.1038/nn.2705.

Burnstock, G., and C. Kennedy. 2011. "Chapter 11 – P2X Receptors in Health and Disease." In: *Advances in Pharmacology*, edited by Kenneth A Jacobson and Joel Linden, 61:333–72. Pharmacology of Purine and Pyrimidine Receptors. Academic Press. doi:10.1016/B978-0-12-385526-8.00011-4.

Burnstock, Geoffrey, and Gillian E. Knight. 2018. "The Potential of P2X7 Receptors as a Therapeutic Target, Including Inflammation and Tumour Progression." *Purinergic Signalling* 14(1): 1–18. doi:10.1007/s11302-017-9593-0.

Chataigneau, Thierry, Damien Lemoine, and Thomas Grutter. 2013. "Exploring the ATP-Binding Site of P2X Receptors." *Frontiers in Cellular Neuroscience* 7:273. doi:10.3389/fncel.2013.00273.

Chaumont, Séverine, Lin-Hua Jiang, Aubin Penna, R. Alan North, and Francois Rassendren. 2004. "Identification of a Trafficking Motif Involved in the Stabilization and Polarization of P2X Receptors." *Journal of Biological Chemistry* 279(28): 29628–38. doi:10.1074/jbc.M403940200.

Coddou, Claudio, Zonghe Yan, Tomas Obsil, J. Pablo Huidobro-Toro, and Stanko S. Stojilkovic. 2011. "Activation and Regulation of Purinergic P2X Receptor Channels." *Pharmacological Reviews* 63(3): 641–83. doi:10.1124/pr.110.003129.

Di Virgilio, Francesco, Diego Dal Ben, Alba Clara Sarti, Anna Lisa Giuliani, and Simonetta Falzoni. 2017. "The P2X7 Receptor in Infection and Inflammation." *Immunity* 47(1): 15–31. doi:10.1016/j.immuni.2017.06.020.

Di Virgilio, Francesco, Anna L. Giuliani, Valentina Vultaggio-Poma, Simonetta Falzoni, and Alba C. Sarti. 2018. "Non-Nucleotide Agonists Triggering P2X7 Receptor Activation and Pore Formation." *Frontiers in Pharmacology* 9: 39. doi:10.3389/fphar.2018.00039.

Donnelly-Roberts, Diana L., Marian T. Namovic, Ping Han, and Michael F. Jarvis. 2009. "Mammalian P2X7 Receptor Pharmacology: Comparison of Recombinant Mouse, Rat and Human P2X7 Receptors." *British Journal of Pharmacology* 157(7): 1203–14. doi:10.1111/j.1476-5381.2009.00233.x.

Dunning, Kate, Adeline Martz, Francisco Andrés Peralta, Federico Cevoli, Eric Boué-Grabot, Vincent Compan, Fanny Gautherat, Patrick Wolf, Thierry Chataigneau, and Thomas Grutter. 2021. "P2X7 Receptors and TMEM16 Channels Are Functionally Coupled with Implications for Macropore Formation and Current Facilitation." *International Journal of Molecular Sciences* 22(12): 6542. doi:10.3390/ijms22126542.

Harkat, Mahboubi, Laurie Peverini, Adrien H. Cerdan, Kate Dunning, Juline Beudez, Adeline Martz, Nicolas Calimet, et al. 2017. "On the Permeation of Large Organic Cations through the Pore of ATP-Gated P2X Receptors." *Proceedings of the National Academy of Sciences of the United States of America* 114(19): E3786–95. doi:10.1073/pnas.1701379114.

Kaczmarek-Hájek, Karina, Eva Lörinczi, Ralf Hausmann, and Annette Nicke. 2012. "Molecular and Functional Properties of P2X Receptors – Recent Progress and Persisting Challenges." *Purinergic Signalling* 8(3): 375–417. doi:10.1007/s11302-012-9314-7.

Karasawa, Akira, and Toshimitsu Kawate. 2016. "Structural Basis for Subtype-Specific Inhibition of the P2X7 Receptor." Edited by Kenton J Swartz. *eLife* 5 : e22153. doi:10.7554/eLife.22153.

Karasawa, Akira, Kevin Michalski, Polina Mikhelzon, and Toshimitsu Kawate. 2017. "The P2X7 Receptor Forms a Dye-Permeable Pore Independent of Its Intracellular Domain but Dependent on Membrane Lipid Composition." *eLife* 6:e31186. doi:10.7554/eLife.31186.

Kawate, Toshimitsu, Jennifer Carlisle Michel, William T. Birdsong, and Eric Gouaux. 2009. "Crystal Structure of the ATP-Gated P2X4 Ion Channel in the Closed State." *Nature* 460(7255): 592–98. doi:10.1038/nature08198.

Khakh, Baljit S., and R. Alan North. 2012. "Neuromodulation by Extracellular ATP and P2X Receptors in the CNS." *Neuron* 76(1): 51–69. doi:10.1016/j.neuron.2012.09.024.

Kopp, Robin, Anna Krautloher, Antonio Ramírez-Fernández, and Annette Nicke. 2019. "P2X7 Interactions and Signaling – Making Head or Tail of It." *Frontiers in Molecular Neuroscience* 12: 183. doi:10.3389/fnmol.2019.00183.

Li, Mufeng, Gilman E. S. Toombes, Shai D. Silberberg, and Kenton J. Swartz. 2015. "Physical Basis of Apparent Pore Dilation of ATP-Activated P2X Receptor Channels." *Nature Neuroscience* 18(11): 1577–83. doi:10.1038/nn.4120.

Mansoor, Steven E., Wei Lü, Wout Oosterheert, Mrinal Shekhar, Emad Tajkhorshid, and Eric Gouaux. 2016. "X-Ray Structures Define Human P2X3 Receptor Gating Cycle and Antagonist Action." *Nature* 538(7623): 66–71. doi:10.1038/nature19367.

McCarthy, Alanna E., Craig Yoshioka, and Steven E. Mansoor. 2019. "Full-Length P2X7 Structures Reveal How Palmitoylation Prevents Channel Desensitization." *Cell* 179(3): 659–670.e13. doi:10.1016/j.cell.2019.09.017.

Moehring, Francie, Ashley M. Cowie, Anthony D. Menzel, Andy D. Weyer, Michael Grzybowski, Thiago Arzua, Aron M. Geurts, Oleg Palygin, and Cheryl L. Stucky. 2018. "Keratinocytes Mediate Innocuous and Noxious Touch via ATP-P2X4 Signaling." Edited by David D Ginty. *eLife* 7 (January): e31684. doi:10.7554/eLife.31684.

Murrell-Lagnado, Ruth D. 2017. "Regulation of P2X Purinergic Receptor Signaling by Cholesterol." *Current Topics in Membranes* 80: 211–32. doi:10.1016/bs.ctm.2017.05.004.

North, R. Alan. 2002. "Molecular Physiology of P2X Receptors." *Physiological Reviews* 82(4): 1013–67. doi:10.1152/physrev.00015.2002.

Peverini, Laurie, Juline Beudez, Kate Dunning, Thierry Chataigneau, and Thomas Grutter. 2018. "New Insights into Permeation of Large Cations Through ATP-Gated P2X Receptors." *Frontiers in Molecular Neuroscience* 11: 265. doi:10.3389/fnmol.2018.00265.

Ralevic, Vera. 2012. "P2X Receptors in the Cardiovascular System." *Wiley Interdisciplinary Reviews: Membrane Transport and Signaling* 1(5): 663–74. doi:10.1002/wmts.58.

Robinson, Lucy E., and Ruth D. Murrell-Lagnado. 2013. "The Trafficking and Targeting of P2X Receptors." *Frontiers in Cellular Neuroscience* 7 (November): 233. doi:10.3389/fncel.2013.00233.

Robinson, Lucy E., Mitesh Shridar, Philip Smith, and Ruth D. Murrell-Lagnado. 2014. "Plasma Membrane Cholesterol as a Regulator of Human and Rodent P2X7 Receptor Activation and Sensitization." *The Journal of Biological Chemistry* 289(46): 31983–94. doi:10.1074/jbc.M114.574699.

Samways, Damien S. K., Zhiyuan Li, and Terrance M. Egan. 2014. "Principles and Properties of Ion Flow in P2X Receptors." *Frontiers in Cellular Neuroscience* 8: 6. doi:10.3389/fncel.2014.00006.

Saul, Anika, Ralf Hausmann, Achim Kless, and Annette Nicke. 2013. "Heteromeric Assembly of P2X Subunits." *Frontiers in Cellular Neuroscience* 7:250. doi:10.3389/fncel.2013.00250.

Syed, Nawazish-i-Husain, and Charles Kennedy. 2012. "Pharmacology of P2X Receptors." *Wiley Interdisciplinary Reviews: Membrane Transport and Signaling* 1(1): 16–30. doi:10.1002/wmts.1.

Ugur, Mehmet, and Özlem Ugur. 2019. "A Mechanism-Based Approach to P2X7 Receptor Action." *Molecular Pharmacology* 95(4): 442–50. doi:10.1124/mol.118.115022.

Vacca, Fabrizio, Michela Giustizieri, Maria Teresa Ciotti, Nicola Biagio Mercuri, and Cinzia Volonté. 2009. "Rapid Constitutive and Ligand-Activated Endocytic Trafficking of P2X Receptor." *Journal of Neurochemistry* 109(4): 1031–41. doi:10.1111/j.1471-4159.2009.06029.x.

Index

Note: Locators in *italics* represent figures and **bold** indicate tables in the text.

For Product Safety Concerns and Information please contact our EU
representative GPSR@taylorandfrancis.com
Taylor & Francis Verlag GmbH, Kaufingerstraße 24, 80331 München, Germany